Progress in Applied Mechanics

MECHANICS OF SURFACE STRUCTURES*

Editors: W.A. Nash and G.Æ. Oravas

Volume 6

P. Seide, Small elastic deformations of thin shells. 1975. ISBN 90-286-0064-7

V. Panc, Theories of elastic plates. 1975. ISBN 90-286-0104-X

T.L. Nowinski, Theory of thermoelasticity with applications. 1978. ISBN 90-286-0457-X

S. Lukasiewicz, Local loads in plates and shells. 1979. ISBN 90-286-0047-7

V. Fiřt, Statics, formfinding and dynamics of air-supported membrane structures. 1983. ISBN 90-247-2672-7

K.Y. Yeh, Progress in applied mechanics. 1987. ISBN 90-247-3249-2

R. Negruţiu, Elastic analysis of slab structures. 1986. ISBN 90-247-3367-7

* The first four titles in this series were published by Sijthoff and Noordhoff International Publishers, Alphen aan den Rijn, The Netherlands, Rockville, Maryland, U.S.A.

Progress in Applied Mechanics

The Chien Wei-zang Anniversary Volume

Edited by

Yeh Kai-yuan

Lanzhou University,
Lanzhou, Gansu 730001
People's Republic of China

1987 **MARTINUS NIJHOFF PUBLISHERS**
a member of the KLUWER ACADEMIC PUBLISHERS GROUP
DORDRECHT / BOSTON / LANCASTER

Distributors

for the United States and Canada: Kluwer Academic Publishers, P.O. Box 358, Accord Station, Hingham, MA 02018-0358, USA
for the UK and Ireland: Kluwer Academic Publishers, MTP Press Limited, Falcon House, Queen Square, Lancaster LA1 1RN, UK
for all other countries: Kluwer Academic Publishers Group, Distribution Center, P.O. Box 322, 3300 AH Dordrecht, The Netherlands

Library of Congress Cataloging in Publication Data

Main entry under title:

Progress in applied mechanics.

(Mechanics of surface structures ; 6)
Bibliography: p.
1. Mechanics, Applied--Addresses, essays, lectures.
2. Ch'ien, Wei-ch'ang. I. Ch'ien, Wei-ch'ang.
II. Yeh, Kai-yuan. III. Series: Mechanics of surface structures ; v. 6.
TA350.3.P77 1986 620.1 85-21715
ISBN-13: 978-94-010-8061-3

ISBN-13: 978-94-010-8061-3 e-ISBN-13: 978-94-009-3487-0
DOI: 10.1007/978-94-009-3487-0

Prof. W. Z. Chien
July 26, 1984

Contents

Composite materials

Finite deformation of plates and shells

Fracture mechanics

Computational mechanics

Foreword

Prof. W. Z. Chien was born on 9 October, 1912 and 1982 saw the 70th anniversary of his birth. Some of his friends, colleagues, and former students prepared this special volume in honour of his outstanding contribution to the field of mechanics. The volume does not contain contributions from all of his students and friends and for this we apologize.

Prof. Chien's family have lived in Qufangquiao Village, Hongshengli, Wuxi County, Jiangsu Province for generations. Many members of his family have been teachers in this village. When he was 14 years old his father died and for a time it appeared necessary to terminate his education but, fortunately, an uncle, Chien Mu, who later became a very famous historian in China, came to his aid and he was able to continue his studies.

In 1931 he took entrance exams and was simultaneously admitted to five prestigious Chinese universities. Of these, he chose to enter Tsing-hau University in Beijing, with major work in physics. He received his baccaulaurate in 1935 and taught at middle school for a time until he was awarded a Sino-British scholarship to study abroad. In the competition for this award, three of the recipients were in the field of mechanics: Prof. C. C. Lin, Prof. Kuo Yung-huai, and Prof. Chien Wei-zang. All three arrived in Toronto in August, 1940 and entered the Department of Applied Mathematics of the University of Toronto to study under Prof. J. L. Synge.

While still a doctoral student, Prof. Chien published in the Th. von Karman Anniversary Volume, his well-known paper "The Intrinsic Theory of Elastic Plates and Shells". After receiving his Ph.D. in 1943, Prof. Chien worked as a research fellow at the Jet Propulsion Laboratory of the California Institute of Technology in Pasadena. At the end of World War II, Prof. Chien returned to the Department of Mechanical Engineering at Tsing-hua University in Beijing. There, he taught many courses in mechanics of solids. His research centred on large deflections of plates and the hydrodynamic theory of lubrication. In 1952 he was appointed Vice President of Tsing-hua University, as well as Vice Chairman of the Institute of Mechanics of the Chinese Academy of Sciences. In 1955 he was elected Academician of the Mathematics–Physics–Chemistry Department and Technical Science Department of the Chinese Academy of Sciences. In 1956 he was honoured by being elected a Foreign Member of the Polish Academy of Sciences. In that same year he was awarded the second rank National Scientific Prize by the Chinese Academy of Sciences for his work on buckling of spherical shells. Later in 1956 he participated in the Ninth International Congress of Theoretical and Applied Mechanics in Brussels, Belgium, and presented results of his recent studies on large deflections of thin elastic plates. The year 1957 saw the presentation in Beijing of special seminars on mechanics led jointly by Prof. Chien and Prof. Tsien Hsue-shen.

In the late 1950's Prof. Chien developed new generalized variational principles in elasticity which, because of the then turbulent conditions in China, remained unpublished. In 1972 he led a Chinese scientific delegation to England, Sweden, Canada, and the United States. A particularly happy event occurred in Toronto, where the visit coincided with the exact date of the award of his doctorate thirty years previously.

Prof. Chien is editor of the Journal of Applied Mathematics and Mechanics, issued in simultaneous Chinese and English editions since 1980. Prof. Chien is a member of the Editorial Board of the International Journal of Engineering Science as well as one of the editors of *Advances in Applied Mathematics*. His recent books have been *Variational Methods and Finite Elements, Book One, Singular Perturbation Theory, Asymptotic Solutions of Plates and Shells*, and, soon to be released, *Variational Methods and Finite Elements, Book Two*, and *Mechanics of Armour Penetration.*

To illustrate his endless energy and enthusiasm, in 1982 Prof. Chien took the position of President of the Shanghai University of Technology as well as serving as Director of its Institute of Applied Mathematics and Mechanics. The authors of this book, as well as the many other friends of Prof. Chien wish him a long, healthy, and happy life.

YEH KAI-YUAN

The main published works of Professor Chien Wei-zang

1. Monographs

1. Chien Wei-zang, Inventions in Chinese History (in Chinese), Chinese Youth Publications, Beijing (1954); translated into Mongolian language and other national languages, 200 pages.
2. Chien Wei-zang and Yeh Kai-yuan, Theory of Elasticity (in Chinese), Science Press, Beijing, 1st Edition (1956); 2nd Edition (1980), 422 pages.
3. Chien Wei-zang, Lin Hung-sun, Hu Hai-chang and Yeh Kai-yuan, Problems of Large Deflection of Elastic Circular Plates (in Chinese), Science Press, Beijing (1954), 152 pages.
4. Chien Wei-zang, Lin Hung-sun, Hu Hai-chang and Yeh Kai-yuan, Torsion of Elastic Columns (in Chinese), Science Press, Beijing (1956), 483 pages.
5. Translation of Monograph 3 in Russian (1957), 207 pages.
6. Chien Wei-zang, Variational Principles and Finite Element Methods, Vol. I (in Chinese), Science Press, Beijing (1980), 599 pages.
7. Chien Wei-zang, Singular Perturbation Theory (in Chinese), to be published in Szechuan People's Publishing House, Chengtu.
8. Chien Wei-zang and Yeh Kai-yuan, Applied Mathematics (in Chinese), to be published in Shanghai Science and Technology Publishing House, Shanghai.
9. Chien Wei-zang, Collected Works of Applied Mathematics and Mechanics (in Chinese), Jiangsu Science and Technology Publishing House, Nanjing (1980), 265 pages.
10. Chien Wei-zang, Armour Penetration Mechanics (in Chinese), to be published in National Defence Publishing House, Beijing.
11. Edited by Chien Wei-zang, Singular Perturbation Theory and Its Application in Mechanics (in Chinese), Science Press (1982), Beijing.
12. Chien Wei-zang, The Development of Zinc Air (oxygen) Cells (published with the name of the Zinc Air Cells Study Group of Tsing Hua University), Science Press, Beijing (1974).
13. Eringen, A. C. (translated by Chien Wei-zang), Tensor Analysis (in Chinese), Jiangsu Science and Technology Publishing House, Nanjing, (1981), 200 pages.

2. Papers of scientific investigation

1. Chien Wei-zang, Measurement of atmospheric electricity in Peiping (in English), presented in the 1935 Annual Meeting of the Chinese Society of Physics (Tsingdao).
2. Chien Wei-zang, The spectrum of double ionized Calcium (Ca III) (in English), Chinese Journal of Physics, 3, 1 (1937), 1–13.
3. Chien Wei-zang, Analysis of the spectrum of singly ionized Cerium (in English), Chinese Journal of Physics, 4 (1939), 86–116.
4. Chien Wei-zang, Highly ionized Potassium and Calcium spectra, (in English), Chinese Journal of Physics, 4 (1939), 117–147.
5. Synge, J. L. and Chien Wei-zang, The intrinsic theory of elastic shells and plates (in English), Applied Mechanics, Theodore von Kàrmàn Anniversary Volume (1940), 103–120.
6. Weinstein, A. and Chien Wei-zang, On the vibration of a clamped plate under tension (in English), Quarterly of Applied Mathematics, I, 1 (1943), 61–68.

7. Chien Wei-zang, The intrinsic theory of thin shells and plates, Part I, General theory (in English), Quarterly of Applied Mathematics, I, 1 (1944), 297–327.
8. Chien Wei-zang, The intrinsic theory of thin shells and plates, Part II, Application to thin plates (in English), Quarterly of Applied Mathematics, II, 2 (1944), 43–59.
9. Chien Wei-zang, The intrinsic theory of thin shells and plates, Part III, Application to thin shells (in English), Quaterly of Applied Mathematics, II, 2 (1944), 120–135.
10. Chien Wei-zang, The resistance of antennae of various shapes and positions in rectangular and circular wave guides (in English), Radio Report No. 5 Special Committee on Applied Mathematics, National Research Council of Canada (1943).
11. Chien Wei-zang, The reactance, matching conditions, matching resistance of a circular wave guide in the case of E_{01} wave (in English), Radio Report, No. 6, Special Committee on Applied Mathematics, National Research Council of Canada (1943).
12. Chien Wei-zang, The trajectories of the missile XF10S1000 (in English), Progress Report, No. 4-1, Jet Propulsion Laboratory, GALCIT (1944).
13. Chien Wei-zang, Correction calculation of missiles trajectories (in English), Progress Report, No. 4-2, Jet Propulsion Laboratory, GALCIT (1944).
14. Chien Wei-zang, The trajectories of Waccorporal (in English), Progress Report, No. 4-5, Jet Propulsion Laboratory, GALCIT (1944).
15. Chien Wei-zang, The trajectories of the missile XF30L20,000 (Corporal) (in English), Progress Report, No. 4-7, Jet Propulsion Laboratory, GALCIT (1945).
16. Chien Wei-zang, Estimated values of the aerodynamic coefficients of the Corporal E (in English), Progress Report, No. 4-20, Jet Propulsion Laboratory, GALCIT (1946).
17. Th. von Kàrmàn and Chien Wei-zang, Torsion with variable twist (in English), J. of Aeronautical Science, 13 (1946), 503–510.
18. Infeld, L., Smith, V. G. and Chien Wei-zang, On some series of Bessel functions (in English), Journal of Mathematics and Physics, XXVI, 1 (1947), 22–28.
19. Chien Wei-zang, Infeld, L., Pounder, J. R., Stevenson, A. F. and Synge, J. L., Contributions to the theory of wave guides (in English), Canadian Journal Research, A27 (1949), 69–129.
20. Chien Wei-zang, Symmetrical conical flow at supersonic speed by perturbation method (in English), The Engineering Reports of National Tsing Hua University, 3 (1947), 1–14.
21. Chien Wei-zang, Large deflection of a circular clamped plate under uniform normal pressure (in English), The Chinese Journal of Physics, 7 (1947), 102–113.
22. Chien Wei-zang, Asymptotic behavior of a thin clamped circular plate under uniform normal pressure at very large deflection (in English), The Science Reports of National Tsing Hua University, 5 (1948), 71–86.
23. Chien Wei-zang and Ho Shui-tsing, Asymptotic method on the problems of thin elastic ring shell with rotational symmetrical load (in English), The Engineering Reports of National Tsing Hua University, 3 (1947), 1–14.
24. Chien Wei-zang, Derivation of the equation of equation of equilibrium of an elastic shell from the general theory of elasticity (in English), The Science Reports of National Tsing Hua University, Ser.A, 5 (1948), 240–251.
25. Chien Wei-zang, The true leaving angle for diaphragm and bucket wheel with curved guides at the discharge end (in English), The Engineering Reports of National Tsing Hua University, 4 (1948), 78–102.
26. Chien Wei-zang, Hydrodynamic theory of lubrication for plane sliders of finite width (in English), Chinese Journal of Physics, 7, 4 (1949), 193–229.
27. Chien Wei-zang and Chen Chih-ta, Theory of rolling (in Chinese), Acta Physics Sinica, 9, 2 (1953), 57–92; (in English), Acta Scientia Sinica, I, 2 (1953), 192–229.
28. Chien Wei-zang, Assumptions in Saint Venant's solution for the torsion of an elastic cylinder (in Chinese), Acta Physica Sinica, 9, 4 (1953), 215–220; (in English), Acta Scientia Sinica, II (1953), 165–170.
29. Chien Wei-zang, Continuous beam with non-uniform stiffness (in Chinese), Acta Physica Sinica, 9, 3 (1953), 170–182; (in English), Acta Scientia Sinica, III, 4 (1953), 116–126.
30. Chien Wei-zang and Yeh Kai-yuan, On the large deflection circular plates (in Chinese), Acta Physica Sinica, 10, 3 (1954), 209–238; (in English), Acta Scientia Sinica, III, 4 (1954), 405–436.

31. Chien Wei-zang and Yeh Kai-yuan, Design data of problem of large deflection of thin circular plate (in Chinese), Chinese Journal Mechanical Engineering, 3, 1 (1955), 15–35.
32. Chien Wei-zang, Preliminary report on the snapping pressure of a thin spherical cap (in English), unpublished, (1945), but is adoped by Hu Hai-chang in his paper "The snapping problem of a spherical shallow thin circular shell" which is included in the above mentioned Monograph 3.
33. Chien Wei-zang, Design formulas for the snapping pressure of a thin cap (in English), unpublished (1945) but most materials of it are adopted by Hu Hai-chang in his paper, ibid.
34. Chien Wei-zang, Building in Ancient China (in Russian), Historical Problems in Science and Technology, No. 1 (1956), 124–136, Moscow, Academy of Science, S.S.S.R.
35. Chien Wei-zang, Problem of large deflection of circular plate (in English), Archiwum Mechaniki Stosowanej, Warsaw, VIII (1956), 1–12.
36. Chien Wei-zang and Yeh Kai-yuan, On the large deflection of rectangular plate (in English), Proc. Ninth Intern. Congr. Appl. Mech., Brussels, 6 (1957), 403–412; presented in this Congress in 1956.
37. Chien Wei-zang and Hu Hai-chang, On the snapping of a thin spherical cap (in English), Proc. Ninth Intern. Congr. Appl. Mech., Brussels (1957); presented in this Congress in 1956.
38. Chien Wei-zang, On generalized variational principle in elasticity and their application in problems of plates and shells (in Chinese) unpublished manuscript.
39. Chien Wei-zang etc., Discussion on "Plane problem of semi-infinite elastic solid subjected to a concentrated force at the midpoint of bottom of a groove" (in Chinese), Acta Mechanica Sinica, No. 3 (1964), 251–259.
40. Chien Wei-zang, Fang Cui-zhang and Jing Shi-dou, theory of cylindrical shell in arbitrary cross-section with tangential reinforced ribs (in Chinese), Research Report of the Faculty of Strength Materials of Tsing Hua University (July 15, 1964).
41. Chien Wei-zang and Jing Shi-dou, Asymptotic solution of a cylindrical shell of arbitrary closed cross-section (elliptical cross-section) with reinforced ribs under uniformly distributed load (in Chinese), Research Report of the Faculty of Strength Materials of Tsing Hua University (Aug. 15, 1964).
42. Study and manufacture of Zinc air (Oxygen) cells (in Chinese) (published with the name of the Zinc Air Cells Study Group of Tsing Hua University), Journal of Tsing Hua University, No. 1 (1973), 37–53.
43. Internal resistance of the cell and the distribution of pole current (in Chinese), Research Report of the Zinc Air Cells Study Group of Tsing Hua University (March 30, 1972).
44. The calculation of temperature rise of cells and accumulators (in Chinese), The 9th Part of the Summary of the Zinc Air Cells Study Group of Tsing Hua University (1972).
45. Theory of heat radiation and temperature rise of cells and accumulators (in Chinese), unpublished (1973).
46. Study, manufacture and experiment of Zinc air cells used in vehicles (in Chinese) (published with the name of the Zinc Air Cells Study Group of Tsing Hua University), Journal of Tsing Hua University (Dec., 1973), 1–10.
47. Zinc air cells used in railway portable beacon light (in Chinese) (published with the name of the Zinc Air Cells Study Group of Tsing Hua University) (Oct., 1974), 167–178.
48. Chien Wei-zang, On the summation of some trigonometric series (in Chinese), Journal of Tsing Hua University, 18, 4 (1978), 53–78; see also Collected Works of Applied Mathematics and Mechanics, Jiangsu Science and Technology Published House (1979), 202–227.
49. Chien Wei-zang, The proof of two integral formulas (in Chinese), Collected Works of Applied Mathematics and Mechanics, Jiangsu Science and Technology Publishing House (1979), 228–232.
50. Ching Wei-zang, Numerical tables of $\Sigma_{k=1}^{\infty} \cos kx/k + 5/m$, $\Sigma_{k=1}^{\infty} \sin kx/k + 5/m$ (in Chinese), Collected Works of Applied Mathematics and Mechanics, Jiangsu Science and Technology Publishing House (1979), 233–265.
51. Chien Wei-zang, Analyses of strokes of Chinese characters and a primary design on a Chinese character typewriter, Yu-Wen-Xian-Dai-Hua or Modernization of the Chinese Language, No. 2 (1980), 159–175.
52. Chien Wei-zang, Studies on generalized variational principles in elasticity and their application in finite element calculation (I) (in Chinese), Mechanics and Practice, 1, 1 (1979), 16–24.

53. Chien Wei-zang, Studies on generalized variational principles in elasticity and their application in finite element calculation (II) (in Chinese), Mechanics and Practice, 1, 2 (1979), 18–27.
54. Chien Wei-zang, Studies on generalized variational principles in elasticity and their application in finite element calculation (in Chinese), Chinese Journal of Mechanical Engineering, 15, 2 (1979), 1–23; presented in the Meeting of Computing Mechanics, held by the Ministry of Education of the People's Republic of China, Dalian, (Oct., 1978); presented in the Meeting of Finite Elements, held by the Chinese Society of Mechanical Engineering, the Chinese Society of Aeronautics and the Chinese Society of Ship Manufacturing, Bangbu, (Dec., 1978).
55. Chien Wei-zang, Studies on convergence problem of power series solutions of ring shell equation and their related theorems on series convergence (in Chinese), Journal of Lanzhou University (Natural Sciences), Special Number of Mechanics, No. 1 (1979), 1–38; Collected Works of Applied Mathematics and Mechanics, Jiang-su Science and Technology Publishing House (1979), 31–68.
56. Chien Wei-zang and Zheng Si-liang, Equations of symmetrical ring shells in complex quantities and their general solutions for slender ring shells (in Chinese), Journal of Tsing Hua University, 19, 1 (1979), 24–47.
57. Chien Wei-zang, Calculations for semi-circular arc type corrugated tube-applications of the theory of slender ring shells (in Chinese), Journal of Tsing Hua University, 19, 1 (1979), 84–99.
58. Chien Wei-zang, Reasonable foundation of the small deflection theory of thin shells (in Chinese), Collected Works of Applied Mathematics and Mechanics, Jiangsu Science and Technology Publishing House (1980), 1–10.
59. Chien Wei-zang, The manufacture, design, experiment and theory of corrugated tubes (in Chinese), Collected Works of Applied Mathematics and Mechanics, Jiangsu Science and Technology Publishing House (1980), 110–126; presented in the Fifth All China Meeting in Elastic Element, Shanghai (Dec., 1978).
60. Chien Wei-zang, Recent development of the finite element method (in Chinese), presented in the All China Meeting of Elasticity and Plasticity, Chengqing (June, 1980); presented in the All China Meeting of Computational Mechanics, Hongzhou (Oct., 1980).
61. Chien Wei-zang, Foreward of Applied Mathematics and Mechanics, Applied Mathematics and Mechanics (Chinese Edition and English Edition), 1, 1 (1980).
62. Chien Wei-zang, Shieh Chih-cheng, Zheng Si-liang and Wang Rui-wu, Shape function of a compatible bending triangle finite element and their related stiffness matrix (in Chinese), Chinese Journal of Mechanical Engineering, 16, 4 (1980), 1–11.
63. Chien Wei-zang, Finite element analysis of axisymmetric elastic body problems, Applied Mathematics and Mechanics, 1, 1 (1980), (Chinese Edition), 25–35; (English Edition), 23–33.
64. Chien Wei-zang, The explicit form of field functions in tetrahedron element with 16 and 20 degree of freedom, Applied Mathematics and Mechanics (Chinese Edition), 1, 2 (Aug., 1980), 153–158; (English Edition), 1, 2 (Dec., 1980), 159–164.
65. Chien Wei-zang, Shieh Chih-cheng, Gu Qiu-ling, Yang Chong fa and Zhou Chun-tian, The superposition of the finite element method on the singularity terms in determining the stress intensity factors (in Chinese), Journal of Tsing Hua University, 20, 2 (1980), 15–24; presented in the All China Meeting of Fracture Mechanics, Wuhan (1980).
66. Chien Wei-zang, Non-homogeneous solution of slender ring shell equations and their applications in instrumental design (in Chinese), Chinese Journal of Instrument and Meter, 1, 1 (Feb., 1980), 89–112.
67. Chien Wei-zang and Zheng Si-liang, General solutions of axial symmetrical ring shells, Applied Mathematics and Mechanics, (Chinese Edition), 1, 3 (Nov., 1980), 287–299; (English Edition), 1, 3 (Dec., 1980), 305–318; presented in the All China Meeting of Elasticity and Plasticity, Chengqing, (June, 1980).
68. Chien Wei-zang and Zheng Si-liang, Calculations for semi-circular arc type corrugated tube – applications of general solutions of ring shell equation, Applied Mathematics and Mechanics, (Chinese Edition), 2, 1 (Feb., 1981), 97–111; (English Edition), 2, 1 (June, 1981), 103–115.
69. Chien Wei-zang and Dai Fu-lung, Effective elastic constants for thick perforated plates (in Chinese), Acta Mechanica Sinica, No. 4 (1981), 364–371.
70. Chien Wei-zang, Wu Ming-de, Shieh Chih-cheng, Zheng Si-liang and Wang Rui-wu, The perturbation method of large deflection problem of corrugated tube (in Chinese), presented in the Sixth All China Meeting in Elastic Element, Xiamen (Nov., 1981).

71. Chien Wei-zang, On information processing in Chinese language (in Chinese), New Asian Life Monthly, The Chinese University of Hongkong (Nov. 15, 1981).
72. Chien Wei-zang, Generalized variational principles (in Chinese), Lectures on Natural Sciences of Gueizhou Province (Sep., 1981).
73. Chien Wei-zang, Nonlinear finite element (in Chinese), Lectures on Natural Sciences of Gueizhou Province (Sep., 1981).
74. Chien Wei-zang, Incompatible plate elements based upon the generalized variational principles (in English), International Conference of Hybrid and Mixed Finite Element in Celebration of the 60th Anniversary of Professor Theordore Pian Hsueh-huang's Birthday, Atlanta, Georgia, U.S.A., April 8–11, 1981; will be published in the Proceedings of this Conference.
75. Chien Wei-zang, Incompatible element and generalized variational principles (in Chinese), International Invitation Workshop Of Finite Element Method, Hefei, Anhui Province, (May 20–24, 1981), will be published in the Proceedings of this Conference and also have the English Edition.
76. Chien Wei-zang, Wang Zhi-zhong, Xu Yin-ge and Chen Shan-lin, The symmetrical deformation of circular membrane under the action of uniformly distributed loads in its central portion, Applied Mathematics and Mechanics, 2, 6 (1981), (Chinese Edition), 599–612; (English Edition), 653–668.
77. Chien Wei-zang, Foreward to the Chinese Edition of "Continuum Physics, edited by Eringen, A. C." (in Chinese), Jiangsu Science and Technology Publishing House (1980).
78. Chien Wei-zang, Foreward to Singular Perturbation Theory and Its Application in Mechanics (in Chinese), Science Press, Beijing (1981).
79. Chien Wei-zang, Foreward to the first number of "Scientific Exploration" (in Chinese) (1981).
80. Chien Wei-zang, Dynamic finite element with diagonalized consistent mass matrix and elastic-plastic impact calculation (in English), Proceedings of the International Conference of Finite Element Methods, Shanghai (Aug. 2–6, 1982); Applied Mathematics and Mechanics, 3, 3 (1982), (Chinese Edition), 281–296; (English Edition), 319–334.
81. Chien Wei-zang, Diagonalized consistent mass matrix and the dynamic finite element analysis of elastic-plastic impact in axisymmetric problems (in English), paper read at Dalian International Symposium on Mixed/Hybrid Finite Element Method, (Aug. 11–29, 1982); Applied Mathematics and Mechanics, 3, 4 (1982), (Chinese Edition), 429–448; (English Edition), 469–489.
82. Chien Wei-zang, Compatible dynamic finite element with diagonalized consistent mass matrix, paper read at Dalian International Symposium on Mixed/Hybrid Finite Element Method (in English), (Aug. 11–29, 1980); Applied Mathematics and Mechanics, 3, 5 (1982), (Chinese Edition), 565–576; (English Edition), 609–622.
83. Chien Wei-zang, The analytical solution of G. I. Taylor's theory of plastic deformation in impact of cylindrical projectiles and its improvement, Applied Mathematics and Mechanics, 3, 6 (1982), (Chinese Edition), 743–755; (English Edition), 801–815.
84. Chien Wei-zang, History, present situation and problems to be solved of armour penetration mechanics (in Chinese), Applied Mechanics, 1, 2 (1982).
85. Chien Wei-zang, Environment pollution in capitalist countries (in Chinese), Environment Protection, No. 1 (1974), 32–35.
86. Chien Wei-zang, Generalized variational principles of three kinds obtained by releasing all the variational restrains of complementary energy principles by high order Lagrange multiplier method and their related equivalent theorems, paper read at Dalian International Symposium on Mixed/Hybrid Finite Element Method (Aug. 11–29, 1982).
87. Chien Wei-zang, On the Kirchhoff-Love's assumptions about the problem of approximation in the classic small deflection theory of shells, Research Report of the Faculty of Strength Materials of Tsing Hua University (1963).
88. Chien Wei-zang, Method of high-order Lagrange multiplier and generalized variational principles of elasticity with more general forms of functionals, Applied Mathematics and Mechanics, (Chinese Edition), 4, 2 (1983), 137–150; (English Edition), 4, 2 (1983), 143–157.
89. Chien Wei-zang and Wu Ming-de, The nonlinear characteristics of U-shaped bellows – calculations by the method of perturbation, Applied Mathematics and Mechanics, 4, 5 (1983) (Chinese Edition), 595–608; (English Edition), 4, 5 (1983), 649–665.

Continuum Mechanics

Variational principles for the linear dynamic theory of micropolar viscoelasticity

TAI TIEN-MIN

Department of Mathematics, Liaoning University, Shenyang, Liaoning, PRC

Abstract. In the present paper, variational principles for the linear dynamic theory of micropolar viscoelasticity are derived by using J. T. Oden and J. N. Reddy's method of the inverse problem of the calculus of variations, M. E. Gurtin's convolution method, and Chien Wei-zang's method of the Lagrange multipliers.

1. Fundamental relations, boundary conditions and initial conditions for the linear dynamic theory of micropolar viscoelasticity

Oden and Reddy (1976) [1] gave the variational principles for the classical linear dynamic theory of viscoelasticity according to the methods of the inverse problem of calculus of variations and convolution due to Gurtin (1963) [2]. Chien Wei-zang (1964) [3] proposed the Lagrange multiplier method which is a universal and convenient one in constructing functions of generalized variation principles. The general theory of micropolar electricity was first established by A. C. Eringen [4].

In this paper, combining the tensor notation in [5] with the operations of tensors and matrices and using the above-mentioned three methods suggested by Oden and Reddy, Gurtin, and Chien, we derive the variational principles for the linear dynamic theory of micropolar viscoelasticity.

The fundamental relations, boundary conditions and initial conditions for the linear dynamic theory of micropolar viscoelasticity have the following component and notation forms:

1.1. *Equation of motion*

$$\begin{aligned} t_{ji,j}(\mathbf{x}, t) + F_i(\mathbf{x}, t) &= \varrho \ddot{u}_i(\mathbf{x}, t) \\ m_{ji,j}(\mathbf{x}, t) + \varepsilon_{ijk} t_{jk}(\mathbf{x}, t) + M_i(\mathbf{x}, t) &= I_{ij}\ddot{\phi}_j(\mathbf{x}, t) \end{aligned} \tag{1.1}$$

where u_i, ϕ_i; t_{ji}, m_{ji}; $F_i = \varrho f_i$, $M_i = \varrho l_i$; ϱ, $I_{ij} = \varrho j_{ji}$ are the components of the macro-linear displacement and micro-angular displacement vectors; the stress and couple stress tensors; the body force and body couple tensors; the density of mass and the components of micro-inertial tensors respectively.

Yeh, K. Y. (ed) Progress in Applied Mechanics
 ISBN-13: 978-94-010-8061-3

If set

$$
\begin{aligned}
\boldsymbol{\sigma} &= \boldsymbol{\sigma}(x, t) = \begin{Bmatrix} t_{ji}(\mathbf{x}, t) \\ m_{ji}(\mathbf{x}, t) \end{Bmatrix}, \qquad \mathbf{f} = \mathbf{f}(\mathbf{x}, t) = \begin{Bmatrix} F_i(\mathbf{x}, t) \\ M_i(\mathbf{x}, t) \end{Bmatrix}, \\
\mathbf{e} &= \begin{bmatrix} \varepsilon_{ijk} & 0 \\ 0 & \varepsilon_{ijk} \end{bmatrix}, \qquad \mathbf{M} = \begin{bmatrix} \varrho\delta_{ij} & 0 \\ 0 & I_{ij} \end{bmatrix}, \\
\mathbf{R} &= \begin{bmatrix} 0 & 1 \\ 0 & 0 \end{bmatrix}, \qquad \mathbf{d} = \mathbf{d}(\mathbf{x}, t) = \begin{Bmatrix} u_i(\mathbf{x}, t) \\ \phi_i(\mathbf{x}, t) \end{Bmatrix},
\end{aligned} \tag{1.2}
$$

then the equations of motion (1.1) can be written in compact symbolic form:

$$\nabla \cdot \boldsymbol{\sigma} + \mathbf{e} : \mathbf{R}^T \boldsymbol{\sigma} + \mathbf{f} = \mathbf{M} \cdot \ddot{\mathbf{d}}. \tag{1.3}$$

1.2. *Strain–displacement relations*

$$
\begin{aligned}
e_{ij}(\mathbf{x}, t) &= u_{j,i}(\mathbf{x}, t) - \varepsilon_{ijk}\phi_k(\mathbf{x}, t) \\
\mu_{ij}(\mathbf{x}, t) &= \phi_{j,i}(\mathbf{x}, t),
\end{aligned} \tag{1.4}
$$

where e_{ij} and μ_{ij} are the components of macro-strain and micro-strain tensors.
Now if set

$$\boldsymbol{\varepsilon} = \boldsymbol{\varepsilon}(\mathbf{x}, t) = \begin{Bmatrix} e_{ij}(\mathbf{x}, t) \\ \mu_{ij}(\mathbf{x}, t) \end{Bmatrix}, \tag{1.5}$$

then the strain–displacement relations (1.4) can be written in the following compact symbolic form

$$\boldsymbol{\varepsilon} = \nabla \mathbf{d} - \mathbf{e} \cdot \mathbf{R}\mathbf{d}. \tag{1.6}$$

1.3. *Constitutive equations*

The constitutive equations are defined by

(a) *Relaxation type*

$$
\begin{aligned}
t_{ij}(\mathbf{x}, t) &= \int_0^t G_{ijkl}(\mathbf{x}, t-s)\frac{\partial e_{kl}(\mathbf{x}, s)}{\partial s}\,\mathrm{d}s + \int_0^t H_{ijkl}(\mathbf{x}, t-s)\frac{\partial \mu_{kl}(\mathbf{x}, s)}{\partial s}\,\mathrm{d}s \\
&= G_{ijkl}(\mathbf{x}, 0)e_{kl}(\mathbf{x}, t) + (\dot{G}_{ijkl} * e_{kl})(\mathbf{x}, t) \\
&\quad + H_{ijkl}(\mathbf{x}, 0)\mu_{kl}(\mathbf{x}, 0) + (\dot{H}_{ijkl} * \mu_{kl})(\mathbf{x}, t), \\
m_{ij}(\mathbf{x}, t) &= \int_0^t H_{ijkl}(\mathbf{x}, t-s)\frac{\partial e_{kl}(\mathbf{x}, s)}{\partial s}\,\mathrm{d}s + \int_0^t I_{ijkl}(\mathbf{x}, t-s)\frac{\partial \mu_{kl}(\mathbf{x}, s)}{\partial s}\,\mathrm{d}s \\
&= H_{ijkl}(\mathbf{x}, 0)e_{kl}(\mathbf{x}, t) + (\dot{H}_{ijkl} * e_{kl})(\mathbf{x}, t) \\
&\quad + I_{ijkl}(\mathbf{x}, 0)\mu_{kl}(\mathbf{x}, t) + (\dot{I}_{ijkl} * \mu_{kl})(\mathbf{x}, t),
\end{aligned} \tag{1.7}
$$

where G_{jikl}, H_{ijkl} and I_{ijkl} can be called the components of the stiffness tensors of the relaxation type, and the symbol "$*$" denotes the convolution.

(b) *Creep type*

$$\begin{aligned}
e_{ij}(\mathbf{x}, t) &= \int_0^t J_{ijkl}(\mathbf{x}, t-s)\frac{\partial t_{kl}(\mathbf{x}, s)}{\partial s}\,\mathrm{d}s + \int_0^t K_{ijkl}(\mathbf{x}, t-s)\frac{\partial m_{kl}(\mathbf{x}, s)}{\partial s}\,\mathrm{d}s \\
&= J_{ijkl}(\mathbf{x}, 0)t_{kl}(\mathbf{x}, t) + (\dot{J}_{ijkl} * t_{kl})(\mathbf{x}, t) \\
&\quad + K_{ijkl}(\mathbf{x}, 0)m_{kl}(\mathbf{x}, t) + (\dot{K}_{ijkl} * m_{kl})(\mathbf{x}, t), \\
\mu_{ij}(\mathbf{x}, 0) &= \int_0^t K_{ijkl}(\mathbf{x}, t-s)\frac{\partial t_{kl}(\mathbf{x}, s)}{\partial s}\,\mathrm{d}s + \int_0^t L_{ijkl}(\mathbf{x}, t-s)\frac{\partial m_{kl}(\mathbf{x}, s)}{\partial s}\,\mathrm{d}s \\
&= K_{ijkl}(\mathbf{x}, 0)t_{kl}(\mathbf{x}, t) + (\dot{K}_{ijkl} * t_{kl})(\mathbf{x}, t) \\
&\quad + L_{ijkl}(\mathbf{x}, 0)m_{kl}(\mathbf{x}, t) + (\dot{L}_{ijkl} * m_{kl})(\mathbf{x}, t),
\end{aligned} \tag{1.8}$$

where J_{ijkl}, K_{ijkl} and L_{ijkl} can be called the components of the flexible tensors of the creep type.

If we further set

$$\begin{aligned}
\mathbf{A} &= \mathbf{A}(\mathbf{x}, 0) = \begin{bmatrix} G_{ijkl}(\mathbf{x}, 0) & H_{ijkl}(\mathbf{x}, 0) \\ H_{ijkl}(\mathbf{x}, 0) & I_{ijkl}(\mathbf{x}, 0) \end{bmatrix}, \\
\dot{\mathbf{B}} &= \dot{\mathbf{B}}(\mathbf{x}, t) = \begin{bmatrix} \dot{G}_{ijkl}(\mathbf{x}, t) & H_{ijkl}(\mathbf{x}, t) \\ \dot{H}_{ijkl}(\mathbf{x}, t) & \dot{I}_{ijkl}(\mathbf{x}, t) \end{bmatrix}, \\
\mathbf{R} &= \mathbf{R}(\mathbf{x}, 0) = \begin{bmatrix} J_{ijkl}(\mathbf{x}, 0) & K_{ijkl}(\mathbf{x}, 0) \\ K_{ijkl}(\mathbf{x}, 0) & L_{ijkl}(\mathbf{x}, 0) \end{bmatrix}, \\
\dot{\mathbf{S}} &= \dot{\mathbf{S}}(\mathbf{x}, t) = \begin{bmatrix} \dot{J}_{ijkl}(\mathbf{x}, t) & \dot{K}_{ijkl}(\mathbf{x}, t) \\ \dot{K}_{ijkl}(\mathbf{x}, t) & \dot{L}_{ijkl}(\mathbf{x}, t) \end{bmatrix},
\end{aligned} \tag{1.9}$$

then the constitutive relations (1.7) and (1.8) can be written in the following compact symbolic form:

$$\boldsymbol{\sigma} = \mathbf{A}:\boldsymbol{\varepsilon} + (\dot{\mathbf{B}} * :\boldsymbol{\varepsilon}), \tag{1.10}$$

$$\boldsymbol{\varepsilon} = \mathbf{R}:\boldsymbol{\sigma} + (\dot{\mathbf{S}} * :\boldsymbol{\sigma}). \tag{1.11}$$

1.4. *Boundary conditions*

$$\left.\begin{aligned}
&\text{On } \partial\Omega_u \times [0, t_0], & u_i(\mathbf{x}, t) &= \hat{u}_i(\mathbf{x}, t), \\
&\text{On } \partial\Omega_\phi \times [0, t_0], & \phi_i(\mathbf{x}, t) &= \hat{\phi}_i(\mathbf{x}, t),
\end{aligned}\right\} \tag{1.12}$$

$$\left.\begin{aligned}
&\text{On } \partial\Omega_t \times [0, t_0], & t_i(\mathbf{x}, t) &= \hat{t}_i(\mathbf{x}, t), \\
&\text{On } \partial\Omega_m \times [0, t_0], & m_i(\mathbf{x}, t) &= \hat{m}_i(\mathbf{x}, t),
\end{aligned}\right\} \tag{1.13}$$

where $\hat{u}_i$, $\hat{\phi}_i$ and $\hat{t}_i$, $\hat{m}_i$ are the functions prescribed on the corresponding boundaries. These boundary conditions can be written in the following compact symbolic forms:

$$\text{On} \quad \partial\Omega_{\mathrm{d}} \times [0, t_0], \qquad \mathbf{d} = \mathbf{d}(\mathbf{x}, t) = \hat{\mathbf{d}}, \tag{1.14}$$

$$\text{On} \quad \partial\Omega_{\mathrm{t}} \times [0, t_0], \qquad \mathbf{t} = \mathbf{t}(\mathbf{x}, t) = \hat{\mathbf{t}}. \tag{1.15}$$

1.5. *Initial conditions*

$$\left.\begin{aligned} u_i(\mathbf{x}, 0) &= u_{i0}(\mathbf{x}), \\ \phi_i(\mathbf{x}, 0) &= \phi_{i0}(\mathbf{x}), \end{aligned}\right\} \quad \mathbf{x} \in \bar{\Omega} \tag{1.16}$$

$$\left.\begin{aligned} \dot{u}_i(\mathbf{x}, 0) &= \dot{u}_{i0}(\mathbf{x}), \\ \dot{\phi}_i(\mathbf{x}, 0) &= \dot{\phi}_{i0}(\mathbf{x}), \end{aligned}\right\} \quad \mathbf{x} \in \bar{\Omega}, \tag{1.17}$$

where u_{i0}, ϕ_{i0} and $\dot{u}_{i0}$, $\dot{\phi}_{i0}$ are the initial displacements and velocities prescribed. These initial conditions can also be written in the following compact symbolic forms:

$$\mathbf{d}(\mathbf{x}, 0) = \mathbf{d}_0(\mathbf{x}) = \mathbf{d}_0, \qquad \mathbf{x} \in \bar{\Omega}, \tag{1.18}$$

$$\dot{\mathbf{d}}(\mathbf{x}, 0) = \dot{\mathbf{d}}_0(\mathbf{x}) = \dot{\mathbf{d}}_0, \qquad \mathbf{x} \in \bar{\Omega}. \tag{1.19}$$

2. Derivation by using Oden and Reddy's method of the inverse problem of the calculus of variation

Assuming $\mathbf{u}(\mathbf{x}, t)$ and $\mathbf{v}(\mathbf{x}, t)$ as well as $\boldsymbol{\alpha}(\mathbf{x}, t)$ and $\boldsymbol{\beta}(\mathbf{x}, t)$ are vectors composed by both vector-value and tensor-value functions, for convenience, the bilinear forms of the vectors are defined as the following:

$$[\mathbf{u}, \mathbf{v}] = \int_\Omega \int_0^t \mathbf{u}^T(\mathbf{x}, \tau) \cdot \mathbf{v}(\mathbf{x}, t - \tau)\, d\tau\, dx = \int_\Omega (\mathbf{u}^T * \mathbf{v})(\mathbf{x}, t)\, dx, \tag{2.1}$$

$$[\boldsymbol{\alpha}, \boldsymbol{\beta}] = \int_\Omega \int_0^t \boldsymbol{\alpha}^T(\mathbf{x}, \tau) : \boldsymbol{\beta}(\mathbf{x}, t - \tau)\, d\tau\, dx = \int_\Omega (\boldsymbol{\alpha}^T * \boldsymbol{\beta})(\mathbf{x}, t)\, dx, \tag{2.2}$$

$$[\mathbf{u}, \mathbf{v}]_{\partial\Omega} = \int_{\partial\Omega} \int_0^t \mathbf{u}^T(\mathbf{x}, \tau) \cdot \mathbf{v}(\mathbf{x}, t - \tau)\, d\tau\, dA \tag{2.3}$$

$$[\mathbf{u}, \mathbf{v}]_0 = \int_\Omega \{\mathbf{u}^T(\mathbf{x}, \tau) \cdot \mathbf{v}(\mathbf{x}, t - \tau)\}\big|_{\tau=0}^{t}\, dx. \tag{2.4}$$

Now let S be a space of the set of the ordered function vectors $\boldsymbol{\lambda}(\mathbf{d}, \boldsymbol{\varepsilon}, \boldsymbol{\sigma})$ for any $\bar{\boldsymbol{\lambda}} = (\bar{\mathbf{d}}, \bar{\boldsymbol{\varepsilon}}, \bar{\boldsymbol{\sigma}}) \in S'$, the dual pair in $S' \times S$ is defined by

$$\begin{aligned} \langle \bar{\boldsymbol{\lambda}}, \boldsymbol{\lambda} \rangle = {} & [\bar{\mathbf{d}}, \mathbf{d}] + [\bar{\boldsymbol{\sigma}}, \boldsymbol{\varepsilon}] + [\bar{\boldsymbol{\varepsilon}}, \boldsymbol{\sigma}] + [\bar{\mathbf{t}}, \mathbf{d}]_{\partial\Omega_{\mathrm{d}}} + [\bar{\mathbf{d}}, \mathbf{t}]_{\partial\Omega_{\mathrm{t}}} \\ & + [\bar{\mathbf{d}}, \mathbf{d}]_0 + [\dot{\bar{\mathbf{d}}}, \mathbf{d}]_0 + [\bar{\mathbf{d}}, \dot{\mathbf{d}}]_0. \end{aligned} \tag{2.5}$$

Now let us introduce the following set of ordered vectors $\mathbf{A}$ and $\mathbf{F}$ composed by vectors of vector-value and tensor-value functions as well as the matrix operator $\mathbf{P}$:

$$\left.\begin{aligned} \mathbf{\Lambda} &= \{\mathbf{d}, \boldsymbol{\varepsilon}, \boldsymbol{\sigma}; \mathbf{d}, \mathbf{t}; \mathbf{d}, \dot{\mathbf{d}}\}^T, \\ \mathbf{F} &= \{\mathbf{f}, 0, 0; \hat{\mathbf{t}}, -\mathbf{d}; \mathbf{M} - \dot{\mathbf{d}}_0, -\mathbf{M}\cdot\mathbf{d}_0\}^T, \end{aligned}\right\} \tag{2.6}$$

$$\mathbf{P} = \begin{bmatrix} \mathbf{M}\cdot\dfrac{\partial^2}{\partial t^2} & \mathbf{0} & -(\nabla\cdot + \mathbf{e}:\mathbf{R}^T) & \mathbf{0} & \mathbf{0} & \mathbf{0} & \mathbf{0} \\ \mathbf{0} & (\mathbf{A}: + \dot{\mathbf{B}}*) & -\mathbf{I} & \mathbf{0} & \mathbf{0} & \mathbf{0} & \mathbf{0} \\ (\nabla - \mathbf{e}\cdot\mathbf{R}) & -\mathbf{I} & \mathbf{0} & \mathbf{0} & \mathbf{0} & \mathbf{0} & \mathbf{0} \\ \mathbf{0} & \mathbf{0} & \mathbf{0} & \mathbf{0} & \mathbf{I} & \mathbf{0} & \mathbf{0} \\ \mathbf{0} & \mathbf{0} & \mathbf{0} & -\mathbf{I} & \mathbf{0} & \mathbf{0} & \mathbf{0} \\ \mathbf{0} & \mathbf{0} & \mathbf{0} & \mathbf{0} & \mathbf{0} & \mathbf{0} & \mathbf{M}\cdot \\ \mathbf{0} & \mathbf{0} & \mathbf{0} & \mathbf{0} & \mathbf{0} & -\mathbf{M}\cdot & \mathbf{0} \end{bmatrix}. \tag{2.7}$$

We may write (1.3), (1.6), (1.10), (1.14), (1.15), (1.18) and (1.19) in the following operator form for the vector or vector-value and tensor-value functions:

$$\mathbf{Q}(\mathbf{\Lambda}) = \mathbf{P}(\mathbf{\Lambda}) - \mathbf{F} = \boldsymbol{\theta} \tag{2.8}$$

where $\boldsymbol{\theta}$ is the column matrix composed by zero vector elements.

According to the fundamental theorem [1] of the inverse problem of the calculus of variations, we may construct the following functional for the variational principles of our problem:

$$\Phi(\mathbf{\Lambda}) = \int_0^1 \langle \mathbf{Q}(s\mathbf{\Lambda}), \mathbf{\Lambda}\rangle \,\mathrm{d}s \tag{2.9}$$

such that

$$\mathbf{Q} = \mathbf{grad}\,\Phi \tag{2.10}$$

Noting (2.6) to (2.8), then from (2.9), integrating for s, we get

$$\begin{aligned} \Phi(\mathbf{\Lambda}) &= \int_0^1 \langle \mathbf{P}(s\mathbf{\Lambda}), \mathbf{\Lambda}\rangle \,\mathrm{d}s - \int_0^1 \langle \mathbf{F}, \mathbf{\Lambda}\rangle \,\mathrm{d}s \\ &= \tfrac{1}{2}[\mathbf{M}\cdot\ddot{\mathbf{d}}, \mathbf{d}] - \tfrac{1}{2}[(\nabla\cdot\boldsymbol{\sigma} + \mathbf{e}:\mathbf{R}^T\boldsymbol{\sigma}), \mathbf{d}] - [\mathbf{f}, \mathbf{d}] \\ &\quad + \tfrac{1}{2}[(\mathbf{A}:\boldsymbol{\varepsilon} + \dot{\mathbf{B}} * \boldsymbol{\varepsilon}); \boldsymbol{\varepsilon}] - \tfrac{1}{2}[\boldsymbol{\sigma}; \boldsymbol{\varepsilon}] + \tfrac{1}{2}[(\nabla\mathbf{d} - \mathbf{e}\cdot\mathbf{R}\mathbf{d}); \boldsymbol{\sigma}] \\ &\quad - \tfrac{1}{2}[\boldsymbol{\varepsilon}; \boldsymbol{\sigma}] + \tfrac{1}{2}[\mathbf{t}, \mathbf{d}]_{\partial\Omega_t} - [\hat{\mathbf{t}}, \mathbf{d}]_{\partial\Omega_t} - \tfrac{1}{2}[\mathbf{d}, \mathbf{t}]_{\partial\Omega_d} + [\hat{\mathbf{d}}, \mathbf{t}]_{\partial\Omega_d} \\ &\quad + \tfrac{1}{2}[\mathbf{M}\cdot\dot{\mathbf{d}}, \dot{\mathbf{d}}]_0 - [\mathbf{M}\cdot\dot{\mathbf{d}}_0, \mathbf{d}]_0 - \tfrac{1}{2}[\mathbf{M}\cdot\mathbf{d}, \dot{\mathbf{d}}]_0 + [\mathbf{M}\cdot\mathbf{d}_0, \dot{\mathbf{d}}]_0. \end{aligned} \tag{2.11}$$

Considering

$$[\mathbf{M}\cdot\ddot{\mathbf{d}}, \mathbf{d}] = [\mathbf{M}\cdot\dot{\mathbf{d}}, \dot{\mathbf{d}}] + [\mathbf{M}\cdot\dot{\mathbf{d}}, \mathbf{d}]_0$$

and

$$[(\nabla\cdot\boldsymbol{\sigma} + \mathbf{e}:\mathbf{R}^T\boldsymbol{\sigma}), \mathbf{d}] = [\mathbf{t}, \mathbf{d}]_{\partial\Omega} - [(\nabla\mathbf{d} - \mathbf{e}\cdot\mathbf{R}\mathbf{d}); \boldsymbol{\sigma}]$$

then we can rewrite (2.11) in the following form:

$$\begin{aligned}\Phi(\boldsymbol{\lambda}) = {} & \tfrac{1}{2}[\mathbf{M}\cdot\dot{\mathbf{d}}, \mathbf{d}] + [\{(\nabla\mathbf{d} - \mathbf{e}\cdot\mathbf{R}\mathbf{d}) - \boldsymbol{\varepsilon}\}; \boldsymbol{\sigma}] \\ & - [\mathbf{f}, \mathbf{d}] + \tfrac{1}{2}[\{(\mathbf{A}:\boldsymbol{\varepsilon} + \dot{\mathbf{B}} * \boldsymbol{\varepsilon}) - \boldsymbol{\sigma}\}; \boldsymbol{\varepsilon}] \\ & - [\mathbf{t}, (\mathbf{d} - \hat{\mathbf{d}})]_{\partial\Omega_d} - [\hat{\mathbf{t}}, \mathbf{d}]_{\partial\Omega_t} \\ & + [\mathbf{M}\cdot(\dot{\mathbf{d}} - \dot{\mathbf{d}}_0), \mathbf{d}]_0 - \tfrac{1}{2}[\mathbf{M}\cdot(\mathbf{d} - 2\mathbf{d}_0), \dot{\mathbf{d}}]_0. \end{aligned} \tag{2.12}$$

Thus, from the expression (2.12), we can derive the functions of the variational principles correspnding to the fundamental relations (1.3), (1.6), (1.10) and the boundary conditions (1.14) and (1.15) as well as initial conditions (1.18) and (1.19), i.e. when $\mathbf{\Lambda}$ is a solution of the problem (1.3), (1.6), (1.10), (1.14), (1.15), (1.18) and (1.19), the functional $\Phi(\boldsymbol{\lambda})$ chooses its stationary value.

THEOREM: Let $\boldsymbol{\lambda} = (\mathbf{d}, \boldsymbol{\varepsilon}, \boldsymbol{\sigma}) \in S$ and Φ: $S \to \mathbb{R}$ has the linear Gâteaux differential for any $\boldsymbol{\lambda} \in \omega \subset S$, where $\Phi(\boldsymbol{\lambda})$ is a functional defined by (2.12), then $\boldsymbol{\lambda}$ is a solution of the problem (1.3), (1.6), (1.10), (1.14), (1.15), (1.18) and (1.19), if and only if the gradient of $\Phi(\boldsymbol{\lambda})$ satisfies the following condition:

$$\operatorname{grad}\Phi(\boldsymbol{\lambda}) \equiv \delta\Phi(\boldsymbol{\lambda}) = \boldsymbol{\theta}, \quad \boldsymbol{\lambda} \in \omega \subset S. \tag{2.13}$$

PROOF: This proof is similar to that of Theorem 5.1 in [1], and therefore it is not necessary to repeat it.

According to (2.12), we can also derive the functionals of the variational principles for some special cases.

3. Derivation of Gurtin's convolution method

When we derive the functional of the variational principles for the linear dynamic theory of micropolar viscoelasticity by Oden and Reddy's method, the initial conditions are obviously contained in the functional. This is different from Gurtin's convolution method.

In order to derive the functional of the variational principles corresponding to the problem by using Gurtin's method, it is necessary to prove the following lemma.

LEMMA: The vectors $\mathbf{d}$ and $\boldsymbol{\sigma}$ of the displacement vector and stress tensor functions satisfy the equations (1.3) of motion and the initial conditions (1.18) and (1.19), if and only if the following equation is satisfied

$$\mathbf{g} * (\nabla\cdot\boldsymbol{\sigma} + \mathbf{e}:\mathbf{R}^T\boldsymbol{\sigma}) + \mathbf{b} = \mathbf{M}\cdot\mathbf{d}, \quad (\mathbf{x}, t) \in \Omega \times [0, t_0], \tag{3.1}$$

where

$$\mathbf{g}(t) = \begin{bmatrix} g(t) & 0 \\ 0 & g(t) \end{bmatrix}, \qquad g(t) = t, \tag{3.2}$$

$$\mathbf{b} = \mathbf{g} * \mathbf{f} + \mathbf{M}\cdot(\mathbf{d}_0 - \mathbf{g}\dot{\mathbf{d}}_0). \tag{3.3}$$

PROOF: The proof is similar to that of the Lemma 5.5 in [1], and therefore it is not necessary to repeat it.

This lemma shows that the equations of motion and the initial conditions are coupled, i.e. the expression (3.1). Thus we may write (3.1), (1.6), (1.10), (1.14) and (1.15) in the following operator form:

$$\mathbf{Q}(\Lambda) \;=\; \mathbf{P}(\Lambda) - \mathbf{F} \;=\; \theta, \tag{3.4}$$

where

$$\left.\begin{aligned} \Lambda &= \{\mathbf{d},\, \boldsymbol{\varepsilon},\, \boldsymbol{\sigma};\, \mathbf{d},\, \mathbf{t}\}^T, \\ \mathbf{F} &= \{\mathbf{f},\, \mathbf{0},\, \mathbf{0};\, \mathbf{g} * \hat{\mathbf{t}},\, -\mathbf{g} * \hat{\mathbf{d}}\}^T, \end{aligned}\right\} \tag{3.5}$$

$$\mathbf{P} =$$

$$\begin{bmatrix} \mathbf{M}\,\cdot & \mathbf{0} & -\mathbf{g} * (\nabla\cdot + \mathbf{e}:\mathbf{R}^T) & \mathbf{0} & \mathbf{0} \\ \mathbf{0} & \mathbf{g} * (\mathbf{A}: + \dot{\mathbf{B}}*) & -\mathbf{g}* & \mathbf{0} & \mathbf{0} \\ \mathbf{g} * (\nabla - \mathbf{e}\cdot\mathbf{R}) & -\mathbf{g}* & \mathbf{0} & \mathbf{0} & \mathbf{0} \\ \mathbf{0} & \mathbf{0} & \mathbf{0} & \mathbf{0} & \mathbf{g}* \\ \mathbf{0} & \mathbf{0} & \mathbf{0} & -\mathbf{g}* & \mathbf{0} \end{bmatrix}. \tag{3.6}$$

Now the bilinear form of the vectors of vector and tensor functions is taken as the following

$$\langle \bar{\lambda}, \lambda \rangle \;=\; [\bar{\mathbf{d}}, \mathbf{d}] + [\bar{\boldsymbol{\sigma}}; \boldsymbol{\varepsilon}] + [\bar{\boldsymbol{\varepsilon}}; \boldsymbol{\sigma}] + [\bar{\mathbf{t}}, \mathbf{d}]_{\partial\Omega_d} + [\bar{\mathbf{d}}, \mathbf{t}]_{\partial\Omega_t}. \tag{3.7}$$

From (3.5) to (3.7), the functional $\Phi(\lambda)$ can be constructed, i.e.,

$$\begin{aligned} \Phi(\lambda) &= \int_0^1 \langle \mathbf{P}(s\Lambda), \Lambda\rangle\, \mathrm{d}s - \int_0^1 \langle \mathbf{F}, \Lambda\rangle\, \mathrm{d}s \\ &= \tfrac{1}{2}[\mathbf{M}\cdot\mathbf{d}, \mathbf{d}] - \tfrac{1}{2}[\mathbf{g} * (\nabla\cdot\boldsymbol{\sigma} + \mathbf{e}:\mathbf{R}^T\boldsymbol{\sigma}), \mathbf{d}] - [\mathbf{b}, \mathbf{d}] \\ &\quad + \tfrac{1}{2}[\mathbf{g} * (\mathbf{A}:\boldsymbol{\varepsilon} - \dot{\mathbf{B}} * \boldsymbol{\varepsilon}); \boldsymbol{\varepsilon}] - \tfrac{1}{2}[\mathbf{g} * \boldsymbol{\sigma}; \boldsymbol{\varepsilon}] \\ &\quad + \tfrac{1}{2}[\mathbf{g} * (\nabla\mathbf{d} - \mathbf{e}\cdot\mathbf{R}\mathbf{d}); \boldsymbol{\sigma}] - \tfrac{1}{2}[\mathbf{g} * \boldsymbol{\varepsilon}; \boldsymbol{\sigma}] \\ &\quad + \tfrac{1}{2}[\mathbf{g} * \mathbf{t}, \mathbf{d}]_{\partial\Omega_t} - [\mathbf{g} * \hat{\mathbf{t}}, \mathbf{d}]_{\partial\Omega_t} \\ &\quad - \tfrac{1}{2}[\mathbf{g} * \mathbf{d}, \mathbf{t}]_{\partial\Omega_d} + [\mathbf{g} * \hat{\mathbf{d}}, \mathbf{t}]_{\partial\Omega_d}. \end{aligned} \tag{3.8}$$

Using the divergence theorem, the above expression can be written in either one of the following forms:

$$\begin{aligned} \Phi(\lambda) &= \tfrac{1}{2}[\mathbf{M}\cdot\mathbf{d}, \mathbf{d}] - [\mathbf{b}, \mathbf{d}] + \tfrac{1}{2}[\mathbf{g} * (\mathbf{A}:\boldsymbol{\varepsilon} + \dot{\mathbf{B}} * \boldsymbol{\varepsilon}); \boldsymbol{\varepsilon}] \\ &\quad + \tfrac{1}{2}[\mathbf{g} * \{(\nabla\mathbf{d} - \mathbf{e}\cdot\mathbf{R}\mathbf{d}) - \boldsymbol{\varepsilon}\}; \boldsymbol{\sigma}] \\ &\quad - [\mathbf{g} * \hat{\mathbf{t}}, \mathbf{d}]_{\partial\Omega_t} - [\mathbf{g} * (\mathbf{d} - \hat{\mathbf{d}}), \mathbf{t}]_{\partial\Omega_d} \end{aligned} \tag{3.9}$$

or

$$\begin{aligned} \Phi(\lambda) &= \int_\Omega \{\tfrac{1}{2}\mathbf{M}\cdot\mathbf{d} * \mathbf{d} - \mathbf{b} * \mathbf{d} + \tfrac{1}{2}\mathbf{g} * (\mathbf{A}:\boldsymbol{\varepsilon} * \boldsymbol{\varepsilon} + \dot{\mathbf{B}} * \boldsymbol{\varepsilon} * \boldsymbol{\varepsilon}) \\ &\quad + \mathbf{g} * \{(\nabla\mathbf{d} - \mathbf{e}\cdot\mathbf{R}\mathbf{d}) - \boldsymbol{\varepsilon}\} * \boldsymbol{\sigma}\}\, \mathrm{d}x - \int_{\partial\Omega_t} \mathbf{g} * \hat{\mathbf{t}} * \mathbf{d}\, \mathrm{d}A \\ &\quad - \int_{\partial\Omega_d} \mathbf{g} * (\mathbf{d} - \hat{\mathbf{d}}) * \mathbf{t}\, \mathrm{d}A. \end{aligned} \tag{3.10}$$

This is the functional of the variational principles for the linear dynamic theory of micropolar viscoelasticity by using Gurtin's convolution method.

4. Derivation using Chien's method of the Lagrange multipliers

As mentioned in the above, using Oden and Reddy's method as well as Gurtin's method, we can obtain the functionals (2.12) and (3.9) (or (3.10)) of the variational principles for the linear dynamic theory of micropolar viscoelasticity; in fact they are the functionals of the Hu-Washizu form for generalized potential energy principles. Consequently we may draw the same conclusions by using Chien's method of Lagrange multipliers. For purposes of comparison, we give only the functional of the generalized potential energy principles corresponding to our problem.

For this reason, assuming ξ and ζ to be undetermined Lagrange multipliers of vectors of tensor-value and vector-value functions, then the functional of the generalized potential energy principles for the linear dynamic theory of micropolar viscoelasticity can be written as the following:

$$\begin{aligned}\Phi(\lambda) \;=\; & \tfrac{1}{2}[\mathbf{M}\cdot\mathbf{d},\,\mathbf{d}] - [\mathbf{b},\,\mathbf{d}] + \tfrac{1}{2}[\mathbf{g} * (\mathbf{A}:\varepsilon + \dot{\mathbf{B}} * \varepsilon);\,\varepsilon] - [\mathbf{g} * \hat{\mathbf{t}},\,\mathbf{d}]_{\partial\Omega_t} \\ & + [\mathbf{g} * \{\varepsilon - (\nabla\mathbf{d} - \mathbf{e}\cdot\mathbf{R}\mathbf{d})\};\,\xi] + [\mathbf{g} * (\mathbf{d} - \hat{\mathbf{d}}),\,\zeta]_{\partial\Omega_d}. \qquad (4.1)\end{aligned}$$

In carrying upon the variational process, $\mathbf{d}$, ε, ξ and ζ are treated as vector variables independent from each other, and therefore when $\Phi(\lambda)$ has reached its stationary value, we have $\delta\Phi(\lambda) = 0$, i.e.,

$$\begin{aligned}\delta\Phi(\lambda) \;=\; & [\mathbf{M}\cdot\mathbf{d},\,\delta\mathbf{d}] - [\mathbf{b},\,\delta\mathbf{d}] + [\mathbf{g} * (\mathbf{A}:\varepsilon + \dot{\mathbf{B}} * \varepsilon);\,\delta\varepsilon] \\ & - [\mathbf{g} * \hat{\mathbf{t}},\,\delta\mathbf{d}]_{\partial\Omega_t} + [\mathbf{g} * \{\varepsilon - (\nabla\mathbf{d} - \mathbf{e}\cdot\mathbf{R}\mathbf{d})\};\,\delta\xi] \\ & - [\mathbf{g} * \delta(\nabla\mathbf{d} - \mathbf{e}\cdot\mathbf{R}\mathbf{d});\,\xi] + [\mathbf{g} * \delta\varepsilon;\,\xi] \\ & + [\mathbf{g} * (\mathbf{d} - \hat{\mathbf{d}}),\,\delta\zeta]_{\partial\Omega_d} + [\mathbf{g} * \delta\mathbf{d},\,\zeta]_{\partial\Omega_d} \;=\; 0. \qquad (4.2)\end{aligned}$$

But from the expression

$$\begin{aligned}[\mathbf{g} * (\nabla\delta\mathbf{d} - \mathbf{e}\cdot\mathbf{R}\,\delta\mathbf{d}),\,\xi] \;=\; & [\mathbf{g} * \mathbf{N}\cdot\xi,\,\delta\mathbf{d}]_{\partial\Omega} \\ & - [\mathbf{g} * (\nabla\xi + \mathbf{e}:\mathbf{R}^T\xi),\,\delta\mathbf{d}].\end{aligned}$$

Therefore equation (4.1) can be written in the following form:

$$\begin{aligned}\delta\Phi(\lambda) \;=\; & [\{\mathbf{M}\cdot\mathbf{d} - \mathbf{b} + \mathbf{g} * (\nabla\xi + \mathbf{e}:\mathbf{R}^T\xi)\},\,\delta\mathbf{d}] \\ & + [\mathbf{g} * \{(\mathbf{A}:\varepsilon + \dot{\mathbf{B}} * \varepsilon) + \xi\};\,\delta\varepsilon] \\ & + [\mathbf{g} * \{\varepsilon - (\nabla\mathbf{d} - \mathbf{e}\cdot\mathbf{R}\mathbf{d})\};\,\delta\xi] \\ & + [\mathbf{g} * (\mathbf{d} - \hat{\mathbf{d}}),\,\delta\xi]_{\partial\Omega_d} + [\mathbf{g} * (\zeta - \mathbf{N}\cdot\xi),\,\delta\mathbf{d}]_{\partial\Omega_d} \\ & - [\mathbf{g} * (\hat{\mathbf{t}} + \mathbf{N}\cdot\xi),\,\delta\mathbf{d}]_{\partial\Omega_t} \;=\; 0. \qquad (4.3)\end{aligned}$$

Considering that $\delta\mathbf{d}$, $\delta\boldsymbol{\varepsilon}$, $\delta\boldsymbol{\xi}$ and $\delta\boldsymbol{\zeta}$ are treated as the variations of independent vector variables, from the previous expression, we get

In $\Omega \times [0, t_0]$

$$\mathbf{M}\cdot\mathbf{d} - \mathbf{b} + \mathbf{g} * (\nabla\boldsymbol{\xi} + \mathbf{e}:\mathbf{R}^T\boldsymbol{\xi}) = 0, \tag{4.4}$$

$$\mathbf{A}:\boldsymbol{\varepsilon} + \dot{\mathbf{B}}\boldsymbol{\varepsilon} = -\boldsymbol{\xi}, \tag{4.5}$$

$$\boldsymbol{\varepsilon} = \nabla\mathbf{d} - \mathbf{e}\cdot\mathbf{R}\mathbf{d}. \tag{4.6}$$

On $\partial\Omega_{\mathbf{d}} \times [0, t_0]$

$$\mathbf{d} = \hat{\mathbf{d}}, \qquad \boldsymbol{\zeta} = \mathbf{N}\cdot\boldsymbol{\xi} \tag{4.7}$$

On $\partial\Omega_{\mathbf{t}} \times [0, t_0]$

$$\hat{\mathbf{t}} = -\mathbf{N}\cdot\boldsymbol{\xi}. \tag{4.8}$$

Making use of the constitutive relations (1.10) of the relaxation form, then we can see the physical meaning of the vector Lagrange multipliers $\boldsymbol{\xi}$ and $\boldsymbol{\zeta}$:

$$\text{In} \quad \Omega \times [0, t_0], \qquad \boldsymbol{\xi} = -\boldsymbol{\sigma} \tag{4.9}$$

$$\text{On} \quad \partial\Omega_{\mathbf{d}} \times [0, t_0], \qquad \boldsymbol{\zeta} = -\mathbf{N}\cdot\boldsymbol{\sigma} = -\mathbf{t}. \tag{4.10}$$

Substituting the confirmed $\boldsymbol{\xi}$ and $\boldsymbol{\zeta}$ into (4.1), the the functional of the generalized potential energy principles for the linear dynamic theory of micropolar visco-elasticity can be obtained as the following:

$$\begin{aligned}\Phi(\boldsymbol{\lambda}) = {} & \tfrac{1}{2}[\mathbf{M}\cdot\mathbf{d}, \mathbf{d}] - [\mathbf{b}, \mathbf{d}] + \tfrac{1}{2}[\mathbf{g} * (\mathbf{A}:\boldsymbol{\varepsilon} + \dot{\mathbf{B}} * \boldsymbol{\varepsilon}); \boldsymbol{\varepsilon}] \\ & + [\mathbf{g} * \{(\nabla\mathbf{d} - \mathbf{e}\cdot\mathbf{R}\mathbf{d}) - \boldsymbol{\xi}\}; \boldsymbol{\sigma}] \\ & - [\mathbf{g} * (\mathbf{d} - \hat{\mathbf{d}}), \mathbf{t}]_{\partial\Omega_{\mathbf{d}}} - [\mathbf{g} * \hat{\mathbf{t}}, \mathbf{d}]_{\partial\Omega_{\mathbf{t}}}.\end{aligned} \tag{4.11}$$

Expanding the above functional into component form, then we get

$$\begin{aligned}\Phi = {} & \int_\Omega [\tfrac{1}{2}\varrho u_i * u_i + \tfrac{1}{2}I_{ij}\phi_j * \phi_i - g * \varrho f_i * u_i - g * \varrho l_i * \phi_i \\ & - \varrho(u_{i0} + t\dot{u}_{i0})u_i - I_{ij}(\phi_{j0} + t\dot{\phi}_{j0})\phi_i \\ & + \tfrac{1}{2}G_{ijkl}g * e_{kl} * e_{ij} + \tfrac{1}{2}g * \dot{G}_{ijkl} * e_{kl} * e_{ij} \\ & + \tfrac{1}{2}H_{ijkl}g * \mu_{kl} * e_{ij} + \tfrac{1}{2}g * \dot{H}_{ijkl} * \mu_{kl} * e_{ij} \\ & + \tfrac{1}{2}H_{ijkl}g * e_{kl} * \mu_{ij} + \tfrac{1}{2}g * \dot{H}_{ijkl} * e_{kl} * \mu_{ij} \\ & + \tfrac{1}{2}I_{ijkl}g * \mu_{kl} * \mu_{ij} + \tfrac{1}{2}g * \dot{I}_{ijkl} * \mu_{kl} * \mu_{ij} \\ & - g * \{e_{ij} - (\tfrac{1}{2}(u_{i,j} + u_{j,i}) - \varepsilon_{ijk}\phi_k)\} * t_{ij} \\ & - g * \{\mu_{ij} - \tfrac{1}{2}(\phi_{i,j} + \phi_{j,i})\} * m_{ij}]\,\mathrm{d}x \\ & - \int_{\partial\Omega_u} g * (u_i - \hat{u}_i) * t_i\,\mathrm{d}A - \int_{\partial\Omega_\phi} g * (\phi_i - \hat{\phi}_i) * m_i\,\mathrm{d}A \\ & - \int_{\partial\Omega_t} g * \hat{t}_i * u_i\,\mathrm{d}A - \int_{\partial\Omega_m} g * \hat{m}_i * \phi_i\,\mathrm{d}A.\end{aligned} \tag{4.12}$$

5. Several special cases

From (4.11) and (4.12), we can give out directly the results of several cases.

5.1. *The functional of the generalized potential energy principle for the linear dynamic theory of micropolar elasticity*

In (4.12), putting $\dot{G}_{ijkl} = \dot{H}_{ijkl} = \dot{I}_{jikl} = 0$, we have

$$\begin{aligned}\Phi_1 = \int_\Omega & [\tfrac{1}{2}\varrho u_i * u_i + \tfrac{1}{2}I_{ij}\phi_j * \phi_i - g * \varrho f_i * u_i \\ & - g * \varrho l_i * \phi_i - \varrho(u_{i0} + t\dot{u}_{i0})u_i - I_{ij}(\phi_{j0} + t\dot{\phi}_{j0})\phi_i \\ & + \tfrac{1}{2}G_{ijkl}g * e_{kl} * e_{ij} + \tfrac{1}{2}H_{ijkl}g * \mu_{kl} * e_{ij} \\ & + \tfrac{1}{2}H_{ijkl}g * e_{kl} * \mu_{ij} + \tfrac{1}{2}I_{ijkl}g * \mu_{kl} * \mu_{ij} \\ & - g * \{e_{ij} - (\tfrac{1}{2}(u_{i,j} + u_{j,i}) - \varepsilon_{ijk}\phi_k)\} * t_{ij} \\ & - g * \{\mu_{ij} - \tfrac{1}{2}(\phi_{i,j} + \phi_{j,i})\} * m_{ij}]\,\mathrm{d}x \\ & - \int_{\partial\Omega_u} g * (u_i - \hat{u}_i) * t_i\,\mathrm{d}A - \int_{\partial\Omega_\phi} g * (\phi_i - \hat{\phi}_i) * m_i\,\mathrm{d}A \\ & - \int_{\partial\Omega_t} g * \hat{t}_i * u_i\,\mathrm{d}A - \int_{\partial\Omega_m} g * \hat{m}_i * \phi_i\,\mathrm{d}A. \end{aligned} \tag{5.1}$$

From (5.1), we can obtain I. & M. Hlaváček's results [7].

5.2. *The functional of the generalized potential energy principle for the linear dynamic theory of viscoelasticity*

Neglecting the effect of micropolarization in (4.12), we have

$$\begin{aligned}\Phi_2 = \int_\Omega & [\tfrac{1}{2}\varrho u_i * u_i - g * \varrho f_i * u_i - \varrho(u_{i0} + t\dot{u}_{i0})u_i \\ & + \tfrac{1}{2}G_{ijkl}g * e_{kl} * e_{ij} + \tfrac{1}{2}g * \dot{G}_{ijkl} * e_{kl} * e_{ij} \\ & - g * \{e_{ij} - \tfrac{1}{2}(u_{i,j} + u_{j,i})\} * t_{ij} \\ & - \int_{\partial\Omega_u} g * (u_i - \hat{u}_i) * t_i\,\mathrm{d}A - \int_{\partial\Omega_t} g * \hat{t}_i * u_i\,\mathrm{d}A \end{aligned} \tag{5.2}$$

These represent Gurtin's [8] as well as Oden and Reddy's results [1].

5.3. *The functional of the generalized potential energy principle for the classical linear dynamic theory of elasticity*

In (5.2), if further stating $\dot{G}_{ijkl} = 0$, we obtain

$$\Phi_3 = \int_\Omega [\tfrac{1}{2}\varrho u_i * u_i - g * \varrho f_i * u_i - \varrho(u_{i0} + t\dot{u}_{i0})u_i$$

$$+ \tfrac{1}{2}G_{ijkl}g * e_{kl} * e_{ij} - g * \{e_{ij} - \tfrac{1}{2}(u_{i,j} + u_{j,i})\} * t_{ij}$$
$$- \int_{\partial\Omega_u} g * (u_i - \hat{u}_i) * t_i \, \mathrm{d}A - \int_{\partial\Omega_t} g * \hat{t}_i * u_i \, \mathrm{d}A.$$

These are Gurtin's results [8].

We can also derive other special cases which are not given here.

References

1. Oden, J. T. and Reddy, J. N., *Variational Methods in Theoretical Mechanics* (1976).
2. Gurtin, M. E., "Variational principles in the linear theory of viscoelasticity", *Arch. Rat. Mech. Anal.* **13** (1963), 179–191.
3. Chien Wei-zang, *Variational Method and Finite Elements, (I)* (in Chinese), Science Press, Beijing (1980).
4. Eringen, A. C., "Linear theory of micropolar viscoelasticity", *Int. J. Engng. Sci.* **5** (1967), 191–204.
5. Guo Zhong-heng, *Non-linear Elastic Theory* (in Chinese), Science Press, Beijing (1980).
6. Hu Hai-chang, *Variational Principles of Elasticity and Its Applications* (in Chinese), Science Press, Beijing (1981).
7. Hlaváček, I. & M., "On the existence and uniqueness of solutions and some variational principles in linear theories of elasticity with couple-stresses", *Aplikace Matematiky* **14** (1969), 387–426.
8. Gurtin, M. E., "Variational principles for linear electrodynamics", *Arch. Rat. Mech. Anal.* **16** (1964), 234–250.

An extended version of the transfer theorem

GUO ZHONG-HENG

Department of Mathematics, Peking University, Beijing, PRC

In Truesdell's presentation (p. 167 in [1]), the following:

LEMMA (Rivlin and Ericksen, Serrin, Noll). Let **F** map symmetric tensors, skew tensors, or orthogonal tensors onto tensors. If for all **A** of one of these kinds and for all orthogonal **Q**

$$\mathbf{F}(\mathbf{QAQ}^*) = \mathbf{QF}(\mathbf{A})\mathbf{Q}^*,$$

then every proper vector of **A** is a proper vector of **F**(**A**).

This plays a central role in the proof of representation theorems of isotropic tensor functions [1–5]. As can be seen, many researchers have contributed to the formulation of this lemma. Gurtin in his recent book called it the transfer theorem (p. 231 in [3]) and stated it as follows:

TRANSFER THEOREM. Let

$$\mathbf{G}: \mathscr{A} \to \mathrm{Lin} \quad (\mathscr{A} \subset \mathrm{Sym})$$

be isotropic. Then every eigenvector of $\mathbf{A} \in \mathscr{A}$ is an eigenvector of **G**(**A**).

Here the eigenvector is understood as the right eigenvector, if we distinguish the right from the left eigenvector of a tensor. In this brief note we show that an argument class, broader then symmetric, skew and orthogonal tensors, has a transfer character.

Let $\mathbb{R}$ be the field of real numbers and $\mathscr{V}$ a 3-dimensional Euclidean space with scalar product **uv** of $\mathbf{u}, \mathbf{v} \in \mathscr{V}$. Lin denotes the space of second-order tensors (or simply tensors) on $\mathscr{V}$, with the unity tensor **I**. If $\{\mathbf{e}_i\}$ is an orthonormal basis of $\mathscr{V}$, then the vectors **u**, **v** and tensors **S**, **T** can be expressed in terms of basis vectors:

$$\mathbf{u} = u_i\mathbf{e}_i, \qquad \mathbf{v} = v_j\mathbf{e}_j,$$

$$\mathbf{S} = S_{ij}\mathbf{e}_i \otimes \mathbf{e}_j, \qquad \mathbf{T} = T_{kl}\mathbf{e}_k \otimes \mathbf{e}_l.$$

We introduce the tensor–vector, vector–tensor and tensor–tensor operations in the following sense:

Yeh, K. Y. (ed) Progress in Applied Mechanics
 ISBN-13: 978-94-010-8061-3

$$\mathbf{Tu} = (T_{ij}\mathbf{e}_i \otimes \mathbf{e}_j)\mathbf{u} = T_{ij}(\mathbf{e}_j\mathbf{u})\mathbf{e}_i = T_{ij}u_j\mathbf{e}_i, \tag{1.1}$$

$$\mathbf{vT} = \mathbf{v}(T_{ji}\mathbf{e}_j \otimes \mathbf{e}_i) = T_{ji}(\mathbf{v}\mathbf{e}_j)\mathbf{e}_i = v_j T_{ji}\mathbf{e}_i, \tag{1.2}$$

$$\mathbf{ST} = (S_{ij}\mathbf{e}_i \otimes \mathbf{e}_j)(T_{kl}\mathbf{e}_k \otimes \mathbf{e}_l) = S_{ij}T_{kl}(\mathbf{e}_j\mathbf{e}_k)\mathbf{e}_i \otimes \mathbf{e}_l$$
$$= S_{ik}T_{kj}\mathbf{e}_i \otimes \mathbf{e}_j. \tag{1.3}$$

We call the first two operations the right-side action of **T** on **u** and the left-side action on **v**, respectively. The results of these two operations are vectors. Thus, a tensor equipped with the action is a linear map from $\mathscr{V}$ into $\mathscr{V}$. In general, the map corresponding to the right-side action differs from that which corresponds to the left-side action, unless the tensor is symmetric. However, according to the definition of transpose, the linear map materialized by the tensor **T** with left-side action can be equivalently materialized by the transpose (already with right-side action)

$$\mathbf{T}^* = T_{ji}\mathbf{e}_i \otimes \mathbf{e}_j$$

of **T**:

$$\mathbf{vT} = v_j T_{ji}\mathbf{e}_i = T_{ji}\mathbf{e}_i(\mathbf{e}_j\mathbf{v}) = (T_{ji}\mathbf{e}_i \otimes \mathbf{e}_j)\mathbf{v} = \mathbf{T}^*\mathbf{v}. \tag{1.4}$$

Therefore, "tensors" and "linear maps" are often used synonymously.

Every tensor **T** possesses at least a right eigenvector **r** ($|\mathbf{r}| = 1$),

$$\mathbf{Tr} = \lambda\mathbf{r}, \tag{1.5}$$

and a left eigenvector **l** ($|\mathbf{l}| = 1$),

$$\mathbf{lT} = \lambda\mathbf{l}, \tag{1.6}$$

λ being the associated eigenvalue. The eigenvalues are equal for the both eigenvectors, because by virtue of (1.4), equation (1.6) may be rewritten as

$$\mathbf{T}^*\mathbf{l} = \lambda\mathbf{l} \quad \text{or} \quad (\mathbf{T} - \lambda\mathbf{I})^*\mathbf{l} = 0$$

and the sufficient and necessary condition for the existence of **l** is

$$\det(\mathbf{T} - \lambda\mathbf{I}) = \det(\mathbf{T} - \lambda\mathbf{I})^* = 0,$$

which is at the same time the characteristic equation for (1.5). When $\mathbf{r} = \mathbf{l}$, we call **r** simply an eigenvector of **T**.

The sets of symmetric, skew, orthogonal and spherical (scalar multiples of **I**) elements in Lin are denoted by Sym, Skw, Orth and Sph, respectively. Every $\mathbf{Q} \in$ Orth possesses (modulo multiplicative constants) only one eigenvector **r**:

$$\mathbf{Qr} = \mathbf{rQ} = (\det\mathbf{Q})\mathbf{r}. \tag{1.7}$$

The results of the tensor–tensor operation (1.3) is a tensor, and may be interpreted as the composition of two (or more) linear maps.

DEFINITION: A mapping **F**: Lin → Lin is isotropic, only if

$$\mathbf{QF(T)Q}^* = \mathbf{F}(\mathbf{QTQ}^*), \quad \forall\mathbf{T} \in \text{Lin}, \quad \mathbf{Q} \in \text{Orth}. \tag{1.8}$$

Now we are able to prove the following

EXTENDED TRANSFER THEOREM: Let the mapping

$$\mathbf{F}\colon \text{Lin} \to \text{Lin}, \quad \mathbf{T} \mapsto \mathbf{F}(\mathbf{T})$$

be isotropic. Then every eigenvector of $\mathbf{T} \in \text{Lin}$ is an eigenvector of $\mathbf{F}(\mathbf{T})$.

PROOF: Let $\mathbf{r}$ be the eigenvector of $\mathbf{T}$:

$$\mathbf{Tr} = \mathbf{rT} = \lambda\mathbf{r}. \tag{1.9}$$

Introducing the orthogonal tensor

$$\mathbf{R} = -\mathbf{I} + 2\mathbf{r} \otimes \mathbf{r}$$

which is a rotation about $\mathbf{r}$ through π, and has $\mathbf{r}$ as its unique eigenvector

$$\mathbf{Rr} = \mathbf{rR} = \mathbf{r},$$

we have

$$\begin{aligned}\mathbf{RTR}^* &= (-\mathbf{I} + 2\mathbf{r} \otimes \mathbf{r})\mathbf{T}(-\mathbf{I} + 2\mathbf{r} \otimes \mathbf{r}) \\ &= \mathbf{T} - 2\mathbf{r} \otimes \mathbf{rT} - 2\mathbf{Tr} \otimes \mathbf{r} + 4\mathbf{r} \otimes \mathbf{rTr} \otimes \mathbf{r} = \mathbf{T},\end{aligned}$$

where equation (1.9) has been used. For this particular orthogonal tensor $\mathbf{R}$, the isotropy condition (1.8) becomes

$$\mathbf{RF}(\mathbf{T})\mathbf{R}^* = \mathbf{F}(\mathbf{T}) \quad \text{or} \quad \mathbf{RF}(\mathbf{T}) = \mathbf{F}(\mathbf{T})\mathbf{R},$$

and, consequently,

$$\begin{aligned}\mathbf{R}[\mathbf{F}(\mathbf{T})\mathbf{r}] &= \mathbf{F}(\mathbf{T})\mathbf{Rr} = \mathbf{F}(\mathbf{T})\mathbf{r}, \\ [\mathbf{rF}(\mathbf{T})]\mathbf{R} &= \mathbf{rRF}(\mathbf{T}) = \mathbf{rF}(\mathbf{T}).\end{aligned}$$

Thus, $\mathbf{F}(\mathbf{T})\mathbf{r}$ and $\mathbf{rF}(\mathbf{T})$ present the right and left eigenvectors (modulo multiplicative constants). Since $\mathbf{R}$ has only one eigenvector, $\mathbf{r}$, $\mathbf{F}(\mathbf{T})\mathbf{r}$ and $\mathbf{rF}(\mathbf{T})$ are co-linear with $\mathbf{r}$. This means that $\mathbf{r}$ is the eigenvector of $\mathbf{F}(\mathbf{T})$:

$$\mathbf{F}(\mathbf{T})\mathbf{r} = \xi\mathbf{r}, \qquad \mathbf{rF}(\mathbf{T}) = \xi\mathbf{r}. \qquad \text{Q.E.D.}$$

In order to show that Sym, Skw and Orth are just subsets of the set Eig of tensors which possess eigenvectors, it is enough to construct an element of Eig, not belonging to Sym, Skw or Orth. It is easily verified, that the sum $\mathbf{Q} + \mathbf{S}$ of any asymmetric $\mathbf{Q} \in \text{Orth}$ and $\mathbf{S} = \alpha\mathbf{I} \in \text{Sph}$ fulfils this condition, because $\mathbf{Q} + \mathbf{S}$ and $\mathbf{Q}$ have the same eigenvector $\mathbf{r}$:

$$(\mathbf{Q} + \mathbf{S})\mathbf{r} = \mathbf{Qr} + \mathbf{Sr} = (\det \mathbf{Q} + \alpha)\mathbf{r} = \mathbf{rQ} + \mathbf{rS} = \mathbf{r}(\mathbf{Q} + \mathbf{S}),$$

where equation (1.7) has been used.

References

1. Truesdell, C., *A First Course in Rational Continuum Mechanics*, Academic Press, New York (1977).
2. Truesdell, C. and Noll, W., "The non-linear field theories of mechanics", *Handbuch der Physik*, Vol. III/3, Berlin, Springer (1965).

3. Gurtin, M. E., *An Introduction to Continuum Mechanics*, Academic Press, New York (1981).
4. Guo Zhong-heng, "The representation theorem for isotropic, linear asymmetric stress–strain relations", *J. Elasticity* **13**, 2 (1983), 121–124.
5. Guo Zhong-heng, "An alternative proof of the representation theorem for isotropic, linear asymmetric stress–strain relations", *Quart. Appl. Math.* **41**, 1 (1983), 119–123.
6. Guo Zhong-heng, "Representations of orthogonal tensors", *SM Arch.* **6**, 4 (1981), 451–466.

General Mechanics

On the application of the screw calculus to the mechanics of spatial mechanisms

YU XIN

Peking Astronomical Observatory, Academia Sinica, Peking, PRC

Abstract. In this paper, the positional, kinematic and dynamic problems of the general five-link spatial mechanism with screw pairs have been solved by means of the Screw Calculus. The solutions are obtained via pure screw operations and the results are expressed in terms of given screws. The positional problem of some specialized five-link spatial mechanisms have already been solved by other authors using different methods; otherwise, the rest of the results are new.

1. Introduction

In 1963, Chace [1] published a paper on the vector analysis of linkages with the explicit purpose of describing the kinematics of mechanisms with direct appeal to geometric intuition, so that the derivation of the results may be interpreted geometrically at each step. Unfortunately, Chace considered only four-link mechanisms, and vector analysis seems to lose much of its power when faced with more complex mechanisms, especially those involving screw-pairs (by which we mean revolute, helical, prismatic or cylindrical). To remedy this, subsequent authors used various other mathematical systems for the analysis of spatial mechanisms; notably the Dual-Number Quaternion Algebra by Yang and Freudenstein [2], the Screw-Coordinate Method by Yuan, Freudenstein and Woo [3], the General Quaternion-Operator Method by Sandor [4], the Stretch-Rotation Tensor Method by Sandor and Bisshop [5] and many others besides. But none of these methods appear to retain the simplicity and geometric clarity of the Gibbsian vector analysis, save for the screw calculus originally developed by Ball [6], Kotelnikov [7], von Mises [8], Brand [9], and first applied to the analysis of Spatial Mechanisms by Dimentberg [10]. Yet few authors use it and those who do (such as Keler [11]) rarely go beyond writing the governing equations in terms of Dual Vectors and the results are usually obtained via some other method, such as the conventional matrices, and in the process much of the geometric significance is lost. In this paper we exploit the possibilities of the Screw Calculus to its fullest advantage in the analysis of Spatial Mechanisms. In particular, we shall demonstrate the power of the screw product $\boldsymbol{\alpha} \wedge \boldsymbol{\beta}$ and the dual scalar product $\boldsymbol{\alpha} \cdot \boldsymbol{\beta}$ in obtaining screw solutions to screw equations, in terms of which the analysis of the Spatial Mechanisms can usually be formulated. In so doing, we uphold much of the geometric spirit which is so

Yeh, K. Y. (ed) Progress in Applied Mechanics
ISBN-13: 978-94-010-8061-3

important in design and also bring into line the notation usually used in mechanics. We do this by presenting the establishment of the (i) Configuration (ii) Kinematics and (iii) Dynamics of the general five-link mechanism with screw pairs. It has been found that the problems usually reduce to the solutions of the simple screw equations of the following types:

$$\boldsymbol{\zeta} \wedge \boldsymbol{\alpha} = \boldsymbol{\beta} \quad (\boldsymbol{\zeta} \text{ unknown}) \tag{1.1}$$

$$\boldsymbol{\zeta} \wedge \boldsymbol{\alpha} + p\boldsymbol{\zeta} = \boldsymbol{\beta} \quad (\boldsymbol{\zeta} \text{ unknown}) \tag{1.2}$$

Moreover, the Jacobi identity

$$\boldsymbol{\alpha} \wedge (\boldsymbol{\beta} \wedge \boldsymbol{\lambda}) + \boldsymbol{\beta} \wedge (\boldsymbol{\lambda} \wedge \boldsymbol{\alpha}) + \boldsymbol{\lambda} \wedge (\boldsymbol{\alpha} \wedge \boldsymbol{\beta}) = 0 \tag{1.3}$$

has been found to possess great powers in the formulation of problems.

2. Some notations

We represent a line by the ordered pair $(\mathbf{a}, \mathbf{m}_0)$ where $\mathbf{a}$ is a unit vector in the direction of the line, the sense of the vector being arbitrary, and

$$\mathbf{m}_0 = \mathbf{r} \wedge \mathbf{a} \tag{2.1}$$

is the moment of the line with respect to the reference point 0 and $\mathbf{r}$ is the position vector of a generic point on the line with respect to 0. On introducing the "Clifford Symbol" which obeys the law

$$\varepsilon^2 = 0 \tag{2.2}$$

we have the one–one mapping,

$$(\mathbf{a}, \mathbf{m}_0) \Leftrightarrow \mathbf{a} + \varepsilon\mathbf{m}_0 \tag{2.3}$$

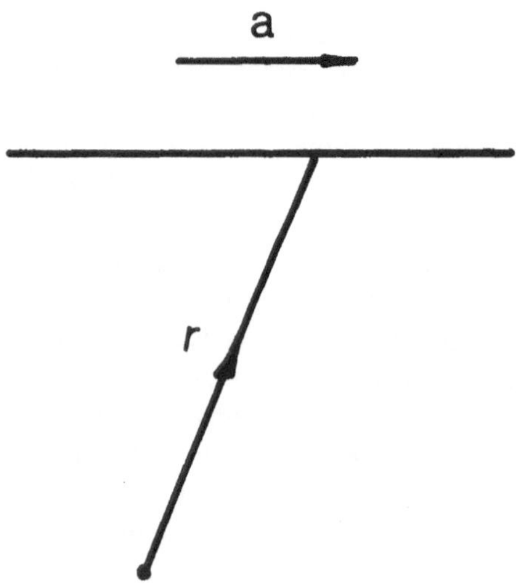

Figure 1

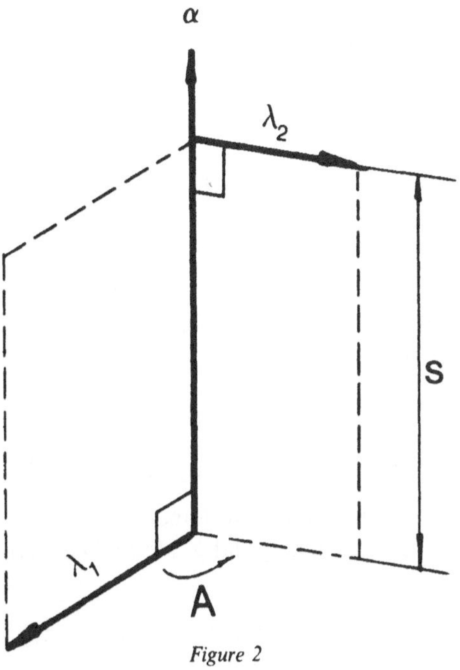

Figure 2

and, in putting

$$\boldsymbol{\lambda} = (\mathbf{a} + \varepsilon \mathbf{m}_0) \tag{2.4}$$

we call $\boldsymbol{\lambda}$ the "dual unit line vector" which represents the line. Evidently, from the definitions (2.2), (2.3) and (2.4) we have,

$$\boldsymbol{\lambda} \cdot \boldsymbol{\lambda} = 1 \tag{2.5}$$

and hence the name.

A screw $\boldsymbol{\phi}$ with intensity R and pitch p on the line is given by,

$$\boldsymbol{\phi} = \mathbf{R}\, e^{\varepsilon p}\, \boldsymbol{\lambda}. \tag{2.6}$$

Now, if A is the angle between the projection of two lines $\boldsymbol{\lambda}_1$ and $\boldsymbol{\lambda}_2$ on a plane parallel to them and S is the perpendicular distance between two lines (Fig. 2), then we can introduce the "dual angle" θ after Study [12], given by

$$\theta = A + \varepsilon S. \tag{2.7}$$

The "dual scalar product" of the given lines is then given by

$$\boldsymbol{\lambda}_1 \cdot \boldsymbol{\lambda}_2 = \cos\theta \tag{2.8}$$

and the "screw product" of the two lines is given by

$$\boldsymbol{\lambda}_1 \wedge \boldsymbol{\lambda}_2 = \sin\theta\, \boldsymbol{\alpha}, \tag{2.9}$$

where $\boldsymbol{\alpha}$ is the dual unit line vector whose line intersects $\boldsymbol{\lambda}_1$ and $\boldsymbol{\lambda}_2$ at right angles.

With these definitions, it has been shown (e.g., [9] and [10]) that there is a complete parallel between the ordinary vector algebra of Gibbs and the algebra of screws. In particular, we have

$$\boldsymbol{\phi}_1 \wedge (\boldsymbol{\phi}_2 \wedge \boldsymbol{\phi}_3) = (\boldsymbol{\phi}_1 \cdot \boldsymbol{\phi}_3)\boldsymbol{\phi}_2 - (\boldsymbol{\phi}_1 \cdot \boldsymbol{\phi}_2)\boldsymbol{\phi}_3. \tag{2.10}$$

3. The configuration of the general five-link mechanism with screw pairs

In Fig. 3 we denote the screw axes by the unit screws $\boldsymbol{\alpha}_i$ ($i = 0, 1, \ldots, 4$); in particular, $\boldsymbol{\alpha}_0$ and $\boldsymbol{\alpha}_4$ are the input and output screw axes respectively. The normal connecting bar between the ith and jth bar are denoted by the unit screws $\boldsymbol{\lambda}_{ij}$. Now, for a given input position of $\boldsymbol{\lambda}_{01}$, the axis $\boldsymbol{\alpha}_1$ is completely determined and so is the mutual normal $\boldsymbol{\lambda}_{41}$ between $\boldsymbol{\alpha}_4$ and $\boldsymbol{\alpha}_1$, since $\boldsymbol{\alpha}_4 \wedge \boldsymbol{\alpha}_1 = \sin \zeta_4 \boldsymbol{\lambda}_{41}$ where ζ_4 is the dual angle between $\boldsymbol{\alpha}_4$ and $\boldsymbol{\alpha}_1$ which is given by $\cos \zeta_4 = \boldsymbol{\alpha}_4 \cdot \boldsymbol{\alpha}_1$. Hence we have the following data:

Given: $\boldsymbol{\alpha}_4, \boldsymbol{\alpha}_1, \boldsymbol{\lambda}_{41}$
Unknowns to be determined: $\boldsymbol{\alpha}_2, \boldsymbol{\alpha}_3, \boldsymbol{\lambda}_{12}, \boldsymbol{\lambda}_{23}, \boldsymbol{\lambda}_{34}$.

However, $\boldsymbol{\lambda}_{12}, \boldsymbol{\lambda}_{23}, \boldsymbol{\lambda}_{34}$ are completely determined once $\boldsymbol{\alpha}_2$ and $\boldsymbol{\alpha}_3$ are, and hence the problem is reduced to determining $\boldsymbol{\alpha}_2$ and $\boldsymbol{\alpha}_3$ alone. We have the following Jacobi identities:

$$\boldsymbol{\alpha}_2 \wedge (\boldsymbol{\alpha}_4 \wedge \boldsymbol{\alpha}_1) + \boldsymbol{\alpha}_4 \wedge (\boldsymbol{\alpha}_1 \wedge \boldsymbol{\alpha}_2) + \boldsymbol{\alpha}_1 \wedge (\boldsymbol{\alpha}_2 \wedge \boldsymbol{\alpha}_4) = 0, \tag{3.1}$$

$$\boldsymbol{\alpha}_3 \wedge (\boldsymbol{\alpha}_4 \wedge \boldsymbol{\alpha}_2) + \boldsymbol{\alpha}_4 \wedge (\boldsymbol{\alpha}_2 \wedge \boldsymbol{\alpha}_3) + \boldsymbol{\alpha}_2 \wedge (\boldsymbol{\alpha}_3 \wedge \boldsymbol{\alpha}_4) = 0. \tag{3.2}$$

Hence

$$[(\boldsymbol{\alpha}_4 \wedge \boldsymbol{\alpha}_1) \wedge (3.1), (\boldsymbol{\alpha}_4 \wedge \boldsymbol{\alpha}_2) \wedge (3.2)] \Rightarrow$$

$$(1 - l_{14}^2)\boldsymbol{\alpha}_2 = V_{412}(\boldsymbol{\alpha}_4 \wedge \boldsymbol{\alpha}_2) + (y - l_{12}l_{41})\boldsymbol{\alpha}_4 + (l_{12} - yl_{14})\boldsymbol{\alpha}_1, \tag{3.3}$$

$$(1 - l_{24}^2)\boldsymbol{\alpha}_3 = V_{423}(\boldsymbol{\alpha}_4 \wedge \boldsymbol{\alpha}_2) + (l_{34} - yl_{23})\boldsymbol{\alpha}_4 + (l_{23} - yl_{34})\boldsymbol{\alpha}_2, \tag{3.4}$$

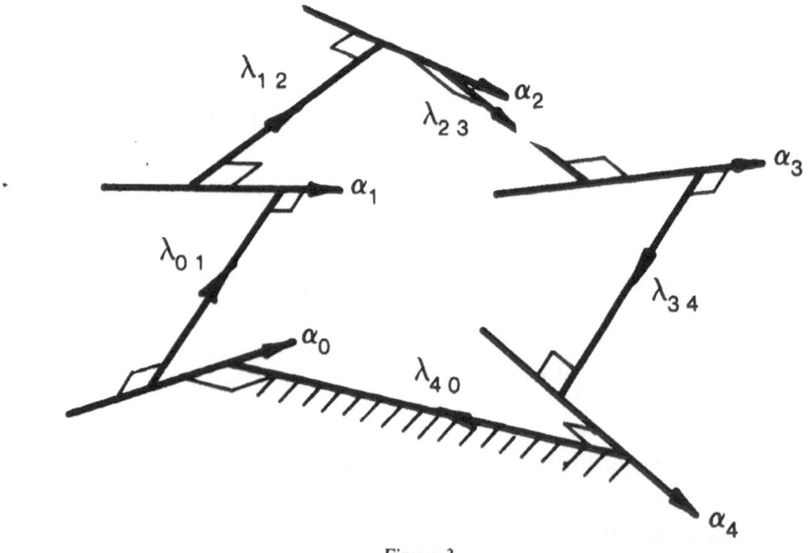

Figure 3

where

$$V_{ijk} = (\boldsymbol{\alpha}_i \wedge \boldsymbol{\alpha}_j) \cdot \boldsymbol{\alpha}_k, \tag{3.5}$$

$$l_{ij} = \boldsymbol{\alpha}_i \cdot \boldsymbol{\alpha}_j, \tag{3.6}$$

$$y = l_{24}. \tag{3.7}$$

However,

$$[\boldsymbol{\alpha}_2 \cdot (3.3),\ \boldsymbol{\alpha}_3 \cdot (3.4)] \Rightarrow$$

$$V_{412} = \pm\sqrt{(1 - l_{41}^2 - l_{12}^2 - y^2 + 2yl_{41}l_{12})}, \tag{3.8}$$

$$V_{423} = \pm\sqrt{(1 - l_{23}^2 - l_{43}^2 - y^2 + 2yl_{23}l_{34})}. \tag{3.9}$$

The sign of V_{ijk} is determined as follows: If the ordered set $(\boldsymbol{\alpha}_i, \boldsymbol{\alpha}_j, \boldsymbol{\alpha}_k)$ form a right-handed set, we take the positive sign; if not, the negative sign. This convention follows from the definition of the screw product $\boldsymbol{\mu} \wedge \boldsymbol{\lambda} = \boldsymbol{\alpha}$, where the right-handed screw rule is followed. Hence, $\boldsymbol{\alpha}_2$ and $\boldsymbol{\alpha}_3$ are completely determined by (3.3) and (3.4) once y is known.

To determine y we state the eight fundamental relations:

$$\boldsymbol{\alpha}_1 \wedge \boldsymbol{\alpha}_2 = g_1\boldsymbol{\lambda}_{12}, \tag{3.10}$$

$$\boldsymbol{\alpha}_2 \wedge \boldsymbol{\alpha}_3 = g_2\boldsymbol{\lambda}_{23}, \tag{3.11}$$

$$\boldsymbol{\alpha}_3 \wedge \boldsymbol{\alpha}_4 = g_3\boldsymbol{\lambda}_{34}, \tag{3.12}$$

$$\boldsymbol{\alpha}_4 \wedge \boldsymbol{\alpha}_1 = g_4\boldsymbol{\lambda}_{41}, \tag{3.13}$$

$$\boldsymbol{\lambda}_{41} \wedge \boldsymbol{\lambda}_{12} = \mu_1\boldsymbol{\alpha}_1, \tag{3.14}$$

$$\boldsymbol{\lambda}_{12} \wedge \boldsymbol{\lambda}_{23} = \mu_2\boldsymbol{\alpha}_2, \tag{3.15}$$

$$\boldsymbol{\lambda}_{23} \wedge \boldsymbol{\lambda}_{34} = \mu_3\boldsymbol{\alpha}_3, \tag{3.16}$$

$$\boldsymbol{\lambda}_{34} \wedge \boldsymbol{\lambda}_{41} = \mu_4\boldsymbol{\alpha}_4, \tag{3.17}$$

where

$$g_i = \sin\zeta_i, \qquad \mu_i = \sin\theta_i. \tag{3.18}$$

On scalar multiplying the consecutive pairs together (i.e. (3.10) · (3.11), . . . , (3.16) · (3.17)) ⇒

$$x = p_1p_2 - g_1g_2v_2, \tag{3.19}$$

$$y = p_1p_3 - g_2g_3v_3, \tag{3.20}$$

$$x = p_3p_4 - g_3g_4v_4 \tag{3.21}$$

$$y = p_4p_1 - g_4g_1v_1, \tag{3.22}$$

$$u = v_1v_2 - p_1\mu_1\mu_2, \tag{3.23}$$

$$v = v_2v_3 - p_2\mu_2\mu_3, \tag{3.24}$$

$$u = v_3v_4 - p_3\mu_3\mu_4, \tag{3.25}$$

$$v = v_4v_1 - p_4\mu_4\mu_1, \tag{3.26}$$

where

$$p_i = \cos \zeta_i, \quad v_i = \cos \theta_i, \tag{3.27}$$

$$u = \boldsymbol{\lambda}_{41} \cdot \boldsymbol{\lambda}_{23}, \quad v = \boldsymbol{\lambda}_{12} \cdot \boldsymbol{\lambda}_{34}, \tag{3.28}$$

and $x = l_{13}$.

The above system constitutes eight relations for the eight unknowns $(x, y, u, v, \theta_1, \ldots, \theta_4)$. However, we shall delete (3.25) and (3.26) and in their place we have,

(3.10) · (3.12), (3.11) · (3.13) ⇒

$$xy - p_2 p_4 = g_1 g_3 v, \tag{3.29}$$

$$xy - p_3 p_1 = g_2 g_4 u. \tag{3.30}$$

Now, $((3.23)^2, (3.24)^2) \Rightarrow$

$$p_1^2(1 - v_1^2)(1 - v_2^2) = (v_1 v_2 - u)^2, \tag{3.31}$$

$$(1 - \quad)(1 - \quad) = (v_2 v_3 - v)^2, \tag{3.32}$$

where we have used the relation,

$$\cos^2\theta + \sin^2\theta = 1.$$

Hence, on replacing v_1, v_2, v_3, u, v in (3.31) and (3.32) by the corresponding functions of x and y given by (3.22), (3.19), (3.20), (3.30) and (3.29) respectively, we obtain two simultaneous equations in x and y in the following form:

$$a_1 x^2 y^2 + b_1 x^2 y + c_1 x y^2 + d_1 x^2 + e_1 y^2 + f_1 xy + g_1 x + h_1 y + k_1 = 0, \tag{3.33}$$

$$a_2 x^2 y^2 + b_2 x^2 y + c_2 x y^2 + d_2 x^2 + e_2 y^2 + f_2 xy + g_2 x + h_2 y + k_2 = 0, \tag{3.34}$$

where $(a_1, \ldots, k_1; a_2, \ldots, k_2)$ are known functions of the given quantities.

Equations (3.33) and (3.34) can be written as:

$$\alpha_1 x^2 + 2\beta_1 x + \gamma_1 = 0 \tag{3.35}$$

$$\alpha_2 x^2 + 2\beta_2 x + \gamma_2 = 0 \tag{3.36}$$

where,

$$\left.\begin{aligned} \alpha_i &= a_i y^2 + b_i y + d_i \\ 2\beta_i &= c_i y^2 + f_i y + g_i \\ \gamma_i &= e_i y^2 + h_i y + k_i \end{aligned}\right\} \quad i = 1, 2. \tag{3.37}$$

Now if (3.35) and (3.36) are to have common roots we must have

$$\frac{-\beta_1 \pm \sqrt{(\beta_1^2 - \alpha_1\gamma_1)}}{\alpha_1} = \frac{-\beta_2 \pm \sqrt{(\beta_2^2 - \alpha_2\gamma_2)}}{\alpha_2}, \tag{3.38}$$

i.e.,

$$\alpha_1\beta_2 - \alpha_2\beta_1 = \pm\alpha_1\sqrt{(\beta_2^2 - \alpha_2\gamma_2)} \pm \alpha_2\sqrt{(\beta_1^2 - \alpha_1\gamma_1)}. \tag{3.39}$$

On squaring both sides of (3.39) and dividing by $\alpha_1\alpha_2$ and then squaring again, we obtain

$$(\alpha_1\gamma_2 + \alpha_2\gamma_1 - 2\beta_1\beta_2)^2 = 4(\beta_2^2 - \alpha_2\gamma_2)(\beta_1^2 - \alpha_1\gamma_1). \tag{3.40}$$

Since $(\alpha_1, \ldots; \alpha_2, \ldots)$ are all quadratic functions in y, (3.40) is a polynomial equation in y of the 8-th degree. The above constitutes a general solution (in principle) for the general five-link mechanism with screw pairs. Special cases have been considered by other authors: e.g. the R–C–R–C–R was solved by Dimentberg, [13] whose solution depends on an equation of the 8-th degree whilst the same mechanism was later solved by Yang [14] who obtained a 4th degree polynomial. The R–R–C–C–R mechanism was considered by Yuan [15] who also obtained an 8-th degree polynomial for its solution.

4. Kinematics of the general five-link mechanism with screw pairs

The problem here is to determine the instantaneous screw

$$\mathbf{\Omega}_i \ (i = 2, 3, 4).$$

From (3.10) and (3.3) respectively, we have

$$\boldsymbol{\alpha}_1 \wedge \boldsymbol{\alpha}_2 = g_1\boldsymbol{\lambda}_{12}, \tag{4.1}$$

$$(1 - l_{14}^2)\boldsymbol{\alpha}_2 = \mathbf{B}, \tag{4.2}$$

where

$$\mathbf{B} = V_{412}\boldsymbol{\alpha}_4 \wedge \boldsymbol{\alpha}_1 + (y - l_{12}l_{41})\boldsymbol{\alpha}_4 + (l_{12} - l_{14}y)\boldsymbol{\alpha}_1 \tag{4.3}$$

Now

$$\left[\frac{\mathrm{d}}{\mathrm{d}t}(4.1)\right] \Rightarrow$$

$$(\mathbf{\Omega}_1 \wedge \boldsymbol{\alpha}_1) \wedge \boldsymbol{\alpha}_2 + \boldsymbol{\alpha}_1 \wedge (\mathbf{\Omega}_2 \wedge \boldsymbol{\alpha}_2) = g_1\mathbf{\Omega}_2 \wedge \boldsymbol{\lambda}_{12}$$

where $\mathbf{\Omega}_i$ is the instantaneous screw of the ith body. The above equation then simplifies to:

$$g_1\boldsymbol{\lambda}_{12} \wedge \mathbf{\Omega}_2 + p_1\mathbf{\Omega}_2 = x\boldsymbol{\alpha}_2 - \mathbf{A}, \tag{4.4}$$

where

$$\mathbf{A} = (\mathbf{\Omega}_1 \wedge \boldsymbol{\alpha}_1) \wedge \boldsymbol{\alpha}_2, \quad x = \mathbf{\Omega}_2 \cdot \boldsymbol{\alpha}_1 \tag{4.5}$$

and $\mathbf{A}$ is known whilst x is not. Thus equation (4.4) is a screw equation in the unknowns $\mathbf{\Omega}_2$ and x. To solve equation (4.4) for $\mathbf{\Omega}_2$ we have,

$$[g_1\boldsymbol{\lambda}_{12} \wedge (4.4)] \Rightarrow$$

$$(\boldsymbol{\lambda}_{12} \cdot \mathbf{\Omega}_2)\boldsymbol{\lambda}_{12} - \mathbf{\Omega}_2 + p_1g_1\boldsymbol{\lambda}_{12} \wedge \mathbf{\Omega}_2 = g_1(x\boldsymbol{\lambda}_{12} \wedge \boldsymbol{\alpha}_2 + \mathbf{A} \wedge \boldsymbol{\lambda}_{12}). \tag{4.6}$$

Also

$$p_1(4.5) \Rightarrow$$

$$p_1 g_1 \boldsymbol{\lambda}_{12} \wedge \boldsymbol{\Omega}_2 + p_1^2 \boldsymbol{\Omega}_2 = x p_1 \boldsymbol{\alpha}_2 - p_1 \mathbf{A}, \tag{4.7}$$

$$[(4.7) - (4.6)] \Rightarrow$$

$$\boldsymbol{\Omega}_2 = x(p_1 \alpha_2 + g_1 \boldsymbol{\alpha}_2 \wedge \boldsymbol{\lambda}_{12}) - (p_1 \mathbf{A} + g_1 \mathbf{A} \wedge \boldsymbol{\lambda}_{12}) + g_1^2(\boldsymbol{\lambda}_{12} \cdot \boldsymbol{\Omega}_2) \tag{4.8}$$

to determine $(\boldsymbol{\Omega}_2 \cdot \boldsymbol{\lambda}_{12})$ we have

$$[\boldsymbol{\lambda}_{12} \cdot (4.4)] \Rightarrow$$

$$\boldsymbol{\Omega}_2 \cdot \boldsymbol{\lambda}_{12} = -\frac{\mathbf{A} \cdot \boldsymbol{\lambda}_{12}}{p_1}$$

and (4.8) becomes

$$\boldsymbol{\Omega}_2 = x(p_1 \boldsymbol{\alpha}_2 + g_1 \boldsymbol{\alpha}_2 \wedge \boldsymbol{\lambda}_{12}) - (p_1 \mathbf{A} + g_1 \mathbf{A} \wedge \boldsymbol{\lambda}_{12}) - \frac{g_1^2(\mathbf{A} \cdot \boldsymbol{\lambda}_{12})\boldsymbol{\lambda}_{12}}{p_1}. \tag{4.9}$$

Or

$$\boldsymbol{\Omega}_2 = x\boldsymbol{\alpha}_2 \cdot \mathbf{T}_1 - \frac{1}{p_1} \mathbf{A} \cdot \mathbf{T}_2, \tag{4.10}$$

$$\mathbf{T}_1 = (p_1 \mathbf{I} + \mathbf{I} \wedge \boldsymbol{\lambda}_{12}), \quad \mathbf{T}_2 = (p_1^2 \mathbf{I} + p_1 g_1 \mathbf{I} \wedge \boldsymbol{\lambda}_{12} + g_1^2 \boldsymbol{\lambda}_{12} \boldsymbol{\lambda}_{12}), \tag{4.11}$$

are the dual tensors and $\mathbf{I}$ is the unit dual tensor. Hence, in order to establish $\boldsymbol{\Omega}_2$ completely, we still must find x.

Now

$$\left[\frac{\mathrm{d}}{\mathrm{d}t}(4.2)\right] \Rightarrow$$

$$(1 - l_{14}^2)\boldsymbol{\Omega}_2 \wedge \boldsymbol{\alpha}_2 = 2 l_{14} \dot{l}_{14} \boldsymbol{\alpha}_2 = \dot{\mathbf{B}}, \tag{4.12}$$

which is a screw equation in the unknown $\boldsymbol{\Omega}_2$. To solve for $\boldsymbol{\Omega}_2$ we have

$$[\boldsymbol{\alpha}_1 \wedge (4.12)] \Rightarrow$$

$$(1 - l_{14}^2)[l_{12}\boldsymbol{\Omega}_2 - x\boldsymbol{\alpha}_2] - 2 l_{14} \dot{l}_{14} \boldsymbol{\alpha}_1 \wedge \boldsymbol{\alpha}_2 = \boldsymbol{\alpha}_1 \wedge \dot{\mathbf{B}}.$$

Hence

$$\boldsymbol{\Omega}_2 = \frac{\boldsymbol{\alpha}_1 \wedge \mathbf{B} + 2 l_{14} \dot{l}_{14} g_1 \boldsymbol{\lambda}_{12}}{p_1(1 - l_{14})} + \frac{x}{p_1} \boldsymbol{\alpha}_2. \tag{4.13}$$

Hence

$$((4.10), (4.13)) \Rightarrow$$

$$x\boldsymbol{\alpha}_2 \cdot \mathbf{T}_1 - \frac{1}{p_1} \mathbf{A} \cdot \mathbf{T}_2 = \frac{\boldsymbol{\alpha}_1 \wedge \dot{\mathbf{B}} + 2 l_{14} \dot{l}_{14} g_1 \boldsymbol{\lambda}_{12}}{p_1(1 - l_{14})} + \frac{x}{p_1} \boldsymbol{\alpha}_2. \tag{4.14}$$

$\therefore \quad [\boldsymbol{\alpha}_2 \cdot (4.14)] \Rightarrow$ (in view of (4.11))

$$p_1 x - (p_1 \mathbf{A} \cdot \boldsymbol{\alpha}_2 + g_1 \mathbf{A} \wedge \boldsymbol{\lambda}_{12} \cdot \boldsymbol{\alpha}_2) = \frac{\boldsymbol{\alpha}_2 \wedge \boldsymbol{\alpha}_1 \cdot \dot{\mathbf{B}}}{p_1(1 - l_{14}^2)} + \frac{x}{p_1}.$$

On transposing terms we have

$$x = \frac{(\boldsymbol{\alpha}_2 \wedge \boldsymbol{\alpha}_1 \cdot \dot{\mathbf{B}}_1)}{(p_1^2 - 1)(1 - l_{14}^2)} + \frac{p_1}{(p_1^2 - 1)}(g_1 \mathbf{A} \wedge \boldsymbol{\lambda}_{12} + p_1 \mathbf{A}) \cdot \boldsymbol{\alpha}_2. \tag{4.15}$$

Or

$$x = \frac{\boldsymbol{\alpha}_2 \wedge \boldsymbol{\alpha}_1 \cdot \dot{\mathbf{B}}}{(p_1^2 - 1)(1 - l_{14}^2)} - \frac{p_1}{(p_1^2 - 1)} \mathbf{A} \cdot \mathbf{T}_1 \cdot \boldsymbol{\alpha}_2. \tag{4.16}$$

Hence $\boldsymbol{\Omega}_2$ is now given by (4.10) and (4.16).

Once $\boldsymbol{\Omega}_2$ is established, $\boldsymbol{\Omega}_3$ and $\boldsymbol{\Omega}_4$ are very easily obtained:

$$\boldsymbol{\Omega}_3 = \boldsymbol{\Omega}_{32} + \boldsymbol{\Omega}_2 \quad \text{and} \quad \boldsymbol{\Omega}_3 = \boldsymbol{\Omega}_{34} + \boldsymbol{\Omega}_4. \tag{4.17}$$

Moreover,

$$\boldsymbol{\Omega}_{32} = \Omega_{32}\boldsymbol{\alpha}_2, \qquad \boldsymbol{\Omega}_{34} = \Omega\boldsymbol{\alpha}_3, \qquad \boldsymbol{\Omega}_4 = \Omega_4\boldsymbol{\alpha}_4. \tag{4.18}$$

Hence, the system (4.17) and (4.18) gives:

$$\boldsymbol{\Omega}_2 = -\Omega_{32}\boldsymbol{\alpha}_2 + \Omega_{34}\boldsymbol{\alpha}_3 + \Omega_4\boldsymbol{\alpha}_4. \tag{4.19}$$

Now

$$\boldsymbol{\alpha}_3 \wedge \boldsymbol{\alpha}_4 \cdot (4.19), \qquad \boldsymbol{\alpha}_4 \wedge \boldsymbol{\alpha}_2 \cdot (4.19), \qquad \boldsymbol{\alpha}_2 \wedge \boldsymbol{\alpha}_3 \cdot (4.19)$$

respectively give

$$\left.\begin{aligned} \Omega_{32} &= -\left(\frac{\boldsymbol{\alpha}_3 \wedge \boldsymbol{\alpha}_2}{V}\right) \cdot \boldsymbol{\Omega}_2, \\ \Omega_{34} &= \left(\frac{\boldsymbol{\alpha}_4 \wedge \boldsymbol{\alpha}_2}{V}\right) \cdot \boldsymbol{\Omega}_2, \\ \Omega_4 &= \left(\frac{\boldsymbol{\alpha}_2 \wedge \boldsymbol{\alpha}_3}{V}\right) \cdot \boldsymbol{\Omega}_2, \end{aligned}\right\} \tag{4.20}$$

where

$$V = \boldsymbol{\alpha}_2 \wedge \boldsymbol{\alpha}_3 \cdot \boldsymbol{\alpha}_4$$

and hence $\boldsymbol{\Omega}_3$ and $\boldsymbol{\Omega}_4$ are now completely established.

5. Dynamics of the R–R–C–C–R mechanism

5.1. *The problem*

The Dynamics of Spatial Mechanisms is a relatively easy matter once the configuration and the Instantaneous Screws of the mechanism are determined. The

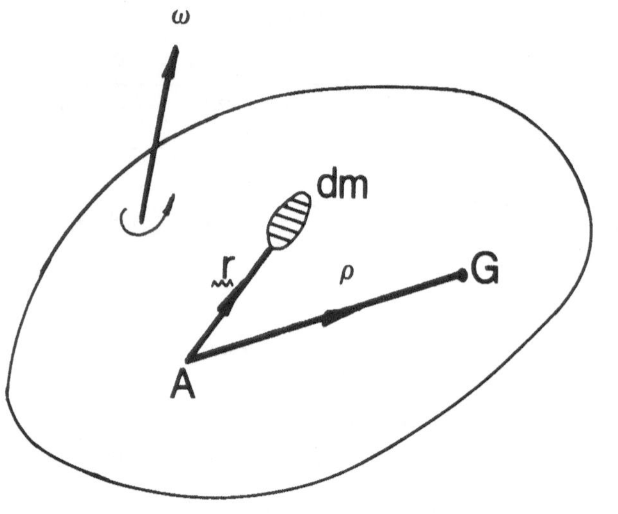

Figure 4

problem here is to determine the input Force Screw (the load) for a given state of motion, and to determine the Reaction Screws at the joints between the successive links.

5.2. *General formulation*

The solution of the problem inevitably depends on the Balance Law for the Momentum Screw of Rigidbody, which may be formulated as follows:

We choose a base point A rigidly connected to the body of mass m, whose centroid is at G. Then the Momentum Screw referred to A is given by

$$\boldsymbol{\mu}(A) = \mathbf{L}(A) + \varepsilon \mathbf{H}_A \tag{5.1}$$

where

$$\mathbf{L}(A) = m(\mathbf{v}_A + \boldsymbol{\omega} \wedge \boldsymbol{\varrho}); \qquad \boldsymbol{\varrho} = \mathbf{AG} \tag{5.2}$$

is the "linear momentum" of the body, and is a line vector through A and

$$\mathbf{H}_A = \int_m (r^2\mathbf{I} - \mathbf{rr}) \cdot \boldsymbol{\omega}\, dm + m(\boldsymbol{\varrho} \wedge \mathbf{v}_A) \tag{5.3}$$

is the "angular momentum" of the body about A whose value depends on the choice of A. In the above formulae, $\mathbf{v}_A$ is the velocity of the point A, $\boldsymbol{\omega}$ is the angular velocity of the rigidbody, and $\mathbf{r}$ is the position vector of a generic particle of the body with respect to A.

The Force Screw acting on the body referred to A is

$$\boldsymbol{\phi}(A) = \mathbf{R}(A) + \varepsilon[p\mathbf{R} + \mathbf{g} \wedge \mathbf{R}], \tag{5.4}$$

where $\mathbf{R}(A)$ is the resultant force acting at A, p is the pitch of the screw and $\mathbf{g}$ is the position vector of A with respect to a point on the screw axis.

Expression (5.4) may, evidently, be written as

$$\boldsymbol{\phi} = R\, e^{\varepsilon p}\, \boldsymbol{\alpha}, \tag{5.5}$$

where

$$\boldsymbol{\alpha} = \mathbf{a} + \varepsilon \mathbf{g} \wedge \mathbf{a}; \qquad \mathbf{a} = \frac{\mathbf{R}}{R} \tag{5.6}$$

represents the screw axis referred to A. Then, if the fixed frame is labelled, and the rigidbody in motion, the Balance Law for the Momentum Screw of the Body is written as

$$\boldsymbol{\phi}(A) = \left(\frac{\mathrm{d}\boldsymbol{\mu}(A)}{\mathrm{d}t}\right)_0 = \dot{\boldsymbol{\mu}}(A), \quad \text{say} \tag{5.7}$$

or

$$\boldsymbol{\phi}(A) = \left(\frac{\mathrm{d}\boldsymbol{\mu}(A)}{\mathrm{d}t}\right)_1 + \boldsymbol{\Omega}_{10} \wedge \boldsymbol{\mu}(A) \tag{5.8}$$

where

$$\boldsymbol{\Omega}_{10} = \boldsymbol{\omega} + \varepsilon \mathbf{v}_A \tag{5.9}$$

is the Instantaneous Screw of the Body.

The Balance Law (5.8) has been known for many years, (see, e.g., A. Gray [16]), but the concise form (5.8) appears to have been first formulated by von Mises [8] and later by Dimentberg [10] and Tang [17].

5.3. *Dynamics of the R–R–C–C–R mechanism*

In order to indicate the method of solution for the Dynamics of a five-link mechanism we choose, in particular, the R–R–C–C–R mechanism, whose configuration and Instantaneous Screw have already been determined.

We label the reaction screws at the i-th screw pair by $(\boldsymbol{\phi}_i, -\boldsymbol{\phi}_i)$, which constitute a null force screw at the point considered. Hence, each link may be regarded as acted on by one force screw at each end, and, on a point along the axis of the connecting pair (Fig. 5). Then, referring all the qualities to a common base point A, we have, applying the Balance Law (5.7) to each link in turn,

$$\boldsymbol{\phi}_0 - \boldsymbol{\phi}_1 = \dot{\boldsymbol{\mu}}_1, \tag{5.10}$$

$$\boldsymbol{\phi}_1 - \boldsymbol{\phi}_2 = \dot{\boldsymbol{\mu}}_2, \tag{5.11}$$

$$\boldsymbol{\phi}_2 - \boldsymbol{\phi}_3 = \dot{\boldsymbol{\mu}}_3, \tag{5.12}$$

$$\boldsymbol{\phi}_3 - \boldsymbol{\phi}_4 = \dot{\boldsymbol{\mu}}_4, \tag{5.13}$$

$$\boldsymbol{\phi}_4 - \boldsymbol{\phi}_0 = 0, \tag{5.14}$$

where $\boldsymbol{\mu}_i$ denotes the momentum screw of the i-th body, and the explicit mention of A has been dropped for convenience. Moreover, we have,

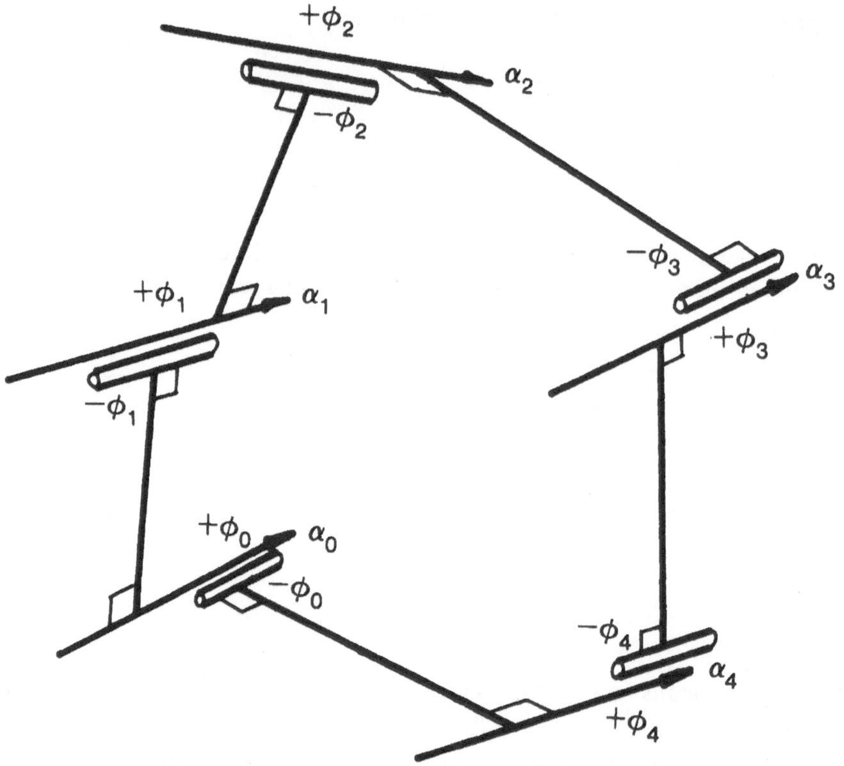

Figure 5

$$\boldsymbol{\phi}_0 \cdot \boldsymbol{\alpha}_2 = F \quad \text{(required)}, \tag{5.15}$$

and

$$\boldsymbol{\phi}_4 \cdot \boldsymbol{\alpha}_4 = G \quad \text{(given)}, \tag{5.16}$$

where F is the required input force along $\boldsymbol{\alpha}_0$ and G is the given output load along $\boldsymbol{\alpha}_4$. Assuming zero friction at the screw pairs we must also have,

$$\boldsymbol{\phi}_2 \cdot \boldsymbol{\alpha}_2 = 0 \tag{5.17}$$

and

$$\boldsymbol{\phi}_3 \cdot \boldsymbol{\alpha}_3 = 0 \tag{5.18}$$

$\boldsymbol{\alpha}_2$ and $\boldsymbol{\alpha}_3$ being cylindrical pairs.

Now we may put,

$$\boldsymbol{\phi}_4 = \lambda\boldsymbol{\alpha}_2 \wedge \boldsymbol{\alpha}_3 + \mu\boldsymbol{\alpha}_3 \wedge \boldsymbol{\alpha}_4 + v\boldsymbol{\alpha}_4 \wedge \boldsymbol{\alpha}_2, \tag{5.19}$$

where λ, μ, and v are dual numbers still to be determined, since we have,

$$(\boldsymbol{\alpha}_2 \wedge \boldsymbol{\alpha}_3) \wedge (\boldsymbol{\alpha}_3 \wedge \boldsymbol{\alpha}_4) \cdot (\boldsymbol{\alpha}_4 \wedge \boldsymbol{\alpha}_2) = (\boldsymbol{\alpha}_2 \wedge \boldsymbol{\alpha}_3 \cdot \boldsymbol{\alpha}_4)^2 \neq 0.$$

Hence

$$\mu = \frac{\boldsymbol{\phi}_4 \cdot \boldsymbol{\alpha}_2}{V}, \qquad \nu = \frac{\boldsymbol{\phi}_4 \cdot \boldsymbol{\alpha}_3}{V}, \qquad \lambda = \frac{\boldsymbol{\phi}_4 \cdot \boldsymbol{\alpha}_4}{V} = G \tag{5.20}$$

where $V = \boldsymbol{\alpha}_2 \wedge \boldsymbol{\alpha}_3 \cdot \boldsymbol{\alpha}_4$. Now

$$[(5.12) + (5.13)] \cdot \boldsymbol{\alpha}_2 \Rightarrow$$

$$-\boldsymbol{\phi}_4 \cdot \boldsymbol{\alpha}_2 = (\dot{\boldsymbol{\mu}}_3 + \dot{\boldsymbol{\mu}}_4) \cdot \boldsymbol{\alpha}_2 \tag{5.21}$$

in view of (5.17), and

$$[(5.13) \cdot \boldsymbol{\alpha}_3] \Rightarrow$$

$$-\boldsymbol{\phi}_4 \cdot \boldsymbol{\alpha}_3 = \dot{\boldsymbol{\mu}}_4 \cdot \boldsymbol{\alpha}_3 \tag{5.22}$$

in view of (5.18). Hence

$$\boldsymbol{\phi}_4 = \frac{1}{V}[G(\boldsymbol{\alpha}_2 \wedge \boldsymbol{\alpha}_3) - (\dot{\boldsymbol{\mu}}_3 + \dot{\boldsymbol{\mu}}_4) \cdot \boldsymbol{\alpha}_4(\boldsymbol{\alpha}_3 \wedge \boldsymbol{\alpha}_4) - \boldsymbol{\alpha}_3 \cdot \dot{\boldsymbol{\mu}}_4(\boldsymbol{\alpha}_1 \wedge \boldsymbol{\alpha}_2)] \tag{5.23}$$

is now completely determined.

The other unknowns follow immediately from the system (5.10), . . . , (5.14):

$$\begin{aligned} \boldsymbol{\phi}_0 &= \boldsymbol{\phi}_4, \\ \boldsymbol{\phi}_1 &= \boldsymbol{\phi}_4 - \dot{\boldsymbol{\mu}}_1, \\ \boldsymbol{\phi}_2 &= \boldsymbol{\phi}_4 - (\dot{\boldsymbol{\mu}}_1 + \dot{\boldsymbol{\mu}}_2), \\ \boldsymbol{\phi}_3 &= \boldsymbol{\phi}_4 + \dot{\boldsymbol{\mu}}_4. \end{aligned} \tag{5.24}$$

The required input force screw is then given by,

$$\begin{aligned} F &= \boldsymbol{\phi}_0 \cdot \boldsymbol{\alpha}_0 = \boldsymbol{\phi}_4 \cdot \boldsymbol{\alpha}_0 \\ &= \frac{\boldsymbol{\alpha}_0}{V}[G(\boldsymbol{\alpha}_2 \wedge \boldsymbol{\alpha}_3) + (\dot{\boldsymbol{\mu}}_3 + \dot{\boldsymbol{\mu}}_4) \cdot \boldsymbol{\alpha}_4(\boldsymbol{\alpha}_3 \wedge \boldsymbol{\alpha}_4) + \dot{\boldsymbol{\mu}}_4 \cdot \boldsymbol{\alpha}_3(\boldsymbol{\alpha}_4 \wedge \boldsymbol{\alpha}_3)] \end{aligned} \tag{5.25}$$

and the problem is completely solved.

For the static case, where all the momentum screws are zero, we have the simple relation between the input and output load:

$$F = \left(\frac{\boldsymbol{\alpha}_2 \wedge \boldsymbol{\alpha}_3 \cdot \boldsymbol{\alpha}_0}{V}\right) G. \tag{5.26}$$

If necessary, the above screw expressions may be resolved into their dual scalar components, but it is here deemed geometrically more meaningful and mathematically more concise, to express the solution for the screws in terms of other (given) screws. Moreover, in avoiding the solution of the problem by first resolving each force screw into its components in the respective reference frames attached to each link, as has been done, e.g. by Yang [17], we have evaded the involvement of the complicated screw matrices in the Balance Equations.

6. Conclusions

The purpose of this paper is not only to demonstrate that the Screw Calculus can be used in the Analysis of Spatial Mechanisms but, above all, how it can be used to the maximum advantage. In particular, the exact parallel between the Screw Algebra and the Gibbsian Vector Algebra has been exploited to the full, so that the Screw Solutions are obtained via pure screw operations without the necessity of having to first resolve the screw equations into their scalar components in some arbitrarily chosen reference frame and then reassembling them back to the screw expressions again. The advantages of pure screw solutions is that the geometry of the solution can be easily visualized and interpreted. Also, in this way, the analysis of spatial mechanisms falls naturally into the traditional scheme of Newtonian mechanics without abrupt changes in notation.

References

1. Chace, M. A., "Vector analysis of linkages", *J. Eng. for Ind., Trans. ASME, Series B* **85** (1963), 289–297.
2. Yang, A. T. and Freudenstein, F., "Application of dual-number quaternion algebra to the analysis of spatial mechanisms", *J. Appl. Mech., Trans. ASME, Series E* **86** (1964), 300–308.
3. Yuan, M. S. C., Freudenstein, F., and Woo, L. S., "Kinematic analysis of spatial mechanisms by means of screw coordinates. Part 2. Analysis of spatial mechanisms", *J. Eng. for Ind., Trans ASME, Series B* **93** (1971), 67–73.
4. Sandor, G. N., "Principles of a general quaternion-operator method of kinematic synthesis", *J. Appl. Mech., Trans. ASME, Series E* **90** (1968), 40–46.
5. Sandor, G. N. and Bisshop, K. E., "On a general method of spatial kinematic synthesis by means of a stretch-rotation tensor", *J. Eng. for Ind., Trans. ASME, Series B* **91** (1969), 115–122.
6. Ball, R. S., *Theory of Screws*, Cambridge University Press (1900).
7. Kotelnikov, A. P., *Screw Calculus and Some Applications to Geometry and Mechanics* (in Russian), Annuals of the Imperial University of Kazan (1895).
8. von Mises, R., "Motorechnung, ein neues Hilfsmittel der Mechanik", *ZAMM* **4** (1924), 155–181, 193–213.
9. Brand, L., *Vector and Tensor Analysis*, J. Wiley, New York (1947).
10. Dimentberg, F. M., *Screw Calculus and its Application to Mechanics* (Russian), IZDOT, Nauka, Moscow (1965).
11. Keler, M. L., "Kinematics and statics including friction in single-loop mechanisms by screw calculus and dual vectors", *J. Eng. for Ind., Trans ASME, Series B* **95** (1973), 471–480.
12. Study, E., *Geometrie der Dynamen*, Teubner Verlag, Leipzig (1903).
13. Dimentberg, F. M., *A General Method for the Investigation of Finite Displacements of Spatial Mechanisms and Certain Cases of Passive Joints*, Purdue Translation, Purdue University, Lafayette, Ind. (May, 1959).
14. Yang, A. T., "Displacement analysis of spatial five-link mechanisms using 3 × 3 matrices with dual number elements", *J. Eng. for Ind., Trans. ASME, Series B* **91** (1969), 152–157.
15. Yuan, M. S. C., "Displacement analysis of the R–R–C–C–R five-link spatial mechanism", *J. Appl. Mech., Trans. ASME, Series E* **37**, (1970), 689–696.
16. Gray, A., *Gyrostatics and Rotational Motion*, Macmillan, New York (1918).
17. Yang, A. T., "Inertia force analysis of spatial mechanisms", *J. Eng. for Ind., Trans. ASME, Series B* **93** (1971), 27–33.

Main properties of root-loci of time-delay control systems and a method of finding the stable interval of gain

WONG CHIA-HO

Teaching Group of Mechanics, Zhejiang University, Hangzhov, Zheijiang, PRC

Abstract. In this paper, we will discuss the extended graphic representation of a characteristic equation of the type

$$1 + KG(s)H(s)\,e^{-\tau s} = 0,$$

where $G(s)H(s)$ is a rational fraction of the complex variable $s\ (\equiv x + iy)$ with real coefficients. Because we have deduced the equation of the root-locus and the expression for the curve (K, x), the main characteristics of such curves are easily described qualitatively. It follows that much unnecessary work can be omitted. When we chose y as the assigned variable, the equation of root-locus of the above expression is an algebraic equation of the $(m + n)$-th degree in x, where $(n + m)$ is the sum of degrees of the numerator and denominator of the fraction $G(s)H(s)$. In order to obtain the stable interval (SI) of gain K, we must solve a transcendental equation of y. Finally, the figures of root-loci and curves (K, x) of the expression $G(s)H(s)$ of the types

$$\frac{1}{s + c}, \quad \frac{1}{s(s + c)}, \quad \frac{s + A}{s + B},$$

are given here for illustration, and a method of finding the SI of K of the expression

$$GH = (1 + T_A s)/s(1 + T_L s)(1 + T_N s)$$

is calculated by a direct numerical method.

1. Introduction

We had discussed the extended graphic representation of polynomials and rational fractions in 1981 [1, 2]. Using the graphs from these sources we can determine the stability of a feed-back system with the characteristic equation

$$1 + G(s)H(s) = 0, \tag{1.1}$$

where

$$G(s)H(s) = Kf^{(d)}(s)/f^{(n)}(s), \tag{1.2}$$

and

$$f^{(n)}(s) = \sum_{i=0}^{n} a_i s^{n-i}, \tag{1.3}$$

Yeh, K. Y. (ed) Progress in Applied Mechanics
 ISBN-13: 978-94-010-8061-3

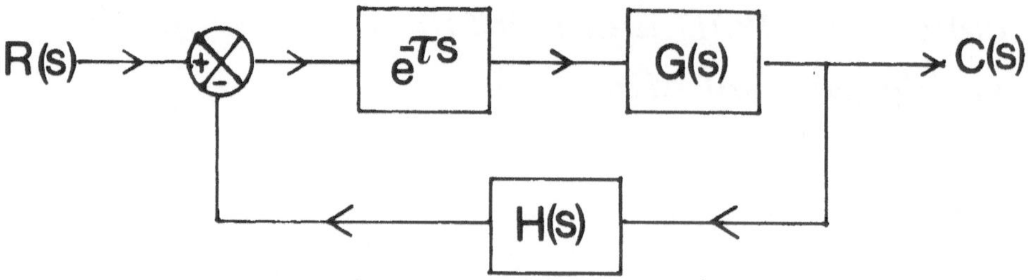

Figure 1

$$f^{(d)}(s) \;=\; \sum_{j=0}^{m} b_j s^{m-j}. \tag{1.4}$$

Now we consider the feed-back system, containing a time delay element as shown in Fig. 1. The transfer function of a time delay element can be written as $e^{-\tau s}$, so that the characteristic equation of this system is

$$1 + G(s)H(s)\,e^{-\tau s} \;=\; 0, \tag{1.5}$$

where $G(s)H(s)$ is taken the same form as (1.2). Hence (1.5) can be changed to the following expression

$$K \;=\; -\frac{a_0 s^n + a_1 s^{n-1} + \cdots + a_{n-1}s + a_n}{b_0 s^m + b_1 s^{m-1} + \cdots + b_{m-1}s + b_m}\,e^{\tau s}. \tag{1.6}$$

Now we will deduce two fundamental formulas of the extended graphic representation of (1.6) for plotting and find the stable interval (SI) of the gain K numerically.

2. Fundamental formulas

Substituting $s = x + iy$ in (1.6) and using the notation as taken in [1], we get

$$K \;=\; -\frac{\psi^{(n)}(x, y^2) + iy\phi^{(n)}(x, y^2)}{\psi^{(d)}(x, y^2) + iy\phi^{(d)}(x, y^2)}\,e^{\tau x}(\cos\tau y + i\sin\tau y), \tag{2.1}$$

where

$$\psi^{(n)}(x, y^2) \;=\; \alpha^{(n)} - y^2\alpha^{(n-2)} + y^4\alpha^{(n-4)} + \cdots \begin{cases} (-y^2)^t\alpha^{(1)} & \text{when } n = 2t+1, \\ (-y^2)^{t+1}\alpha^{(0)} & \text{when } n = 2t+2, \end{cases} \tag{2.2}$$

$$\phi^{(n)}(x, y^2) \;=\; \alpha^{(n-1)} - y^2\alpha^{(n-3)} + y^4\alpha^{(n-5)} + \cdots \begin{cases} (-y^2)^t\alpha^{(0)} & \text{when } n = 2t+1, \\ (-y^2)^t\alpha^{(1)} & \text{when } n = 2t+2, \end{cases} \tag{2.3}$$

$$\alpha^{(n)} \;\equiv\; f^{(n)}(x) \;\equiv\; \sum_{i=0}^{n} a_i x^{n-i} \tag{2.4}$$

and

$$\alpha^{(t)} = \frac{1}{n-t}\frac{\mathrm{d}}{\mathrm{d}x}\alpha^{(t+1)}, \quad t < n. \tag{2.5}$$

From (1.4), we can write similar expressions of $\psi^{(d)}(x, y^2)$ and $\phi^{(d)}(x, y^2)$.

Multiplying the factor $\psi^{(d)} - \mathrm{i}y\phi^{(d)}$ by the denominator as well as the numerator of (2.1), separating it into two parts; the real part and the imaginary part, and considering the gain K restricted to be real, we have

$$K = -\mathrm{e}^{\tau x}\frac{F_1(x, y^2)\cos\tau y - yF_2(x, y^2)\sin\tau y}{(\psi^{(d)})^2 + y^2(\phi^{(d)})^2} \tag{2.6}$$

and

$$yF_2(x, y^2)\cos\tau y + F_1(x, y^2)\sin\tau y = 0, \tag{2.7}$$

where

$$F_1(x, y^2) \equiv \psi^{(n)}\psi^{(d)} + y^2\phi^{(n)}\phi^{(d)}. \tag{2.8}$$

and

$$F_2(x, y^2) \equiv \psi^{(d)}\phi^{(n)} - \psi^{(n)}\phi^{(d)}. \tag{2.9}$$

Equations (2.6) and (2.7) are two fundamental formulas of extended graphic representation of (1.6). These two equations express a space curve in three dimensions (K, x, iy). This curve is the intersection of the two surfaces (2.6), (2.7), and the latter contains two variables x and y, and is therefore a cylindrical surface. We know that the equation of the projection of this space curve upon the plane (x, iy) is the same as the equation of this cylinder. Equation (2.7) is therefore the root-locus equation of (1.6). In the case of $y = 0$, (2.6) becomes the real curve

$$K_r = -\frac{a_0x^n + a_1x^{n-1} + \cdots + a_{n-1}x + a_n}{b_0x^m + b_1x^{m-1} + \cdots + b_{m-1}x + b_m}\mathrm{e}^{\tau x}. \tag{2.10}$$

This curve is expressed by the solid line in figure (K, x). For $y \neq 0$, (2.6) expresses a set of complex root curves, which are expressed by dashed lines in the figure (K, x). An important property of the figure (K, x) is that we can observe the stable interval of K.

3. Essential properties of root-loci

If we change the sign of y in (2.7), which will be kept the same form as before by means of

$$\cos\tau y = \cos(-\tau y), \qquad \sin\tau y = -\sin(-\tau y).$$

F_1 and F_2 are invariants with the transformation $y \to -y$. Hence we have proved.

THEOREM 1. The root-loci of a time-delay system are symmetrical with respect to the x-axis also.

The extremes of K_r are determined by the condition

$$\frac{dK_r}{dx} = -e^{\tau x}\frac{1}{[\psi^{(d)}(x,0)]^2}[\psi^{(d)}(x,0)\phi^{(n)}(x,0) - \psi^{(n)}(x,0)\phi^{(d)}(x,0)$$
$$+ \tau\psi^{(n)}(x,0)\psi^{(d)}(x,0)] = 0. \tag{3.1}$$

On the other hand, when $y = 0$, (2.7) can be written as

$$\psi^{(d)}(x,0)\phi^{(n)}(x,0) - \psi^{(n)}(x,0)\phi^{(d)}(x,0) + \tau\psi^{(n)}(x,0)\psi^{(d)}(x,0) = 0$$

by using the formula

$$\lim_{y\to 0}\frac{\sin \tau y}{y} = \tau \tag{3.2}$$

and

$$\frac{d}{dx}\psi(x,0) = \phi(x,0).$$

Therefore we have proved the theorem.

THEOREM 2. The abscissas of intersections of root-loci at the x-axis are the corresponding abscissas of the extrema of the real curve K_r.

The highest power of x in (2.7) is contained in the term $\psi^{(n)}(x, y^2)\psi^{(d)}(x, y^2)$. Evidently, (2.7) is an algebraic equation of $(m + n)$-th degree with respect to the variable x, but it is also a transcendental equation with respect to the variable y. Hence we may solve (2.7) for x by assigning it the value of y. In Fig. 3a, the straight line y = const will be cutting the root-loci at $(m + n)$ points or fewer. This depends on the form of $G(s)H(s)$.

When $y = k\pi$, where k is positive or negative integer but not zero, (2.7) becomes

$$\psi^{(d)}(x, y^2)\phi^{(n)}(x, y^2) - \psi^{(n)}(x, y^2)\phi^{(d)}(x, y^2) = 0. \tag{3.3}$$

The above equation is of $(m + n - 1)$-th degree in x. Comparing it with (2.7), we conclude that (3.3) lacks a root of $x = +\infty$ or $x = -\infty$. Let ε be a small positive quantity. Then

$$\tau y = k\pi - \varepsilon \to k\pi \qquad x \to +\infty,$$

$$\tau y = k\pi + \varepsilon \to k\pi \qquad x \to -\infty.$$

The above conclusion cannot apply to the case $G(x)H(s) = (s + A)/(s + B)$, because (3.3) becomes $B - A = 0$ (cf. Example 3).

From the above discussion, the time-delay system (1.5) has an infinite number of horizontal asymptotes, $\tau y = k\pi$.

4. Figure (K, x)

Because $e^{\tau x} > 0$ and $\lim_{x\to-\infty} e^{\tau x} = 0$, we conclude that the real curve K_r has the same zeros and poles as $[1/G(x)H(x)]$ besides the zero point of $x = -\infty$.

Evidently, the x-axis in the plane (K, x) is the asymptote of the real curve K_r. Let $[1/G(x)H(x)]$ have h real poles $B_j (h \leqslant m, j = 1, 2, \ldots, h)$, then K_r has h vertical asymptotes, $x = B_j$. Since

$$\frac{\mathrm{d}K_r}{\mathrm{d}x} = -\mathrm{e}^{\tau x}\left[\tau \frac{1}{G(x)H(x)} + \frac{\mathrm{d}}{\mathrm{d}x}\left(\frac{1}{G(x)H(x)}\right)\right]$$

the extremes of K_r are not the extremes of $[1/G(x)H(x)]$, besides the double point of $1/G(x)H(x)$.

Equation (2.6) for K_c may be rewritten in two other forms as is shown below. Solving (2.7) for $\sin \tau y$, and substituting in (2.6), we have

$$K_c = -\mathrm{e}^{\tau x} \frac{[F_1(x, y^2)]^2 + y^2[F_2(x, y^2)]^2}{[(\psi^{(d)})^2 + y^2(\phi^{(d)})^2]F_1(x, y^2)} \cos \tau y \tag{4.1}$$

Using

$$\cos \tau y = \pm \frac{1}{\sqrt{1 + \tan^2 y}} = \pm 1 \Bigg/ \sqrt{1 + \left(y \frac{F_2(x, y^2)}{F_1(x, y^2)}\right)^2} \tag{4.2}$$

we obtain

$$K_c = -\mathrm{e}^{\tau x}\left(\pm \sqrt{\frac{(\psi^{(n)})^2 + y^2(\phi^{(n)})^2}{(\psi^{(d)})^2 + y^2(\phi^{(d)})^2}}\right), \tag{4.3}$$

where $\pm$ in the parenthesis must be taken as the same sign as $\cos \tau y$ in (4.2).

Eliminating F_1 between (2.7) and (2.6), we get

$$K_c = \mathrm{e}^{\tau x} \frac{y F_2(x, y^2)}{[(\psi^{(d)})^2 + y^2(\phi^{(d)})^2] \sin \tau y}. \tag{4.4}$$

Because $\sin \tau y$ occurs in the denominator of (4.4), we divide τy in intervals

$$L_k\colon \quad (k - 1)\pi \leqslant \tau y \leqslant k\pi, \quad k = 1, 2, \ldots .$$

Assigning the value of τy, we can solve (2.7) for x, then we can calculate K_c from (4.4). In fact, given $H(s)G(s)$, we can determine many main properties of root-loci and the curve (K, x) by applying mathematical analysis to (2.7) and (4.4), in order that we can determine the SI of gain conveniently.

5. Examples

EXAMPLE 1: $\quad K = -a_0(s + c)\,\mathrm{e}^{\tau s}$. (5.1)

Fundamental formula: (2.7) and (4.4) become

Root-locus $\quad y \cos \tau y + (x + c) \sin \tau y = 0$ (5.2)

Curve K_r: $\quad K_r = -a_0(x + c)\,\mathrm{e}^{\tau x}$ (5.3)

Curve K_c: $\quad K_c = a_0\,\mathrm{e}^{\tau x} y/\sin \tau y$ (5.4)

From (5.2), we get

$$x = -c - y \cot \tau y. \tag{5.5}$$

From the above formula, we can find x by assigning the value of y. Because root-loci are symmetrical with respect to the x-axis, we may plot the $y \geqslant 0$ half plane only. In the interval L_1, the locus start from $x_A = -c - (1/\tau)$ on the x-axis, then x is increased with y increasing. After passing through a point B on y-axis, the locus extends to the line $\tau y = \pi$ asymptotically. In the interval L_k, a monotonic locus is passing from $x = -\infty$ to $x = +\infty$ (cf. Fig. 2(a)).

In Fig. 2(b) we first draw the curve K_r, which starts from $(x, K) = (-\infty, 0)$ to a maximum of K_r, $x_A = -c - (1/\tau)$, $K_r(\max) = (1/\tau)a_0\,\mathrm{e}^{-(c\tau+1)}$, then decreases to $x = -c$, $K_r = 0$, next to $x = 0$, $K_r = -a_0 c$, and $x \to +\infty$, $K_r \to -\infty$ at last (cf. Fig. 2(b)).

From (5.4), we can decide that $K_c(k)$ are positive for the interval L_k for odd integer k, and are negative for even k.

PRACTICAL EXAMPLE: $K = -0.4(s + 2.5)\,\mathrm{e}^{0.2s}$ [3].

To find the SI of K, we set $x = 0$ into (5.5), which becomes $y + 2.5 \tan 0.2y = 0$. Solving this equation, we get $y_B = 9.18$. From (5.4), we have $K_F = 3.8$, since $K_r(0) = -1$, and the SI is $(-1, 3.8)$.

EXAMPLE 2: $$K = -a_0 x(x + c)\,\mathrm{e}^{\tau s} \tag{5.6}$$

Root-locus: $$y(2x + c) \cos \tau y + (x^2 + cx - y^2) \sin \tau y = 0 \tag{5.7}$$

K_r: $$K_r = -a_0 x(x + c)\,\mathrm{e}^{\tau x} \tag{5.8}$$

K_c: $$K_c = a_0\,\mathrm{e}^{\tau x}(2x + c)/\sin \tau y. \tag{5.9}$$

From (5.7),

$$x = -\frac{c}{2} - \left(\frac{y}{\sin \tau y}\right) \cos \tau y \pm \sqrt{\left(\frac{y}{\sin \tau y}\right)^2 + \frac{c^2}{4}}. \tag{5.10}$$

By the above expression, we have two real roots of x with respect to any real value of y. On the x-axis, we have

$$x_{A_1} = -\frac{1}{\tau} - \frac{c}{2} + \sqrt{\frac{1}{\tau^2} + \frac{c^2}{4}}, \qquad x_{A_2} = -\frac{1}{\tau} - \frac{c}{2} - \sqrt{\frac{1}{\tau^2} + \frac{c^2}{4}}.$$

The locus of starting from A_1 will be extended asymptotically to $\tau y = \pi$, and then from A_2 to $\tau y = 2\pi$ (cf. Fig. 3(a)). Also, x_{A_1} and x_{A_2} are the abscissas of $K_r(\max)$ and $K_r(\min)$ respectively as shown in Fig. 3(b). To find the SI, we must first solve $c - y \tan \tau y = 0$, and substitute y_B in $K = a_0 cy/\sin \tau y$.

PRACTICAL EXAMPLE: For $\tau = 1$, $c = 1$, $a_0 = 1$ then one may find $y_B = 0.86033$, $K_F = 1.1349$, and SI is $(0, 1.1349)$.

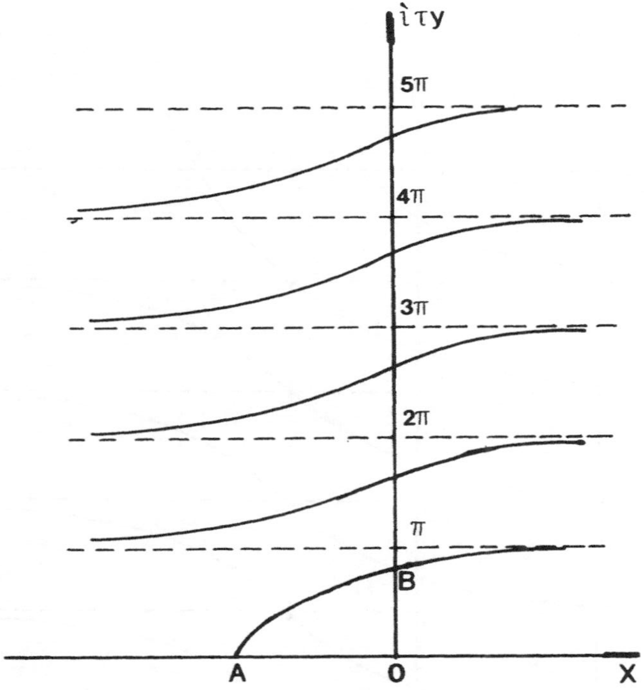

Figure 2a

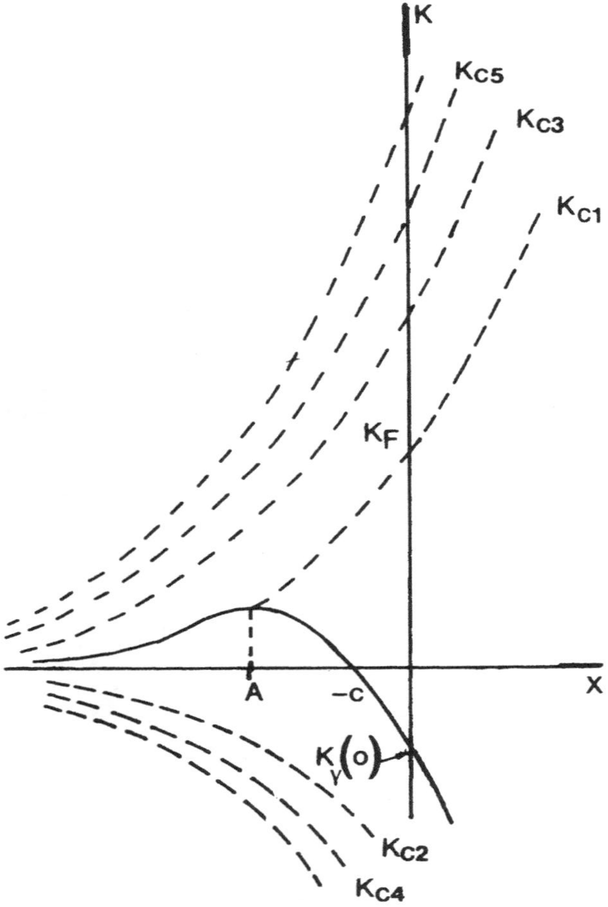

Figure 2b

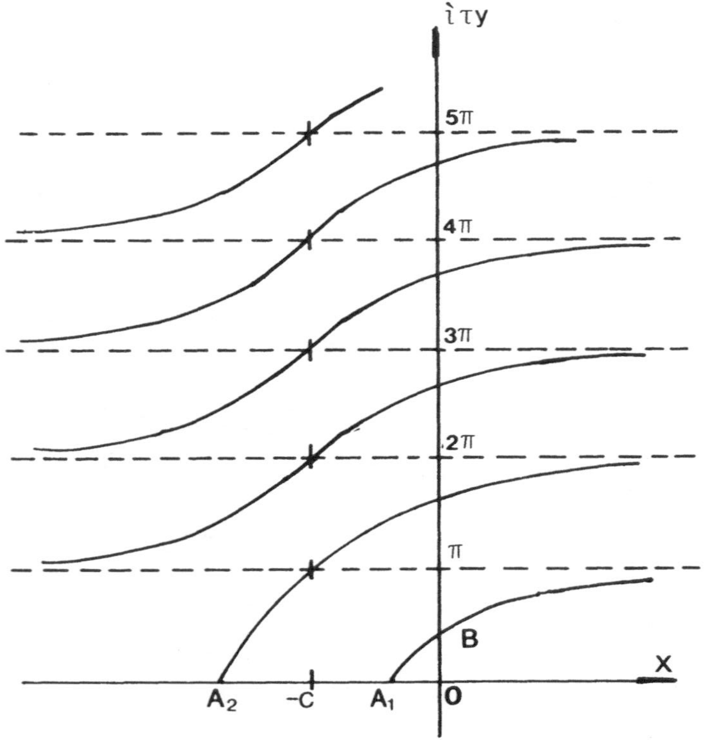

Figure 3a

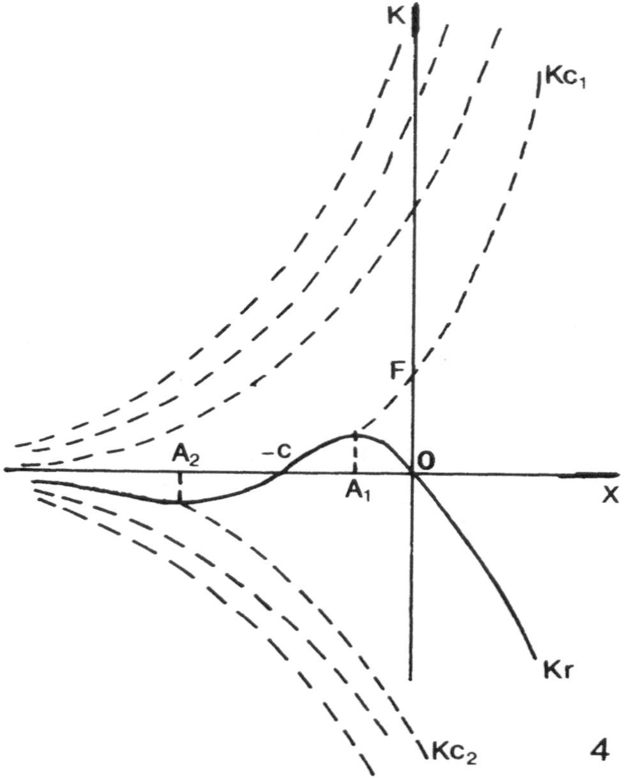

Figure 3b

EXAMPLE 3: $$K = -\frac{s + A}{s + B} e^{\tau s} \qquad (5.11)$$

Root-locus: $$[(x + A)(x + B) + y^2] \sin \tau y + y(B - A) \cos \tau y = 0 \qquad (5.12)$$

K_r: $$K_r = -\frac{x + A}{x + B} e^{\tau x} \qquad (5.13)$$

K_c: $$K_c = e^{\tau x} \frac{(B - A)y}{[(x + B)^2 + y^2] \sin \tau y} \qquad (5.14)$$

or: $$K_c = \pm e^{\tau x} \frac{\sqrt{[(x + A)(x + B) + y^2]^2 + y^2(B - A)^2}}{(x + B)^2 + y^2}. \qquad (5.15)$$

From (5.12),

$$x = -\tfrac{1}{2}[A + B \pm \sqrt{(A - B)^2 - 4[y^2 + y(B - A) \cot \tau y]}]. \qquad (5.16)$$

We know that $(s + A)/(s + B)$ has no complex root-locus [2], but it may have a complex root-locus after multiplying by $e^{\tau s}$, because (5.16) contains $\cot \tau y$ which has the value from $-\infty$ to $+\infty$. The root-loci will intercept the x-axis if the discriminant of (5.16) is positive when $y = 0$, i.e.,

$$B - A \geqslant \frac{4}{\tau}.$$

Now we will classify this problem in the following three cases:

(i) $A > B$ } root-loci are intercepted with
(ii) $B > A$, further more $B - A \geqslant 4/\tau$ } x-axis at two points
(iii) $B > A$, but $B - A < 4/\tau$. root-loci do not intercept x-axis.

From (5.16), we see that all the root-loci have an extreme point on the line $x = -(A + B)/2$, as it becomes a double point. Figures 4, 5 and 6 give us the main characteristic curves of three cases for this problem.

6. Method of finding the SI of *K*

First of all, we must solve (2.7) for y when $x = 0$, i.e.

$$\tan \tau y = -\frac{yF_2(0, y^2)}{F_1(0, y^2)}. \qquad (6.1)$$

Equation (4.4), for the case $x = 0$, becomes

$$K_c = \frac{yF_2(0, y^2)}{\{[\psi^{(d)}(0, y^2)]^2 + y^2[\phi^{(d)}(0, y^2)]^2\} \sin \tau y} \qquad (6.2)$$

or

$$K_c = \pm \sqrt{\frac{[\psi^{(n)}(0, y^2)]^2 + y^2[\phi^{(n)}(0, y^2)]^2}{[\psi^{(d)}(0, y^2)]^2 + y^2[\phi^{(d)}(0, y^2)]^2}}. \qquad (6.3)$$

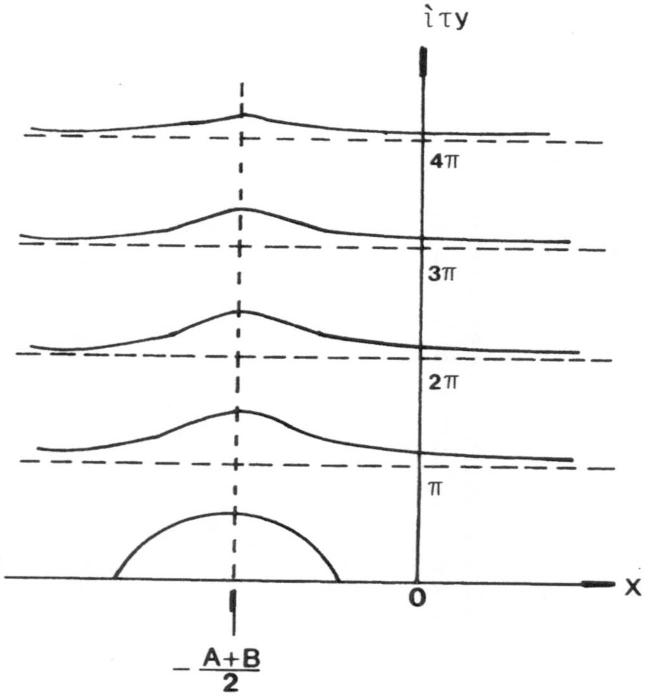

Figure 4a

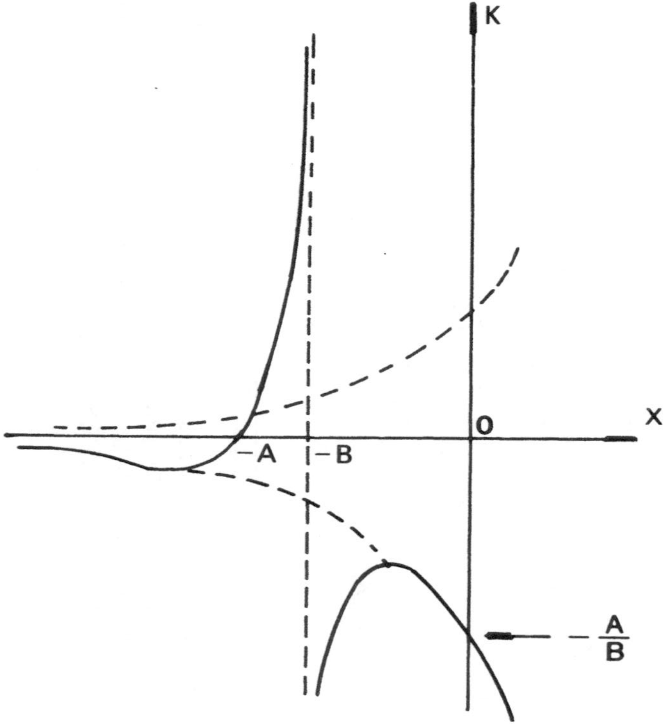

Figure 4b

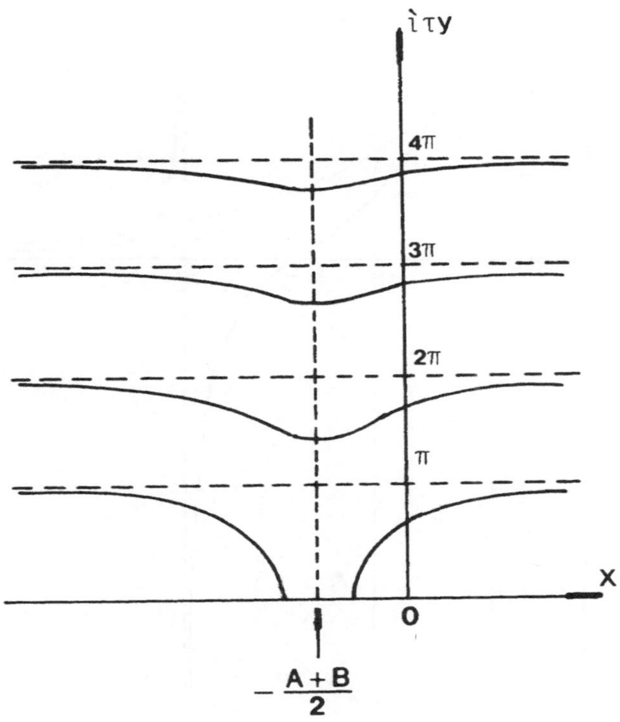

Figure 5a

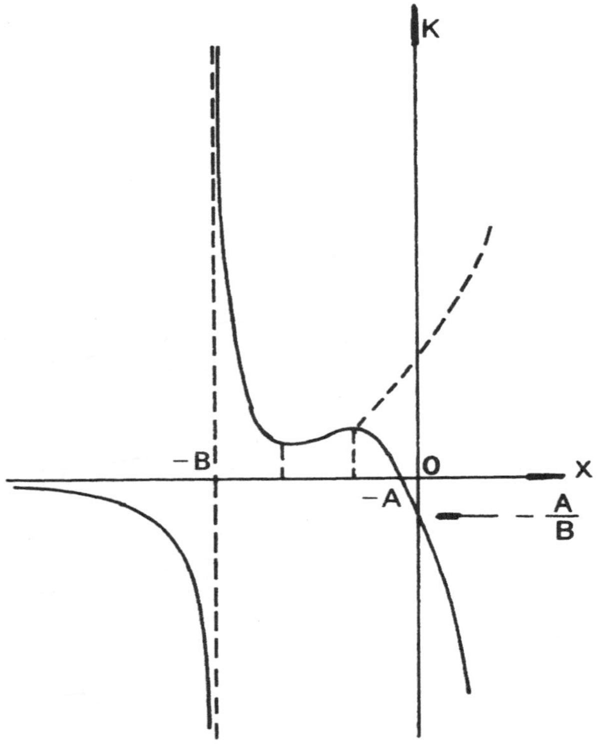

Figure 5b

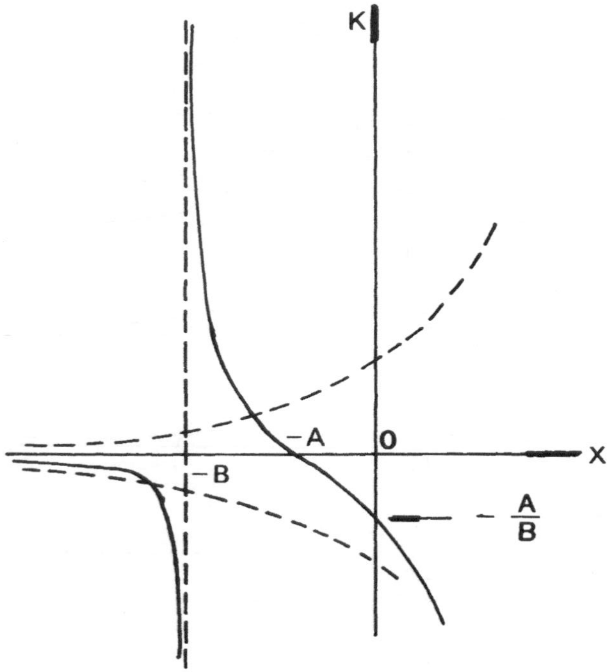

Figure 6a

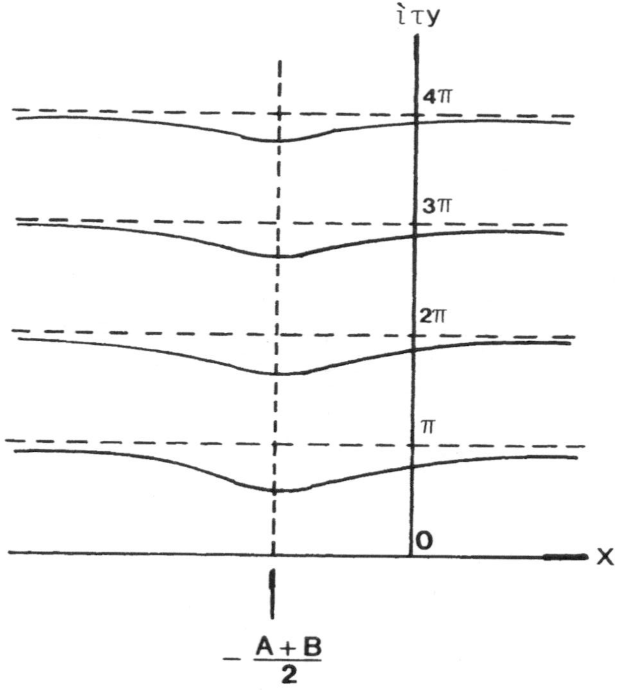

Figure 6b

Let the smallest value of positive K_c be $K_{min}(+)$ and the largest of negative K_c and $K_r(0)$ be $K_{max}(-)$, then the SI is $(K_{max}(-), K_{min}(+))$. From (2.10), we have

$$K_r(0) \;=\; -a_n/b_m. \tag{6.4}$$

To find the SI of Example 3, we have

$$(AB + y^2)\tan y + y(B - A) \;=\; 0, \tag{6.5}$$

$$K_c \;=\; \pm\sqrt{\frac{y^2 + A^2}{y^2 + B^2}} \quad \text{and} \quad K_r(0) \;=\; -A/B. \tag{6.6}$$

(i) $A > B,\ \ y^2 \to +\infty,\ \ K_c \to \pm 1.$

Hence SI is $(-1, 1)$.

(ii) $B > A,\ \ K_{max}(-) \;=\; K_r(0) \;=\; -\dfrac{A}{B}.$

Let y_B be the smallest positive root of (6.5).
Then SI is $(-A/B, \sqrt{y_B^2 + A^2}/\sqrt{y_B^2 + B^2})$.

EXAMPLE 4: To find the SI of [3],

$$1 + \frac{K(1 + T_A s)}{s(1 + T_L s)(1 + T_N s)}\,\mathrm{e}^{-Ds} \;=\; 0.$$

SOLUTION: Since

$$f^{(n)}(s) \;=\; T_L T_N s^3 + (T_L + T_N)s^2 + s,$$

$$f^{(d)}(s) \;=\; 1 + T_A s,$$

it follow that

$$\psi^{(n)} \;=\; T_L T_N x^3 + (T_L + T_N)x^2 + x - y^2(3T_L T_N x + T_L + T_N),$$

$$\phi^{(n)} \;=\; 3T_L T_N x^2 + 2(T_L + T_N)x + 1 - y^2 T_L T_N,$$

$$\psi^{(d)} \;=\; 1 + T_A x$$

and

$$\phi^{(d)} \;=\; T_A.$$

Equations (6.1) and (6.3) become

$$\tan \tau y \;=\; \frac{1 + y^2(T_L T_A + T_N T_A - T_L T_N)}{y[T_L + T_N - T_A) + T_L T_N T_A y^2]}, \tag{6.7}$$

$$K_c \;=\; y\sqrt{\frac{1 + y^2(T_L^2 + T_N^2) + y^4 T_L^2 T_N^2}{1 + y^2 T_A^2}}. \tag{6.8}$$

PRACTICAL EXAMPLE: $D = 0.2\,\text{sec}$, $T_A = 1\,\text{sec}$, $T_L = 10\,\text{sec}$, $T_N = 0.1\,\text{sec}$.

$$\tan 0.2y = (1 + 9.1y^2)/y(9.1 + y^2) \tag{6.9}$$

and

$$K_c = y\sqrt{(1 + 100.01y^2 + y^4)/(1 + y^2)}. \tag{6.10}$$

Solving (6.9), we get $y = 4.713$, substituting it in (6.10), we have $K_c = 50.9789$, $K_r(0) = 0$. Therefore, SI is (0, 50.9789).

References

1. Wong, Chia-ho, "Extended graphical representation of polynomials with applications to cybernetics", *Applied Mathematics and Mechanics* (English Edition), **2**, 3 (1981), 305–318.
2. Wong, Chia-ho, "Extended graphical representation of rational fractions with application to cybernetics", *Applied Mathematics and Mechanics* (English Edition), **2**, 4 (1981), 419–428.
3. Stanley, M. S., *Modern Control System Theory and Applications* (2nd Ed.), pp. 292, 357, Addison-Wesley, New York (1978).

Fluid Mechanics

Concerning the singularity of non-linear vibration of ideal liquid in communication tubes with different end cross sections

LIU HSIEN-CHIH

49 Nanxinjie, Jinan, Shandong, PRC

Abstract. With this paper, we have, in the main, set up the differential equation of motion of the ideal column in an elliptical arc communication tube with different end cross sections; and for the purpose of comparison, those for a circular arc and straight arms as well as with a solid vibrating piston have been cited. For all these cases displacement properties and dependence of the singularity point with the initial displacement of the liquid column are found.

1. Introduction

Conventionally, when one gives an account of the vibrations of mass-points systems, one seldom fails to mention the harmonic vibrations of a liquid column in a U-tube, which represents only the simplest case of communication-tubing systems. In the normal routines of daily life as well as in modern science, the natural phenomena of communication-tube liquid motion often occur in both small and large measure, e.g., a tea pot may be an unnoticed example of the former; the flow between the Mediterranean and the Atlantic ocean through Gibraltar, a portrait of the latter. Whence, a careful observer may perceive that these vibrations seldom obey the simple harmonic law, as in a communication-tube with constant cross section.

In aeroplanes, an artificial gyroscope "horizon" indicator is installed to establish a pseudo-equilibrium position [1], by which one utilizes an annexed communication tube to achieve this effect [1]. Large steamers are sometimes equipped with two huge communication water tanks to attenuate the destabilizing effect of ship's roll [3, 17–19]. The micro-pressure meter in a wind tunnel test is another example of this [4]. The above cases fall in the category of comparatively simple configuration, in which the velocity and the flow direction possess certain period and regularity. But there also exist large scale cases, e.g., the canals connecting the water reservoirs and the sea in aiming to exploit the energy of the tide water.

Finally let us draw attention to the water power station. Here there exists a communication-tube system between water reservoirs, surge tower and the water-turbine with continuously varying flow to accommodate the power output

Yeh, K. Y. (ed) Progress in Applied Mechanics
ISBN-13: 978-94-010-8061-3

fluctuation [6, 8]. Ignorance of flow kinematics has, in the past, been responsible for defective analyses [9] which has led in turn to harmful effects.

Furthermore, in connecting-tube systems for pump stations [13], sensitive pressure meters [14], and the fuel injection system for internal combustion engines [16], there exists communication-liquid motion effects.

This paper sets up the differential equations and finds general solution for several common systems, without recourse to any concrete problem analysis.

2. The equation of motion of the ideal liquid in a communication tube with elliptic arcs

This case has been discussed, by the author, in a previous paper [10]. Since the publication was brief we would like to take this opportunity of offering some supplementary information and to give an account of some new finding.

Referring to Fig. 1, let us set up the differential equation of motion for the ideal liquid in this subsection, wherein θ_1 signifies the eccentric angle, ϕ_1 the supplementary one to θ_1, and φ_1 the so-called central angle. Thus we can write down the following helpful relations:

$$\theta_1 + \phi_1 = \pi/2, \quad \theta_1 \geqslant \varphi_1, \quad (0 \leqslant \varphi_1 \leqslant \pi/2), \tag{2.1}$$

$$\left.\begin{aligned} x &= a \sin \phi_1 = a \cos \theta_1 = \sqrt{a^2 \sin^2\phi_1 + b^2 \cos^2\phi_1} \cdot \cos \varphi_1 \\ &= \sqrt{a^2 \cos^2\theta_1 + b^2 \sin^2\theta_1} \cdot \cos \varphi_1, \\ y &= b \cos \phi_1 = b \sin \theta_1 = \sqrt{a^2 \sin^2\phi_1 + b^2 \cos^2\phi_1} \cdot \sin \varphi_1 \\ &= \sqrt{a^2 \cos^2\theta_1 + b^2 \sin^2\theta_1} \cdot \sin \varphi_1. \end{aligned}\right\} \tag{2.2}$$

By virtue of $\mathrm{d}x = a \cos \phi_1 \, \mathrm{d}\phi_1$, $\mathrm{d}y = -b \sin \phi_1 \, \mathrm{d}\phi_1$ the elliptic arc equation reads

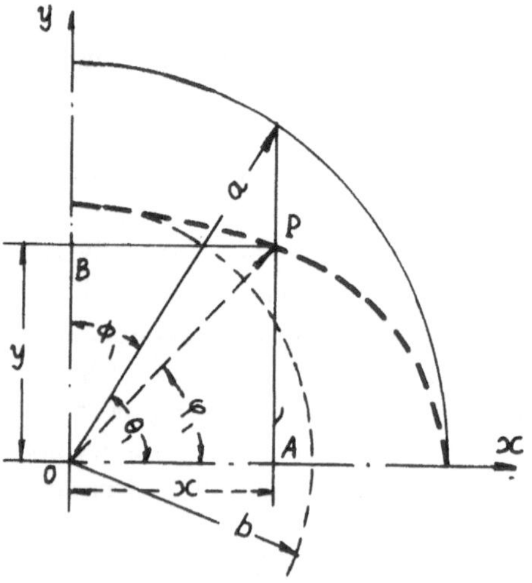

Figure 1. The relation between the three angles ϕ_1, θ_1 and φ_1.

$$\delta = \int_{\phi_{11}}^{\phi_{12}} \sqrt{a^2 \cos^2\phi_1 + b^2 \sin^2\phi_1}\, d\phi_1 = -\int_{\theta_{11}}^{\theta_{12}} \sqrt{a^2 \sin^2\theta_1 + b^2 \cos^2\theta_1}\, d\theta_1. \tag{2.3}$$

Moreover, we can write the equations

$$\theta_1 = \tan^{-1}(ay/bx), \qquad \varphi_1 = \tan^{-1}(y/x)$$

to obtain

$$\delta_1 = \theta_1 - \varphi_1 = \tan^{-1}\frac{(1 - b/a)\tan\varphi_1}{b/a + \tan^2\varphi_1}. \tag{2.4}$$

Again, since

$$d\delta_1/d\varphi_1 = \frac{\left(1 - \frac{b}{a}\right)\left(\frac{b}{a} - \tan^2\varphi_1\right)\sec^2\varphi_1}{\left(\frac{b}{a} + \tan^2\varphi_1\right)^2 + \left(1 - \frac{b}{a}\right)^2 \tan^2\varphi_1},$$

then we get

$$\left(\frac{d\delta_1}{d\varphi_1}\right)_{\varphi_1=0} = \frac{a}{b} - 1 \gtreqless 0, \quad (a \gtreqless b); \quad \left(\frac{d\delta_1}{d\varphi_1}\right)_{\varphi_1=\pi/2} = \frac{b}{a} - 1 \lesseqgtr 0, \quad (a \gtreqless b).$$

For the maximum value of the angle difference δ_1, the condition $d\delta_1/d\varphi_1 = 0$ yields

$$(\varphi_1)_{\delta_1=\max} = \tan^{-1}\sqrt{b/a}. \tag{2.5}$$

For obtaining an intuitive expression for $\delta_{1\max}$, we take $b/a = 2/3$, thus get Table 1 and Fig. 2, and in turn the relation

$$\delta_{1\max} = \tan^{-1}\left(\frac{a - b}{b/\tan\varphi_1 + a\tan\varphi_1}\right) = \tan^{-1}0.204 = 11°31'13''.$$

We perceive from Fig. 2 that in the major part of the interval $0 \leqslant \varphi_1 \leqslant \pi/2$, we have $\delta_1 < 7°$, therefore for acquisition of an approximate solution of our related differential equation, we may use φ_1 to replace θ_1. If $b/a < 2/3$, then $\delta_1 \ll 7°$. Based on this fact, we then get

Table 1. Angle difference.

x	0	$a/4$	$a/2$	$3a/4$	a
y	b	$\frac{\sqrt{\pi}}{4}a$	$\frac{\sqrt{3}}{2}a$	$\frac{\sqrt{7}}{4}a$	0
y/x	∞	$\sqrt{15}$	$\sqrt{3}$	$\frac{\sqrt{7}}{3}$	0
ay/bx	∞	$\frac{3\sqrt{15}}{2}$	$\frac{3\sqrt{3}}{2}$	$\frac{\sqrt{7}}{2}$	0
φ_1	$\frac{\pi}{2}$	75°31′	60°	41°25′	0
θ_1	$\frac{\pi}{2}$	80°14′	68°58′	53°4′	0
δ_1	0	4°43′	8°58′	11°38′	0

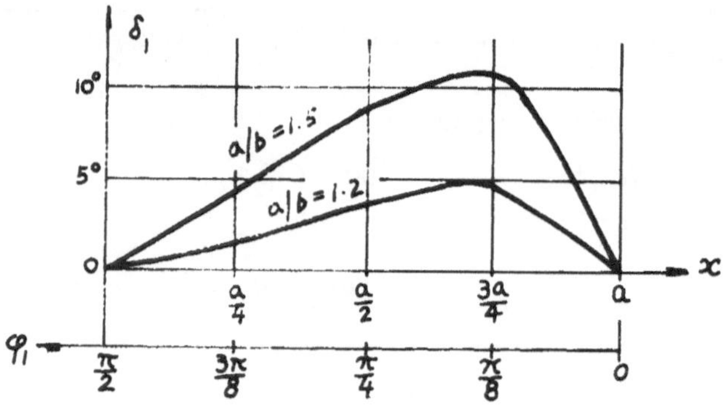

Figure 2. δ_1 depends on x and φ_1.

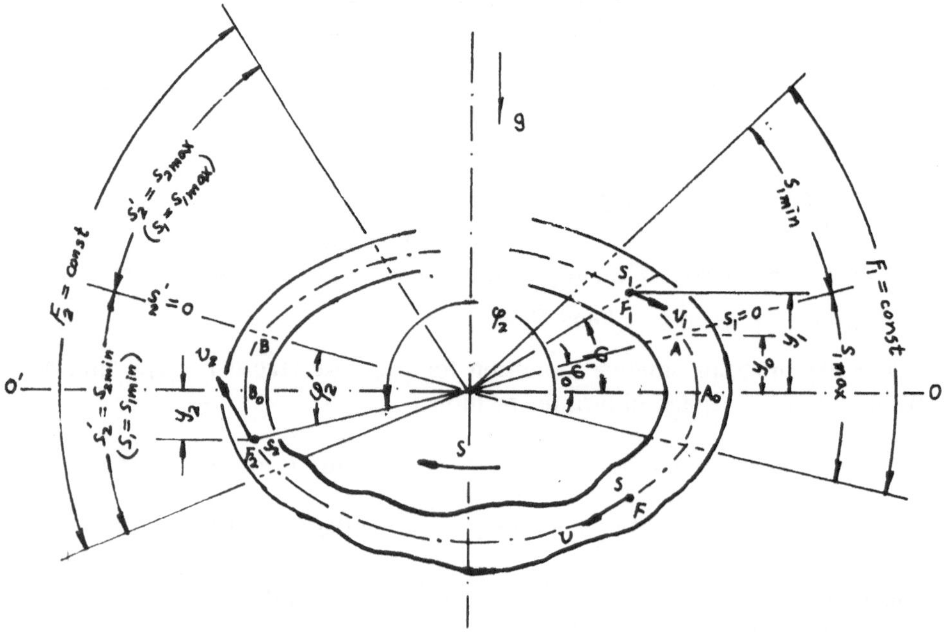

Figure 3. Communication-tube with elliptic arcs.

$$x = \sqrt{a^2 \cos^2 \varphi_1 + b^2 \sin^2 \varphi_1} \cos \varphi_1, \quad y \approx \sqrt{a^2 \cos^2 \varphi_1 + b^2 \sin^2 \varphi_1} \sin \varphi_1, \tag{2.6}$$

$$s \approx -\int_{\varphi_{11}}^{\varphi_{12}} \sqrt{a^2 \sin^2 \varphi_1 + b^2 \cos^2 \varphi_1}\, d\varphi_1. \tag{2.7}$$

Making reference to the geometrical configuration in Fig. 3, we may use the generalized Bernoulli equation

$$\frac{v_1^2 - v_2^2}{2g} - \int_{s_{11}}^{s_{12}} \frac{1}{g} \frac{\partial v}{\partial t} ds + y_1 - y_2 = 0 \tag{2.8}$$

to deduce the related equation of motion, where v_i is the velocity of the liquid particle, t the time, g the acceleration of gravity, and y_i the distance from the liquid surface to the equilibrium position.

For the elliptic arc communication tube, let us first examine the displacement heights y_1 and y_2 in (2.8). Let y_1 (y_2) indicate the vertical distance from the point s_1 (s_2) to the line 00′. The arc lengths are $s_{21} = \widehat{BB_0}$ and $s_{22} = \widehat{B_0 s_2}$; hence we get

$$s_{21} + s_{22} = \eta_2 \widehat{s_1 A} = \eta_2 \int_{\varphi_0}^{\varphi_1} \Phi \, d\varphi_1,$$

where $\Phi = \sqrt{a^2 \sin^2\varphi_1 + b^2 \cos^2\varphi_1}$, $\eta_2 = F_1/F_2$. Thus we obtain

$$s_{22} = \eta_2 \int_{\varphi_0}^{\varphi_1} \Phi \, d\varphi_1 - s_{21} = \eta_2 \int_0^{\varphi_1} \Phi \, d\varphi_1 - (1 + \eta_2) \int_0^{\varphi_0} \Phi \, d\varphi_1 .$$

The two integrals yield the arc length s_{22} and in turn to yield $y_2 = \Psi(s_{22})$, and again in turn

$$y_1 - y_2 = \sqrt{a^2 \cos^2\varphi_1 + b^2 \sin^2\varphi_1} \cdot \sin\varphi_1 - y_2, \quad (y_2 < 0). \tag{2.9}$$

With $\varphi_0 = 0$ and $y_2 = -\eta_2 y_1$, we get

$$y_1 - y_2 = (1 + \eta_2) \sqrt{a^2 \cos^2\varphi_1 + b^2 \sin^2\varphi_1} \cdot \sin\varphi_1 . \tag{2.10}$$

With $\eta_2 = 1$, $\varphi_0 = 0$ and $y_1 = b \sin\theta_1 \approx b \sin\varphi_1$, then reads

$$y_1 - y_2 = 2y_1 \approx 2b \sin\varphi_1 . \tag{2.11}$$

Moreover, from (2.7) we get $v_1 = \dot{s}_1 = -\sqrt{a^2 \sin^2\varphi_1 + b^2 \cos^2\varphi_1} \cdot \dot{\varphi}_1$ and

$$v_1^2 - v_2^2 = v_1^2(1 - \eta_2) = (1 - \eta_2^2)[a^2 \sin^2\varphi_1 + b^2 \cos^2\varphi_1]\dot{\varphi}_1^2 .$$

By using the relations

$$\frac{\partial v_1}{\partial t} = \frac{dv_1}{dt} = \frac{\partial^2 s_1}{\partial \varphi_1^2}\left(\frac{\partial \varphi_1}{\partial t}\right)^2 + \frac{\partial s_1}{\partial \varphi_1} \cdot \frac{d^2 \varphi_1}{dt^2},$$

$$\frac{\partial s_1}{\partial \varphi_1} = -\sqrt{a^2 \sin^2\varphi_1 + b^2 \cos^2\varphi_1},$$

$$\frac{\partial^2 s_1}{\partial \varphi_1^2} = -[(a^2 - b^2) \sin\varphi_1 \cos\varphi_1]/\sqrt{a^2 \sin\varphi_1 + b^2 \cos\varphi_1},$$

$$L = F_1 \int_{s_1}^{s_2} ds/F$$

and

$$y_2 = \Psi(s_{22}) = \Psi\left\{\eta_2 \int_0^{\varphi_1} \Phi \, d\varphi_1 - (1 + \eta_2) \int_0^{\varphi_0} \Phi \, d\varphi_1\right\}.$$

we then succeed in setting up the following differential equation

$$\{a^2 \sin^2\varphi_1 + b^2 \cos^2\varphi_1\}\ \ddot{\varphi}_1 + \Bigg\{(a^2 - b^2) \sin \varphi_1 \cos \varphi_1$$

$$+ \frac{1 - \eta_2^2}{2L} (a^2 \sin^2\varphi_1 + b^2 \cos^2\varphi_1)^{3/2}\Bigg\}\ \dot{\varphi}_1^2$$

$$+ \frac{g}{L} \sqrt{a^2 \sin^2\varphi_1 + b^2 \cos^2\varphi_1} \Bigg\{\sqrt{a^2 \cos^2\varphi_1 + b^2 \sin^2\varphi_1} \cdot \sin \varphi_1$$

$$- \Psi \left[\eta_2 \int_0^{\varphi_1} \Phi\, d\varphi_1 - (1 + \eta_2)\int_0^{\varphi_0} \Phi\, d\varphi_1\right]\Bigg\} = 0, \quad (\varphi_0 \neq 0, \eta_2 \neq 1). \tag{2.12}$$

With $\varphi_0 = 0$ and $\eta_2 \neq 1$, then we get

$$\{a^2 \sin^2\varphi_1 + b^2 \cos^2\varphi_1\}\ \ddot{\varphi}_1 + \Bigg\{(a^2 - b^2) \sin \varphi_1 \cos \varphi_1$$

$$+ \frac{1 - \eta_2^2}{2L} (a^2 \sin^2\varphi_1 + b^2 \cos^2\varphi_1)^{3/2}\Bigg\}\ \dot{\varphi}_1^2 + \frac{g}{L} \sqrt{a^2 \sin^2\varphi_1 + b^2 \cos^2\varphi_1}$$

$$\times \left\{\sqrt{a^2 \cos^2\varphi_1 + b^2 \sin^2\varphi_1} \cdot \sin \varphi_1 - \Psi\left(\eta_2 \int_0^{\varphi_1} \Phi\, d\varphi_1\right)\right\} = 0$$

$$(\varphi_0 = 0, \eta_2 \neq 1). \tag{2.13}$$

With $\varphi_0 = 0$, $\eta_2 = 1$ and $\varphi_1 = \varphi$, then

$$\sqrt{a^2 \cos^2\varphi_1 + b^2 \sin^2\varphi_1} \cdot \sin \varphi_1 = y_1 = -y_2 = y = \Psi\left(\eta_2 \int_0^{\varphi_1} \Phi\, d\varphi\right)$$

will convert (3.13) into

$$(a^2 \sin^2\varphi + b^2 \cos^2\varphi)\ \ddot{\varphi} + (a^2 - b^2) \sin \varphi \cos \varphi\ \dot{\varphi}^2$$

$$+ \frac{2g}{L} y \sqrt{a^2 \sin^2\varphi + b^2 \cos^2\varphi} = 0 \quad (\varphi_0 = 0, \eta_2 = 1, y_i = y, \varphi_i = \varphi). \tag{2.14}$$

If we let $y \approx b \sin \varphi$, the above equation simplifies to

$$(a^2 \sin^2\varphi + b^2 \cos^2\varphi)\ \ddot{\varphi} + (a^2 - b^2) \sin \varphi \cos \varphi\ \dot{\varphi}^2$$

$$+ \frac{2gb}{L} \sqrt{a^2 \sin^2\varphi + b^2 \cos^2\varphi} \cdot \sin \varphi = 0 \tag{2.15}$$

which is the equation set up by the author in an earlier paper [10].

Setting $a = b$, we then get the reasonable outcome

$$\ddot{\varphi} + \frac{2g}{L} \sin \varphi = 0$$

for the case only of circular arc and constant cross-section, which is equation (15) in [10].

3. For the case of a communication-tube with circular arcs

We have reported a detailed account of its derivation in [10], and the essential results will be cited here only for the purpose of comparison. The essential mathematical expressions are:

$$\frac{v_1^2}{2g}(1-\eta_2^2) + R\{\sin\varphi_1 + \cos\varphi_0 \sin\eta_2(\varphi_1-\varphi_0) - \sin\varphi_0\cos\eta_2(\varphi_1-\varphi_0)\}$$

$$= \frac{1}{g}\int_{s_1}^{s_2}\frac{\partial v}{\partial t}\,\mathrm{d}s, \tag{3.1}$$

$$L\ddot{s}_1 + \tfrac{1}{2}(1-\eta_2^2)\,\dot{s}_1^2 + gR\,\{\sin\varphi_1 + \cos\varphi_0\sin\eta_2(\varphi_1-\varphi_0)$$

$$-\sin\varphi_0\cos\eta_2(\varphi_1-\varphi_0)\} = 0, \tag{3.2}$$

$$\{L_0 + \lambda R(\varphi_0-\varphi_1)\}\,\ddot{\varphi}_1 - \tfrac{1}{2}\lambda R\dot{\varphi}_1^2$$

$$+g\,\{\sin\varphi_1 + \sin[\eta_2\varphi_1 - (1+\eta_2)\varphi_0]\} = 0. \tag{3.3}$$

The first solution of (3.3) is

$$\eta_2\{L_0 + \lambda R(\varphi_0-\varphi_1)\}\,\dot{\varphi}_1^2 + 2g\,\{\eta_2(\cos\varphi_{10} - \cos\varphi_1)$$

$$+\cos[\eta_2\varphi_{10} - (1+\eta_2)\varphi_0] - \cos[\eta_2\varphi_1 - (1+\eta_2)\varphi_0]\} = 0 \tag{3.4}$$

or written into the form

$$\sqrt{\frac{2g}{\eta_2}}\int_{t_1}^{t_2}\mathrm{d}t = \int_{\vartheta_1}^{\vartheta_2}\frac{\Psi(\varphi_1)}{\Phi(\varphi_1)}\,\mathrm{d}\varphi_1, \tag{3.5}$$

where

$$\Psi(\varphi_1) = \sqrt{L_0 + \lambda R(\varphi_0-\varphi_1)}$$

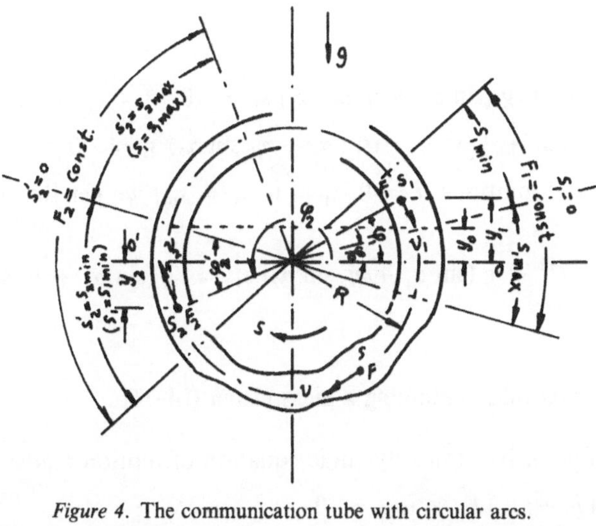

Figure 4. The communication tube with circular arcs.

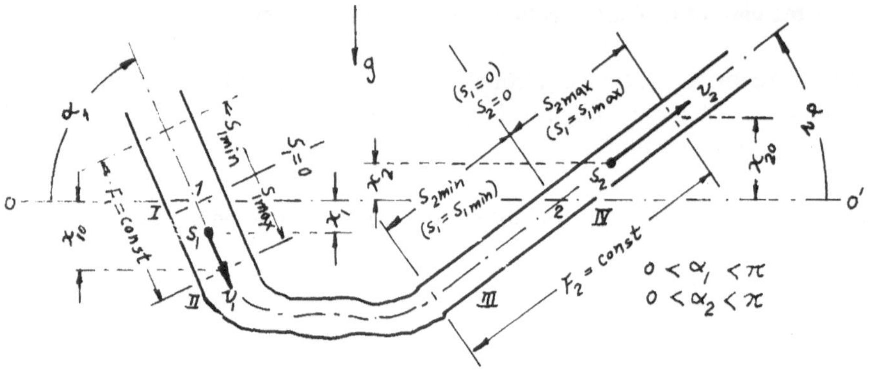

Figure 5. Slanting angles of the two arms are α_1 and α_2.

and

$$\Phi(\varphi_1) \; = \; \sqrt{\eta_2(\cos\varphi_1 - \cos\varphi_{10}) + \cos[\eta_2\varphi_1 - (1+\eta_2)\varphi_0] - \cos[\eta_2\varphi_{10} - (1+\eta_2)\varphi_0]}.$$

Equation (3.3) discloses that at points which are equidistant from the equilibrium position, there exist unequal accelerations. The right side of (3.5) is an odd function of φ_1; this means that vibrations corresponding to opposite directions consume different times.

4. Communication-tube with slanted arms with unequal cross sections

For brevity we quote only the following equations from [11]:
From

$$v_1^2/2g - s_1 \sin\alpha_1 \;=\; v_2^2/2g + s_2 \sin\alpha_2 + \frac{1}{g}\int_{s_1}^{s_2} (\partial v/\partial t)\, \mathrm{d}s \tag{4.1}$$

we deduce

$$L\ddot{s}_1 + \tfrac{1}{2}\lambda\dot{s}_1^2 + g(\sin\alpha_1 + \eta_2 \sin\alpha_2)\, s_1 \;=\; 0, \tag{4.2}$$

$$(L_0 + \lambda s_1)\, \ddot{s}_1 + \tfrac{1}{2}\lambda\dot{s}_1^2 + g(\sin\alpha_1 + \eta_2\sin\alpha_2)\, s_1 \;=\; 0. \tag{4.3}$$

With the initial conditions: $t = 0$, $\dot{s}_1 = 0$, $s_1 = s_{10}$, we get the first integral of (4.3) as

$$(L_0 + \lambda s_1)\dot{s}_1^2 + g(\sin\alpha_1 + \eta_2\sin\alpha_2)\, s_1^2 \;=\; g(\sin\alpha_1 + \eta_2\sin\alpha_2)\, s_{10}^2. \tag{4.4}$$

5. Communication-tube containing a mass piston [14–16]

Referring to Fig. 6, Newton's dynamic equation of motion reads:

$$m\ddot{s}_1 + (p_1 - p_0) F_1 + c s_1 \;=\; 0, \tag{5.1}$$

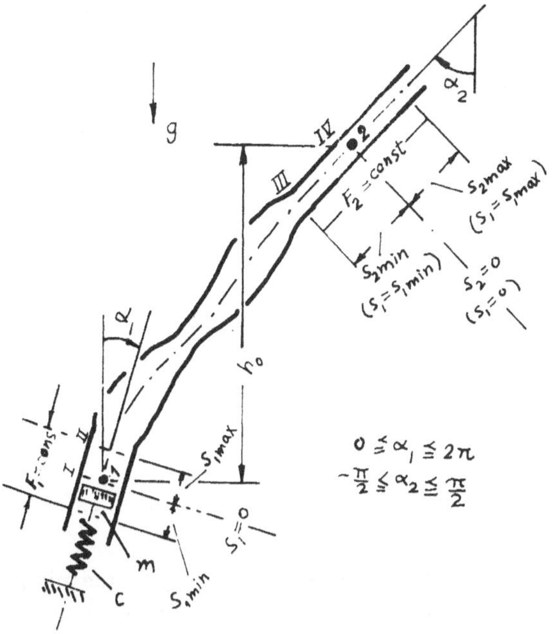

Figure 6. A mass piston joins the vibration of the liquid.

where m is the piston mass, c the spring constant, and $p_1 - p_0 = p_1 - p_2$, then we have the Bernoulli equation

$$p_1 - p_0 = \gamma h + \varrho(v_2^2 - v_1^2)/2 + \varrho L \ddot{s}_1. \tag{5.2}$$

The combination of (5.1) and (5.2) yields in sequence

$$\left(m + \varrho F_1 \int_{s_1}^{s_2} \eta \, \mathrm{d}s\right) \ddot{s}_1 + \tfrac{1}{2} \varrho F_1 (\eta_1^2 - 1) \dot{s}_1^2 + c s_1 + \gamma F_1 h = 0, \tag{5.3}$$

$$\{m + \varrho F_1 L_0 + \varrho F_1(\eta_2^2 - 1) s_1\} \ddot{s}_1 + \tfrac{1}{2} \varrho F_1 (\eta_2^2 - 1) \dot{s}_1^2$$
$$+ \{\gamma F_1(\eta_2 \cos \alpha_2 - \cos \alpha_1) + c\} s_1 + \gamma F_1 h = 0, \tag{5.4}$$

$$(m_1 + 2a\xi_1)\ddot{\xi}_1 + a\dot{\xi}_1^2 + c_1\xi_1 = 0, \tag{5.5}$$

$$m_1\dot{\xi}_1^2 + 2a\xi_1\dot{\xi}_1^2 + c_1\xi_1^2 = c_1\xi_{10}^2, \tag{5.6}$$

wherein $\gamma F_1(\eta_2 \cos \alpha_2 - \cos \alpha_1) + c = c_1$, $m + \varrho F_1 h - 2a\gamma F_1 h_0/c_1 = m_1$, $\varrho F_1(\eta_2^2 - 1) = 2a$, $s_1 = \xi_1 - \gamma F_1 h_0/c_1$ and with the initiative $t = 0$, $\xi_1 = \xi_{10}$, $\dot{\xi}_1 = 0$.

6. The singularity location of the ideal liquid motion in the communication-tube with elliptic arcs

In the preceding two sections we have set up four differential equations (2.12) to (2.15) for the motion of the ideal liquid in communication-tube with elliptic arcs

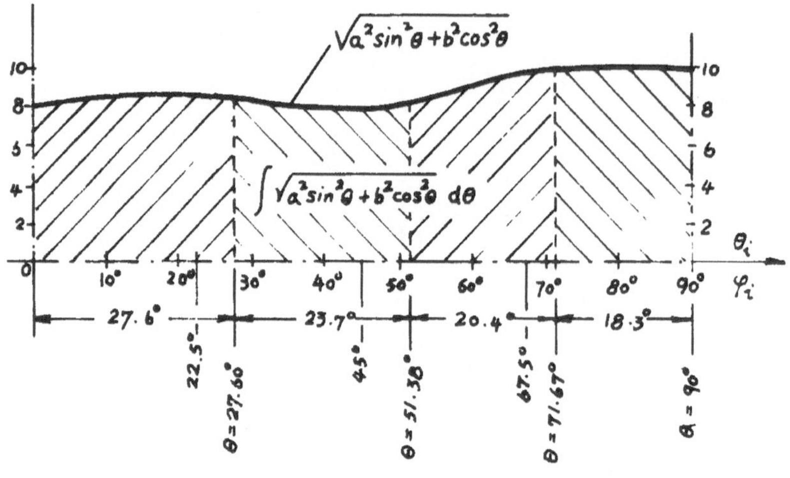

Figure 7. The graphical method of $\int\sqrt{a^2\sin^2\varphi + b^2\cos^2\varphi}\,d\varphi$.

of variable cross section. The solution of the former three might be quite involved, nevertheless, through graphical means, we may get an approximate result. In this procedure, the kernel point lies in the preliminary integration of the integral $\int\sqrt{a^2\sin^2\theta + b^2\cos^2\theta}\,d\theta$. For this goal first we need to divide the first quadrant into equal parts with angle φ_i; then utilize the relation (2.4) to calculate the angles θ_i; but the angles θ_i do not lead to equal angular sections. We write down the points θ_i on the abscissa in Fig. 7, in which the slanted parallel lines correspond to the integral $\int\sqrt{a^2\sin^2\varphi + b^2\cos^2\varphi}\,d\varphi$ with appropriate upper and lower integral limits, i.e., the related points on the elliptic arc.

In the communication-tube we should assign some initial displacement (i.e., some initial arc lengths), find the related angle by the graphical method, and finally acquire the arc lengths of the divided sections. With $a = 10$ and $b = 8$, then through (2.4) to get Table 2. The φ_i lay on equal segments, but they lead to unequal $\theta_{i+1} - \theta_i$, where $\Theta_i = \sqrt{a^2\sin^2\theta_i + b^2\cos^2\theta_i}$.

Here we do not intend to solve the former three equations (2.12) to (2.14), since they pertain in the main to involved concrete configurations of the tube, but without any barrier. We will treat (2.15) only in order to attain some general properties which might enable us to say something about the former three.

Equation (2.15) can be written in the form

$$\{\alpha^2\sin^2\varphi + b^2\cos^2\varphi\}\,\ddot\varphi + (a^2 - b^2)\sin\varphi\cos\varphi\,\dot\varphi^2$$

$$+ \frac{2gab}{L_0}\left\{1 - \sum_{n=1}^{\infty}\frac{M}{2n-1}\alpha^n\cos^{2n}\varphi\right\}\sin\varphi = 0,$$

$$\frac{d}{dt}\left\{\frac{a^2\sin^2\varphi + b^2\cos^2\varphi}{2}\cdot\dot\varphi^2\right.$$

$$\left. - \frac{2gab}{L_0}\left(\cos\varphi - \sum_{n=1}^{\infty}\frac{M}{4n^2-1}\alpha^n\cos^{2n+1}\varphi\right)\right\} = 0,$$

Table 2. With $a = 10$, $b = 8$; the values of $\int_{\theta_i}^{\theta_{i+1}} \Theta_i \, d\theta$.

φ_i	0		$\pi/8$		$\pi/4$		$3\pi/8$		$\pi/2$
$\varphi_i[°]$	0		22.5		45		67.5		90
$\theta_i - \varphi_i = \varphi_i' = \tan^{-1} \dfrac{2 + \tan \varphi_i}{8 + 10 \tan^2 \varphi_i}$	0		5.10°		6.30°		4.17°		0
$\theta_i = \varphi_i + \varphi_i'$	0		27.60°		51.38°		71.67°		90°
$\theta_{i+1} - \theta_i$		27.60°		23.70°		20.37°		18.33°	
Θ_i	8.02		8.46		8.27		9.81		10
$\int_{\theta_i}^{\theta_{i+1}} \Theta_i \, d\theta$		10.26		8.42		8.35		7.96	

wherein

$$\alpha = 1 - b^2/a^2, \qquad M = \frac{1 \cdot 3 \cdot 5 \ldots (2n-1)}{2 \cdot 4 \cdot 6 \ldots 2n}.$$

Then we get the solution of (2.15)

$$\frac{a^2 \sin^2\varphi + b^2 \cos^2\varphi}{2} \dot{\varphi}^2 - \frac{2gab}{L_0}\left(\cos \varphi - \sum_{n=1}^{\infty} \frac{M}{4n^2 - 1} \alpha^n \cos^{2n+1} \varphi\right) = C,$$

wherein with the initial conditions $t = 0$, $\varphi = \varphi_0$, $\dot{\varphi} = 0$ we get

$$C = -\frac{2gab}{L_0}\left\{\cos \varphi_0 - \sum_{n=1}^{\infty} \frac{M}{4n^2 - 1} \alpha^n \cos^{2n+1} \varphi_0\right\}.$$

From the relation $\dot{\varphi} = \phi$, in the state plane, we can get

$$\frac{d\phi}{d\varphi} = -\frac{L_0(a^2 - b^2) \sin \varphi \cos \varphi \, \phi^2 + 2gb\sqrt{a^2 \sin^2\varphi + b^2 \cos^2\varphi} \cdot \sin \varphi}{L_0[a^2 \sin^2\varphi + b^2 \cos^2\varphi]\, \phi}. \tag{6.1}$$

Then we can utilize

$$\phi^2 = \left[2C + \frac{2gab}{L_0}\left(\cos \varphi - \sum_{n=1}^{\infty} \frac{M}{4n^2 - 1} \alpha^n \cos^{2n+1}\right)\right] \Big/ (a^2 \sin^2\varphi + b^2 \cos^2\varphi)$$

and

$$U(\varphi) = \int_{\varphi_0}^{\varphi} \phi \, (d\phi/d\varphi) \, d\varphi$$

to acquire the potential function

$$U(\varphi) = \frac{2gab}{L_0} \cdot \frac{\cos \varphi_0 - \cos \varphi + \sum_{n=1}^{\infty} \frac{M}{4n^2 - 1} \alpha^n (\cos^{2n+1} \varphi_0 - \cos^{2n+1} \varphi)}{a^2 \sin^2\varphi + b^2 \cos^2\varphi} \geqslant 0$$

$$(\varphi \geqslant \varphi_0) \tag{6.2}$$

$$[U(\varphi)]_{\min} = -\frac{2ga}{L_0 b}\left\{1 - \cos\varphi_0 + \sum_{n=1}^{\infty}\frac{M}{4n^2-1}\alpha^n(1-\cos^{2n+1}\varphi_0)\right\},$$

$$(\varphi = 0) \tag{6.3}$$

$$U(\varphi) = U(-\varphi), \tag{6.4}$$

The above results indicate that $\varphi = 0$ is the stable singular point of the potential function. The position of this point depends on the initial angular displacement φ_0; accordingly, we do *not have here a fixed singular point.*

Although we have not dealt directly with a communication-tube of varying cross sections, the complexity caused by section variation may allow us to conclude that their singularities are also *movable in nature.*

7. The singularity location of the ideal liquid motion in communication-tube with circular arcs

Here, we limit the initial conditions to: $t = 0$, $\varphi_1 = \varphi_{10}$, $\dot{\varphi}_1 = 0$ and use $\dot{\varphi}_1 = \phi_1$, then we deduce the differential equation for the motion trajectory in the state plane

$$\frac{d\phi_1}{d\varphi_1} = \frac{\lambda R\phi_1^2 - 2g\{\sin\varphi_1 + \sin[\eta_2\varphi_1 - (1+\eta_2)\varphi_0]\}}{2[L_0 + \lambda R(\varphi_0 - \varphi_1)]\,\phi_1}. \tag{7.1}$$

We complete the integration, after tedious manipulation of

$$U(\varphi_1) = -\int_0^{\varphi_1}\phi_1\,(d\phi_1/d\varphi_1)\,d\varphi_1, \tag{7.2}$$

there then ensues

$$U(\varphi_1) = \frac{g}{2\eta_2} \times \frac{\eta_2(\cos\varphi_{10} - \cos\varphi_1) + \cos[\eta_2\varphi_{10} - (1+\eta_2)\varphi_0]\cos[\eta_2\varphi_1 - (1+\eta_2)\varphi_0]}{L_0 + \lambda R(\varphi_0 - \varphi_1)} \tag{7.3}$$

wherein, after discarding an immaterial constant

$$A = \frac{g}{2\eta_2}\,\frac{(2+\cos\varphi_0)/2 + 2\cos(1+\eta_2)\varphi_0 + \cos[\eta_2\varphi_{10} - (1+\eta_2)\varphi_0]}{L_0 + \lambda R\varphi_0}.$$

Moreover, the following expressions can be obtained

$$\left.\begin{aligned} U(\pm\varphi_{10}) &= 0 \\ [U(\varphi_1)]_{\min} &= \frac{g}{2\eta_2} \\ &\times \frac{\eta_2(\cos\varphi_{10} - 1) + \cos[\eta_2\varphi_{10} - (1+\eta_2)\varphi_0] - \cos(1+\eta_2)\varphi_0}{L_0 + \lambda R\varphi_0} \end{aligned}\right\} \tag{7.4}$$

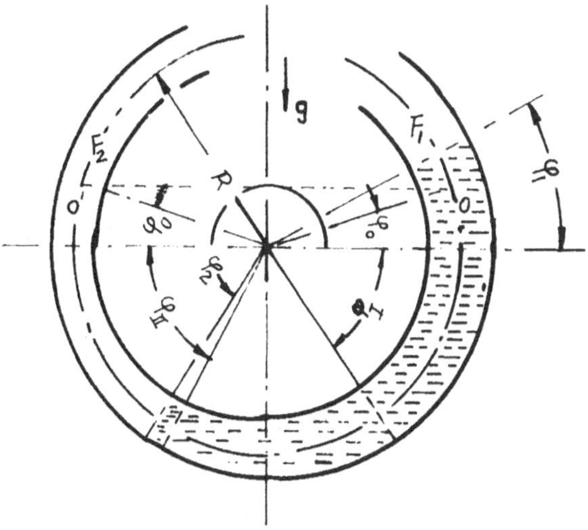

Figure 8. Communication tube with circular arcs.

$$U(\varphi_1) \neq U(-\varphi_1). \tag{7.5}$$

For brevity, let $\varphi_0 = 0$; corresponding to Fig. 8, we have taken $R = 10\,\text{cm}$, $\eta_2 = 2$, $L_0 = R[\varphi_{10} + \eta_2(\pi - \varphi_1)] = 57.5\,\text{cm}$. Then with $\varphi_{10} = 10°, 20°, 30°$ to calculate it yields the curves in Fig. 9, with three unequal displacements. Under these one finds the related trajectories in the state plane. The lowest points of the potential function are connected to the corresponding singular points with dotted lines, respectively. The lowest points in Fig. 9 were not obtained through analysis, since $\mathrm{d}U/\mathrm{d}\varphi_1 = 0$ yields a very involved equation.

$$\frac{\lambda R}{g} C_{\varphi_{10}} - \eta_2 [L_0 + \lambda R(\varphi_0 - \varphi_1)] \{\sin \varphi_1 + \sin [\eta_2\varphi_1 - (1 + \eta_2)\varphi_0]\}$$

$$+ \lambda R\eta_2 \cos \varphi_1 + \lambda R \cos [\eta_2\varphi_1 - (1 + \eta_2)\varphi_0] = 0. \tag{7.6}$$

Since $C_{\varphi_{10}}$ contains the integration initial values, the location of the singular points are dependant upon these initial values.

The lowest point of the potential function shifts along the φ_1-axis with variation of the parameter φ_{10}. The three closed curves $(\dot{\varphi}_{1i} - \varphi_{1i})$ have their individual, single singular point, then they are also of *movable nature*. Besides, the singular point always falls on the half axis $\varphi_1 > 0$, i.e., in the direction where the value $L = \int \eta \, \mathrm{d}\varphi_1$ tends to diminish.

Moreover, one envisions from (3.4) that, with the conditions $t = 0$, $\varphi_1 = \varphi_{10}$ and $\dot{\varphi}_1 = 0$, the integration constant will be $C\varphi_{10} = g\{\eta_2\varphi_{10} + \cos [\eta_2\varphi_{10} - (1 + \eta_2)\varphi_0]\}$. It can be verified that $-(\varphi_{10} - 2\varphi_0)$ is the symmetrical point to φ_{10} with respect to the central position φ_0. Herewith, we get

$$C_{-(\varphi_{10}-2\varphi_0)} = g\{\eta_2 \cos [-(\varphi_{10} - 2\varphi_0)]$$

$$+ \cos [-\eta_2(\varphi_{10} - 2\varphi_0) - (1 + \eta_2)\varphi_0]\}.$$

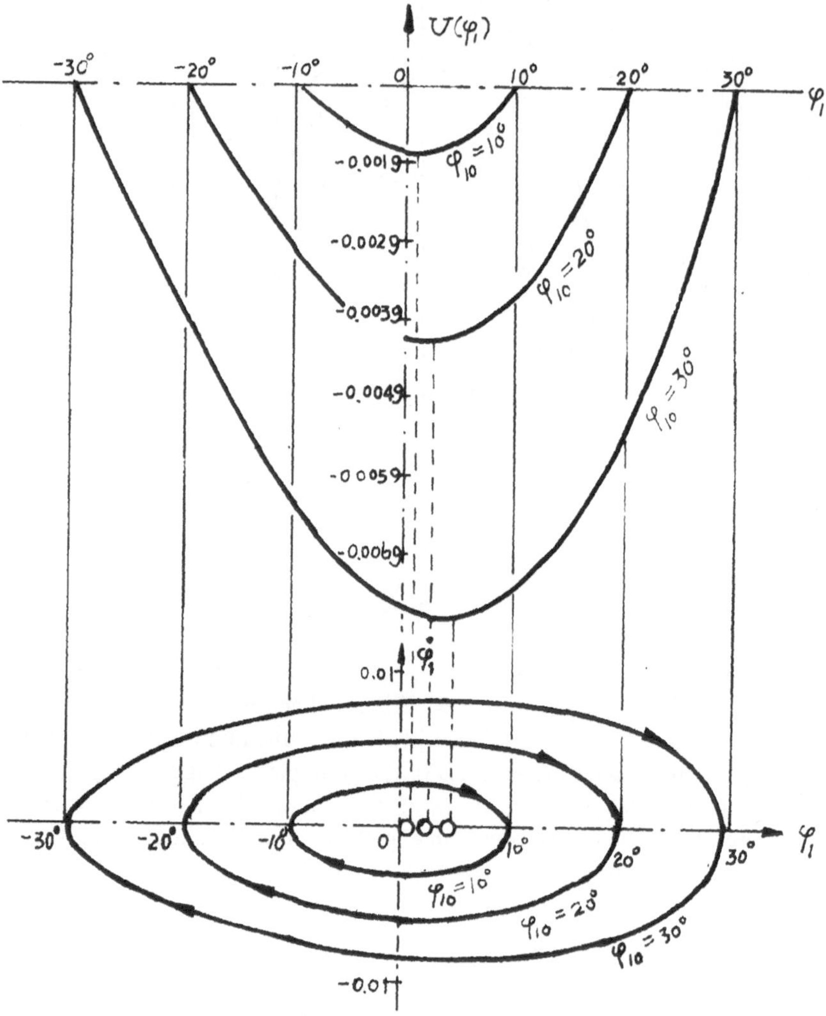

Figure 9. Curves for potential function and motion trajectories in the state plane for communication tube with circular arcs.

Accordingly, we can conclude that

$$\{C_{\varphi_{10}}\}_{\varphi_{10}\to\varphi_0} = \{C_{-(\varphi_{10}-2\varphi_0)}\}_{\varphi_{10}\to\varphi_0} = (1+\eta_2)g\cos\varphi_0, \tag{7.7}$$

$$\{\delta\}_{\varphi_{10}\neq\varphi_0} = \{C_{\varphi_{10}} - C_{-(\varphi_{10}-2\varphi_0)}\}_{\varphi_{10}\neq\varphi_0} \neq 0. \tag{7.8}$$

If we use $\varphi_{10} = 10°, 20°, 30°, 40°, 41°, 42°$ to calculate, then we get Table 3, from which we find, for the case of $\varphi_0 \neq 0$, that its differential equation is fairly difficult to solve. One should therefore endeavor to avoid this in practice.

For the calculation of

$$L_{01} = \int_0^{0'} \frac{F_1}{F}\,ds \quad \text{and} \quad \int_{0'}^{0} \frac{F_2}{F}\,ds,$$

Table 3. $\delta = C_{\varphi_{10}} - C_{-(\varphi_{10}-2\varphi_0)}$ depends on φ_{10}.

$\varphi_{10}[^\circ]$	10	20	30	40	41	42
δ	0	−0.001	−0.016	−0.048	−0.052	−0.056

we can refer to Fig. 8, to get

$$L_{01} = R(\varphi_I + \varphi_{10}) + \eta_2 R(\varphi_{II} + \varphi_{10}) + RF_1 \int_{\varphi_I}^{\pi-\varphi_{II}} \frac{d\varphi_1}{F}, \tag{7.9}$$

$$L_{02} = -\frac{R}{\eta_2}\{(1+\eta_2)\varphi_{10} + \varphi_I + \eta_2\varphi_{II}\} - R\frac{F_1}{F_2}\int_{\pi-\varphi_{II}}^{\varphi_I} \frac{d\varphi_2}{F}, \quad (\eta_2 = F_1/F_2). \tag{7.10}$$

If we take the section variation as $d = c_1 + c_2\varphi_1$ for the conversion length of the intermediate tube piece, then with the boundary equations $d_1 = c_1 + c_2\varphi_I$ and $d_2 = c_1 + c_2(\pi - \varphi_{II})$ we get

$$\int_{\varphi_I}^{\pi-\varphi_{II}} \frac{d\varphi_1}{F} = \frac{4}{\pi c_2}\left\{\frac{1}{c_1 + c_2\varphi_I} - \frac{1}{c_1 + c_2(\pi - \varphi_{II})}\right\} \tag{7.11}$$

and the constraints for the angles φ_I and φ_{II}

$$\varphi_I \leqslant \frac{\pi - 2(1+\eta_2)\varphi_0}{2\eta_2}, \qquad \varphi_{II} \leqslant \frac{\pi}{2} - 2\varphi_0 \tag{7.12}$$

with the constants

$$c_1 = d_1 - \frac{(d_1 - d_2)\varphi_I}{\pi - \varphi_I - \varphi_{II}}, \qquad c_2 = \frac{d_2 - d_1}{\pi - \varphi_I - \varphi_{II}}. \tag{7.13}$$

8. The properties of the singular points for a communication tube with straight arms

For this case with $\dot{s}_1 = \xi_1$, the differential equation of the trajectory in the state plane is $d\xi_1/ds_1 = -[\lambda\xi_1^2 + 2cs_1]/(L_{01} + \lambda s_1)\xi_1$, where $c = g(\sin\alpha_1 + \eta_2 \sin\alpha_2)$. For calculating the potential function

$$V(s_1) = \int_0^{s_1} \xi_1(d\xi_1/ds_1)\, ds_1,$$

we should put $\xi_1^2 = (A - cs_1^2)/(L_{01} + \lambda s_1)$ and $A = cs_{10}^2$, then get

$$V(s_1) = \frac{cs_{10}^2}{2L_{01}} - \frac{c(s_{10}^2 - s_1^2)}{2(L_{01} + \lambda s_1)}. \tag{8.1}$$

Accordingly, $dV(s_1)/ds_1 = 0$ yields

$$(s_1)_{V=\min} = -\frac{L_{01}}{\lambda} \pm \sqrt{\frac{L_{01}^2}{\lambda^2} - s_{10}^2}, \quad (\lambda \gtrless 0). \tag{8.2}$$

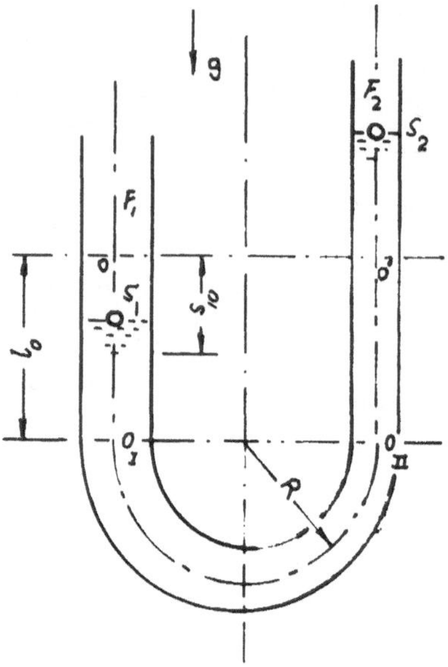

Figure 10. Straight arm communication tube for numerical example.

The above relation shows that the position of the singular point varies with the initial value s_{10}, and again is of a *movable nature.*

Referring to the configuration in Fig. 10, we have

$$L_{01} = \int_{s_1}^{s_2} (F_1/F)\,\mathrm{d}s = \left\{\int_{s_1}^{0} + \int_{0}^{0'} + \int_{0'}^{s_2}\right\} \eta\,\mathrm{d}s = l_0 + F_1 \int_{0_\mathrm{I}}^{0_\mathrm{II}} \frac{\mathrm{d}s}{F} + \eta_2 l_0.$$

Using the variable $d = c_1 + c_2\varphi$ we then get

$$\int_{0_\mathrm{I}}^{0_\mathrm{II}} \frac{\mathrm{d}s}{F} = \frac{4R}{\pi}\int_0^{\pi} \frac{\mathrm{d}\varphi}{(c_1 + c_2\varphi)^2} = \frac{4R}{\pi c_1 c_2}\left\{1 - \frac{1}{1 + \pi c_2/c_1}\right\}$$

where $c_1 = d_1$, $c_2 = (d_2 - d_1)/\pi$. Accordingly we get

$$L_{01} = (1 + \eta_2)\,l_0 + \pi R \eta_2. \qquad (8.3)$$

With $R = 5\,\mathrm{cm}$, $l_0 = 10\,\mathrm{cm}$, $\eta_2 = 2$, $s_{10} \leqslant l_0/\eta = 5\,\mathrm{cm}$, $d_1 = d_2 - \pi/2$, $L_{01} = 61.42\,\mathrm{cm}$, we can use the potential function (8.1) to get the curves in Fig. 11 discarding an immaterial constant.

Figure 11 discloses that the lower points of the potential function are compatible with the requirement of (8.2). Every trajectory in the state plane has its individual singular point, and the limit relation is

$$(s_1)_{V=\min} = \lim_{s_{10}\to 0}\left\{-\frac{L_{01}}{\lambda} \pm \sqrt{\frac{L_{01}^2}{\lambda^2} - s_{10}^2}\right\} = 0, \quad (\lambda \gtrless 0).$$

For the unequal cross section straight arm tube, which is similar to the tube with elliptic arcs, *the differential equation of motion* is also non-linear; hence its singular point is also of a *movable nature*.

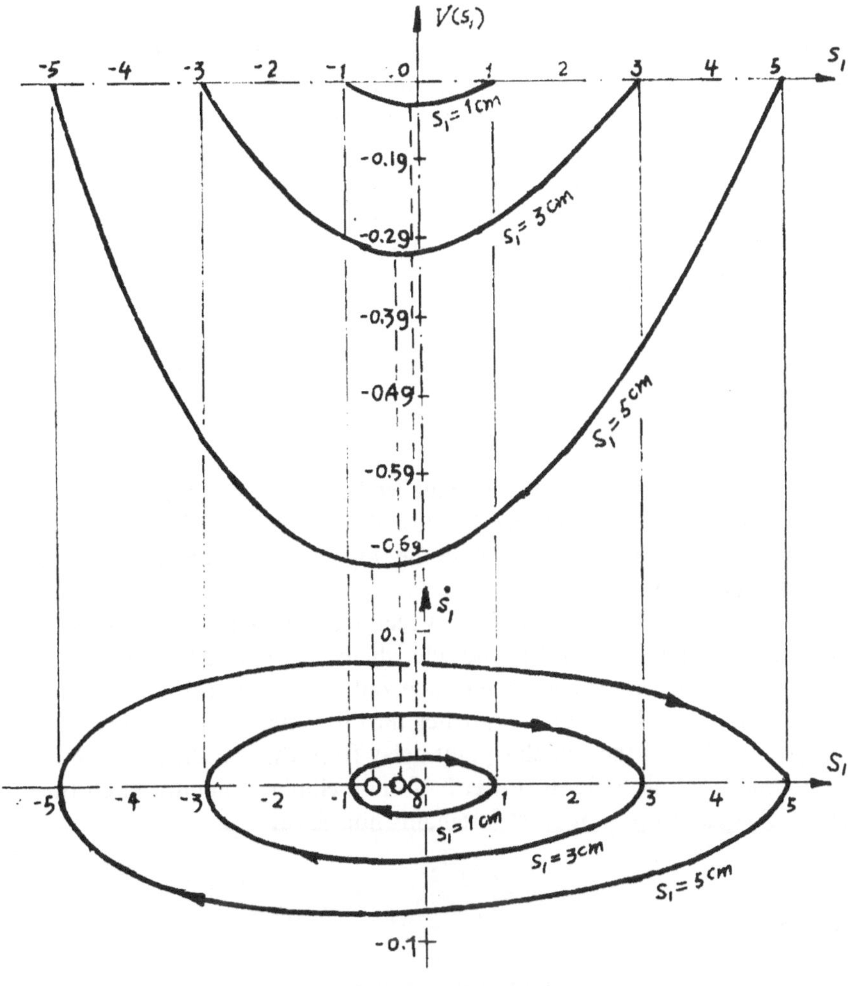

Figure 11. Potential function and trajectories for the case of Fig. 10.

9. The case with a mass piston entering the vibration

For this case, if our analysis uses the state plane method with $d\xi_1/dt = \zeta$, then we acquire from (5.5)

$$\zeta \, d\zeta/d\xi_1 = -(2aA - ca\xi_1^2)/(m_1 + 2a\xi_1)^2 - c_1\xi_1/(m_1 + 2a\xi_1)$$

and the potential function

$$V(\xi_1) = A\int_0^{\xi_1} \frac{2ad\xi_1}{(m_1 + 2a\xi_1)^2} - \frac{c_1}{2}\int_0^{\xi_1} \frac{2a\xi_1^2 \, d\xi_1}{(m_1 + 2a\xi_1)^2} + c_1\int_0^{\xi_1} \frac{\xi_1 \, d\xi_1}{m_1 + 2a\xi_1}$$

$$= \frac{A}{m_1} - \frac{c_1(\xi_{10}^2 - \xi_1^2)}{2(m_1 + 2a\xi_1)}. \tag{9.1}$$

The equations (9.1) and (8.1) possess the same structure. This case pertains to two applications. The first concerns the design of the pressure meter [14, 15], since in it there exists just a mass piston, a spring and a liquid column, and in general the cross section of the tubing is not constant. The mass of the piston, the spring constant and the reduced length of the liquid column cooperatively determine the pressure specification of the meter. The second application comprises a simple installation in which we can get vibration curves which give us the reduced length of the liquid column in some irregular shaped tubing system, by experiment. In addition to time-saving it gives more exact results, and for this we refer again to [2] and [15].

10. Conclusion

When the communication-tube has unequal end cross sections, it possesses *singular points of a movable nature*, and their location depends on the initial displacements. When the initial displacements are small the singular points then strive toward the equilibrium position.

For the tubing with unequal end cross sections, singular points fall away from the coordinate origin, but shift to the side where the reduced length of the liquid column is diminishing; the shift magnitude and the initial displacements increase or decrease jointly. Moreover, the greater the difference of the end cross section, the larger the shift of the singular point away from the coordinate origin.

When the liquid fills only the lower half of an elliptic communication tube, the differential equation of motion still remains non-linear.

References

1. Schuler, M., *Einführung der Mechanik*, Teil II, Wolfenbruttel (1951).
2. Grammel, R., *Der Kreisel seine Theorie und seine Anwendung*, zweiter Band, Springer-Verlag, Berlin (1950).
3. Bragovishenski, S. N., *Oscillation of the Ships* (in Russian), Sudpromgiz (National Ship Industry Press) (1954).
4. Prandtl, L., *Führer durch die Strömungslehre*, Braunschweig (1942).
5. Popov, S. G., *Some Problems and Methods of Experimental Aeromechanics* (in Russian), Moscow (1958).
6. Sideriades, L., "Méthodes de Topologie qualitative: Applications a l'étude descheminées d'équilibre", *Akad. Nauk USSR*, Kiev (1961).
7. Gvazava, G. N. and Kanjelaki, N. A., "On a nonlinear vibration problem for the appendix to calculate the unstable motion with liquid mass in the press-shock system of the hydropower station" (in Russian), *Akad. Nauk USSR*, Kiev (1961).
8. Prasil, F., "Das Wassenschlossproblem der Wasserkraftstationen", *Schweizerische Bauzeitung* (1908).
9. Wittenbauer, F., *Aufgaben aus der Technischen Mechanik*, Berlin (1921).
10. Liu Hsien-chih, Über Schwingungen einer idealen Flüssigkeit in ellipsen- und kreisbogenförmigen kommunizierender Röhren, *Ingenieur-Archiv*, 20. Band, 5. Heft (1952).
11. Liu Hsien-chih, "Beitrag zur Kenntnis der Eigenschwingung einer idealen Flussigkeit in kommunitierenden Röhren", *ZAMP*, Schweiz, Vol. IV (1953).

12. Liu Hsien-chih, "Eigenschwingungen idealen Flüssigkeitin Rohrleitungen mit verschiedenen Endquerschnitten", *Ingenieur-Archiv*, Band 2 (1952).
13. Lehr, E., *Schwingungstechnik*, Band 2, Berlin (1948).
14. Lojanski, L. T. and Lurie, A. I., *A Course in the Theoretical Mechanics* (in Russian), Vol. II, Moscow (1955).
15. den Hartog, J. P., *Mechanical Vibrations*, McGraw-Hill, New York (1947).
16. Cordier, O., *Gemischbildung und Verbrennung in Dieselmotor*, Springer-Verlag, Wien (1956).
17. Schuler, M., "Schlingertank as ship rolling absorber", *Proc. 2nd International Congress for Applied Mechanics*, Zurich (1926), 219 and "Werft, Reederei, Hafen", Vol. 9 (1928).
18. Frahm, H., *Neuartige Schiffsbautechnik Ges.*, Vol. 12 (1911).
19. Arnold, R. N. and Maunder, L., *Gyrodynamics and its Engineering Applications*, New York and London (1961).

Magnetohydrodynamic flow in elliptic ducts

JIANG FU-RU

Shanghai University of Technology, Shanghai, PRC

Abstract. In this paper we will study the flow of an incompressible conducting fluid along an elliptic duct situated in a uniform magnetic field in the case of when the Hartmann number is large. The asymptotic expansion of the solution up to an order of small magnitude has been constructed by means of the method of multiple scales. The flow along a duct of any uniform smooth cross-section can be studied by the same method.

1. Introduction

In 1965, Hunt and Stewartson [1] studied the flow of an incompressible conducting fluid along a rectangular duct under an imposed transverse uniform magnetic field. In 1967, Roberts [2] studied the same flow along an insulating circular duct, and constructed the asymptotic approximation of the solution up to order M^{-1} by means of the method of matched asymptotic expansions, where M is the Hartmann number of flow. In this paper we study the flow along an insulating duct with elliptic cross-section by a new perturbation method offered by the author [3]. By comparing these two methods we see that our method is simpler than the former, and it can give the asymptotic approximation of the solution up to any order of a small number, and can be applied to study the flow along a duct with any uniform smooth cross-section, and under any boundary condition. Here for comparison, we restrict ourselves to study the flow along an insulating elliptic duct.

Let a duct of elliptic cross-section be situated in a uniform magnetic field $\mathbf{B}_0$. The axis of the duct is perpendicular to the direction of $\mathbf{B}_0$. Take the centre of the duct as the origin of coordinates (x, y, z), and let the x-axis be parallel to the direction of flow, and the z-axis be parallel to the direction of $\mathbf{B}_0$ (cf. Fig. 1). Let A be the long radius of elliptic cross-section, which is laid on the z-axis, and B be the short radius laid on the y-axis. Take A as the unit length of a scale on the coordinate axes. From [1] and [2] we know that the normalized velocity $v(y, z)$ of flow and the induced magnetic field $b(y, z)$ (both in the x-direction) are governed by the differential equations:

$$\frac{\partial^2 v}{\partial y^2} + \frac{\partial^2 v}{\partial z^2} + M\frac{\partial b}{\partial z} = -1, \tag{1.1}$$

$$\frac{\partial^2 b}{\partial y^2} + \frac{\partial^2 b}{\partial z^2} + M\frac{\partial v}{\partial z} = 0, \tag{1.2}$$

Yeh, K. Y. (ed) Progress in Applied Mechanics
 ISBN-13: 978-94-010-8061-3

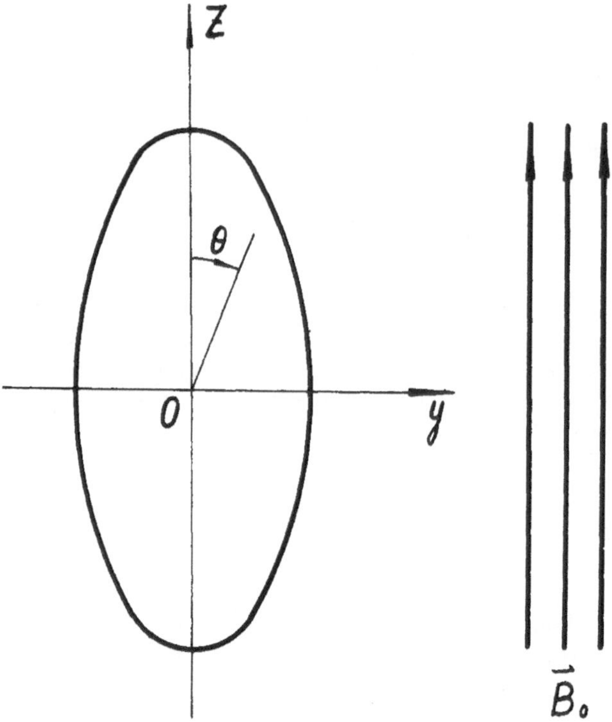

Figure 1. Cross-section of duct.

where $M = AB_0(\sigma/\varrho\nu)^{1/2}$ is the Hartmann number of flow; ϱ, ν and σ are the density, kinematic viscosity, and electrical conductivity of the fluid respectively; and B_0 is the flux density of the imposed magnetic field.

Since the wall of the duct is insulated, we have the following boundary conditions:

$$v = b = 0, \quad \text{on the ellipse } \frac{y^2}{p^2} + z^2 = 1, \tag{1.3}$$

where $p = B/A$.

Let us define

$$X(y, z) = v(y, z) + b(y, z). \tag{1.4}$$

From (1.1)–(1.3) we know that it is governed by

$$\varepsilon\left(\frac{\partial^2 X}{\partial y^2} + \frac{\partial^2 X}{\partial z^2}\right) + \frac{\partial X}{\partial z} = -\varepsilon, \tag{1.5}$$

$$X|_{z=\pm\sqrt{1-y^2/p^2}} = 0, \tag{1.6}$$

where $\varepsilon = 1/M \ll 1$.

Again, define

$$\tilde{X}(y, z) = v(y, z) - b(y, z), \tag{1.7}$$

it is easily shown that

$$\tilde{X}(y, z) = X(y, -z). \tag{1.8}$$

In case X is known, we can find $\tilde{X}$ by (1.8), and the unknowns v and b can be found by the following formulas:

$$v = \tfrac{1}{2}(X + \tilde{X}) = \tfrac{1}{2}[X(y, z) + X(y, -z)],$$

$$b = \tfrac{1}{2}(X - \tilde{X}) = \tfrac{1}{2}[X(y, z) - X(y, -z)].$$

So we need only to find the solution of the boundary value problem (1.5)–(1.6).

2. Outer solution

Let the outer expansion of the solution be

$$X = \varepsilon[w_0(y, z) + \varepsilon w_1(y, z) + \cdots + \varepsilon^i w_i(y, z) + \cdots]. \tag{2.1}$$

Substituting X into (1.5), and equating the coefficients of like powers of ε, we obtain the recursive equations for w_i $(i = 0, 1, \ldots)$,

$$\frac{\partial w_0}{\partial z} = -1, \tag{2.2}$$

$$\frac{\partial w_i}{\partial z} = -\Delta w_{i-1}, \quad (i = 1, 2, \ldots). \tag{2.3}$$

They are equations of the first order, and we cannot therefore prescribe boundary conditions on the whole edge of the ellipse. From the results of [4] we know that the boundary condition for w_i should be given on the upper half-edge of the ellipse, i.e., on that segment of the boundary through which the characteristics of the degenerated equation cross over the region. Substituting (2.1) into boundary condition (1.6), and equating the coefficient of each power of ε to zero, we obtain the boundary condition for w_i $(i = 0, 1, \ldots)$,

$$w_i\big|_{z=\sqrt{1-y^2/p^2}} = 0, \quad (i = 0, 1, \ldots). \tag{2.4}$$

From (2.2) and (2.4) we find that

$$w_0 = -z + \sqrt{1 - (y^2/p^2)}. \tag{2.5}$$

Substituting w_0 into (2.3) (with $i = 1$), and considering the boundary condition (2.4), we find that

$$w_1 = -\frac{1}{p^2}[-z + \sqrt{1 - (y^2/p^2)}][1 - (y^2/p^2)]^{-3/2}. \tag{2.6}$$

*Formula (9) in [2] should be corrected as

$$v(y, z) = v(-y, z) = \tfrac{1}{2}[X(y, z) + X(y, -z)],$$

$$b(y, z) = b(-y, z) = \tfrac{1}{2}[X(y, z) - X(y, -z)].$$

Progressing similarly, we have

$$w_i = \int_{\sqrt{1-(y^2/p^2)}}^{z} (-\Delta w_{i-1})\, dz, \quad (i = 2, 3, \ldots). \tag{2.7}$$

From the above we see that the expansion constructed cannot, in general, satisfy the boundary condition (1.6) on the lower half-edge of the ellipse. We call it "outer solution" of the problem, and denote it by $X^{(0)}$, describing the main stream of flow.* In order to replenish $X^{(0)}$ to satisfy the boundary condition (1.6) on the lower half-edge of the ellipse, we will construct the boundary layer correction along this segment in the next section.

3. Boundary layer correction

In the neighbourhood of the boundary let us introduce local coordinates (ϱ, θ):

$$y = (1 - \varrho)p \sin \theta, \qquad z = (1 - \varrho) \cos \theta, \quad (1 > \varrho \geqslant 0).$$

Equation (1.5) and boundary condition (1.6) are transformed into

$$L_\varepsilon(X) \equiv \varepsilon \left(a(\theta) \frac{\partial^2}{\partial \varrho^2} + 2b(\varrho, \theta) \frac{\partial^2}{\partial \varrho \partial \theta} + c(\varrho, \theta) \frac{\partial^2}{\partial \theta^2} + d(\varrho, \theta) \frac{\partial}{\partial \varrho} + e(\varrho, \theta) \frac{\partial}{\partial \theta} \right) X + \left(\alpha(\theta) \frac{\partial}{\partial \varrho} + \beta(\varrho, \theta) \frac{\partial}{\partial \theta} \right) X = -\varepsilon, \tag{3.1}$$

$$X|_{\varrho=0} = 0, \tag{3.2}$$

where

$$a(\theta) = \frac{\sin^2\theta}{p^2} + \cos^2\theta, \qquad b(\varrho, \theta) = \frac{\sin\theta \cos\theta}{1 - \varrho}[1 - (1/p^2)],$$

$$c(\varrho, \theta) = \frac{1}{(1-\varrho)^2}\left(\frac{\cos^2\theta}{p^2} + \sin^2\theta\right),$$

$$d(\varrho, \theta) = -\frac{1}{1-\varrho}\left(\frac{\cos^2\theta}{p^2} + \sin^2\theta\right),$$

$$e(\varrho, \theta) = \frac{\sin 2\theta}{(1-\varrho)^2}[1 - (1/p^2)],$$

$$\alpha(\theta) = -\cos, \theta, \qquad \beta(\varrho, \theta) = -\frac{\sin\theta}{1-\varrho}.$$

Here, we apply the method of multiple scales offered by the author in [3] to construct a more accurate boundary layer correction than that obtained by the method of matched asymptotic expansions and the Lyusternik–Vishik Method. In

*For the special case of a circular duct, the expansion obtained above is in agreement with that derived by Gold [4], Shercliff [5] and Roberts [2] from a different point of view. Further, they only obtained the asymptotic expansion up to order $0(\varepsilon^2)$.

the neighbourhood of the lower half-edge of the ellipse we introduce the variables with multiple scales

$$f = \frac{g(\varrho, \theta)}{\varepsilon}, \quad \xi = \varrho, \quad \eta = \theta, \tag{3.3}$$

and replace the partial derivatives with respect to ϱ and θ by

$$\frac{\partial}{\partial \varrho} = \varepsilon^{-1}\left(g_\varrho \frac{\partial}{\partial f} + \varepsilon \frac{\partial}{\partial \xi}\right), \quad \frac{\partial}{\partial \theta} = \varepsilon^{-1}\left(g_\theta \frac{\partial}{\partial f} + \varepsilon \frac{\partial}{\partial \eta}\right),$$

$$\frac{\partial^2}{\partial \varrho \partial \theta} = \varepsilon^{-2}\left[g_\varrho g_\theta \frac{\partial^2}{\partial f^2} + \varepsilon\left(g_\theta \frac{\partial^2}{\partial f \partial \xi} + g_\varrho \frac{\partial^2}{\partial f \partial \eta} + g_{\varrho,\theta} \frac{\partial}{\partial f}\right) + \varepsilon^2 \frac{\partial^2}{\partial \xi \partial \eta}\right],$$

The differential operator L_ε can be expanded as

$$L_\varepsilon \equiv \varepsilon^{-1}(K_0 + \varepsilon K_1 + \varepsilon^2 K_2), \tag{3.4}$$

where

$$K_0 \equiv (a g_\varrho^2 + 2b g_\varrho g_\theta + c g_\theta^2) \frac{\partial^2}{\partial f^2} + (\alpha g_\varrho + \beta g_\theta) \frac{\partial}{\partial f},$$

$$K_1 \equiv a\left(2 g_\varrho \frac{\partial^2}{\partial f \partial \xi} + g_{\varrho,\varrho} \frac{\partial}{\partial f}\right) + 2b\left(g_\theta \frac{\partial^2}{\partial f \partial \xi} + g_\varrho \frac{\partial^2}{\partial f \partial \eta} + g_{\varrho,\theta} \frac{\partial}{\partial f}\right)$$

$$+ c\left(2 g_\theta \frac{\partial^2}{\partial f \partial \eta} + g_{\theta,\theta} \frac{\partial}{\partial f}\right) + d g_\varrho \frac{\partial}{\partial f} + e g_\theta \frac{\partial}{\partial f} + \alpha \frac{\partial}{\partial \xi} + \beta \frac{\partial}{\partial \eta},$$

$$K_2 \equiv a \frac{\partial^2}{\partial \xi^2} + 2b \frac{\partial^2}{\partial \xi \partial \eta} + c \frac{\partial^2}{\partial \eta^2} + d \frac{\partial}{\partial \xi} + e \frac{\partial}{\partial \eta}.$$

Let the boundary layer term have an expansion of the form

$$V = \varepsilon[v_0(f, \xi, \eta) + \varepsilon v_1(f, \xi, \eta) + \cdots + \varepsilon^i v_i(f, \xi, \eta) + \cdots]. \tag{3.5}$$

Substituting this into the homogeneous equation of (3.1), and equating the coefficient of each power of ε to zero, we obtain the recursive equations for v_i $(i = 0, 1, \ldots)$

$$(a g_\varrho^2 + 2b g_\varrho g_\theta + c g_\theta^2) \frac{\partial^2 v_0}{\partial f^2} + (\alpha g_\varrho + \beta g_\theta) \frac{\partial v_0}{\partial f} = 0, \tag{3.6}$$

$$(a g_\varrho^2 + 2b g_\varrho g_\theta + c g_\theta^2) \frac{\partial^2 v_i}{\partial f^2} + (\alpha g_\varrho + \beta g_\theta) \frac{\partial v_i}{\partial f}$$

$$= -[K_1(v_{i-1}) + K_2(v_{i-2})], \quad (i = 1, 2, \ldots), \tag{3.7}$$

the letters with negative subscript are taken as zero here. Since the boundary layer term V is needed to eliminate the value of $X^{(0)}$ on the lower half-edge of ellipse, we have the boundary conditions

$$v_0|_{\varrho=0} = -\tilde{w}_0(\varrho, \theta)|_{\varrho=0} = 2\cos\theta, \quad (|\theta| > \pi/2), \tag{3.8}$$

$$v_i|_{\varrho=0} = -\tilde{w}_i(\varrho, \theta)|_{\varrho=0}, \quad (|\theta| > \pi/2), \quad (i = 1, 2, \ldots), \tag{3.9}$$

where $\tilde{w}_i$ is w_i in local coordinates (ϱ, θ).

We take $g(\varrho, \theta)$ in (3.3) to be the solution of the boundary value problem

$$\left.\begin{aligned} ag_\varrho^2 + 2bg_\varrho g_\theta + cg_\theta^2 &= \alpha g_\varrho + \beta g_\theta \\ g|_{\varrho=0} &= 0, \quad (|\theta| > \pi/2) \\ g > 0, \quad (\varrho > 0, |\theta| > \pi/2). \end{aligned}\right\} \tag{3.10}$$

Equation (3.6) for v_0 reduces to

$$\frac{\partial^2 v_0}{\partial f^2} + \frac{\partial v_0}{\partial f} = 0, \tag{3.11}$$

and has a solution

$$v_0 = C_0(\xi, \eta)\,\mathrm{e}^{-f} = C_0(\varrho, \theta)\,\mathrm{e}^{-g(\varrho,\theta)/\varepsilon}. \tag{3.12}$$

decreasing exponentially in the neighbourhood of lower half-edge of the ellipse, where C_0 is an arbitrary function determined later. Substituting v_0 into (3.7) (with $i = 1$), and setting its right-hand side equal to zero, we obtain the differential equation governing C_0:

$$(\alpha - 2ag_\varrho - 2bg_\theta)\frac{\partial C_0}{\partial \xi} + (\beta - 2bg_\varrho - 2cg_\theta)\frac{\partial C_0}{\partial \eta}$$
$$-(ag_{\varrho,\varrho} + 2bg_{\varrho,\theta} + cg_{\theta\theta} + dg_\varrho + eg_\theta)C_0 = 0. \tag{3.13}$$

From (3.8), we have the boundary condition for C_0:

$$C_0|_{\xi=0} = 2\cos\theta, \qquad (|\theta| > \pi/2). \tag{3.14}$$

From the Cauchy problem (3.13)–(3.14) we can determine C_0. Meanwhile, equation (3.7) (with $i = 1$) governing v_1 is reduced to a homogeneous equation with constant coefficients as that governing v_0

$$\frac{\partial^2 v_1}{\partial f^2} + \frac{\partial v_1}{\partial f} = 0 \tag{3.15}$$

which gives an exponentially decreasing solution

$$v_1 = C_1(\xi, \eta)\,\mathrm{e}^{-f} = C_1(\varrho, \theta)\,\mathrm{e}^{-g(\varrho,\theta)/\varepsilon}, \tag{3.16}$$

where C_1 is the arbitrary function determined in a way similar to that of C_0. Finally, we have an asymptotic expansion of the solution of the form

$$X = \begin{cases} X^{(0)} \sim \varepsilon[-z + \sqrt{1 - (y^2/p^2)}] + \varepsilon^2 \dfrac{-1}{p^2}[-z + \sqrt{1 - (y^2/p^2)}] \\ \quad \times [1 - (y^2/p^2)]^{-3/2} + \cdots + \varepsilon^i \displaystyle\int_{\sqrt{1-(y^2/p^2)}}^{z} (-\Delta w_{i-1})\,\mathrm{d}z + \cdots, \\ \text{for} \quad -\sqrt{1 - (y^2/p^2)} + \delta_2 \leqslant z \leqslant \sqrt{1 - (y^2/p^2)}, \\ \quad -p + \delta_1 \leqslant y \leqslant p - \delta_1, \\ X^{(0)} + V \sim X^{(0)} + \varepsilon[C_0(\varrho, \theta) + \varepsilon C_1(\varrho, \theta) \\ \quad + \cdots + \varepsilon^i C_i(\varrho, \theta) + \cdots]\,\mathrm{e}^{-g(\varrho,\theta)/\varepsilon}, \\ \text{for} \quad -\sqrt{1 - (y^2/p^2)} \leqslant z \leqslant -\sqrt{1 - (y^2/p^2)} + \delta_2, \\ \quad -p + \delta_1 \leqslant y \leqslant p - \delta_1, \end{cases}$$

where δ_1 is a small positive number less than p, and $\delta_2 < \sqrt{1 - (p - \delta_1)^2/p^2}$.

In case we only need to obtain an asymptotic expansion of the solution up to order $O(\varepsilon)$, we may take

$$g(\varrho, \theta) = \int_0^{\varrho} \frac{\alpha(\theta)}{a(\theta)} \mathrm{d}\varrho = -\varrho \frac{p^2 \cos\theta}{\sin^2\theta + p^2\cos^2\theta}, \quad (|\theta| > \pi/2)$$

as the approximate solution of the boundary value problem (3.10), and set

$$L_\varepsilon \equiv \varepsilon^{-1} K_0,$$

$$K_0 \equiv a g_\varrho^2 \frac{\partial^2}{\partial f^2} + \alpha g_\varrho \frac{\partial}{\partial f} = \frac{\alpha^2}{a}\left(\frac{\partial^2}{\partial f^2} + \frac{\partial}{\partial f}\right),$$

in (3.4). Then we have

$$X \sim \begin{cases} \varepsilon[-z + \sqrt{1 - (y^2/p^2)}] + O(\varepsilon^2), \\ \text{for } -\sqrt{1 - (y^2/p^2)} + \delta_2 \leqslant z \leqslant \sqrt{1 - (y^2/p^2)}, \\ -p + \delta_1 \leqslant y \leqslant p - \delta_1, \\ \varepsilon\Bigg[-(1 - \varrho)\cos\theta + \sqrt{1 - (1 - \varrho)^2 \sin^2\theta} + 2\cos\theta \\ \times \exp\left(\dfrac{\varrho}{\varepsilon} \dfrac{p^2 \cos\theta}{\sin^2\theta + p^2\cos^2\theta}\right)\Bigg] + O(\varepsilon^2), \\ \text{for } 0 \leqslant \varrho \leqslant \delta_3, \quad |\theta| > \pi/2 + \delta_4, \end{cases}$$

where δ_3 and δ_4 are certain small positive numbers.

For the special case of a circular cross-section, $p = 1$, the above formula is in agreement with that obtained in [2].

From the above, we see that the asymptotic expansion of the solution would be singular at points $(\pm p, 0)$, where the lines of magnetization are tangent to the boundary, and the thickness of the boundary layer would approach infinity at $\theta \to \pm\pi/2$. Therefore, the expansion obtained above is invalid in the neighbourhoods of these two points. In the following, we shall apply the method offered in [6] to construct the asymptotic expansion which is valid in the neighbourhoods of these two points.

4. Singular point layer correction

In order to make use of the results of [6], we must set $\tilde{z} = -z$, $\tilde{\theta} = (3\pi/2) = \theta$. Figure 1 is transformed into Fig. 2, and the boundary value problem (3.1)–(3.2) takes the form

$$\tilde{L}_\varepsilon[X] \equiv \varepsilon\left[\tilde{a}(\tilde{\theta}) \frac{\partial^2}{\partial\varrho^2} + 2\tilde{b}(\varrho, \tilde{\theta}) \frac{\partial^2}{\partial\varrho\partial\tilde{\theta}} + \tilde{c}(\varrho, \tilde{\theta}) \frac{\partial^2}{\partial\tilde{\theta}^2} + \tilde{d}(\varrho, \tilde{\theta}) \frac{\partial}{\partial\varrho} + \tilde{e}(\varrho, \tilde{\theta}) \frac{\partial}{\partial\tilde{\theta}}\right] X - \left[\tilde{\alpha}(\tilde{\theta}) \frac{\partial}{\partial\varrho} + \tilde{\beta}(\varrho, \tilde{\theta}) \frac{\delta}{\partial\tilde{\theta}}\right] X = -\varepsilon, \quad (4.1)$$

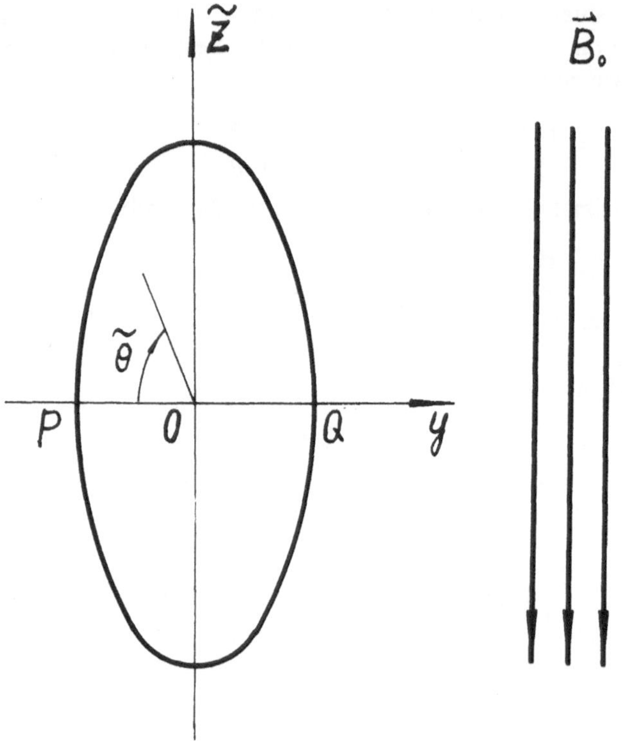

Figure 2. Cross-section of transformed duct.

$$X|_{\varrho=0} = 0 \tag{4.2}$$

where

$$\tilde{a}(\tilde{\theta}) = \frac{\cos^2\tilde{\theta}}{p^2} + \sin^2\tilde{\theta}, \qquad \tilde{b}(\varrho, \tilde{\theta}) = \frac{\sin\tilde{\theta}\cos\tilde{\theta}}{1-\varrho}\left(\frac{1}{p^2} - 1\right),$$

$$\tilde{c}(\varrho, \tilde{\theta}) = \frac{1}{(1-\varrho)^2}\left(\frac{\sin^2\tilde{\theta}}{p^2} + \cos^2\tilde{\theta}\right),$$

$$\tilde{d}(\varrho, \tilde{\theta}) = \frac{-1}{1-\varrho}\left(\frac{\sin^2\tilde{\theta}}{p^2} + \cos^2\tilde{\theta}\right),$$

$$\tilde{e}(\varrho, \tilde{\theta}) = \frac{\sin 2\tilde{\theta}}{(1-\varrho)^2}\left(\frac{1}{p^2} - 1\right),$$

$$\tilde{\alpha}(\tilde{\theta}) = -\sin\tilde{\theta}, \qquad \tilde{\beta}(\varrho, \tilde{\theta}) = \frac{\cos\tilde{\theta}}{1-\varrho}.$$

We will only construct the expansion of the solution in the neighbourhood of point $P(0, 0)$. Similarly we can get the expansion of the solution in the neighbourhood of the point $Q(0, \pi)$.

In the neighbourhood of point P, we introduce local coordinates

$$\xi = \frac{\varrho}{\varepsilon_1^2}, \qquad \eta = \frac{\tilde{\theta}}{\varepsilon_1}, \qquad (\varepsilon_1 = \varepsilon^{1/3}), \tag{4.3}$$

and expand the coefficients of equation (3.1) with respect to ε_1 (for simplicity, we find only the expansion up to order $O(\varepsilon_1^3)$)

$$\tilde{a}(\varepsilon, \eta) = \frac{1}{p^2} + \varepsilon_1^2 \frac{(p^2 - 1)\eta^2}{p^2} + \cdots,$$

$$\tilde{b}(\varepsilon_1^2\xi, \varepsilon_1\eta) = \varepsilon_1 \frac{(1 - p^2)\eta}{p^2} + \cdots$$

$$\tilde{c}(\varepsilon_1^2\xi, \varepsilon_1\eta) = 1 + \cdots, \qquad \tilde{d}(\varepsilon_1^2\xi, \varepsilon_1\eta) = -1 + \cdots$$

$$\tilde{\alpha}(\varepsilon_1\eta) = -\varepsilon_1\eta + \varepsilon_1^3 \frac{\eta^3}{3!} + \cdots, \qquad \tilde{\beta}(\varepsilon_1^2\xi, \varepsilon_1\eta) = 1 + \varepsilon_1^2 \frac{2\xi - \eta^2}{2} + \cdots$$

The differential operator $\tilde{L}_\varepsilon$ can be expanded as

$$\tilde{L}_\varepsilon \equiv \varepsilon_1^{-1}(E_0 + \varepsilon_1 E_1 + \varepsilon_1^2 E_2 + \cdots) \tag{4.4}$$

where

$$E_0 \equiv \frac{1}{p^2}\frac{\partial^2}{\partial \xi^2} + \eta \frac{\partial}{\partial \xi} - \frac{\partial}{\partial \eta}, \qquad E_1 \equiv 0,$$

$$E_2 \equiv \frac{p^2 - 1}{p^2}\eta^2 \frac{\partial^2}{\partial \xi^2} + 2\frac{(1 - p^2)}{p^2}\eta \frac{\partial^2}{\partial \xi \partial \eta} + \frac{\partial^2}{\partial \eta^2}$$

$$- \frac{1}{6}\eta^3 \frac{\partial}{\partial \xi} - \frac{\partial}{\partial \xi} - \frac{2\xi - \eta^2}{2}\frac{\partial}{\partial \eta}.$$

Suppose that the solution of the boundary value problem (1.5)–(1.6) has the expansion of the form:

$$Y = \varepsilon_1^3(y_0 + \varepsilon_1 y_1 + \varepsilon_1^2 y_2 + \cdots). \tag{4.5}$$

Substituting Y into (4.1)–(4.2), and equating the coefficients of like powers of ε, we obtain the differential equations and boundary conditions for $y_0, y_1, \ldots$:

$$\frac{1}{p^2}\frac{\partial^2 y_0}{\partial \xi^2} + \eta \frac{\partial y_0}{\partial \xi} - \frac{\partial y_0}{\partial \eta} = 0, \qquad y_0\big|_{\xi=0} = 0, \tag{4.6}$$

$$\frac{1}{p^2}\frac{\partial^2 y_1}{\partial \xi^2} + \eta \frac{\partial y_1}{\partial \xi} - \frac{\partial y_1}{\partial \eta} = -1, \qquad y_1\big|_{\xi=0} = 0, \tag{4.7}$$

$$\frac{1}{p^2}\frac{\partial^2 y_2}{\partial \xi^2} + \eta \frac{\partial y_2}{\partial \xi} - \frac{\partial y_2}{\partial \eta} = -E_2(y_0), \quad y_2\big|_{\xi=0} = 0. \tag{4.8}$$

. . . .

Considering (4.6), we may take

$$y_0 = 0. \tag{4.9}$$

It is difficult to find the exact solution of the boundary value problem (4.7), but it can be approximated by means of the method offered in [6]. From Theorem 2 of [6] we have

$$y_1 \sim y_{1,0} + y_{1,1} + \cdots = \hat{w}_{0,1} + \hat{w}_{1,1} + \cdots, \quad \text{for } \eta \leqslant -1, \tag{4.10}$$

where $\hat{w}_{0,1}$, $\hat{w}_{1,1}$ ... are the coefficients of the powers of ε_1 in the expansions of $\tilde{w}_0(\varepsilon_1^2\xi, \varepsilon_1\eta)$, $\tilde{w}_1(\varepsilon_1^2\xi, \varepsilon_1\eta)$, ... with respect to ε_1 respectively,

$$\hat{w}_{0,1} = \eta + \sqrt{\eta^2 + 2\xi}. \tag{4.11}$$

For $-1 \leqslant \eta \leqslant 1$, we may take y_1 to be the solution of the boundary value problem for the parabolic equation:

$$\left.\begin{aligned} &\frac{1}{p^2}\frac{\partial^2 y_1}{\partial \xi^2} + \eta\frac{\partial y_1}{\partial \xi} - \frac{\partial y_1}{\partial \eta} = -1, \\ &y_1\Big|_{\xi=0} = 0, \\ &y_1\Big|_{\eta=-1} = (y_{1,0} + y_{1,1} + \cdots)\Big|_{\eta=-1} = (-1 \pm \sqrt{1 + 2\xi}) + \cdots. \end{aligned}\right\} \tag{4.12}$$

For $\eta > 1$, from Theorem 3 of [6] we have

$$\begin{aligned} y_1 &\sim y_{1,0} + y_{1,1} + \cdots + (\tilde{y}_{1,0} + \tilde{y}_{1,1} + \cdots) \\ &= \hat{w}_{0,1} + \hat{w}_{1,1} + \cdots + (\hat{v}_{0,1} + \hat{v}_{1,1} + \cdots), \end{aligned}$$

where $\hat{v}_{0,1}$, $\hat{v}_{1,1}$, ... are the coefficients of powers of ε_1 in the expansions of $\tilde{v}_0(\varepsilon_1^2\xi, \varepsilon_1\eta)$, $\tilde{v}_1(\varepsilon_1^2\xi, \varepsilon_1\eta)$, ... with respect to ε_1 respectively,

$$\hat{v}_{0,1} = -2\eta\, e^{-p^2\xi\eta}. \tag{4.14}$$

Hence, the expansion of the solution in the neighbourhood of point P is

$$Y \sim \begin{cases} \varepsilon_1^4(\eta + \sqrt{\eta^2 + 2\xi}) = \varepsilon(\tilde{\theta} + \sqrt{\tilde{\theta}^2 + 2\varrho}), \quad \text{for } \tilde{\theta} \leqslant -\varepsilon^{1/3}, \\ \text{the approximate solution of (4.12),} \quad \text{for } -\varepsilon^{1/3} \leqslant \tilde{\theta} \leqslant \varepsilon^{1/3}, \\ \varepsilon_1^4(\eta + \sqrt{\eta^2 + 2\xi}) - 2\varepsilon_1^4\eta\, e^{-p^2\xi\eta} \\ \quad = \varepsilon[(\tilde{\theta} + \sqrt{\tilde{\theta}^2 + 2\varrho}) - 2\tilde{\theta}\, e^{-p^2\tilde{\theta}\varrho/\varepsilon}], \quad \text{for } \tilde{\theta} > \varepsilon^{1/3}. \end{cases}$$

The expansion of the solution in the neighbourhood of point Q is obtained by setting $\eta = (\pi - \tilde{\theta})/\varepsilon_1$ in Y.

For the special case of a duct with circular cross-section ($p = 1$), the expansion of the solution in the neighbourhood of point, say, Q is

$$Y \sim \begin{cases} \varepsilon_1^4\left[\eta + \eta\left(1 + \dfrac{\xi}{\eta^2}\right) + O(\eta^{-2})\right], \quad \text{as } \eta \to -\infty, \\ \varepsilon_1^4\left\{\left[\eta + \eta\left(1 + \dfrac{\xi}{\eta^2}\right) + O(\eta^{-2})\right] - 2\, e^{\xi\eta}[\eta + O(\eta^{-1})]\right\}, \\ \text{as } \eta \to +\infty, \\ \varepsilon_1^4\left[\eta + 2^{1/2}\xi^{1/2} + \dfrac{\eta^2}{2^{3/2}\xi^{1/2}} + O(\xi^{-1})\right], \quad \text{as } \xi \to +\infty. \end{cases}$$

The foregoing is in agreement with the results derived from [2] by a complicated process.

References

1. Hunt, J. C. R. and Stewartson, K., "Magnetohydrodynamic flow in rectangular ducts, II", *J. Fluid Mech.* **23**, 3 (1965), 563–581.
2. Roberts, P. H., "Singularities of Hartmann layers", *Proc. Royal Soc., Series A*, **300**, 22 (1967), 94–107.
3. Jiang Fu-ru, "On the boundary layer methods", *Applied Mathematics and Mechanics* (English Edition), **2**, 5 (1981), 505–518.
4. Gold, R. R., "Magnetohydrodynamic pipe flow", *J. Fluid Mech.* **13**, 4 (1962), 505–512.
5. Shercliff, L. A., "Magnetohydrodynamic pipe flow", Part 2, *J. Fluid Mech.* **13**, 4 (1962), 513–518.
6. Jiang, Fu-ru, "On the Dirichlet problem for a quasilinear elliptic equation with a small parameter", *Applied Mathematics and Mechanics* (English Edition), **2**, 1 (1981), 21–47.

Flow around a cone at supersonic speed

HUANG DUN

Department of Mathematics, Peking University, Beijing, PRC

Abstract. In memory of the educational efforts of Prof. Chien Wei-zang, the author presents this thesis in its original form, written when he was a senior student of Tsinghua University under the guidance of Prof. Chien in the period 1947–1948. The inviscid supersonic conical flow around a cone at zero angle of attack was analysed by the method of asymptotic expansion. The two kinds of parameter used are both small difference of angles. Asymptotic expansions up to the fourth order were obtained.

To make the paper more complete, some formulas were added recently, and a comparison with the *Table of Conical Flow* edited by Kopal [5] is added. The table was not available when the thesis was being written. Comparison shows that for adequately high Mach number of incoming flow, the accuracy of the result of this paper is satisfactory, since it is correct to four significant figures. Some possible applications for the results of this paper are indicated.

1. The work accomplished in the years 1947–1948

Prof. Chien analysed symmetrical supersonic conical flow in 1946 by the method of perturbations [1]. He used the linearized solution of the paper written jointly by von Kàrmàn and Moore [2] as his first approximation. From the winter of 1947 Prof. Chien guided the author of this paper for his thesis. We retain the original notation but present the work briefly.

Consider the air as an inviscid perfect gas. For the supersonic case and an infinite cone the flow is conical. The shock wave starts from the apex of the right cone at zero incidence. The only independence variable for the description of flow is the angle θ. The flow behind the shock wave is irrotational.

Let $\bar{u}$, $\bar{v}$ denote the components of the velocity of fluid particle (Fig. 1).

By Bernoulli's equation and shock wave conditions, we introduce a constant c

$$c = U\left[1 + \frac{2}{\gamma - 1}\left(\frac{a_1}{U}\right)^2\right]^{1/2} = a_1\left(M^2 + \frac{2}{\gamma - 1}\right)^{1/2}$$

$$= \left(\frac{2}{\gamma - 1} a^2 + \bar{u}^2 + \bar{v}^2\right)^{1/2}, \qquad a_1 = \left(\gamma \frac{p_1}{\varrho_1}\right)^{1/2} \tag{1.1}$$

where U, a_1, ϱ_1, p_1 and $M = U/a_1$ are the velocity, sound speed, density, pressure and Mach number of the incoming flow respectively, while a, $\bar{u}$, and $\bar{v}$ are sound speed and velocity components in the disturbed region behind the shock wave.

For irrotational flow $\bar{v} = d\bar{u}/d\theta$. We choose c to make $\bar{u}$, $\bar{v}$ dimensionless, i.e., $u = \bar{u}/c$, $v = \bar{v}/c$. The equation satisfied by u takes the form:

Yeh, K. Y. (ed) Progress in Applied Mechanics
 ISBN-13: 978-94-010-8061-3

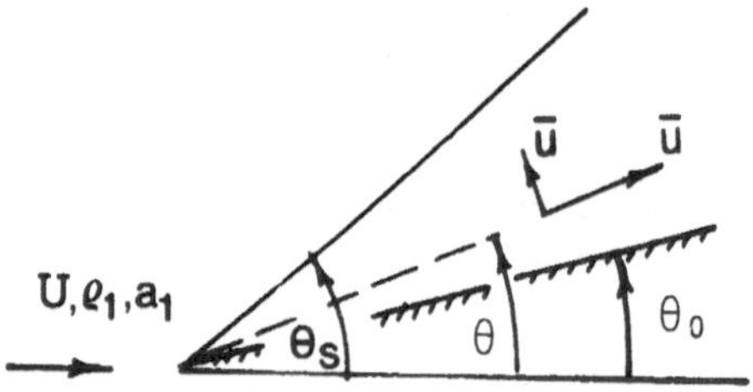

Figure 1

$$\frac{d^2u}{d\theta^2}\left[\frac{\gamma+1}{2}\left(\frac{du}{d\theta}\right)^2 - \frac{\gamma-1}{2}(1-u^2)\right] = (\gamma-1)u(1-u^2) - \gamma u\left(\frac{du}{d\theta}\right)^2$$

$$-\frac{\gamma-1}{2}\cot\theta\left(\frac{du}{d\theta}\right)^3 + \frac{\gamma-1}{2}\frac{du}{d\theta}\cot\theta(1-u^2) \tag{1.2}$$

The domain of integration is $0 < \theta_0 \leqq \theta \leqq \theta_s < \pi/2$.
The boundary condition along the body surface is

$$u_0' \equiv \left(\frac{du}{d\theta}\right)_{\theta=\theta_0} = v_{\theta=\theta_0} = \left(\frac{\bar{v}}{c}\right)_{\theta=\theta_0} = 0. \tag{1.3}$$

The conditions along the shock wave are:

(i) continuity of tangential velocity:

$$U\cos\theta_s = cu_s, \tag{1.4}$$

(ii) conservation of mass:

$$\varrho_1 U\sin\theta_s = \varrho_s(-v_s c) \equiv m, \tag{1.5}$$

(iii) conservation of momentum:

$$p_s - p_1 = m(U\sin\theta_s + v_s c), \tag{1.6}$$

(iv) Hugoniot relation:

$$\varrho_s/\varrho_1 = [(\gamma+1)p_s + (\gamma-1)p_1]/[(\gamma+1)p_1 + (\gamma-1)p_s]. \tag{1.7}$$

The subscript s denotes the quantities just behind the shock wave, while subscript 0 is used for values at the surface of solid cone.

For a given set of values of θ_0, M, a_1, γ, ϱ_1 integration gives u, v, θ_s, and then the pressure on the cone can be evaluated by Bernoulli's equation. The main difficulty in integration lies in the determination of the half apex angle θ_s of the canonical shock wave and the fulfilment of shock conditions, whereas the differential equation (1.2) is nonlinear.

The mathematical problem in finding $u(\theta)$ can be formulated as: to integrate the equation (1.2) for a given set of values of M, γ and θ_0, for $\theta_0 \leqq \theta \leqq \theta_s$, while θ_s is to be determined during integration. The boundary conditions to be satisfied are equation (1.3) and

$$u_s = \frac{v}{c}\cos\theta_s = \frac{M\cos\theta_s}{[M^2 + 2/(\gamma-1)]^{1/2}} \tag{1.8}$$

derived from (1.4), and a condition derived from (1.5), (1.6), (1.7)

$$\left(\frac{\mathrm{d}u}{\mathrm{d}\theta}\right)_{\theta=\theta_s} \equiv u_s' = -\frac{K-1+M^2\sin^2\theta_s}{KM\sin\theta_s(M^2+K-1)^{1/2}}, \tag{1.9}$$

where

$$K = (\gamma+1)/(\gamma-1), \qquad K-1 = 2/(\gamma-1).$$

G. I. Taylor [3] and von Kàrmàn [2] investigated the above problem by different methods after their discussion in the year 1930. von Kàrmàn linearized the problem for a small half apex angle θ_0 of the solid cone and for M greater but near 1. The linearized solution is simple to use but the accuracy is not high, while it is applicable for bodies of revolution other than the cone. G. I. Taylor [3] integrated numerically the exact non-linear equation (equivalent to (1.2) of this paper). The result is of high accuracy but Taylor considered only the case $\gamma = 1.405$ and only three typical values of θ_0, say 10°, 20° and 30°.

Prof. Chien used a small parameter, $\varepsilon = \cot\theta_s \tan\theta_0$, in [1]. This corresponds to the case when $\theta_s \gg \theta_0$, and therefore M is sufficiently near 1.

In 1947 Prof. Chien suggested that the author work out the solution by the method of perturbation using $\theta_s - \theta$ or $\theta - \theta_0$ as a small parameter. The asymptotic expansion takes the form

$$u = u_s - u_s'(\theta_s-\theta) + \frac{u_s''}{2}(\theta_s-\theta)^2 - \frac{u_s'''}{6}(\theta_s-\theta)^3 + \frac{u_s''''}{24}(\theta_s-\theta)^4 + O((\theta_s-\theta)^5), \tag{1.10}$$

or

$$u = u_0 + u_0'(\theta-\theta_0) + \frac{u_0''}{2}(\theta-\theta_0)^2 + \frac{u_0'''}{6}(\theta-\theta_0)^3 + \frac{u_0''''}{24}(\theta-\theta_0)^4 + O((\theta-\theta_0)^5). \tag{1.11}$$

The expression for v, i.e., $\mathrm{d}u/\mathrm{d}\theta$, has a similar expression.

In (1.10), u_s'', u_s''', u_s'''' denote the second to fourth derivatives just behind the shock wave, while u_0'', u_0''', u_0'''' are derivatives at $\theta = \theta_0$. The author of this paper obtained these derivatives by differentiating (1.2), and then making use of the boundary conditions (1.8), (1.9) and (1.3). The results obtained in the thesis are

$$\left(\frac{\mathrm{d}^2u}{\mathrm{d}\theta^2}\right)_{\theta=\theta_s} \equiv u_s'' = -\frac{\cos\theta_s\left[\left(2-\frac{1}{K^2}\right)M^2\sin^2\theta_s - \frac{K-1}{K^2}\right]}{M\sin^2\theta_s(M^2+K-1)^{1/2}}, \tag{1.12}$$

$$u_s''' = -u_s' - \frac{(K-1)\left[M^4\sin^4\theta_s\left(2-\frac{1}{K}\right)\left(1+\frac{1}{K}\right) + M^2\sin^2\theta_s\left(\frac{1}{K^0}+\frac{2}{K^2}\right) - \left(\frac{1}{K}+\frac{1}{K^2}\right)\right]}{KM(M^2+K-1)^{1/2}(M^2\sin^2\theta_s-1)\tan^2\theta_s\sin\theta_s}. \tag{1.13}$$

The expression for the fourth derivative is rather long. Denoting by $\phi = 1 - u^2 - K(u')^2$ equation (1.2) gives

$$\frac{d^4u}{d\theta^4} \equiv u'''' = \left(-u''' + 2\cot\theta u'' - \frac{2u'}{\sin^2\theta}\right)\cot\theta$$

$$+ \frac{1}{\phi}\Big\{2(u'')^2[Ku'' + u(K+2) + 3\cot\theta u'] - 2uu'(K+3)$$

$$+ u''[6u'(u\cot\theta - K) + 4u^2 + 2(K-3)(\cot\theta u')^2$$

$$+ (u')^2(1+5K)] + 2(u')^2\left[\cot\theta\left(1 - \frac{K-1}{\sin^2\theta}\right)u'\right.$$

$$\left.\left. + \frac{K-3}{2}\cot\theta + 2u\left(3 - \frac{1}{\sin^2\theta}\right)\right]\right\}. \qquad (1.14)$$

The u_s'''' can be evaluated from (1.14) setting $\theta = \theta_s$, and making use of (1.8), (1.9), (1.12), and (1.13).

The derivatives on the solid cone ($\theta = \theta_0$) can be found similarly:

$$u = u_0, \quad u_0' = \left(\frac{du}{d\theta}\right)_{\theta=\theta_0} = v_0 = 0, \quad \left(\frac{d^2u}{d\theta^2}\right)_{\theta=\theta_0} = u_0'' = -2u_0, \qquad (1.15)$$

$$u_0''' = 2u_0\cot\theta_0, \quad \left(\frac{d^4u}{d\theta^4}\right)_{\theta=\theta_0} = u_0'''' = -2u_0\left[3\cot^2\theta_0 + \frac{4u_0^2}{1-u_0^2}(K-1)\right]. \qquad (1.16)$$

θ_s in (1.8), (1.9), (1.12), and (1.13) is unknown, while θ_0 is unknown in (1.15). Prof. Chien and Zheng Zhe-min checked formula (1.12). The derivation of (1.13) and (1.14) is long but elementary and is omitted here for brevity. In my thesis the result of this paper is compared with the results of G. I. Taylor and J. W. Maccoll [3]. At that time Kopal's table [5] was unknown in China. To calculate θ_s only simplified formula (2.4) below was used.

Above we have presented the main results of the thesis written in 1948 under the guidance and encouragement of Prof. Chien. The final report is still preserved, but the sheets of paper have now become yellow and fragile. It records the educational work of Prof. Chien.

2. Recent supplement

Nearly 40 years have passed since the original thesis was produced. The question may be posed: Has this paper any practicle significance now? Can these results be applied in engineering applications? We now supplement the thesis with some recent discussion.

It is well known that the tables of trigonometric function can be constructed in different ways. To reduce the size of calculator or algorithm, people study algorithms even now. The study of conical flow is time consuming, and the formulas of this paper will enable us to utilize a short algorithm with sufficient accuracy. The formulae can give flow variables for arbitrary values of the ratio of specific heats γ and for a wide range of Mach number M and cone angle θ_0.

It is necessary for us to interpret the accuracy of the numerical results obtainable from our formulas. For this we must draw a comparison with Kopal's table, which was available in China in the late 1950's. Kopal's table was obtained through four years of machine computation under the guidance of a group of experts. It gives numerical values with five significant values in general, but only for $\gamma = 1.405$. As well known, many gases have quite different values of γ under different conditions.

To show the accuracy of our formulas, first consider the case of an infinite Mach number. Such a case is applicable for strong shock waves, say for $M^2 \gg 2/(\gamma - 1)$. It gives at least an interesting qualitative tendency.

As $M \to \infty$ many formulas simplify:

$$\left(M^2 + \frac{2}{\gamma - 1}\right)^{1/2} \to M, \quad (M^2 \sin^2\theta_s - 1) \to M^2 \sin^2\theta_s, \quad u_s \to \cos\theta_s,$$

$$u_s' \to -\frac{\sin\theta_s}{K}, \quad u_s'' \to -\left(2 - \frac{1}{K^2}\right)\cos\theta_s, \tag{2.1}$$

$$u_s''' \to -u_s' - \left(2 - \frac{1}{K}\right)\left(1 - \frac{1}{K^2}\right)\cos^2\theta_s/\sin\theta_s, \tag{2.2}$$

and for the particular value of $\gamma = 1.405$

$$u_s'''' \to -45.6717\cos\theta_s\cot^2\theta_s - 2.24341\cos\theta_4$$
$$+4.83051\cot\theta_s/\sin\theta_s. \tag{2.3}$$

For $M \to \infty$ and $\theta_0 \leqq 35°$, $(\theta_s - \theta_0)$ is small, and (1.3) can be written as

$$v_{\theta=\theta_0} = \left(\frac{du}{d\theta}\right)_{\theta=\theta_0} = u_0' = 0 = u_s' - u_s''(\theta_s - \theta_0) + O((\theta_s - \theta_0)^2),$$

which gives approximately for $M \to \infty$,

$$0 < \theta_s - \theta_0 \doteq \frac{u_s'}{u_s''} = \frac{\tan\theta_s}{K\left(2 - \frac{1}{K^2}\right)} < 0.17\,\text{rad}. \tag{2.4}$$

This approximate formula gives θ_s readily for given values of θ_0 and γ. It also gives a value coincident with the values of Kopal's table at least with three significant figures for $5° \leqq \theta_0 \leqq 35°$. Notice that the half apex angle $\theta_0 \leqq 35°$ is a sufficiently wide range of values.

For large M we can calculate the value u_0 and the pressure on the solid cone neglecting the term $O((\theta_s - \theta_0)^5)$ in formula (1.10). For this (2.1), (2.2), and (2.3)

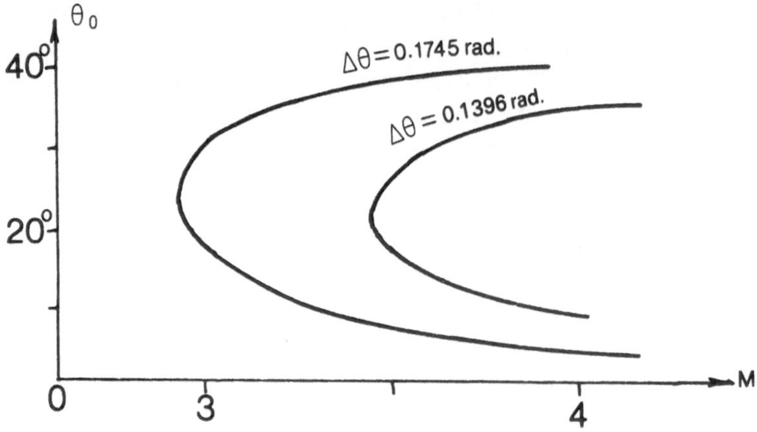

Figure 2

should be used and the numerical results coincide with Kopal's table to four or even five significant figures for $\theta_0 \leqq 35°$. The procedure to obtain θ_s and M for a given set of values of θ_0 and γ in [3–5] requires a lot of labour.

For moderate values of M the formulae of this paper are still applicable. In Fig. 2 we show two curves. To the right of the right curve $\Delta\theta = \theta_s - \theta_0 < 0.1396$, the expansion up to $(\Delta\theta)^4$ gives a result of high accuracy; while to the right of the left curve $\Delta\theta = \theta_s - \theta_0 < 0.1745$ rad, the accuracy is still satisfactory.

To apply the formulae of this paper to practical problems one should at first find θ_s and u_0 for a given set of values of γ, θ_0 and M. If θ_s is known by other means, the calculation is straightforward. In most cases θ_s is not known, so that we can derive two equations which determine simultaneously θ_s and u_0.

Denoting by $\Delta\theta = \theta_s - \theta_0$, for $\Delta\theta < 0.174$,

$$0.58 < u_0 = \frac{\left(1 + \dfrac{2}{\gamma - 1}\dfrac{1}{M^2}\right)^{-1/2}\left[\cos\theta_s + \dfrac{\Delta\theta}{4K}\dfrac{\sin^2\theta_s + \dfrac{2}{\gamma-1}M^{-2}}{\sin\theta_s}\right]}{1 - \frac{1}{2}\Delta\theta^2 + \dfrac{\cot\theta_s}{12}\Delta\theta^3 + O(\Delta\theta^5)} < 1, \tag{2.5}$$

$$\frac{\cos\theta_s}{\left(1 + \dfrac{2}{\gamma - 1}M^{-2}\right)^{1/2}} = u_0\left\{1 - \Delta\theta^2 + \frac{\cot\theta_0}{3}\Delta\theta^3 - \left[\frac{\cot^2\theta_0}{4} + \frac{K-1}{3}\frac{u_0^2}{1-u_0^2}\right]\Delta\theta^4 + O(\Delta\theta^5)\right\}. \tag{2.6}$$

When we solve (2.5) and (2.6) simultaneously by iteration, equation (2.4) can be used to obtain a first approximation for $\Delta\theta$. For brevity, we omit large amounts of numerical examples for the special case of $\gamma = 1.405$, and the comparison with the values from [5].

References

1. Chien Wei-zang, "Symmetrical conical flow at supersonic speeds by perturbation method", *Report of Tsunghua Univ.*(1946) (in Chinese) see also reference [163] of the book of Courant, R. and Friedrichs, K. O., *Supersonic Flow and Shock Waves*, Springer-Verlag, (1976), 403, 449.
2. von Kàrmàn, Th. and Moore, N. B., "Resistance of slender bodies moving with supersonic velocities, with special reference to projectiles", *ASME Trans.* **54** (1932), 303.
3. Taylor, G. I. and Maccoll, J. W., "The air pressure on a cone moving at high speeds", *Proc. Royal Soc.* (*A*) **139** (1933), 278–311.
4. Huang Dun, "Conical flow at supersonic speed", Thesis of senior, Mechanical Eng. Dept. Tsinghua Univ. (1948) (in Chinese).
5. Kopal, Z., *Tables of Supersonic Flow Around Cones*, MIT (1947).

Elasticity and Plasticity

A Note on biharmonic functions

FAN WEI-XUN

Nanjing Aeronautical Institute, Nanjing, PRC

Abstract. The condition that four complex functions $\phi_1(z)$, $\phi_2(\bar{z})$, $\psi_1(z)$, $\psi_2(\bar{z})$ are two pairs of mutually conjugate ones $\phi_2(\bar{z}) = \overline{\phi_1(z)}$, $\psi_2(\bar{z}) = \overline{\psi_1(z)}$ is a sufficient but not necessary one for

$$U = \phi_1(z) + \phi_2(\bar{z}) + \bar{z}\psi_1(z) + z\psi_2(\bar{z})$$

to be a real function. This paper presents a necessary and sufficient condition.

Thesis

In the plane theory of elasticity, it is seen that (in the absence of body forces) there always exists some function called an Airy stress function $U(x, y)$ by the help of which the three stress components may be expressed in the following manner:

$$X_x = \frac{\partial^2 U}{\partial y^2}, \quad X_y = -\frac{\partial^2 U}{\partial x \partial y}, \quad Y_y = \frac{\partial^2 U}{\partial x^2}. \tag{1.1}$$

The function $U(x, y)$ has continuous derivatives up to and including the fourth order and these derivatives, from the second order onwards, must be single-valued functions throughout the region S, occupied by the body.

For a homogeneous and isotropic body in the absence of body forces, the function $U(x, y)$ must also satisfy the biharmonic equation

$$\Delta\Delta U = \frac{\partial^4 U}{\partial x^4} + 2\frac{\partial^4 U}{\partial x^2 \partial y^2} + \frac{\partial^4 U}{\partial y^4} = 0 \tag{1.2}$$

the solutions of which are called biharmonic functions.

Muskhelishvili (1954) [1] quoted the method of deduction from Goursat's paper (1898), i.e., introducing instead of x and y the new variables $z = x + iy$ and $\bar{z} = x - iy$ from the fact that biharmonic function

$$U = \phi_1(z) + \phi_2(\bar{z}) + \bar{z}\psi_1(z) + z\psi_2(\bar{z}) \tag{1.3}$$

is a real, one must put

$$\phi_2(\bar{z}) = \overline{\phi_1(z)}, \quad \psi_2(\bar{z}) = \overline{\psi_1(z)}. \tag{1.4}$$

We suppose that condition (1.4) is a sufficient but not necessary one for (1.3) to be a real function. We deduce a necessary and sufficient condition as follows.

Yeh, K. Y. (ed) Progress in Applied Mechanics
 ISBN-13: 978-94-010-8061-3

The sum of any complex function and its conjugate must be real:

$$\phi_2(\bar{z}) + \overline{\phi_2(\bar{z})} + z\psi_2(\bar{z}) + \bar{z}\overline{\psi_2(\bar{z})} = \text{a real function,} \tag{1.5}$$

where

$$\overline{\phi_2(\bar{z})} = \bar{\phi}_2(z), \quad \overline{\psi_2(\bar{z})} = \bar{\psi}_2(z), \tag{1.6}$$

are analytic functions in the region S. The difference between (1.3) and (1.5) must also be real.

$$[\phi_1(z) - \bar{\phi}_2(z)] + \bar{z}[\psi_1(z) - \bar{\psi}_2(z)] = \text{a real function,} \tag{1.7}$$

where the two square brackets

$$[\phi_1(z) - \bar{\phi}_2(z)] = u(x, y) + iv(x, y), \tag{1.8a}$$

$$[\psi_1(z) - \bar{\psi}_2(z)] = u_1(x, y) + iv_1(x, y), \tag{1.8b}$$

are analytic functions in the region S. Their real parts and imaginary parts are all harmonic functions, and satisfy the Cauchy–Riemann conditions. The imaginary part of (1.7) must be equal to zero:

$$v + xv_1 - yu_1 = 0, \tag{1.9}$$

$$\Delta(v + xv_1 - yu_1) = \Delta v + x\Delta v_1 - y\Delta u_1 + 2\frac{\partial v_1}{\partial x} - 2\frac{\partial u_1}{\partial y}$$

$$= 2\left(\frac{\partial v_1}{\partial x} - \frac{\partial u_1}{\partial y}\right) = 0.$$

From the Cauchy–Riemann condition

$$\frac{\partial u_1}{\partial y} = -\frac{\partial v_1}{\partial x} = \frac{\partial v_1}{\partial x} = 0,$$

the two functions

$$u_1(x, y) = f(x), \tag{1.10a}$$

$$v_1(x, y) = g(y), \tag{1.10b}$$

are functions of a single variable. From the Cauchy–Riemann condition

$$\frac{\partial u_1}{\partial x} = \frac{\partial v_1}{\partial y} = f'(x) = g'(y),$$

$$u_1 = Cx + C_1, \tag{1.11a}$$

$$v_1 = Cy + C_2. \tag{1.11b}$$

Substituting into (1.9)

$$v = yu_1 - xv_1 = C_1 y - C_2 x. \tag{1.12a}$$

From the Cauchy–Riemann conditions

$$\frac{\partial u}{\partial x} = \frac{\partial v}{\partial y} = C_1, \quad \frac{\partial u}{\partial y} = -\frac{\partial v}{\partial x} = C_2,$$

$$u = C_1 x + C_2 y + C_3, \tag{1.12b}$$

where C, C_1, C_2, C_3 are arbitrary real constants. Substituting into (1.8)

$$\phi_1(z) = \bar{\phi}_2(z) + (C_1 - \mathrm{i}C_2)z + C_3, \tag{1.13}$$

$$\psi_1(z) = \bar{\psi}_2(z) + Cz + (C_1 + \mathrm{i}C_2). \tag{1.14}$$

Substituting into (1.3)

$$U = \phi_1^*(z) + \phi_2^*(\bar{z}) + \bar{z}\psi_1^*(z) + z\psi_2^*(\bar{z}), \tag{1.15}$$

where

$$\left.\begin{aligned} \phi_1^*(z) &= \bar{\phi}_2(z) + (C_1 - \mathrm{i}C_2)z + C_3/2, \\ \phi_2^*(\bar{z}) &= \phi_2(\bar{z}) + (C_1 + \mathrm{i}C_2)\bar{z} + C_3/2, \\ \psi_1^*(z) &= \bar{\psi}_2(z) + (C/2)z, \\ \psi_2^*(\bar{z}) &= \psi_2(\bar{z}) + (C/2)\bar{z}, \end{aligned}\right\} \tag{1.16}$$

are two pairs of mutually conjugate functions

$$\phi_2^*(\bar{z}) = \overline{\phi_1^*(z)}, \quad \psi_2^*(\bar{z}) = \overline{\psi_1^*(z)}, \tag{1.17}$$

i.e., the expression (1.15) is of the same form as (1.3) satisfying condition (1.4), but the four complex functions $\phi_1^*(z)$, $\phi_2^*(\bar{z})$, $\psi_1^*(z)$, $\psi_2^*(\bar{z})$, in general, are not equal to the four functions $\phi_1(z)$, $\phi_2(\bar{z})$, $\psi_1(z)$, $\psi_2(\bar{z})$ respectively; if, and only if, the four arbitrary real constants C, C_1, C_2, C_3 are equal to zero simultaneously, then the differences between two pairs of the four functions vanish.

Note: It is apparent that the expression (1.15) is of the form of (1.3) satisfying condition (1.4). In fact, any real function must be equal to its conjugate,

$$\begin{aligned} U &= \phi_1(z) + \phi_2(\bar{z}) + \bar{z}\psi_1(z) + z\psi_2(\bar{z}) \\ &= \bar{\phi}_1(\bar{z}) + \bar{\phi}_2(z) + z\bar{\psi}_1(\bar{z}) + \bar{z}\bar{\psi}_2(z) \\ &= \phi_1^*(z) + \phi_2^*(\bar{z}) + \bar{z}\psi_1^*(z) + z\psi_2^*(\bar{z}), \end{aligned} \tag{1.18}$$

where

$$\left.\begin{aligned} \phi_1^*(z) &= (1/2)[\phi_1(z) + \bar{\phi}_2(z)], \\ \phi_2^*(\bar{z}) &= (1/2)[\bar{\phi}_1(\bar{z}) + \phi_2(\bar{z})], \\ \psi_1^*(z) &= (1/2)[\psi_1(z) + \bar{\psi}_2(z)], \\ \psi_2^*(\bar{z}) &= [(1/2)[\bar{\psi}_1(\bar{z}) + \psi_2(\bar{z})]. \end{aligned}\right\} \tag{1.19}$$

These four complex functions obviously satisfy the conditions (1.17).

Reference

1. Muskhelishvili, N. I., *Several Fundamental Problems on Mathematical Elasticity*, Science Academy Press, U.S.S.R. (1954).

The effect of transverse shear deformations on ring-shaped elastic plates

CHENG CHANG-JUN, SHANG XIN-CHUN, DUAN DAI-MING and ZHU JIAN-GANG

Department of Mathematics and Mechanics, Lanzhou University, Lanzhou, Gansu 730001, PRC

Abstract. In this paper, the difficulty of resolving the fundamental equations of E. Reissner's theory of the bending of finite thick plates is overcome by means of the theory of generalized functions and properties of the δ-function. The Green's functions of bending problems of ring-shaped plates under a concentrated load or a concentrated moment are obtained. On this basis, the possibilities of generalizing our method are discussed in three respects, i.e., elastic properties of the plate, distributive cases of loads, and supported conditions.

1. Introduction

For treating widely practical problems, the Kirchhoff's classical theory of the bending of thin plates is sufficiently accurate. However, for some problems, such as those where stress is concentrated at the edge of a hole in a bent plate, the results obtained by the classical theory do not agree with that of experiments. Also, according to the classical theory, there will be concentrated reaction forces at the corners on multi-angular plates with simply supported edges, but this also disagrees with our knowledge. All these are due to neglecting the effect of the transverse shear deformations and to the statement of boundary conditions which are not compatible with the theory of elasticity. Many investigators have tried to find a better physical model of the bending of plates. The theory rendered by E. Reissner [1, 2] is widely used. Using this theory to solve some technically important problems, satisfactory results coinciding with experiments are obtained [1].

The system of fundamental equations of E. Reissner's theory is one consisting of three second-order coupled partial differential equations which contain the deflection w and two shearing forces V_r, V_θ of plates. The general solution of the system of the equations may be finally represented by two fundamental variables, namely, the deflection w and the stress function Φ. Here w and Φ satisfy respectively a fourth-order and a second-order partial differential equation. In general, this leads to a sixth-order equation. Hence, to determine the solution of a problem, three boundary conditions must be satisfied at each edge of plates. The statement of boundary conditions differs from that of the classical theory, and this coincides with the theory of elasticity and reflects the physical reality of the bending of plates.

Yeh, K. Y. (ed) Progress in Applied Mechanics
 ISBN-13: 978-94-010-8061-3

But, because of the formidable difficulty caused by the complexity of mathematics, solutions obtained are still few in number [3, 4]. The general solutions for the bending of ring-shaped plates under concentrated loads, in particular, are not yet available.

In this paper, we first discuss the structure of the general solution of fundamental equations of E. Reissner's theory of isotropic plates in a polar coordinate system (r, θ). Then, by means of the theory of generalized functions and the properties of the δ-function, the difficulty of resolving the equations of E. Reissner's theory is overcome and the Green's functions of the bending of ring-shaped plates under the actions of concentrated loads are obtained. Thereby, from the superposition principle, the general solutions under the actions of arbitrary transversely distributive forces and distributive moments can be obtained, regardless of boundary conditions. Thus, the problem may be reduced to a system of sixth-order linear algebraic equations in six undetermined integral constants. Finally, we will discuss some possibilities of generalizing our method, particularly the structure of general solutions of fundamental equations of a variety of E. Reissner's plates which contain transversely isotropic plates, sandwich plates and non-homogeneous plates along their thickness. Only by bringing elastic constants into suitable correspondence with each other, can our method be applied to solve the above problems.

2. Statement of the problem

In the polar coordinates (r, θ) of the middle plane of the plate, the fundamental equations of E. Reissner's theory have the following forms:

$$\frac{\partial(rV_r)}{\partial r} + \frac{\partial V_\theta}{\partial \theta} = -rp, \tag{2.1}$$

$$V_r - \frac{h^2}{10}\left(\Delta V_r - \frac{2}{r^2}\frac{\partial V_\theta}{\partial \theta} - \frac{1}{r^2}V_r\right) = -D\frac{\partial \Delta w}{\partial r} - \frac{h^2}{10(1-\mu)}\frac{\partial p}{\partial r}, \tag{2.2a}$$

$$V_\theta - \frac{h^2}{10}\left(\Delta V_\theta + \frac{2}{r^2}\frac{\partial V_r}{\partial \theta} - \frac{1}{r^2}V_\theta\right) = -D\frac{\partial \Delta w}{r\partial \theta} - \frac{h^2}{10(1-\mu)}\frac{\partial p}{r\partial \theta}, \tag{2.2b}$$

$$M_r = -D\left(\frac{\partial^2 w}{\partial r^2} + \frac{\mu}{r}\frac{\partial w}{\partial r} + \frac{\mu}{r^2}\frac{\partial^2 w}{\partial \theta^2}\right) + \frac{h^2}{5}\frac{\partial V_r}{\partial r} - \frac{\mu h^2}{10(1-\mu)}p, \tag{2.3a}$$

$$M_\theta = -D\left(\frac{1}{r}\frac{\partial w}{\partial r} + \frac{1}{r^2}\frac{\partial^2 w}{\partial \theta^2} + \mu\frac{\partial^2 w}{\partial r^2}\right) + \frac{h^2}{5}\left(\frac{1}{r}\frac{\partial V_\theta}{\partial \theta} + \frac{V_r}{r}\right) - \frac{\mu h^2}{10(1-\mu)}p, \tag{2.3b}$$

$$H_{r\theta} = -(1-\mu)D\frac{\partial}{\partial r}\left(\frac{1}{r}\frac{\partial w}{\partial \theta}\right) + \frac{h^2}{10}\left(\frac{1}{r}\frac{\partial V_r}{\partial \theta} + r\frac{\partial}{\partial r}\left(\frac{V_\theta}{r}\right)\right), \tag{2.3c}$$

$$\varphi_r = -\frac{\partial w}{\partial r} + \frac{12(1+\mu)}{5Eh}V_r, \qquad \varphi_\theta = -\frac{1}{r}\frac{\partial w}{\partial \theta} + \frac{12(1+\mu)}{5Eh}V_\theta, \tag{2.4a, b}$$

where w, V_r, V_θ, M_r, M_θ, $H_{r\theta}$, φ_r and φ_θ are the deflection, the transverse shearing forces, the bending moments, the twisting moment and the angles of rotation of the plate respectively. Here, p is the transverse load. The convention of the notations is the same as that in [1, 2]. The flexural rigidity D of the plate is defined by

$$D = \frac{Eh^3}{12(1-\mu^2)} \tag{2.5}$$

where h is the thicknenss, and E, μ are the Young's modulus and Poisson's ratio, and Δ is the Laplacian

$$\Delta = \frac{\partial^2}{\partial r^2} + \frac{1}{r}\frac{\partial}{\partial r} + \frac{1}{r^2}\frac{\partial^2}{\partial \theta^2}. \tag{2.6}$$

According to the Theorem of Minimum Complementary Energy and the method of the Lagrange multiplier, the natural boundary conditions on an edge $r =$ constant developed by E. Reissner should be of the following forms:

$$\begin{cases} w = \bar{w} \\ \text{or} \quad V_r = \bar{V}_r \end{cases}, \quad \begin{cases} \varphi_r = \bar{\varphi}_r \\ \text{or} \quad M_r = \bar{M}_r \end{cases}, \quad \begin{cases} \varphi_\theta = \bar{\varphi}_\theta \\ \text{or} \quad H_{r\theta} = \bar{H}_{r\theta} \end{cases}, \tag{2.7}$$

where the quantities with a bar represent given functions of θ. A variety of boundary conditions is possible provided the existence and uniqueness theorem of the solutions is not violated. Particularly,

(1) for a free edge, we have

$$V_r = 0, \quad M_r = 0, \quad H_{r\theta} = 0 \tag{2.8}$$

(2) for a fixed edge, we have

$$w = 0, \quad \varphi_r = 0, \quad \varphi_\theta = 0 \tag{2.9}$$

(3) for a simply supported edge, we have

$$w = 0, \quad M_r = 0, \quad \varphi_\theta = 0, \tag{2.10}$$

consequently, we see that fundamentally unknown quantities are w and V_r, V_θ. Once they are obtained from (2.1) and (2.2), then (2.3) and (2.4), M_r, M_θ, $H_{r\theta}$ and φ_r, φ_θ can be obtained by differentiation.

Eliminating V_r and V_θ from (2.1) and (2.3), we obtain an equation for w:

$$D\Delta^2 w = p - \frac{kh^2}{10}\Delta p, \tag{2.11}$$

$$k = \frac{2-\mu}{1-\mu}. \tag{2.12}$$

According to [3], if w is known, we can easily show that V_r and V_θ will be represented as

$$V_r = -\left(D\frac{\partial \Delta w}{\partial r} + \frac{kh^2}{10}\frac{\partial p}{\partial r}\right) + \frac{\partial \Phi}{r\partial \theta}, \tag{2.13a}$$

$$V_\theta = -\left(D\frac{\partial \Delta w}{r\partial \theta} + \frac{kh^2}{10}\frac{\partial p}{r\partial \theta}\right) - \frac{\partial \Phi}{\partial r}. \tag{2.13b}$$

The first part in (2.13) is a particular solution of (2.1) and (2.2), and the second part is the general solution of the corresponding homogeneous equation. Φ is called the stress function and satisfies the equation

$$\Delta\Phi - \frac{10}{h^2}\Phi = 0. \tag{2.14}$$

Obviously, the general solution of the equation is

$$\Phi(r, \theta) = \sum_{m=0}^{\infty} \left\{ \left[B_{1m} I_m \left(\frac{\sqrt{10}}{h} r \right) + B_{2m} K_m \left(\frac{\sqrt{10}}{h} r \right) \right] \cos m\theta \right.$$
$$\left. + \left[\tilde{B}_{1m} I_m \left(\frac{\sqrt{10}}{h} r \right) + \tilde{B}_{2m} K_m \left(\frac{\sqrt{10}}{h} r \right) \right] \sin m\theta \right\} \tag{2.15}$$

in which $I_m(\varrho)$ and $K_m(\varrho)$ are a modification of Bessel's functions of integer-order of the first and second kinds respectively.

Therefore, our main task is to solve (2.11). Once w is obtained, then V_r and V_θ can be obtained from formulae (2.13) and (2.15), and from (2.3) and (2.4), M_r, M_θ, $H_{r\theta}$ and φ_r, φ_θ can also be obtained. Finally, if these quantities satisfy simultaneously prescribed boundary conditions, then the solution of a given problem will be obtained.

3. General solution of deflection w under a concentrated load

Now we will consider three types of concentrated loads:

(1) a concentrated force P;

(2) a concentrated tangential moment L_θ;

(3) a concentrated radial moment L_r.

We assume that any one of these loads is applied at a point (ξ, η) on the ring-shaped plate.

Using the δ-function, we can represent the densities $p(r, \theta)$ of these loads as the following forms:

$$(1)\ p(r, \theta) = P\delta(r - \xi)\delta(\theta - \eta), \tag{3.1a}$$

$$(2)\ p(r, \theta) = L_\theta \delta'(r - \xi)\delta(\theta - \eta), \tag{3.1b}$$

$$(3)\ p(r, \theta) = L_r \xi^{-1} \delta(r - \xi)\delta'(\theta - \eta). \tag{3.1c}$$

Now we expand $p(r, \theta)$ into Fourier series:

$$p(r, \theta) = \sum_{m=0}^{\infty} (q_m(r) \cos m\theta + \tilde{q}_m(r) \sin m\theta), \tag{3.2}$$

where

$$\left.\begin{aligned} q_0(r) &= \frac{1}{2\pi} \int_0^{2\pi} p(r, \theta)\, d\theta, \\ q_m(r) &= \frac{1}{\pi} \int_0^{2\pi} p(r, \theta) \cos m\theta\, d\theta, \quad m \geqslant 1, \\ \tilde{q}_m(r) &= \frac{1}{\pi} \int_0^{2\pi} p(r, \theta) \sin m\theta\, d\theta. \end{aligned}\right\} \tag{3.3}$$

According to the properties of δ-function, we can write $q_m(r)$ and $\tilde{q}_m(r)$ of these loads in (3.1) in the unified forms:

$$\begin{aligned} q_m(r) &= a_{m1}\delta(r-\xi) + a_{m2}\delta'(r-\xi), \\ \tilde{q}_m(r) &= \tilde{a}_{m1}\delta(r-\xi) + \tilde{a}_{m2}\delta'(r-\xi), \end{aligned} \tag{3.4}$$

where a_{mi} and $\tilde{a}_{mi}$ ($i = 1, 2$) are given in Table 1. For $m = 0$, $\tilde{a}_{0i} = 0$ and $2a_{0i}$ are equal to a_{mi} with $m = 0$ in Table 1.

Now we expand $w(r, \theta)$ into the Fourier series:

$$w = \sum_{m=0}^{\infty} (R_m(r) \cos m\theta + \tilde{R}_m(r) \sin m\theta). \tag{3.5}$$

Substituting formulae (3.2) and (3.5) into (2.11), we obtain two sets of the ordinary differential equations in $R_m(r)$ and $\tilde{R}_m(r)$ for all integral numbers $m \geqslant 0$.

$$\begin{aligned} \left(\frac{\mathrm{d}^2}{\mathrm{d}r^2} + \frac{1}{r}\frac{\mathrm{d}}{\mathrm{d}r} - \frac{m^2}{r^2}\right)^2 R_m(r) &= \sum_{j=0}^{3} h_{jm}(r)\delta^{(j)}(r-\xi), \quad m \geqslant 0, \\ \left(\frac{\mathrm{d}^2}{\mathrm{d}r^2} + \frac{1}{r}\frac{\mathrm{d}}{\mathrm{d}r} - \frac{m^2}{r^2}\right)^2 \tilde{R}_m(r) &= \sum_{j=0}^{3} \tilde{h}_{jm}(r)\delta^{(j)}(r-\xi), \quad m \geqslant 1, \end{aligned} \tag{3.6a, b}$$

where $h_{jm}(r)$ and $\tilde{h}_{jm}(r)$ ($j = 0, 1, 2, 3$) are given in Table 2.

For $m = 0$, $\tilde{h}_{j0}(r) \equiv 0$ and $2h_{j0}(r)$ are equal to $h_{jm}(r)$ with $m = 0$ in Table 2.

The equations (3.6) are ordinary differential equations with variable coefficients. On the right-sides of these equations, the δ-function and its derivatives are contained, but their coefficients now are analytic functions of r. According to the theory of generalized functions. We can now easily show the following equations:

$$\begin{aligned} f(r)\delta(r-\xi) &= f(\xi)\delta(r-\xi), \\ f(r)\delta'(r-\xi) &= f(\xi)\delta'(r-\xi) - f'(\xi)\delta(r-\xi), \\ f(r)\delta''(r-\xi) &= f(\xi)\delta''(r-\xi) - 2f'(\xi)\delta'(r-\xi) + f''(\xi)\delta(r-\xi), \\ f(r)\delta'''(r-\xi) &= f(\xi)\delta'''(r-\xi) - 3f'(\xi)\delta''(r-\xi) \\ &\quad + 3f''(\xi)\delta'(r-\xi) - f'''(\xi)\delta(r-\xi). \end{aligned} \tag{3.7}$$

Therefore, using (3.7), the equations (3.6) may be reduced to the equivalent forms:

$$\begin{aligned} \left(\frac{\mathrm{d}^2}{\mathrm{d}r^2} + \frac{1}{r}\frac{\mathrm{d}}{\mathrm{d}r} - \frac{m^2}{r^2}\right)^2 R_m(r) &= \sum_{j=0}^{s} K_{jm}\delta^{(j)}(r-\xi), \quad m \geqslant 0, \\ \left(\frac{\mathrm{d}^2}{\mathrm{d}r^2} + \frac{1}{r}\frac{\mathrm{d}}{\mathrm{d}r} - \frac{m^2}{r^2}\right)^2 \tilde{R}_m(r) &= \sum_{j=0}^{s} \tilde{K}_{jm}\delta^{(j)}(r-\xi), \quad m \geqslant 1, \end{aligned} \tag{3.8a, b}$$

where we assume that s is a fixed integer greater than 4 without loss of generality. In fact, it is true, if we let $K_{jm} = \tilde{K}_{jm} = 0$ ($s \geqslant j \geqslant 4$) in (3.8). Now K_{jm} and $\tilde{K}_{jm}$ ($j = 1, \ldots, s$) are constants. They are given in Table 3.

Table 1. Expressions of a_{mi} and $\tilde{a}_{mi}$ for $m \geqslant 1$.

	Case (1)	Case (2)	Case (3)
a_{m1}	$p\pi^{-1}\cos m\eta$	0	$L_r m(\pi\xi)^{-1}\sin m\eta$
a_{m2}	0	$L_\theta \pi^{-1}\cos m\eta$	0
$\tilde{a}_{mi}$	$a_{mi}\tan m\eta$	$a_{mi}\tan m\eta$	$-a_{mi}c\tan m\eta$

Table 2. Expressions of $h_{jm}(r)$ and $\tilde{h}_{jm}(r)$ for $m \geqslant 1$.

	Case (1)	Case (2)	Case (3)
$h_{0m}(r)$	$\frac{P}{\pi D}\left(1+\frac{kh^2}{10}\frac{m^2}{r^2}\right)\cos m\eta$	0	$\frac{L_r}{\pi D\xi}\left(1+\frac{kh^2}{10}\frac{m^2}{r^2}\right)m\sin m\eta$
$h_{1m}(r)$	$\frac{-Pkh^2}{10\pi Dr}\cos m\eta$	$\frac{L_\theta}{\pi D}\left(1+\frac{kh^2}{10}\frac{m^2}{r^2}\right)\cos m\eta$	$\frac{-L_r kh^2}{10\pi D\xi r}m\sin m\eta$
$h_{2m}(r)$	$\frac{-Pkh^2}{10\pi D}\cos m\eta$	$\frac{-L_\theta kh^2}{10\pi Dr}\cos m\eta$	$\frac{-L_r kh^2}{10\pi D\xi}m\sin m\eta$
$h_{3m}(r)$	0	$\frac{-L_\theta kh^2}{10\pi D}\cos m\eta$	0
$\tilde{h}_{jm}(r)$	$h_{jm}(r)\tan m\eta$	$h_{jm}(r)\tan m\eta$	$-h_{jm}(r)c\tan m\eta$

For $m = 0$, $\tilde{K}_{j0} = 0$ and $2K_{j0}$ are equal to K_{jm} with $m = 0$ in Table 3.

The equations (3.8) are the ordinary differential equations stated in [5]. When the system of fundamental solutions of the homogeneous equations (3.8) are known, the solutions of (3.8) can be obtained by solving two systems of linear algebraic equation. According to [6], the fundamental solutions $R_{im}(r) \equiv \tilde{R}_{im}(r)$ $(i = 1, 2, 3, 4)$ of the homogeneous equations of (3.8) are given in Table 4.

Thus, the general solutions of (3.8) can be written as

$$R_m(r) = \sum_{i=1}^{4}(A_{im} + I(r-\xi)g_{im})R_{im}(r) + \sum_{i=4}^{s} X_{im}\delta^{(i-4)}(r-\xi),$$
$$\tilde{R}_m(r) = \sum_{i=1}^{4}(\tilde{A}_{im} + I(r-\xi)\tilde{g}_{im})\tilde{R}^{im}(r) + \sum_{i=4}^{s} \tilde{X}_{im}\delta^{(i-4)}(r-\xi). \quad (3.9a, b)$$

Hence, the k-th derivatives of $R_m(r)$ and $\tilde{R}_m(r)$ can be written as

$$R_m^{(k)}(r) = \sum_{i=1}^{4}(A_{im} + I(r-\xi)g_{im})R_{im}^{(k)}(r) + \sum_{i=4-k}^{s} X_{im}\delta^{(i-4+k)}(r-\xi),$$
$$\tilde{R}_m(r) = \sum_{i=1}^{4}(\tilde{A}_{im} + I(r-\xi)\tilde{g}_{im})\tilde{R}_{im}^{(k)}(r) + \sum_{i=4-k}^{s} \tilde{X}_{im}\delta^{(i-4+k)}(r-\xi), \quad (3.10a, b)$$

Table 3. Expressions of K_{jm} and $\tilde{K}_{jm}$ for $m \geqslant 1$.

	Case (1)	Case (2)	Case (3)
K_{0m}	$\frac{P}{\pi D}\left[1 + \frac{kh^2}{10}\frac{(m^2-1)}{\xi^2}\right]\cos m\eta$	$\frac{L_\theta kh^2}{5\pi D}\frac{(m^2-1)}{\xi^2}\cos m\eta$	$\frac{L_r m}{\pi D\xi}\left[1 + \frac{kh^2}{10}\frac{(m^2-1)}{\xi^2}\right]\sin m\eta$
K_{1m}	$-\frac{Pkh^2}{10\pi D}\frac{1}{\xi}\cos m\eta$	$\frac{L_\theta}{\pi D}\left[1 + \frac{kh^2}{10}\frac{(m^2-1)}{\xi^2}\right]\cos m\eta$	$-\frac{L_r kh^2}{10\pi D}\frac{m}{\xi^2}\sin m\eta$
K_{2m}	$-\frac{Pkh^2}{10\pi D}\cos m\eta$	$-\frac{L_\theta kh^2}{10\pi D}\frac{1}{\xi}\cos m\eta$	$-\frac{L_r kh^2}{10\pi D}\frac{m}{\xi}\sin m\eta$
K_{3m}	0	$-\frac{L_\theta kh^2}{10\pi D}\cos m\eta$	0
K_{jm} $s \geqslant j \geqslant 4$	0	0	0
$\tilde{K}_{jm}$	$K_{jm}\tan m\eta$	$K_{jm}\tan m\eta$	$-K_{jm}c\tan m\eta$

Table 4. Expressions for $R_{jm}(r) \equiv \tilde{R}_{im}(r)$.

	R_{1m}	R_{2m}	R_{3m}	R_{4m}
$m = 0$	1	r^2	$\ln r$	$r^2 \ln r$
$m = 1$	r	r^3	r^{-1}	$r \ln r$
$m \geqslant 2$	r^m	r^{-m}	r^{m+2}	r^{-m+2}

where $I(r - \xi)$ is Heaviside's function

$$I(r - \xi) = \begin{cases} 1, & r \geqslant \xi, \\ 0, & r < \xi, \end{cases} \tag{3.11}$$

and X_{im} and $\tilde{X}_{im}$ are the solutions of the following systems of linear algebraic equations:

$$\left.\begin{aligned} \sum_{k=0}^{4} f_{km}(\xi) X_{k+im} &= K_{im} \\ \sum_{k=0}^{4} f_{km}(\xi) \tilde{X}_{k+im} &= \tilde{K}_{im} \end{aligned}\right\} \quad i = 0, 1, \ldots, s, \tag{3.12}$$

where $f_{km}(\xi)$ are the values at $r = \xi$ of the coefficients $f_{km}(r)$ of $R_m^{(4-k)}(r)$ and $\tilde{R}_m^{(4-k)}(r)$ on the left-sides of (3.8), namely,

$$\begin{aligned} f_{0m}(\xi) &= 1, \quad f_{1m}(\xi) = 2\xi^{-1}, \quad f_{2m}(\xi) = -(2m^2 + 1)\xi^{-2} \\ f_{3m}(\xi) &= (2m^2 + 1)\xi^{-3}, \quad f_{4k}(\xi) = m^2(m^2 - 4)\xi^{-4}, \quad (m \geqslant 0). \end{aligned} \tag{3.13}$$

Substituting (3.13) into (3.12), after carrying on some algebraic operations, we can get the unique solutions of X_{im} and $\tilde{X}_{im}$ ($i = 0, 1, \ldots, s$). Their expressions are given in Table 5. For $m = 0$, $\tilde{X}_{i0} = 0$ and $2X_{i0}$ are equal to X_{im} with $m = 0$ in Table 5.

The constants g_{im} and $\tilde{g}_{im}$ ($i = 1, 2, 3, 4$) in expressions (3.9) and (3.10) will be given by the solutions of the following systems of linear algebraic equations:

$$\left.\begin{aligned} \sum_{i=0}^{4} R_{im}^{(k)}(\xi) g_{im} &= X_{4-k-1m} \\ \sum_{i=0}^{4} \tilde{R}_{im}^{(k)}(\xi) \tilde{g}_{im} &= \tilde{X}_{4-k-1m} \end{aligned}\right\} \quad k = 0, 1, 2, 3. \tag{3.14}$$

The results of g_{im} and $\tilde{g}_{im}$ are listed in Table 6.

Therefore, from Table 5, we see that the general solutions of (3.8) (hence equivalent (3.6)) are

$$\begin{aligned} R_m(r) &= \sum_{i=1}^{4} (A_{im} + I(r - \xi) g_{im}) R_{im}(r), \\ \tilde{R}_m(r) &= \sum_{i=1}^{4} (\tilde{A}_{im} + I(r - \xi) \tilde{g}_{im}) \tilde{R}_{im}(r). \end{aligned} \tag{3.15}$$

Table 5. Expressions of X_{im} and $\tilde{X}_{im}$ for $m \geqslant 1$.

	Case (1)	Case (2)	Case (3)
X_{0m}	$\frac{P}{\pi D}\left[1 - \frac{kh^2(m^2+4)}{10\xi^2}\right]\cos m\eta$	$-\frac{2L_\theta}{\pi D\xi}\left[1 - \frac{kh^2(4m^2+5)}{10\xi^2}\right]\cos m\eta$	$\frac{L_r m}{\pi D\xi}\left[1 - \frac{kh^2(m^2+4)}{10\xi^2}\right]\sin m\eta$
X_{1m}	$\frac{P}{\pi D}\frac{kh^2}{10\xi}\cos m\eta$	$\frac{L_\theta}{\pi D}\left[1 - \frac{kh^2(m^2+5)}{10\xi^2}\right]\cos m\eta$	$\frac{L_r kh^2}{10\pi D}\frac{m}{\xi^2}\sin m\eta$
X_{2m}	$-\frac{P}{\pi D}\frac{kh^2}{10\xi}\cos m\eta$	$\frac{L_\theta}{\pi D}\frac{kh^2}{10\xi}\cos m\eta$	$-\frac{L_r kh^2}{10\pi D}\frac{m}{\xi}\sin m\eta$
X_{3m}	0	$-\frac{L_\theta kh^2}{10\pi D}\cos m\eta$	0
X_{im} $4 \leqslant i \leqslant s$	0	0	0
$\tilde{X}_{im}$	$X_{im}\tan m\eta$	$X_{im}\tan m\eta$	$-X_{im}c\tan m\eta$

Table 6. Expressions of g_{im} and $\tilde{g}_{im}$.

	g_{1m}	g_{2m}	g_{3m}	g_{4m}
$m = 0$	$\frac{\xi}{4}[4\xi^{-1}X_{30} - (2 + \ln\xi)X_{20} + \xi\ln\xi X_{10} + \xi^2(1 - \ln\xi)X_{00}]$	$\frac{1}{4}[\xi^{-1}(2 + \ln\xi)X_{20} - \ln\xi X_{10} - \xi(1 + \ln\xi)X_{00}]$	$\frac{\xi}{4}(X_{20} - \xi X_{10} + \xi^2 X_{00})$	$\frac{1}{4}(-\xi^{-1}X_{20} + X_{10} + \xi^{-1}X_{00})$
$m = 1$	$\frac{1}{4}[3(1 + \ln\xi)\xi^{-1}X_{31} + (1 - 3\ln\xi)X_{21} - \xi X_{11} + \xi^2\ln\xi X_{01}]$	$\frac{1}{16}[\xi^{-3}X_{31} - \xi^{-2}X_{21} + 2\xi^{-1}X_{11} + X_{01}]$	$\frac{\xi}{16}[3X_{31} - 3\xi X_2] + 2\xi^2 X_{11} - \xi^3 X_{01}]$	$\frac{1}{4}[-3\xi^{-1}X_{31} + 3X_{21} - \xi^2 X_{01}]$
$m \geqslant 2$	$\frac{\xi^{1-m}}{8m(m-1)}[m(m^2-4)\xi^{-1}X_{3m} + (m^2 + 3m - 1)X_{2m} - (m-1)\xi X_{1m} - \xi^2 X_{0m}]$	$\frac{\xi^{1+m}}{8m(m+1)}[-m(m^2-4)\xi^{-1}X_{3m} + (m^2 - 3m - 1)X_{2m} + (m+1)\xi X_{1m} - \xi^2 X_{0m}]$	$\frac{\xi^{-1-m}}{8m(m+1)}[-m^2(m-2)\xi^{-1}X_{3m} - (m^2 - m + 1)X_{2m} + (m+1)\xi X_{1m} + \xi^2 X_{0m}]$	$\frac{\xi^{-1+m}}{8m(m-1)}[m^2(m+2)\xi^{-1}X_{3m} - (m^2 + m + 1)X_{2m} - (m-1)\xi X_{1m} + \xi^2 X_{0m}]$

Remarks: Substituting X_{im} for $\tilde{X}_{im}$ in the corresponding expressions of g_{im}, $\tilde{g}_{im}$ can be obtained.

Thus, the general solution of (2.11) is

$$w = \sum_{m=0}^{\infty} \left\{ \left[\sum_{i=1}^{4} (A_{im} + I(r - \xi) g_{im}) R_{im}(r) \right] \cos m\theta + \left[\sum_{i=1}^{4} (\tilde{A}_{im} + I(r - \xi) \tilde{g}_{im}) \tilde{R}_{im}(r) \right] \sin m\theta \right\}, \tag{3.16}$$

in which A_{im} and $\tilde{A}_{im}$ ($i = 1, 2, 3, 4$) together with B_{im} and $\tilde{B}_{im}$ ($i = 1, 2$) in expression (2.15) are integral constants, they will be determined by the boundary conditions of a given problem.

4. General solutions of V_r, V_θ, φ_r, φ_θ, M_r, M_θ and $H_{r\theta}$

4.1. *General solutions of V_r and V_θ*

Now, we expand V_r and V_θ into the Fourier series:

$$V_r = \sum_{m=0}^{\infty} (Q_{rm}(r) \cos m\theta + \tilde{Q}_{rm}(r) \sin m\theta),$$
$$V_\theta = \sum_{m=0}^{\infty} (Q_{\theta m}(r) \cos m\theta + \tilde{Q}_{\theta m}(r) \sin m\theta). \tag{4.1a, b}$$

Substituting expressions (2.15), (3.2) and (3.5) into (2.13a, b), and comparing the resulting expressions with (4.1a, b), we can obtain

$$\begin{aligned}
Q_{rm} &= -\left[D \frac{\mathrm{d}}{\mathrm{d}r} (\Delta_m R_m(r)) + \frac{kh^2}{10} \frac{\mathrm{d}q_m}{\mathrm{d}r} \right] + \frac{m}{r} \left[\tilde{B}_{1m} I_m \left(\frac{\sqrt{10}}{h} r \right) + \tilde{B}_{2m} K_m \left(\frac{\sqrt{10}}{h} r \right) \right] \\
\tilde{Q}_{rm} &= -\left[D \frac{\mathrm{d}}{\mathrm{d}r} (\Delta_m \tilde{R}_m(r)) + \frac{kh^2}{10} \frac{\mathrm{d}\tilde{q}_m}{\mathrm{d}r} \right] - \frac{m}{r} \left[B_{1m} I_m \left(\frac{\sqrt{10}}{h} r \right) + B_{2m} K_m \left(\frac{\sqrt{10}}{h} r \right) \right] \\
Q_{\theta m} &= -\frac{m}{r} \left[D \Delta_m \tilde{R}_m(r) + \frac{kh^2}{10} \tilde{q}_m \right] - \left[B_{1m} \frac{\mathrm{d}}{\mathrm{d}r} I_m \left(\frac{\sqrt{10}}{h} r \right) + B_{2m} \frac{\mathrm{d}}{\mathrm{d}r} K_m \left(\frac{\sqrt{10}}{h} r \right) \right] \\
\tilde{Q}_{\theta m} &= \frac{m}{r} \left[D \Delta_m R_m(r) + \frac{kh^2}{10} q_m \right] - \left[\tilde{B}_{1m} \frac{\mathrm{d}}{\mathrm{d}r} I_m \left(\frac{\sqrt{10}}{h} r \right) + \tilde{B}_{2m} \frac{\mathrm{d}}{\mathrm{d}r} K_m \left(\frac{\sqrt{10}}{h} r \right) \right]
\end{aligned} \tag{4.2}$$

where

$$\Delta_m \equiv \frac{d^2}{dr^2} + \frac{1}{r}\frac{d}{dr} - \frac{m^2}{r^2}. \tag{4.3}$$

Substituting expressions (3.4) and (3.5) into (4.2), and using (3.7) and (3.10), we have

$$Q_{rm} = -\left\{\sum_{i=1}^{4}(A_{im} + I(r-\xi)g_{im})F^m_{v_r}[R_{im}(r)] - (\tilde{B}_{1m}\chi^m_{v_r} + \tilde{B}_{2m}\psi^m_{v_r}) + \sum_{i=0}^{2} e^m_{v_r i}\delta^{(i)}(r-\xi)\right\}, \tag{4.4a}$$

$$\tilde{Q}_{rm} = -\left\{\sum_{i=1}^{4}(\tilde{A}_{im} + I(r-\xi)\tilde{g}_{im})F^m_{v_r}[R_{im}(r)] + (B_{1m}\chi^m_{v_r} + B_{2m}\psi^m_{v_r}) + \sum_{i=0}^{2} \tilde{e}^m_{v_r i}\delta^{(i)}(r-\xi)\right\}, \tag{4.4b}$$

$$Q_{\theta m} = -\left\{\sum_{i=1}^{4}(\tilde{A}_{im} + I(r-\xi)\tilde{g}_{im})F^m_{v_\theta}[R_{im}(r)] + (B_{1m}\chi^m_{v_\theta} + B_{2m}\psi^m_{v_\theta}) + \sum_{i=0}^{1} \tilde{e}^m_{v_\theta i}\delta^{(i)}(r-\xi)\right\}, \tag{4.4c}$$

$$\tilde{Q}_{\theta m} = -\left\{\sum_{i=1}^{4}(A_{im} + I(r-\xi)g_{im})F^m_{v_\theta}[R_{im}(r)] - (\tilde{B}_{1m}\chi^m_{v_\theta} + \tilde{B}_{2m}\psi^m_{v_\theta}) + \sum_{i=0}^{1} e^m_{v_\theta i}\delta^{(i)}(r-\xi)\right\}, \tag{4.4d}$$

in which the operators $F^m_{v_r}[R_{im}]$ and $F^m_{v_\theta}[R_{im}]$ are defined by

$$F^m_{v_r}[\cdot] \equiv D\frac{d}{dr}(\Delta_m[\cdot]), \qquad F^m_{v_\theta}[\cdot] \equiv D\frac{m}{r}(\Delta_m[\cdot]). \tag{4.5}$$

The expressions of $F^m_{v_r}[R_{im}]$ and $F^m_{v_\theta}[R_{im}]$ are given in Tables 7(a–c). The expressions of $\chi^m_{v_r}$, $\psi^m_{v_r}$, $\chi^m_{v_\theta}$ and $\psi^m_{v_\theta}$ are given in Table 8, and in Table 9(a), we give $e^m_{v_r i}$, $\tilde{e}^m_{v_r i}$ $(i = 0, 1, 2)$ and $e^m_{v_\theta i}$, $\tilde{e}^m_{v_\theta i}$ $(i = 0, 1)$.

4.2. *General solutions for φ_r and φ_θ*

Next, we will expand φ_r and φ_θ into the Fourier series:

$$\varphi_r = \sum_{m=0}^{\infty}(\beta_{rm}(r)\cos m\theta + \tilde{\beta}_{rm}(r)\sin m\theta),$$
$$\varphi_\theta = \sum_{m=0}^{\infty}(\beta_{\theta m}(r)\cos m\theta + \tilde{\beta}_{\theta m}(r)\sin m\theta). \tag{4.6a, b}$$

Table 7a. Expressions of $F^m[R_{im}]$ for $m = 0$.

	R_{10}	R_{20}	R_{30}	R_{40}
$F^0_{v_r}[\cdot]$	0	0	0	0
$F^0_{v_\theta}[\cdot]$	0	0	0	0
$F^0_{\varphi_r}[\cdot]$	0	$2r$	r^{-1}	$(1 + 2\ln r)r + 4k_1h^2r^{-1}$
$F^0_{\varphi_\theta}[\cdot]$	0	0	0	0
$F^0_{M_r}[\cdot]$	0	$2(1+\mu)D$	$-(1-\mu)r^{-2}D$	$2(1+\mu)D\ln r + (3+\mu)D - \frac{4}{5}h^2Dr^{-2}$
$F^0_{M_\theta}[\cdot]$	0	$2(1+\mu)D$	$(1-\mu)r^{-2}D$	$(1+3\mu)D + 2(1+\mu)D\ln r + \frac{4}{5}h^2Dr^{-2}$
$F^0_H[\cdot]$	0	0	0	0

Table 7b. Expressions of $F^m[R_{im}]$ for $m = 1$.

	R_{11}	R_{21}	R_{31}	R_{41}
$F^1_{v_r}[\cdot]$	0	$8D$	0	$-2Dr^{-2}$
$F^1_{v_\theta}[\cdot]$	0	$8D$	0	$2Dr^{-2}$
$F^1_{\varphi_r}[\cdot]$	1	$8k_1h^2 + 3r^2$	$-r^2$	$1 + \ln r - 2k_1h^2r^{-2}$
$F^1_{\varphi_\theta}[\cdot]$	1	$8k_1h^2 + r^2$	r^{-2}	$\ln r + 2k_1h^2r^{-2}$
$F^1_{M_r}[\cdot]$	0	$2(3+\mu)Dr$	$2(1-\mu)Dr^{-3}$	$(1+\mu)Dr^{-1} + \frac{4}{5}h^2Dr^{-3}$
$F^1_{M_\theta}[\cdot]$	0	$2(1+3\mu)Dr$	$-2(1-\mu)Dr^{-3}$	$(1+\mu)Dr^{-1} - \frac{4}{5}h^2Dr^{-3}$
$F^1_H[\cdot]$	0	$-2(1-\mu)Dr$	$2(1-\mu)Dr^{-3}$	$-(1-\mu)Dr^{-1} + \frac{4}{5}h^2Dr^{-3}$

Substituting expressions (3.15) and (4.2) into (2.4a, b), and using (3.7) and (3.10), and then comparing the resulting expressions with (4.6), we find

$$\begin{aligned}
\beta_{rm} &= -\left\{\sum_{i=1}^{4}(A_{im} + I(r-\xi)g_{im})F^m_{\varphi_r}[R_{im}(r)] - (\tilde{B}_{1m}\chi^m_{\varphi_r} + \tilde{B}_{2m}\psi^m_{\varphi_r})\right.\\
&\qquad \left. + \sum_{i=0}^{2} e^m_{\varphi_r i}\delta^{(i)}(r-\xi)\right\},\\
\tilde{\beta}_{rm} &= -\left\{\sum_{i=1}^{4}(\tilde{A}_{im} + I(r-\xi)\tilde{g}_{im})F^m_{\varphi_r}[R_{im}(r)] + (B_{1m}\chi^m_{\varphi_r} + B_{2m}\psi^m_{\varphi_r})\right.\\
&\qquad \left. + \sum_{i=0}^{2} \tilde{e}^m_{\varphi_r i}\delta^{(i)}(r-\xi)\right\},\\
\beta_{\theta m} &= -\left\{\sum_{i=1}^{4}(\tilde{A}_{im} + I(r-\xi)\tilde{g}_{im})F^m_{\varphi_\theta}[R_{im}(r)] + (B_{1m}\chi^m_{\varphi_\theta} + B_{2m}\psi^m_{\varphi_\theta})\right.\\
&\qquad \left. + \sum_{i=0}^{1} \tilde{e}^m_{\varphi_\theta i}\delta^{(i)}(r-\xi)\right\},\\
\tilde{\beta}_{\theta m} &= \sum_{i=1}^{4}(A_{im} + I(r-\xi)g_{im})F^m_{\varphi_\theta}[R_{im}(r)] - (\tilde{B}_{1m}\chi^m_{\varphi_\theta} + \tilde{B}_{2m}\psi^m_{\varphi_\theta})\\
&\qquad + \sum_{i=0}^{1} e^m_{\varphi_\theta i}\delta^{(i)}(r-\xi),
\end{aligned} \tag{4.7}$$

Table 7c. Expressions of $F^m[R_{im}]$ for $m \geqslant 2$.

	R_{1m}	R_{2m}	R_{3m}	R_{4m}
$F^m_{v_r}[\cdot]$	0	0	$4m(m+1)Dr^{m-1}$	$4m(m-1)Dr^{-m-1}$
$F^m_{v_\theta}[\cdot]$	0	0	$4m(m+1)Dr^{m-1}$	$-4m(m-1)Dr^{-m-1}$
$F^m_{\varphi_r}[\cdot]$	mr^{m-1}	$-mr^{-m-1}$	$(m+2)r^{m+1}+4k_1h^2m(m+1)r^{m-1}$	$-(m-2)r^{-m+2}+4k_1h^2m(m-1)r^{-m-1}$
$F^m_{\varphi_\theta}[\cdot]$	mr^{m-1}	mr^{-m-1}	$mr^{m+1}+4k_1h^2m(m+1)r^{m-1}$	$mr^{-m+1}-4k_1h^2m(m-1)r^{-m-1}$
$F^m_{M_r}[\cdot]$	$(1-\mu)m$ $\times(m-1)Dr^{m-2}$	$(1-\mu)m$ $\times(m+1)Dr^{-m-2}$	$(m+1)[(1-\mu)m$ $+2(1+\mu)]Dr^m$ $+\frac{4}{5}h^2m(m^2-1)Dr^{m-2}$	$(m-1)[(1-\mu)m$ $-2(1+\mu)]Dr^{-m}$ $-\frac{4}{5}h^2m(m^2-1)Dr^{-m-2}$
$F^m_{M_\theta}[\cdot]$	$-(1-\mu)m$ $\times(m-1)Dr^{m-2}$	$-(1-\mu)m$ $\times(m+1)Dr^{-m-2}$	$-(m+1)[(1-\mu)m$ $+2(1+\mu)]Dr^m$ $-\frac{4}{5}h^2m(m^2-1)Dr^{m-2}$	$-(m-1)[(1-\mu)m$ $+2(1+\mu)]Dr^{-m}$ $+\frac{4}{5}h^2m(m^2-1)Dr^{-m-2}$
$F^m_H[\cdot]$	$-(1-\mu)m$ $\times(m-1)Dr^{m-2}$	$(1-\mu)m$ $\times(m+1)Dr^{-m-2}$	$-(1-\mu)m(m+1)Dr^m$ $-\frac{4}{5}h^2m(m^2-1)Dr^{m-2}$	$(1-\mu)m(m-1)Dr^{-m}$ $-\frac{4}{5}h^2m(m^2-1)Dr^{-m-2}$

in which the operators $F^m_{\varphi_r}[\cdot]$ and $F^m_{\varphi_\theta}[\cdot]$ are defined by

$$F^m_{\varphi_r}[\cdot] \equiv \frac{\mathrm{d}}{\mathrm{d}r}[\cdot] + k_1 h^2 \frac{\mathrm{d}}{\mathrm{d}r}(\Delta_m[\cdot]),$$
$$F^m_{\varphi_\theta}[\cdot] \equiv \frac{m}{r}([\cdot] + k_1 h^2 \Delta_m[\cdot]). \tag{4.8a, b}$$

Here we had set

$$k_1 = \frac{1}{5(1-\mu)}. \tag{4.9}$$

The expressions of $F^m_{\varphi_r}[R_{im}]$ and $F^m_{\varphi_\theta}[R_{im}]$ are given in Tables 7(a–c). The expressions of $\chi^m_{\varphi_r}$, $\psi^m_{\varphi_r}$, $\chi^m_{\varphi_\theta}$ and $\psi^m_{\varphi_\theta}$ are given in Table 8, and in Table 9(b), we give $e^m_{\varphi_r i}$, $\tilde{e}_{\varphi_r i}$ $(i = 0, 1, 2)$ and $e^m_{\varphi_\theta i}$, $\tilde{e}^m_{\varphi_\theta i}$, $(i = 0, 1)$.

4.3. *General solutions for M_r and M_θ*

We now expand M_r and M_θ into the Fourier series:

$$M_r = \sum_{m=0}^{\infty} (T_{rm}(r)\cos m\theta + \tilde{T}_{rm}(r)\sin m\theta),$$
$$M_\theta = \sum_{m=0}^{\infty} (T_{\theta m}(r)\cos m\theta + \tilde{T}_{\theta m}(r)\sin m\theta). \tag{4.10a, b}$$

Substituting expressions (3.2), (3.4), (3.5), (3.15) and (4.2) into (2.3), and using (3.7) and (3.10), then using the resulting expressions with (4.10a, b), we find

$$T_{rm} = -\left\{\sum_{i=1}^{4} (A_{im} + I(r-\xi)g_{im})F^m_{M_r}[R_{im}(r)] + (\tilde{B}_{1m}\chi^m_{M_r} + \tilde{B}_{2m}\psi^m_{M_r}) + \sum_{i=0}^{3} e^m_{M_r i}\delta^{(i)}(r-\xi)\right\}, \tag{4.11a}$$

$$\tilde{T}_{rm} = -\left\{\sum_{i=1}^{4} (\tilde{A}_{im} + I(r-\xi)\tilde{g}_{im})F^m_{M_r}[R_{im}(r)] - (B_{1m}\chi^m_{M_r} + B_{2m}\psi^m_{M_r}) + \sum_{i=0}^{3} \tilde{e}^m_{M_r i}\delta^{(i)}(r-\xi)\right\}, \tag{4.11b}$$

$$T_{\theta m} = -\left\{\sum_{i=1}^{4} (A_{im} + I(r-\xi)g_{im})F^m_{M_\theta}[R_{im}(r)] - (\tilde{B}_{1m}\chi^m_{M_\theta} + \tilde{B}_{2m}\psi^m_{M_\theta}) + \sum_{i=0}^{2} e^m_{M_\theta i}\delta^{(i)}(r-\xi)\right\}, \tag{4.11c}$$

$$\tilde{T}_{\theta m} = -\left\{\sum_{i=1}^{4} (\tilde{A}_{im} + I(r-\xi)g_{im})F^m_{M_\theta}[R_{im}(r)] + (B_{1m}\chi^m_{M_\theta} + B_{2m}\psi^m_{M_\theta}) + \sum_{i=0}^{2} \tilde{e}^m_{M_\theta i}\delta^{(i)}(r-\xi)\right\}, \tag{4.11d}$$

Table 8. Expressions of $\chi^m_\cdot$ and $\psi^m_\cdot$.

	$\chi^m_\cdot$	$\psi^m_\cdot$
V_r	$\frac{m}{r} I_m\left(\frac{\sqrt{10}}{h} r\right)$	$\frac{m}{r} K_m\left(\frac{\sqrt{10}}{h} r\right)$
V_θ	$\frac{m}{r} I_m\left(\frac{\sqrt{10}}{h} r\right) + \frac{\sqrt{10}}{\mathrm{h}} I_{m+1}\left(\frac{\sqrt{10}}{h} r\right)$	$\frac{m}{r} K_m\left(\frac{\sqrt{10}}{h} r\right) - \frac{\sqrt{10}}{h} K_{m+1}\left(\frac{\sqrt{10}}{h} r\right)$
φ_r	$\frac{12(1+\mu)}{5Eh} \frac{m}{r} I_m\left(\frac{\sqrt{10}}{h} r\right)$	$\frac{12(1+\mu)}{5Eh} \frac{m}{r} K_m\left(\frac{\sqrt{10}}{h} r\right)$
φ_θ	$\frac{12(1+\mu)}{5Eh}\left[\frac{m}{r} I_m\left(\frac{\sqrt{10}}{h} r\right) + \frac{\sqrt{10}}{h} I_{m+1}\left(\frac{\sqrt{10}}{h} r\right)\right]$	$\frac{12(1+\mu)}{5Eh}\left[\frac{m}{r} K_m\left(\frac{\sqrt{10}}{h} r\right) - \frac{\sqrt{10}}{h} K_{m+1}\left(\frac{\sqrt{10}}{h} r\right)\right]$
M_r	$-\frac{h^2}{5}\frac{m}{r}\left[\frac{m-1}{r} I_m\left(\frac{\sqrt{10}}{h} r\right) + \frac{\sqrt{10}}{h} I_{m+1}\left(\frac{\sqrt{10}}{h} r\right)\right]$	$-\frac{h^2}{5}\frac{m}{r}\left[\frac{m-1}{r} K_m\left(\frac{\sqrt{10}}{h} r\right) - \frac{\sqrt{10}}{h} K_{m+1}\left(\frac{\sqrt{10}}{h} r\right)\right]$
M_θ	$-\frac{h^2}{5}\frac{m}{r}\left[\frac{m-1}{r} I_m\left(\frac{\sqrt{10}}{h} r\right) + \frac{\sqrt{10}}{h} I_{m+1}\left(\frac{\sqrt{10}}{h} r\right)\right]$	$-\frac{h^2}{5}\frac{m}{r}\left[\frac{m-1}{r} K_m\left(\frac{\sqrt{10}}{h} r\right) - \frac{\sqrt{10}}{h} K_{m+1}\left(\frac{\sqrt{10}}{h} r\right)\right]$
H	$\frac{2m(m-1)}{r^2} I_m\left(\frac{\sqrt{10}}{h} r\right) + \frac{2m}{r} I_{m+1}\left(\frac{\sqrt{10}}{h} r\right) + \frac{10}{h^2} I_{m+2}\left(\frac{\sqrt{10}}{h} r\right)$	$\frac{2m(m-1)}{r^2} K_m\left(\frac{\sqrt{10}}{h} r\right) - \frac{2m}{r} K_{m+1}\left(\frac{\sqrt{10}}{h} r\right) + \frac{10}{h^2} K_{m+2}\left(\frac{\sqrt{10}}{h} r\right)$

Table 9a. Expressions of $e^m_{v_r i}$ and $e^m_{v_\theta i}$.

	$i = 0$	$i = 1$	$i = 2$	$i = 3$
$e^m_{v_r i}$	$D[X_{1m} - \xi^{-1}X_{2m} - m^2\xi^{-2}X_{3m}]$	$D(X_{2m} + \xi^{-1}X_{3m}) - \frac{kh^2}{10}a_{m1}$	$DX_{3m} + \frac{kh^2}{10}a_{m2}$	0
$e^m_{v_\theta i}$	$Dm\xi^{-1}(X_{2m} + 2\xi^{-1}X_{3m}) + \frac{kh^2}{10}(a_{m1} + \xi^{-1}a_{m2})$	$Dm\xi^{-1}X_{3m} + \frac{kh^2}{10}a_{m2}$	0	0

Remarks: Substituting X_{im} and a_{mi} for $\tilde{X}_{im}$ and $\tilde{a}_{mi}$ in corresponding expressions of $e^m_{[\cdot]i}$, $\tilde{e}_{[\cdot]i}$ can be obtained.

Table 9b. Expressions of $e^m_{\varphi_r i}$ and $e^m_{\varphi_\theta i}$.

	$i = 0$	$i = 1$	$i = 2$	$i = 3$
$e^m_{\varphi_r i}$	$(1 - k_1h^2m^2\xi^{-2})X_{3m} + k_1h^2(\xi^{-1}X_{2m} + X_{1m})$	$k_1h^2(\xi^{-1}X_{3m} + X_{2m}) + \frac{6(1+\mu)}{25E}kha_{m1}$	$k_1h^2X_{3m} + \frac{6(1+\mu)}{25E}kha_{m2}$	0
$e^m_{\varphi_\theta i}$	$k_1h^2m\xi^{-1}\left[(2\xi^{-1}X_{3m} + X_{2m}) + \frac{kh^2}{100}(a_{m1} + \xi^{-1}a_{m2})\right]$	$k^1h^2m\xi^{-1}\left(X_{3m} + \frac{kh^2}{100}a_{m2}\right)$	0	0

Remarks: Substituting X_{im} and a_{mi} for $\tilde{X}_{im}$ and $\tilde{a}_{mi}$ in corresponding expressions of $e^m_{[\cdot]i}$, then $\tilde{e}_{[\cdot]i}$ can be obtained.

in which the operators $F^m_{M_r}[\cdot]$ and $F^m_{M_\theta}[\cdot]$ are defined by

$$F^m_{M_r}[\cdot] \equiv D\left\{(1-\mu)\frac{\mathrm{d}^2}{\mathrm{d}r^2}[\cdot] + \mu\Delta_m[\cdot] + \frac{h^2}{5}\frac{\mathrm{d}^2}{\mathrm{d}r^2}(\Delta_m[\cdot])\right\},$$

$$F^m_{M_\theta}[\cdot] \equiv D\left\{-(1-\mu)\frac{\mathrm{d}^2}{\mathrm{d}r^2}[\cdot] + \Delta_m[\cdot] + \frac{h^2}{5}\left[\frac{1}{r}\frac{\mathrm{d}}{\mathrm{d}r}(\Delta_m[\cdot]) - \frac{m^2}{r^2}\Delta_m[\cdot]\right]\right\}. \tag{4.12a, b}$$

Here we had set

$$k_2 = \frac{\mu}{1-\mu}. \tag{4.13}$$

The expressions of $F^m_{M_r}[R_{im}]$ and $F^m_{M_\theta}[R_{im}]$ are given in Tables 7(a–c). The expressions of $\chi^m_{M_r}$, $\psi^m_{M_r}$, $\chi^m_{M_\theta}$ and $\psi^m_{M_\theta}$ are given in Table 8, and in Table 9(c) we give $e^m_{M_r i}$, $\tilde{e}^m_{M_r i}$ $(i = 0, 1, 2, 3)$ and $e^m_{M_\theta i}$, $\tilde{e}^m_{M_\theta i}$ $(i = 0, 1, 2)$.

4.1. *General solution for $H_{r\theta}$*

Finally, we expand $H_{r\theta}$ into the Fourier series:

$$H_{r\theta} = \sum_{m=0}^{\infty}(h_m(r)\cos m\theta + \tilde{h}_m(r)\sin m\theta). \tag{4.14}$$

Table 9c. Expressions of $e^m_{M_r i}$, $e^m_{M_\theta i}$ and e^m_{Hi}.

	$i = 0$	$i = 1$	$i = 2$	$i = 3$
$e^m_{M_r i}$	$D\left[\left(\mu\xi^{-1} + \frac{h^2}{10} m^2\xi^{-3}\right) X_{3m} + \left(1 - \frac{h^2}{5}(m^2 + 1)\xi^{-1}\right) X_{2m} + \frac{h^2}{5}(\xi^{-1}X_{1m} + X_{0m})\right] + \frac{k_2 h^2}{10} a_{m1}$	$D\left[\left(1 - \frac{h^2}{5} m^2\xi^{-2}\right) X_{3m} + \frac{h^2}{5}(\xi^{-1}X_{2m} + X_{1m})\right] + \frac{k_2 h^2}{10} a_{m2}$	$\frac{h^2}{5} D(\xi^{-1}X_{3m} + X_{2m}) + \frac{kh^4}{50} a_{m1}$	$\frac{h^2}{5} DX_{3m} + \frac{kh^2}{50} a_{m2}$
$e^m_{M_\theta i}$	$D\left\{\xi^{-1}\left[1 - \frac{h^2}{5}(4m^2 - 3)\xi^{-2}\right] X_{3m} + \left[\mu - \frac{h^2}{5}(m^2 - 2)\xi^{-2}\right] X_{2m} + \frac{h^2}{5}\xi^{-1}X_{1m}\right\} - \frac{kh^4}{25}(m^2 - 1)\xi^{-3} a_{m2} + \frac{h^2}{10}\left[k_2 - \frac{kh^2}{5}(m^2 - 1)\xi^{-2}\right] a_{m1}$	$D\left\{\left[\mu - \frac{h^2}{5}(m^2 - 3)\xi^{-2}\right] X_{3m} + \frac{h^2}{5}\xi^{-1}X_{2m}\right\} + \frac{kh^4}{50}\xi^{-1}a_{m1} + \frac{h^2}{10}\left[k_2 - \frac{kh^2}{5}(m^2 - 2)\xi^{-2}\right] a_{m2}$	$\frac{h^2}{5} D\xi^{-1}X_{3m} + \frac{kh^4}{50}\xi^{-1}a_{m2}$	0
e^m_{Hi}	$-Dm\xi^{-1}\left\{[(1 - \mu) - m^2\xi^{-2}]X_{3m} + \frac{h^2}{5}(\xi^{-1}X_{2m} + X_{1m})\right\}$	$-\frac{h^2}{5} Dm\xi^{-1}(2\xi^{-1}X_{3m} + X_{2m}) - \frac{kh^4}{50} m\xi^{-1}(\xi^{-1}a_{m2} + a_{m1})$	$-\frac{h^2}{5} Dm\xi^{-1}X_{3m} - \frac{kh^4}{50} m\xi^{-1}a_{m2}$	0

Remarks: Substituting X_{im} and a_{mi} for $\tilde{X}_{im}$ and $\tilde{a}_{mi}$ in corresponding expressions of $e^m_{[\cdot]i}$, then $\tilde{e}^m_{[\cdot]i}$ can be obtained.

Substituting expressions (3.2), (3.3), (3.5), (3.15) and (4.2) into (2.3c), and using (3.7) and (3.10), then comparing the result obtained with (4.14), we find

$$h_m(r) = \left\{\sum_{i=1}^{4} (\tilde{A}_{im} + I(r-\xi)\tilde{g}_{im})F_H^m[R_{im}(r)] - (B_{1m}\chi_H^m + B_{2m}\psi_H^m) + \sum_{i=0}^{2} e_{Hi}^m \delta^{(i)}(r-\xi)\right\}, \tag{4.15a}$$

$$\tilde{h}_m(r) = \left\{\sum_{i=1}^{4} (A_{im} + I(r-\xi)g_{im})F_H^m[R_{im}(r)] + (\tilde{B}_{1m}\chi_H^m + B_{2m}\psi_H^m) + \sum_{i=0}^{2} \tilde{e}_{Hi}^m \delta^{(i)}(r-\xi)\right\}, \tag{4.15b}$$

in which the operator $F_H^m[\cdot]$ is defined by

$$F_H^m[\cdot] \equiv D\left\{-(1-\mu)\frac{m}{r}\left[\frac{\mathrm{d}}{\mathrm{d}r}[\cdot] - \frac{1}{r}[\cdot]\right] + \frac{h^2}{5}\frac{m}{r}\left[\frac{1}{r}\Delta_m[\cdot] - \frac{\mathrm{d}}{\mathrm{d}r}(\Delta_m[\cdot])\right]\right\}. \tag{4.16}$$

The expressions of $F_H^m[R_{im}]$ are given in Tables 7(a–c). The expressions of χ_H^m and ψ_H^m are given in Table 8, and in Table 9(c) we give e_{Hi}^m and $\tilde{e}_{Hi}^m$ $(i = 0, 1, 2)$.

5. Determination of integral constants A_{im}, $\tilde{A}_{im}$ and B_{im}, $\tilde{B}_{im}$

From E. Reissner's theory, three boundary conditions must be satisfied on each edge of a bent plate. By the aid of expressions w, V_r, V_θ, φ_r, φ_θ, M_r, M_θ and $H_{r\theta}$, the boundary conditions can now be reduced to two sets. For example:

(1) if the inner and outer edges of a ring-shaped plate are built-in, we have

$$\begin{aligned} R_m(r)\big|_{r=a,b} &= 0, \quad \beta_{rm}(r)\big|_{r=a,b} = 0, \quad \beta_{\theta m}(r)\big|_{r=a,b} = 0, \\ \tilde{R}_m(r)\big|_{r=a,b} &= 0, \quad \tilde{\beta}_{rm}(r)\big|_{r=a,b} = 0, \quad \tilde{\beta}_{\theta m}(r)\big|_{r=a,b} = 0. \end{aligned} \tag{5.1}$$

(2) if the inner and outer edges of a ring-shaped plate are simply supported, we have

$$\begin{aligned} R_m(r)\big|_{r=a,b} &= 0, \quad T_{rm}(r)\big|_{r=a,b} = 0, \quad \beta_{\theta m}(r)\big|_{r=a,b} = 0, \\ \tilde{R}_m(r)\big|_{r=a,b} &= 0, \quad \tilde{T}_{rm}(r)\big|_{r=a,b} = 0, \quad \tilde{\beta}_{\theta m}(r)\big|_{r=a,b} = 0,. \end{aligned} \tag{5.2}$$

(3) if the inner and outer edges of a ring-shaped plate are free, we have

$$\begin{aligned} T_{rm}(r)\big|_{r=a,b} &= 0, \quad h_m(r)\big|_{r=a,b} = 0, \quad Q_{rm}(r)\big|_{r=a,b} = 0, \\ \tilde{T}_{rm}(r)\big|_{r=a,b} &= 0, \quad \tilde{h}_m(r)\big|_{r=a,b} = 0, \quad \tilde{Q}_{rm}(r)\big|_{r=a,b} = 0. \end{aligned} \tag{5.3}$$

Obviously, other combinations of boundary conditions are often encountered. However, a well-posed theory requires that the statement of boundary conditions cannot violate the existence and uniqueness theorem of solutions of fundamental boundary-value problems. Under each combination of boundary conditions, we shall obtain a set of twelfth-order linear algebraic equations in twelve integral constants, namely, A_{im}, $\tilde{A}_{im}$ $(i = 1, 2, 3, 4)$ and B_{im}, $\tilde{B}_{im}$ $(i = 1, 2)$. But we emphasize that the set of twelve algebraic equations can be divided practically into two independent sets, and each of them only depends on six integral constants, respectively. Moreover, we point out that it is sufficient to solve any one of them. The solution of another set can be obtained by a suitable substitution of the coefficients of another set for one in the obtained formulae of solutions. Owing to the limitations of space, we will not pursue these lengthy algebraic operations and results any further.

6. Discussion and generalization

We will now discuss the possibility of generalizing our method in the following three ways.

6.1. *On the elastic properties of plate*

(a) *Transversely isotropic plates.* For plates with this elastic property, the fundamental equations have been derived by E. Reissner [2] (cf. see Eqs. (36) in [2]). Corresponding to (2.1) and (2.2) in the present paper we have

$$\frac{\partial(rV_r)}{\partial r} + \frac{\partial V_\theta}{\partial \theta} = -rp \tag{6.1}$$

$$V_r - \frac{(1-\mu)D}{2C_s}\left[\Delta V_r - \frac{2}{r^2}\frac{\partial V_\theta}{\partial \theta} - \frac{V_r}{r^2}\right] = -D\frac{\partial \Delta w}{\partial r} - (1+\mu)D\left(\frac{1}{2C_s} - \frac{1}{C_n}\right)\frac{\partial p}{\partial r} \tag{6.2a}$$

$$V_\theta - \frac{(1-\mu)D}{2C_s}\left[\Delta V_\theta + \frac{2}{r^2}\frac{\partial V_r}{\partial \theta} - \frac{V_\theta}{r^2}\right] = -D_r\frac{\partial \Delta w}{\partial \theta} - (1+\mu)D\left(\frac{1}{2C_s} - \frac{1}{C_n}\right)\frac{1}{r}\frac{\partial p}{\partial \theta} \tag{6.2b}$$

Eliminating V_r and V_θ from these equations, we get

$$\Delta^2 w = \frac{p}{D} - \left(\frac{1}{C_s} - \frac{1+\mu}{C_n}\right)\Delta p \tag{6.3}$$

in which the constants D, C_s and C_n depend on the properties of materials of the plates and on the case of stress distributions across the thickness. For so-called transversely isotropic plates, the constants are

$$D = \frac{Eh^3}{12(1-\mu^2)}, \quad C_n = \frac{5}{6}\frac{E_z h}{\mu_z}, \quad C_s = \tfrac{5}{6}G_z h \tag{6.4}$$

where E, μ and E_z, μ_z are respectively Young's modulus and Poisson's ratio in the plane of a plate and in the direction normal to the plane of the plate, and G_z is another independent elastic constant.

If the solution of (6.3) is obtained under a transverse load $p(r, \theta)$, then we can show that the general solution of (6.1) and (6.2) may be represented as

$$\begin{aligned} V_r &= -D\frac{\partial \Delta w}{\partial r} - \left(\frac{1}{C_s} - \frac{1+\mu}{C_n}\right) D\frac{\partial p}{\partial r} + \frac{1}{r}\frac{\partial \Phi}{\partial \theta}, \\ V_\theta &= -D\frac{1}{r}\frac{\partial \Delta w}{\partial \theta} - \left(\frac{1}{C_s} - \frac{1+\mu}{C_n}\right) D\frac{1}{r}\frac{\partial p}{\partial \theta} - \frac{\partial \Phi}{\partial r}, \end{aligned} \tag{6.5}$$

where the sum of the first two parts is a particular solution of (6.1) and (6.2), and the last part is the general solution of the corresponding homogeneous equations. The stress function Φ satisfies the equation

$$\Delta\Phi - \frac{2C_s}{(1-\mu)D}\Phi = 0. \tag{6.6}$$

Clearly, if the elastic constants in (6.2), (6.3) and (6.5), (6.6) correspond with those in suitable equations of isotropic plates, then we can give solutions for the bending problems of transversely isotropic plates.

(b) *Sandwich plates.* The present method can be applied to resolving the fundamental equations of sandwich plates as established by E. Reissner [2]. In fact, according to [2], it is sufficient to set

$$D = \frac{t(h+t)^2}{2(1-\mu^2)}E_f, \quad C_n = \infty, \quad C_s = G_c h, \tag{6.7}$$

in (6.1) to (6.6). Here h and t are the thicknesses of core layers and face layers respectively, E_f and μ are Young's modulus and Poisson's ratio of the face layers, and G_c is the shearing modulus of the core layers.

(c) *Plates non-homogeneous across the thickness.* Here we will assume that the plate is isotropic and non-homogeneous across its thickness. We will also assume that Young's modulus and Poisson's ratio of the plate satisfy $E(z) = E(-z)$, $\mu(z) = \mu(-z)$.

Using the same method as used in deriving (3.6) in [2], we can obtain the equations which formally are the same as (3.6) in [2], except that in the present case the constants D, C_n and C_s should be respectively

$$D = \frac{4K_1}{K_2(4K_1 - K_2)}, \quad C_n = \frac{2}{K_3}, \quad C_s = \frac{1}{K_4} \tag{6.8}$$

and Poisson's ratio μ is replaced by $\bar{\mu}$. Here $\bar{\mu}$ is defined by

$$\bar{\mu} \equiv \frac{K_2}{2K_1} - 1. \tag{6.9}$$

The constants K_i ($i = 1, 2, 3, 4$) in (6.8) and (6.9) are represented as

$$
\begin{aligned}
K_1 &= \left(\frac{12}{h^3}\right)^2 \int_{-h/2}^{h/2} \frac{z^2}{E(z)}\,\mathrm{d}z, \\
K_2 &= 2\left(\frac{12}{h^3}\right)^2 \int_{-h/2}^{h/2} \frac{1+\mu(z)}{E(z)}\,z^2\,\mathrm{d}z, \\
K_3 &= \frac{1}{2}\left(\frac{12}{h^3}\right)^2 \int_{-h/2}^{h/2} \frac{\mu(z)}{E(z)}\,z^2\left[\frac{1}{2} - \frac{3}{4}\left(\frac{2z}{h}\right) - \frac{1}{4}\left(\frac{2z}{h}\right)^3\right]\mathrm{d}z, \\
K_4 &= \frac{1}{2}\left(\frac{3}{h}\right)^2 \int_{-h/2}^{h/2} \frac{1+\mu(z)}{E(z)}\left[1 - \left(\frac{2z}{h}\right)^2\right]^2 \mathrm{d}z.
\end{aligned}
\tag{6.10}
$$

Obviously, when E and μ are constants, we have $\bar{\mu} \equiv \mu$, and in this case, our equations coincide exactly with those of homogeneous isotropic plates.

6.2. *On cases of load distribution*

In the preceding articles the solutions obtained for the bending of ring-shaped plates under the action of a concentrated load constitute Green's functions of the general solutions of fundamental equations of E. Reissner's theory. If therefore, some load of intensity $p(\xi, \eta)$ distributed over an area A of the plate is given, then corresponding solutions may be obtained. In fact, if we apply an elementary load $p(\xi, \eta)\xi\, \mathrm{d}\xi\, \mathrm{d}\eta$ at $r = \xi, \theta = \eta$ using the principle of superposition, then integrating for ξ and η over the loaded area A, the expected results can be gained.

6.3. *On constraint conditions of plates*

(a) *Plates of generalized displacements and forces prescribed on edges.* In this case, if we first expand the given displacements and forces into Fourier series of θ and make the values of displacements and forces on the edges of plates equal to the given values, then compare them with the coefficients of corresponding terms on both sides in resulting equalities, we will obtain a set of linear algebraic equations in the undetermined integral constants.

(b) *Plates supported at several points* [7]. Now, except where some boundary conditions on the two edges $r = a$ and $r = b$ are satisfied, we also assume that the plate is supported at several points. Let the polar coordinates of the points of support be (ξ_i, η_i) ($i = 1, 2, \ldots, n$). The points of support may be any one of the following four kinds and/or a combination of all of them.

(i) *Fully fixed point of support.* In this case, the constraint conditions at (ξ_i, η_i) are

$$
w\big|_{(\xi_i,\eta_i)} = 0, \quad \varphi_r\big|_{(\xi_i,\eta_i)} = 0, \quad \varphi_\theta\big|_{(\xi_i,\eta_i)} = 0. \tag{6.11}
$$

(ii) *Radially fixed point of support.* In this case, the corresponding conditions at (ξ_i, η_i) are

$$w\Big|_{(\xi_i,\eta_i)} = 0, \quad \varphi_r\Big|_{(\xi_i,\eta_i)} = 0, \quad H_{r\theta}\Big|_{(\xi_i,\eta_i)} = 0. \tag{6.12}$$

(iii) *Tangentially fixed point of support.* In this case, the corresponding conditions at (ξ_i, η_i) are

$$w\Big|_{(\xi_i,\eta_i)} = 0, \quad M_r\Big|_{(\xi_i,\eta_i)} = 0, \quad \varphi_\theta\Big|_{(\xi_i,\eta_i)} = 0. \tag{6.13}$$

(iv) *Free point of support.* In this case, the corresponding conditions at (ξ_i, η_i) are

$$w\Big|_{(\xi_i,\eta_i)} = 0, \quad M_r\Big|_{(\xi_i,\eta_i)} = 0, \quad H_{r\theta}\Big|_{(\xi_i,\eta_i)} = 0. \tag{6.14}$$

Clearly, (ii), (iii), and (iv) are special cases of (i).

When we treat problems of the bending of plates supported at some points, we first remove the supports at (ξ_i, η_i) $(i = 1, 2, \ldots, n)$ and denote the concentrated reacting forces and moments by P_i^*, L_{ri}^* and $L_{\theta i}^*$. According to the method in this paper, we can obtain the general solutions of the bending of ring-shaped plates under the actions of the reactions. Using the constraint conditions at the point (ξ_i, η_i) of support, we shall obtain the equations determining the reacting forces P_i^*, L_{ri}^* and $L_{\theta i}^*$. Since the number of constraint conditions is just equal to the number of unknown reactions, the equations that determine the reactions are sufficient.

References

1. Reissner, E., "The effect of transverse shear deformations on the bending of elastic plates", *J. Appl. Mech. ASME* **67** (1945), A69–A77.
2. Reissner, E., "On bending of elastic plates", *Quart. Appl. Mech.* **5**, 1 (1947), 55.
3. Salerno, V. L. and Goldberg, M. A., "Effect of shear deformations on the bending of rectangular plates", *J. Appl. Mech.* **27** (1960), 54.
4. Koeller, R. C. and Essenburg, F., "Shear deformations in rectangular plates", *Proc. 4th U.S. Nat. Cong. Appl. Mech.* **1** (1962), 555.
5. Cheng Chang-jun, "Bending of plates with a concentrated load", *Appl. Math. Mech.* (English Edition), **2**, 5 (1981), 529–537.
6. Timoshenko, S. and Woinowsky-Krieger, S., *Theory of Plates and Shells*, McGraw-Hill, New York (1959).
7. Yeh, Kai-yuan and Hsin, Chien-yun, "General solutions of bending of ring-shaped elastic thin plates under arbitrary lateral loads and several practical problems", *Journal of Lanzhou University, Natural Sciences, Special Number of Mechanics*, No. 1 (1979), 202–225.

Generalized variational principles

Sub-region generalized variational principles in elastic thin plates

LONG YU-QIU

Tsing Hua University, Beijing, PRC

Abstract. The variational principle in elasticity was stated systematically in [1], [2] and [3]. The variational principle of non-conforming elements in elastic thin plates was discussed in [4] and [5]. This paper is a supplement based on [4].

(1) The multi-region mixed energy generalized variational principle of elastic thin plates is proposed. As a consequence of this, it is possible to consider the generalized variational principle of multi-region potential energy and multiregion complementary energy as special cases in this paper.

(2) In the literature, the number of unknown variables in different regions is usually the same. Now this unitary sub-region system is generalized into an arbitrary mixture of different types of sub-region, i.e., an arbitrary mixture of a three-variable-region, a two-variable-region and single-variable-region.

(3) The transform relation between the sub-region generalized potential energy and sub-region generalized complementary energy contributed in this paper allows the possibility of transforming the functionals of different variational principles directly.

A similar principle in the case of three-dimensional continuum and thick plates was discussed in [6] and [8].

1. Sub-region three-variable generalized mixed variational principle

An elastic thin plate is assumed to be divided into two sub-regions a and b (Fig. 1). The areas of the sub-regions a and b are Ω_a and Ω_b respectively. The external boundaries C_a and C_b of the sub-regions a and b are composed of three parts respectively:

$$C_a = C_{1a} + C_{2a} + C_{3a},$$

$$C_b = C_{1b} + C_{2b} + C_{3b},$$

where C_{1a} and C_{1b} are fixed boundaries (the given values of deflection and normal slope are $\bar{w}$ and $\bar{\psi}_n$ respectively), C_{2a} and C_{2b} are simply supported boundaries (the given values of deflection and normal bending moment are $\bar{w}$ and $\bar{M}_n$ respectively), and C_{3a} and C_{3b} are free boundaries (the given values of normal bending moment and effective shear force are $\bar{M}_n$ and $\bar{V}_n$ respectively).

The corner points A_a and A_b in the external boundaries of sub-regions a and b are composed of two kinds of corner points:

$$A_a = A_{1a} + A_{2a},$$

$$A_b = A_{1b} + A_{2b},$$

Yeh, K. Y. (ed) Progress in Applied Mechanics
 ISBN-13: 978-94-010-8061-3

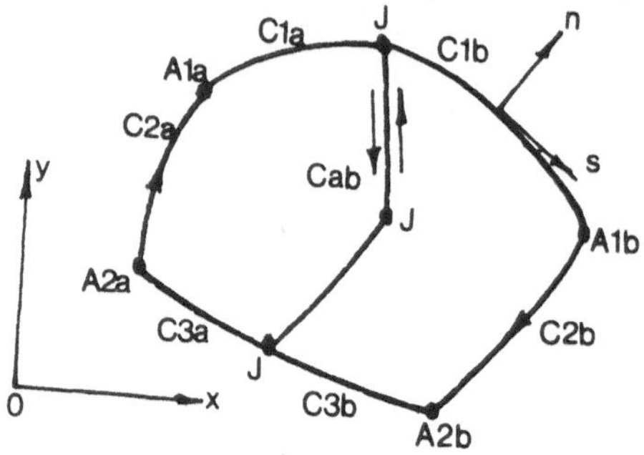

Figure 1

in which A_{1a} and A_{1b} are corner points of given deflection $\bar{w}$, A_{2a} and A_{2b} are corner points of given concentrated force $\bar{R}$.

The common line between two sub-regions is C_{ab}. The nodal points J consist of two kinds of node:

$$J = J_1 + J_2$$

where J_1 are nodes of given deflection $\bar{w}$, and J_2 are nodes of given concentrated force $\bar{R}$. x and y are rectangular coordinates of the middle plane of the plate, and n and s are the external normal and tangential directions respectively. The positive direction of s is shown in Fig. 1.

The main points of the sub-region three-variable generalized mixed variational principle are as follows:

1.1. *Unknown variables*

In sub-regions a and b, there are three kinds of variables respectively:

deflection:

$$w^{(a)}, \; w^{(b)}$$

bending and twisting moments:

$$\{M\}^{(a)} = \lfloor M_x M_y M_{xy} \rfloor^{T(a)}, \qquad \{M\}^{(b)} = \lfloor M_x M_y M_{xy} \rfloor^{T(b)}$$

curvature and twist:

$$\{k\}^{(a)} = \lfloor k_x k_y k_{xy} \rfloor^{T(a)}, \qquad \{k\}^{(b)} = \lfloor k_x k_y k_{xy} \rfloor^{T(b)}$$

These variables can vary independently in the sub-regions and on the boundaries and common lines.

1.2. *Definition of the functional*

Let sub-region a be specified as a potential energy region and sub-region b be specified as a complementary energy region. The functional is defined as

$$\pi_3 = \pi_{3p}^{(a)} - \pi_{3c}^{(b)} + H_{pc} + G_{1pc} + G_{2pc} \tag{1.1}$$

in which, $\pi_{3p}^{(a)}$ is the three-variable generalized potential energy of sub-region a (interline C_{ab} and nodes J are not included):

$$\pi_{3p}^{(a)} = \iint_{\Omega_a}\left[A(\{k\}) - qw - \left(\frac{\partial^2 w}{\partial x^2} + k_x\right)M_x - \left(\frac{\partial^2 w}{\partial y^2} + k_y\right)M_y\right.$$

$$\left. -2\left(\frac{\partial^2 w}{\partial x \partial y} + k_{xy}\right)M_{xy}\right]\mathrm{d}x\,\mathrm{d}y - \int_{C_{1a}+C_{2a}}\left(Q_n + \frac{\partial M_{ns}}{\partial s}\right)(w - \bar{w})\,\mathrm{d}s$$

$$- \int_{C_{3a}} \bar{V}_n w\,\mathrm{d}s + \int_{C_{1a}} M_n\left(\frac{\partial w}{\partial n} - \bar{\psi}_n\right)\mathrm{d}s + \int_{C_{2a}+C_{3a}} \bar{M}_n \frac{\partial w}{\partial n}\,\mathrm{d}s$$

$$- \sum_{A_{1a}} \Delta M_{ns}(w - \bar{w}) - \sum_{A_{2a}} \bar{R}w \tag{1.2}$$

where q is the load intensity, A is the strain energy density

$$A = \frac{D}{2}[(k_x + k_y)^2 + 2(1 - \mu)(k_{xy}^2 - k_x k_y)] \tag{1.3}$$

and D is the bending rigidity of the plate. Also, μ is Poisson's ratio of the material; M_n, M_{ns}, Q_n are normal bending moment, twisting moment and shearing force on the boundary respectively; and ΔM_{ns} is the increment of twisting moment on both sides of the corner point on the boundary.

$\pi_{3c}^{(b)}$ is the three-variable generalized complementary energy of sub-region b (interline C_{ab} and nodes J are not included):

$$\pi_{3c}^{(b)} = \iint_{\Omega_b}\left[M_x k_x + M_y k_y + 2M_{xy}k_{xy} - A(\{k\}) + \left(\frac{\partial^2 M_x}{\partial x^2} + \frac{\partial^2 M_y}{\partial y^2}\right.\right.$$

$$\left.\left. +2\frac{\partial^2 M_{xy}}{\partial x \partial y} + q\right)w\right]\mathrm{d}x\,\mathrm{d}y - \int_{C_{1b}+C_{2b}}\left(Q_n + \frac{\partial M_{ns}}{\partial s}\right)\bar{w}\,\mathrm{d}s$$

$$- \int_{C_{3b}}\left(Q_n + \frac{\partial M_{ns}}{\partial s} - \bar{V}_n\right)w\,\mathrm{d}s + \int_{C_{1b}} M_n \bar{\psi}_n\,\mathrm{d}s$$

$$+ \int_{C_{2b}+C_{3b}} (M_n - \bar{M}_n)\frac{\partial w}{\partial n}\,\mathrm{d}s - \sum_{A_{1b}} \Delta M_{ns}\bar{w} - \sum_{A_{2b}} (\Delta M_{ns} - \bar{R})w. \tag{1.4}$$

Also, H_{pc}, G_{1pc} and G_{2pc} are additional energy items of the common line C_{ab} and nodes J_1, J_2:

$$H_{pc} = \int_{C_{ab}}\left[M_n^{(b)}\left(\frac{\partial w}{\partial n}\right)^{(a)} + \left(Q_n + \frac{\partial M_{ns}}{\partial s}\right)^{(b)} w^{(a)}\right]\mathrm{d}s, \tag{1.5}$$

$$G_{1pc} = \sum_{J_1}[-(\Delta M_{ns})^{(a)}(w^{(a)} - \bar{w}) + (\Delta M_{ns}^{(b)})\bar{w}], \tag{1.6}$$

$$G_{2pc} = \sum_{J_2}[(\Delta M_{ns})^{(b)} - \bar{R}]w^{(a)}. \tag{1.7}$$

1.3. *The stationary condition*

The stationary condition of the functional

$$\delta\pi_3 = \delta\pi_{3p}^{(a)} - \delta\pi_{3c}^{(b)} + \delta H_{pc} + \delta G_{1pc} + \delta G_{2pc} = 0 \tag{1.8}$$

is equivalent to all the field equations, boundary conditions, conjunction conditions on the common line, corner and nodal conditions of the whole sub-region system of the thin plate. They are:

(1) the field equations in Ω_a and Ω_b:

$$\left.\begin{aligned} &M_x = D(k_x + \mu k_y),\ M_y = D(k_y + \mu k_x),\ M_{xy} = D(1-\mu)k_{xy}, \\ &k_x = -\frac{\partial^2 w}{\partial x^2},\quad k_y = -\frac{\partial^2 w}{\partial y^2},\quad k_{xy} = -\frac{\partial^2 w}{\partial x \partial y}, \\ &\frac{\partial^2 M_x}{\partial x^2} + \frac{\partial^2 M_y}{\partial y^2} + 2\frac{\partial^2 M_{xy}}{\partial x \partial y} + q = 0. \end{aligned}\right\} \tag{1.9}$$

(2) the boundary conditions on C_a and C_b:

$$\left.\begin{aligned} &Q_n + \frac{\partial M_{ns}}{\partial s} = \bar{V}_n \quad \text{on } C_{3a} + C_{3b}, \\ &w = \bar{w} \quad \text{on } C_{1a} + C_{2a} + C_{1b} + C_{2b}, \\ &M_n = \bar{M}_n \quad \text{on } C_{2a} + C_{3a} + C_{2b} + C_{3b}, \\ &\frac{\partial w}{\partial n} = \bar{\psi}_n \quad \text{on } C_{1a} + C_{1b}. \end{aligned}\right\} \tag{1.10}$$

(3) the conjunction conditions on the interline C_{ab}:

$$\left.\begin{aligned} &M_n^{(a)} = M_n^{(b)}, \\ &\left(Q_n + \frac{\partial M_{ns}}{\partial s}\right)^{(a)} = -\left(Q_n + \frac{\partial M_{ns}}{\partial s}\right)^{(b)}, \\ &\left(\frac{\partial w}{\partial n}\right)^{(a)} = -\left(\frac{\partial w}{\partial n}\right)^{(b)}, \\ &w^{(a)} = w^{(b)}. \end{aligned}\right\} \tag{1.11}$$

(4) the corner conditions on the boundary:

$$\left.\begin{aligned} &w = \bar{w} \quad \text{at } A_{1a} + A_{1b}, \\ &\Delta M_{ns} = \bar{R} \quad \text{at } A_{2a} + A_{2b}. \end{aligned}\right\} \tag{1.12}$$

(5) the nodal conditions on the common line:

$$\left.\begin{aligned} &w^{(a)} = \bar{w} \quad \text{at } J_1, \\ &w^{(b)} = \bar{w} \quad \text{at } J_1, \\ &w^{(a)} = w^{(b)} \quad \text{at } J_2, \\ &(\Delta M_{ns})^{(a)} + (\Delta M_{ns})^{(b)} = \bar{R} \quad \text{at } J_2. \end{aligned}\right\} \tag{1.13}$$

The derivation of the above equivalent equations from the functional stationary condition (1.8) is shown in Appendix A.

2. The sub-region three-variable generalized principle of potential energy and complementary energy

2.1. *The transform relation between* $\pi_{3p}^{(a)}$ *and* $\pi_{3c}^{(a)}$

Between the three-variable generalized potential energy $\pi_{3p}^{(a)}$ and generalized complementary energy $\pi_{3c}^{(a)}$ of sub-region a (the common line C_{ab} and nodes J are not included) there exists the following transform relation:

$$\pi_{3p}^{(a)} + \pi_{3c}^{(a)} = \int_{C_{ab}} \left[-M_n^{(a)} \left(\frac{\partial w}{\partial n} \right)^{(a)} + \left(Q_n + \frac{\partial M_{ns}}{\partial s} \right)^{(a)} w^{(a)} \right] \mathrm{d}s + \sum_{J_1+J_2} w^{(a)} (\Delta M_{ns})^{(a)}. \tag{2.1}$$

This relation can be proved as follows:
From (1.2) and (1.4) (in (1.4), (a) is replaced by (b)), we obtain

$$\begin{aligned} \pi_{3p}^{(a)} + \pi_{3c}^{(a)} = {} & \iint_{\Omega_a} \left[-\left(\frac{\partial^2 w}{\partial x^2} M_x + \frac{\partial^2 w}{\partial y^2} M_y + 2 \frac{\partial^2 w}{\partial x \partial y} M_{xy} \right) \right. \\ & \left. + \left(\frac{\partial^2 M_x}{\partial x^2} + \frac{\partial^2 M_y}{\partial y^2} + 2 \frac{\partial^2 M_{xy}}{\partial x \partial y} \right) w \right] \mathrm{d}x \, \mathrm{d}y \\ & - \int_{C_{1a}+C_{2a}+C_{3a}} \left[\left(Q_n + \frac{\partial M_{ns}}{\partial s} \right) w - M_n \frac{\partial w}{\partial n} \right] \mathrm{d}s \\ & - \sum_{A_{1a}+A_{2a}} (\Delta M_{ns}) w, \end{aligned} \tag{2.2}$$

with integration by parts, we have

$$\begin{aligned} & \iint_{\Omega_a} \left(\frac{\partial^2 M_x}{\partial x^2} + \frac{\partial^2 M_y}{\partial y^2} + 2 \frac{\partial^2 M_{xy}}{\partial x \partial y} \right) w \, \mathrm{d}x \, \mathrm{d}y \\ & = \iint_{\Omega_a} \left(M_x \frac{\partial^2 w}{\partial x^2} + M_y \frac{\partial^2 w}{\partial y^2} + 2 M_{xy} \frac{\partial^2 w}{\partial x \partial y} \right) \mathrm{d}x \, \mathrm{d}y \\ & - \int_{C_{1a}+C_{2a}+C_{3a}+C_{ab}} \left[M_n \frac{\partial w}{\partial n} - \left(Q_n + \frac{\partial M_{ns}}{\partial s} \right) w \right] \mathrm{d}s \\ & + \sum_{A_{1a}+A_{2a}+J_1+J_2} w (\Delta M_{ns}). \end{aligned} \tag{2.3}$$

Substituting (2.3) into (2.2), (2.1) is then obtained.

If the whole region is not divided into sub-regions, then C_{ab}, J_1 and J_2 do not exist, and then we have

$$\pi_{3p}^{(a)} + \pi_{3c}^{(a)} = 0. \tag{2.4}$$

2.2. *Sub-region three-variable generalized potential energy principle*

In the functional expression (1.1) of the sub-region three-variable generalized mixed variational principle, sub-region a is specified as a potential energy region and sub-region b is specified as a complementary energy region. Now let sub-region b be specified as a potential energy region. Then from (2.1), we have

$$\pi_{3c}^{(b)} = -\pi_{3p}^{(b)} + \int_{C_{ab}} \left[-M_n^{(b)} \left(\frac{\partial w}{\partial n}\right)^{(b)} + \left(Q_n + \frac{\partial M_{ns}}{\partial s}\right)^{(b)} w^{(b)} \right] \mathrm{d}s + \sum_{J_1+J_2} w^{(b)} (\Delta M_{ns})^{(b)}.$$

Substituting this into (1.1), we obtain

$$\pi_3 = \pi_{3p}^{(a)} + \pi_{3p}^{(b)} + H_{pp} + G_{1pp} + G_{2pp}, \tag{2.5}$$

where H_{pp}, G_{1pp} and G_{2pp} are additional potential energy items of the common line C_{ab} and nodes J_1, J_2:

$$H_{pp} = \int_{C_{ab}} \left\{ M_n^{(b)} \left[\left(\frac{\partial w}{\partial n}\right)^{(a)} + \left(\frac{\partial w}{\partial n}\right)^{(b)} \right] + \left(Q_n + \frac{\partial M_{ns}}{\partial s}\right)^{(b)} (w^{(a)} - w^{(b)}) \right\} \mathrm{d}s, \tag{2.6a}$$

$$G_{1pp} = -\sum_{J_1} [(\Delta M_{ns})^{(a)} (w^{(a)} - \bar{w}) + (\Delta M_{ns})^{(b)} (w^{(b)} - \bar{w})], \tag{2.7}$$

$$G_{2pp} = \sum_{J_2} [(\Delta M_{ns})^{(b)} (w^{(a)} - w^{(b)}) - \bar{R} w^{(a)}]. \tag{2.8a}$$

Equations (2.5), (2.6a), (2.7) and (2.8a) define the functional of sub-region three-variable generalized potential energy principle. It can be proved that the stationary condition of this functional is equivalent to all the field equations, boundary conditions, conjunction conditions on the common line, corner and nodal conditions of the whole sub-region system of the thin plate.

If a and b in (2.6a) and (2.8a) are interchanged, then we obtain another expression for H_{pp} and G_{2pp}:

$$H_{pp} = \int_{C_{ab}} \left\{ M_n^{(a)} \left[\left(\frac{\partial w}{\partial n}\right)^{(a)} + \left(\frac{\partial w}{\partial n}\right)^{(b)} \right] + \left(Q_n + \frac{\partial M_{ns}}{\partial s}\right)^{(a)} (w^{(b)} - w^{(a)}) \right\} \mathrm{d}s, \tag{2.6b}$$

$$G_{2pp} = \sum_{J_2} [(\Delta M_{ns})^{(a)} (w^{(b)} - w^{(a)}) - \bar{R} w^{(b)}]. \tag{2.8b}$$

If the displacement conjunction conditions on the common line C_{ab} and nodes J_1, J_2 are already satisfied, then from (2.6), (2.7) and (2.8), we obtain

$$H_{pp} = 0,$$

$$G_{1pp} = 0,$$

$$G_{2pp} = -\sum_{J_2} \bar{R} w^{(a)} \quad \text{or} \quad -\sum_{J_2} \bar{R} w^{(b)}.$$

2.3. *Sub-region three-variable generalized complementary energy principle*

Now let both sub-regions a and b be specified as complementary energy region. Then substituting (2.1) into (1.1), we obtain

$$\pi_3 = -\pi_{3c}^{(a)} - \pi_{3c}^{(b)} + H_{cc} + G_{1cc} + G_{2cc}, \tag{2.9}$$

where H_{cc}, G_{1cc} and G_{2cc} are additional complementary energy items of the common line C_{ab} and nodes J_1, J_2:

$$H_{cc} = \int_{C_{ab}} \left\{ (M_n^{(b)} - M_n^{(a)}) \left(\frac{\partial w}{\partial n}\right)^{(a)} + \left[\left(Q_n + \frac{\partial M_{ns}}{\partial s}\right)^{(b)} + \left(Q_n + \frac{\partial M_{ns}}{\partial s}\right)^{(a)} \right] w^{(a)} \right\} \mathrm{d}s, \tag{2.10a}$$

$$G_{1cc} = \sum_{J_1} [(\Delta M_{ns})^{(b)} + (\Delta M_{ns})^{(a)}] \bar{w}, \tag{2.11}$$

$$G_{2cc} = \sum_{J_2} \{[(\Delta M_{ns})^{(b)} + (\Delta M_{ns})^{(a)} - \bar{R}] w^{(a)}\}. \tag{2.12a}$$

Equations (2.9), (2.10a), (2.11) and (2.12a) define the functional of the sub-region three-variable generalized complementary energy principle. If a and b in equations (2.10a) and (2.12a) are interchanged, then we obtain another expression for H_{cc} and G_{2cc}:

$$H_{cc} = \int_{C_{ab}} \left\{ (M_n^{(a)} - M_n^{(b)}) \left(\frac{\partial w}{\partial n}\right)^{(b)} + \left[\left(Q_n + \frac{\partial M_{ns}}{\partial s}\right)^{(a)} + \left(Q_n + \frac{\partial M_{ns}}{\partial s}\right)^{(b)} \right] w^{(b)} \right\} \mathrm{d}s, \tag{2.10b}$$

$$G_{2cc} = \sum_{J_2} \{[(\Delta M_{ns})^{(b)} + (\Delta M_{ns})^{(a)} - \bar{R}] w^{(b)}\}. \tag{2.12b}$$

If the stress conjunction conditions on the common line C_{ab} and nodes J_1, J_2 are already satisfied, then from (2.10) and (2.12) we have

$$H_{cc} = 0, \qquad G_{2cc} = 0.$$

3. Sub-region two-variable and single-variable generalized variational principles

3.1. *Sub-region two-variable generalized variational principle*

Using the following relation between strain energy density $A(\{k\})$ and complementary strain energy density $B(\{M\})$:

$$B(\{M\}) = M_x k_x + M_y k_y + 2M_{xy} k_{xy} - A(\{k\}) \tag{3.1}$$

we can eliminate the variable $\{k\}$ in three-variable generalized potential energy $\pi_{3p}^{(a)}$ and generalized complementary energy $\pi_{3c}^{(a)}$ of sub-region a and consequently we may obtain two-variable (displacement w, internal moment $\{M\}$) generalized potential energy $\pi_{2p}^{(a)}$ and generalized complementary energy $\pi_{2c}^{(a)}$ of sub-region a as follows:

$$
\begin{aligned}
\pi_{2p}^{(a)} = & \iint_{\Omega_a} \left[-\left(\frac{\partial^2 w}{\partial x^2} M_x + \frac{\partial^2 w}{\partial y^2} M_y + 2 \frac{\partial^2 w}{\partial x \partial y} M_{xy} \right) \right. \\
& \left. - B(\{M\}) - qw \right] \mathrm{d}x\, \mathrm{d}y - \int_{C_{1a}+C_{2a}} \left(Q_n + \frac{\partial M_{ns}}{\partial s} \right) (w - \bar{w})\, \mathrm{d}s \\
& - \int_{C_{3a}} \bar{V}_n w\, \mathrm{d}s + \int_{C_{1a}} M_n \left(\frac{\partial w}{\partial n} - \bar{\psi}_n \right) \mathrm{d}s + \int_{C_{2a}+C_{3a}} \bar{M}_n \frac{\partial w}{\partial n} \mathrm{d}s \\
& - \sum_{A_{1a}} \Delta M_{ns}(w - \bar{w}) - \sum_{A_{2a}} \bar{R} w,
\end{aligned} \tag{3.2}
$$

$$
\begin{aligned}
\pi_{2c}^{(a)} = & \iint_{\Omega_a} \left[B(\{M\}) + \left(\frac{\partial^2 M_x}{\partial x^2} + \frac{\partial^2 M_y}{\partial y^2} + 2 \frac{\partial^2 M_{xy}}{\partial x \partial y} + q \right) w \right] \mathrm{d}x\, \mathrm{d}y \\
& - \int_{C_{1a}+C_{2a}} \left(Q_n + \frac{\partial M_{ns}}{\partial s} \right) \bar{w}\, \mathrm{d}s - \int_{C_{3a}} \left(Q_n + \frac{\partial M_{ns}}{\partial s} - \bar{V}_n \right) w\, \mathrm{d}s \\
& + \int_{C_{1a}} M_n \bar{\psi}_n\, \mathrm{d}s + \int_{C_{2a}+C_{3a}} (M_n - \bar{M}_n) \frac{\partial w}{\partial n} \mathrm{d}s \\
& - \sum_{A_{1a}} \Delta M_{ns} \bar{w} - \sum_{A_{2a}} (\Delta M_{ns} - \bar{R}) w.
\end{aligned} \tag{3.3}
$$

According to (1.1), (2.5), (2.9), the functionals of sub-region two-variable generalized mixed, potential and complementary energy principles can be expressed as follows:

$$\pi_2 = \pi_{2p}^{(a)} - \pi_{2c}^{(b)} + H_{pc} + G_{1pc} + G_{2pc}, \tag{3.4}$$

$$\pi_2 = \pi_{2p}^{(a)} + \pi_{2p}^{(b)} + H_{pp} + G_{1pp} + G_{2pp}, \tag{3.5}$$

$$\pi_2 = -\pi_{2c}^{(a)} - \pi_{2c}^{(b)} + H_{cc} + G_{1cc} + G_{2cc}, \tag{3.6}$$

where H_{pc}, H_{pp} and H_{cc} are still given by (1.5), (2.6) and (2.10). G_{1pc}, G_{2pc}, G_{1pp}, G_{2pp}, G_{1cc}, G_{2cc} are still given by (1.6), (1.7), (2.7), (2.8), (2.11) and (2.12).

3.2. *Sub-region single-variable variational principle*

Let us consider the case where each sub-region is specified as a single-variable region.

If sub-region a is specified as a potential energy region, then the displacement w is taken as the independent variable. In this case, $\pi_{3p}^{(a)}$ of (1.2) or $\pi_{2p}^{(a)}$ of (3.2) transforms into the single variable potential energy $\pi_{1p}^{(a)}$ of sub-region a as follows:

$$
\begin{aligned}
\pi_{1p}^{(a)} = & \iint_{\Omega_a} [A(w) - qw]\, \mathrm{d}x\, \mathrm{d}y - \int_{C_{1a}+C_{2a}} \left(Q_n + \frac{\partial M_{ns}}{\partial s} \right) (w - \bar{w})\, \mathrm{d}s \\
& - \int_{C_{3a}} \bar{V}_n w\, \mathrm{d}s + \int_{C_{1a}} M_n \left(\frac{\partial w}{\partial n} - \bar{\psi}_n \right) \mathrm{d}s + \int_{C_{2a}+C_{3a}} \bar{M}_n \frac{\partial w}{\partial n} \mathrm{d}s \\
& - \sum_{A_{1a}} \Delta M_{ns}(w - \bar{w}) - \sum_{A_{2n}} \bar{R} w,
\end{aligned} \tag{3.7a}
$$

where $[Q_n + (\partial M_{ns}/\partial s)]$, M_n and ΔM_{ns} can be expressed as functions of displacement w, or can be regarded as Lagrange multipliers at the boundary or the corner points. $A(w)$ is the strain energy density expressed with displacement w:

$$A = \frac{D}{2}\left\{\left(\frac{\partial^2 w}{\partial x^2} + \frac{\partial^2 w}{\partial y^2}\right)^2 + 2(1-\mu)\left[\left(\frac{\partial^2 w}{\partial x \partial y}\right)^2 - \frac{\partial^2 w}{\partial x^2}\frac{\partial^2 w}{\partial y^2}\right]\right\}.$$

If displacement w already satisfies the geometric boundary and corner conditions, then we have

$$\pi_{1p}^{(a)} = \iint_{\Omega_a} [A(w) - qw]\, dx\, dy - \int_{C_{3a}} \bar{V}_n w\, ds + \int_{C_{2a}+C_{3a}} \bar{M}_n \frac{\partial w}{\partial n} ds - \sum_{A_{2a}} \bar{R}w. \tag{3.7b}$$

If sub-region a is specified as a complementary energy region, then the internal moment $\{M\}$ is taken as the independent variable. Moreover, $\{M\}$ must satisfy the equilibrium differential equation:

$$\frac{\partial^2 M_x}{\partial x^2} + \frac{\partial^2 M_y}{\partial y^2} + 2\frac{\partial^2 M_{xy}}{\partial x \partial y} + q = 0.$$

In this case, $\pi_{3c}^{(a)}$ of (1.4) or $\pi_{2c}^{(a)}$ of (3.3) transforms into single-variable complementary energy as follows:

$$\begin{aligned}\pi_{1c}^{(a)} = &\iint_{\Omega_a} B(\{M\})\, dx\, dy - \int_{C_{1a}+C_{2a}} \left(Q_n + \frac{\partial M_{ns}}{\partial s}\right)\bar{w}\, ds \\ &- \int_{C_{3a}} \left(Q_n + \frac{\partial M_{ns}}{\partial s} - \bar{V}_n\right) w\, ds + \int_{C_{1a}} M_n \bar{\psi}_n\, ds \\ &+ \int_{C_{2a}+C_{3a}} (M_n - \bar{M}_n)\frac{\partial w}{\partial n} ds - \sum_{A_{1a}} \Delta M_{ns}\bar{w} - \sum_{A_{2a}} (\Delta M_{ns} - \bar{R})w,\end{aligned} \tag{3.8a}$$

where w and $\partial w/\partial n$ can be regarded as Lagrange multipliers at the boundary or corner points.

If $\{M\}$ still satisfies the stress boundary and corner conditions, then we have

$$\begin{aligned}\pi_{1c}^{(a)} = &\iint_{\Omega_a} B(\{M\})\, dx\, dy - \int_{C_{1a}+C_{2a}} \left(Q_n + \frac{\partial M_{ns}}{\partial s}\right)\bar{w}\, ds \\ &+ \int_{C_{1a}} M_n \bar{\psi}_n\, ds - \sum_{A_{1a}} \Delta M_{ns}\bar{w}.\end{aligned} \tag{3.8b}$$

According to equations (1.1), (2.5), (2.9) or equations (3.4), (3.5), (3.6), the functionals of sub-region single-variable mixed, potential and complementary energy principles can be expressed as follows:

$$\pi_1 = \pi_{1p}^{(a)} - \pi_{1c}^{(b)} + H_{pc} + G_{1pc} + G_{2pc}, \tag{3.9}$$

$$\pi_1 = \pi_{1p}^{(a)} + \pi_{1p}^{(b)} + H_{pp} + G_{1pp} + G_{2pp}, \tag{3.10}$$

$$\pi_1 = -\pi_{1c}^{(a)} - \pi_{1c}^{(b)} + H_{cc} + G_{1cc} + G_{2cc}, \tag{3.11}$$

where the expressions of H_{pc}, G_{1pc}, G_{2pc}, H_{pp}, G_{1pp}, G_{2pp}, H_{cc}, G_{1cc}, G_{2cc} are the same as above.

4. The general form of sub-region generalized variational principle

From the above discussion, the general form of the sub-region generalized variational principle of elastic thin plates can be established.

The elastic thin plate is assumed to be divided into several sub-regions. Each sub-region may be specified arbitrarily as a single-variable, two-variable or three-variable potential energy region (as regions Ω_{p1}, Ω_{p2}, Ω_{p3} in Fig. 2) or complementary energy region (as regions Ω_{c1}, Ω_{c2}, Ω_{c3} in Fig. 2).

There are three kinds of common line of the neighbouring sub-regions, i.e., C_{pc}, C_{pp}, C_{cc} (Fig. 2): on one side of C_{pc} is the potential energy region, on the other side is the complementary energy region. Both sides of C_{pp} are the potential energy regions and both sides of C_{cc} are the complementary energy regions.

There are two kinds of nodes of the neighbouring sub-regions, i.e., J_1 and J_2 (Fig. 2): J_1 are nodes with given displacement w, and J_2 are nodes with given concentrated force $\bar{R}$. Around node J there are γ_p potential energy elements e_p, and γ_c complementary energy elements e_c.

The general form of the functional of sub-region generalized variational principle may be given as follows:

$$\pi = \sum_{\Omega_p} \pi_p - \sum_{\Omega_c} \pi_c + \sum_{C_{pc}} H_{pc} + \sum_{C_{pp}} H_{pp} + \sum_{C_{cc}} H_{cc} + \sum_{J_1} G_1 + \sum_{J_2} G_2. \quad (4.1)$$

The terms on the right part of (4.1) are described as follows:

The first term is the sum of the potential energy or generalized potential energy π_p of each potential energy region Ω_p. π_p may be π_{1p}, π_{2p} or π_{3p} which are given by (3.7), (3.2), and (1.2) respectively.

The second term is the sum of the complementary energy or generalized complementary energy π_c of each complementary energy region Ω_c. π_c may be π_{1c}, π_{2c} or π_{3c} which are given by (3.8), (3.3), and (1.4) respectively.

The third term represents the sum of the additional mixed energy item H_{pc} on the common line C_{pc}. H_{pc} is given by (1.5).

The fourth term represents the sum of the additional potential energy item H_{pp} on the common line C_{pp}. H_{pp} is given by (2.6).

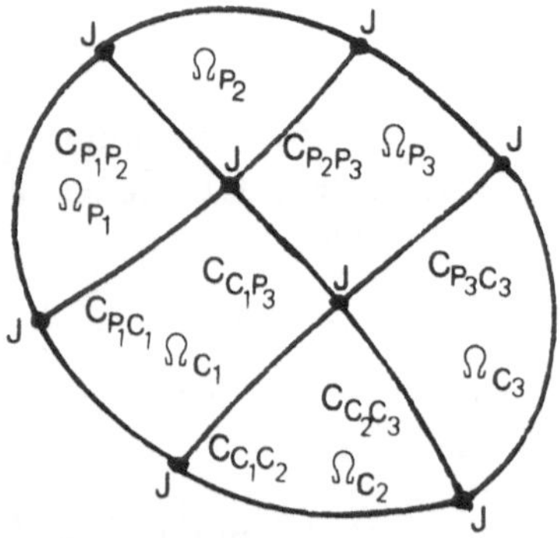

Figure 2

The fifth term represents the sum of the additional complementary energy item H_{cc} on the common line C_{cc}. H_{cc} is given by (2.10).

The sixth term represents the sum of the additional energy item G_1 at node J_1 with given displacement. G_1 is given by the following equation:

$$G_1 = -\sum_{e_p} (\Delta M_{ms})^{(e_p)}(w^{(e_p)} - \bar{w}) + \sum_{e_c} (\Delta M_{ns})^{(e_c)}\bar{w}. \tag{4.2}$$

On the right side of the above equation, the first term represents the sum for all potential energy elements e_p around node J_1; the second term represents the sum for all complementary energy elements e_c around node J_1.

The seventh term represents the sum of additional energy item G_2 at node J_2 with a given concentrated force. G_2 is given by the following equation:

$$G_2 = \left[\sum_{e} (\Delta M_{ns})^{(e)} - \bar{R}\right] w^{(a)} - \sum_{e_p} (\Delta M_{ns})^{(e_p)} w^{(e_p)}. \tag{4.3}$$

On the right side of the above equation, Σ_e in the first term represents the sum for all elements e (including all e_p and all e_c) around node J_2; Σ_{e_p} in the second term represents the sum for all potential energy elements e_p around node J_2.

G_{1pc} of (1.6), G_{1pp} of (2.7) and G_{1cc} of (2.11) are special forms of G_1 of (4.2). G_{2pc} of (1.7), G_{2pp} of (2.8) and G_{2cc} of (2.12) are special forms of G_2 of (4.3).

It can be proved that the stationary condition of functional π in (4.1)

$$\delta\pi = 0 \tag{4.4}$$

is equivalent to all the field equations, boundary conditions, conjunction conditions on the common line, corner and nodal conditions of the multi-region system of the thin plate. The derivation of nodal condition for node J from the stationary condition (4.4) is shown in Appendix B.

If all sub-regions are specified as potential energy regions, the functional of the principle of sub-region potential energy or sub-region generalized potential energy can be obtained from (4.1) as follows:

$$\pi = \sum_{\Omega_p} \pi_p + \sum_{C_{pp}} H_{pp} + \sum_{J_1} G_{1pp} + \sum_{J_2} G_{2pp}, \tag{4.5}$$

where G_{1pp} and G_{2pp} can be obtained from (4.2), (4.3) as follows:

$$G_{1pp} = -\sum_{e_p} (\Delta M_{ns})^{(e_p)}(w^{(e_p)} - \bar{w}), \tag{4.6}$$

$$\begin{aligned} G_{2pp} &= \left[\sum_{e_p} (\Delta M_{ns})^{(e_p)} - \bar{R}\right] w^{(a)} - \sum_{e_p} (\Delta M_{ns})^{(e_p)} w^{(e_p)} \\ &= -\sum_{e_p} (\Delta M_{ns})^{(e_p)}(w^{(e_p)} - w^{(a)}) - \bar{R}w^{(a)}. \end{aligned} \tag{4.7}$$

Equations (2.5), (3.5) and (3.10) are special forms of (4.5). In the special case, of which all sub-regions are specified as single-variable regions and the term π_p in (4.5) is specified as π_{1p} in (3.7a), then from (4.5) we have

$$\pi = \sum_{\Omega_p} \pi_{1p} + \sum_{C_{pp}} H_{pp} + \sum_{J_1} G_{1pp} + \sum_{J_2} G_{2pp}. \tag{4.8}$$

The functional π defined by (4.8) is equivalent to the functional π^{**}_{IIIA} defined by (11.6) in [4].

If all sub-regions are specified as complementary energy regions, the functional of the principle of sub-region complementary energy or sub-region generalized complementary energy can be obtained from (4.1) as follows:

$$\pi = -\sum_{\Omega_c} \pi_c + \sum_{C_{cc}} H_{cc} + \sum_{J_1} G_{1cc} + \sum_{J_2} G_{2cc}, \tag{4.9}$$

where

$$G_{1cc} = \sum_{e_c} (\Delta M_{ns})^{(e_c)} \bar{w}, \tag{4.10}$$

$$G_{2cc} = \left[\sum_{e_c} (\Delta M_{ns})^{(e_c)} - \bar{R}\right] w^{(a)}. \tag{4.11}$$

Equations (2.9), (3.6) and (3.11) are special forms of (4.9). In the special case, of which all sub-regions are specified as two-variable regions, and the term π_c in (4.9) is specified as π_{2c} in (3.3), then from (4.9) we have

$$\pi = -\sum_{\Omega_c} \pi_{2c} + \sum_{C_{cc}} H_{cc} + \sum_{J_1} G_{1cc} + \sum_{J_2} G_{2cc}. \tag{4.12}$$

The functional π defined by (4.12) is equivalent to the functional π^{**}_{IVf} defined by (14.3) in [4].

Appendix A. The equivalent equation of the stationary condition (1.8)

In order to obtain the equivalent equation of the stationary condition (1.8), the expression of functional variation $\delta\pi_3$ is derived first. In the derivation, the formula (2.3) for integration by parts is used.

The double integral terms in $\delta\pi_3$ are

$$\iint_{\Omega_a+\Omega_b} \{[D(k_x + \mu k_y) - M_x]\delta k_x + [D(k_y + \mu k_x) - M_y]\delta k_y$$
$$+ 2[D(1-\mu)k_{xy} - M_{xy}]\delta k_{xy}\}\,\mathrm{d}x\,\mathrm{d}y - \iint_{\Omega_a+\Omega_b}\left[\left(\frac{\partial^2 w}{\partial x^2} + k_x\right)\delta M_x\right.$$
$$\left. + \left(\frac{\partial^2 w}{\partial y^2} + k_y\right)\delta M_y + 2\left(\frac{\partial^2 w}{\partial x \partial y} + k_{xy}\right)\delta M_{xy}\right]\mathrm{d}x\,\mathrm{d}y$$
$$- \iint_{\Omega_a+\Omega_b}\left(\frac{\partial^2 M_x}{\partial x^2} + \frac{\partial^2 M_y}{\partial y^2} + 2\frac{\partial^2 M_{xy}}{\partial x \partial y} + q\right)\delta w\,\mathrm{d}x\,\mathrm{d}y.$$

The line integral terms in $\delta\pi_3$ are

$$\int_{C_{3a}+C_{3b}}\left(Q_n + \frac{\partial M_{ns}}{\partial s} - \bar{V}_n\right)\delta w\,\mathrm{d}s$$

$$- \int_{C_{1a}+C_{1b}+C_{2a}+C_{2b}} (w - \bar{w})\delta\left(Q_n + \frac{\partial M_{ns}}{\partial s}\right)\mathrm{d}s$$

$$- \int_{C_{2a}+C_{2b}+C_{3a}+C_{3b}} (M_n - \bar{M}_n) \frac{\partial \delta w}{\partial n} \, ds + \int_{C_{1a}+C_{1b}} \left(\frac{\partial w}{\partial n} - \bar{\psi}_n \right) \delta M_n \, ds$$

$$+ \int_{C_{ab}} \left\{ (M_n^{(b)} - M_n^{(a)}) \frac{\partial \delta w^{(a)}}{\partial n} + \left[\left(Q_n + \frac{\partial M_{ns}}{\partial s} \right)^{(b)} + \left(Q_n + \frac{\partial M_{ns}}{\partial s} \right)^{(a)} \right] \delta w^{(a)} \right.$$

$$\left. + \left[\left(\frac{\partial w}{\partial n} \right)^{(b)} + \left(\frac{\partial w}{\partial n} \right)^{(a)} \right] \delta M_n^{(b)} - (w^{(b)} - w^{(a)}) \delta \left(Q_n + \frac{\partial M_{ns}}{\partial s} \right)^{(b)} \right\} ds.$$

The corner and nodal terms in $\delta\pi_3$ are

$$- \sum_{A_{1a}+A_{1b}} (w - \bar{w}) \Delta\delta M_{ns} + \sum_{A_{2a}+A_{2b}} (\Delta M_{ns} - \bar{R}) \delta w$$

$$- \sum_{J_1} [(w^{(b)} - \bar{w})(\Delta\delta M_{ns})^{(b)} + (w^{(a)} - \bar{w})(\Delta\delta M_{ns})^{(a)}]$$

$$+ \sum_{J_2} \{[(\Delta M_{ns})^{(a)} + (\Delta M_{ns})^{(b)} - \bar{R}] \delta w^{(a)} - (w^{(b)} - w^{(a)})(\Delta\delta M_{ns})^{(b)}\}.$$

Since all independent variables can vary arbitrarily, it follows from the stationary condition (1.8) that the integral terms and the corner nodal terms in $\delta\pi_3$ must each be zero respectively. From the condition in which the double integral terms are zero, the field (1.9) in Ω_a and Ω_b are then obtained. From the condition in which the line integral terms are zero, the boundary condition (1.10) on boundaries C_a and C_b and the conjunction condition (1.11) on the common line C_{ab} are then obtained. From the condition in which the corner and nodal terms are zero, the corner condition (1.12) on the boundary, and the nodal condition (1.13) on the common line are then obtained.

Appendix B. Derivation of nodal condition from stationary condition (4.4)

Consider two different cases for nodes J_1 and J_2, and from the stationary condition (4.4), $\delta\pi = 0$, the corresponding nodal conditions are derived respectively.

First, consider the case of supported nodes J_1. The corresponding nodal terms for node J_1 in $\delta\pi$ are composed of three terms:

(1) In the potential energy variation $\Sigma_{e_p} \delta\pi_p^{(e_p)}$ for all potential energy elements e_p around node J_1, the corresponding nodal terms are $\Sigma_{e_p} (\Delta M_{ns})^{(e_p)} \delta w^{(e_p)}$.

(2) In the complementary energy variation $-\Sigma_{e_c} \delta\pi_c^{(e_c)}$ for all complementary energy elements e_c around node J_1, the corresponding nodal terms are $-\Sigma_{e_c} w^{(e_c)} (\Delta\delta M_{ns})^{(e_c)}$.

(3) From δG_1, the corresponding nodal terms are

$$- \sum_{e_p} [(w^{(e_p)} - \bar{w})(\Delta\delta M_{ns})^{(e_p)} + (\Delta M_{ns})^{(e_p)} \delta w^{(e_p)}] + \sum_{e_c} \bar{w} (\Delta\delta M_{ns})^{(e_c)}.$$

Adding up all terms, we obtain the corresponding nodal terms for node J_1 in $\delta\pi$ as follows:

$$- \sum_{e} (w^{(e)} - \bar{w})(\Delta\delta M_{ns})^{(e)},$$

which represents the sum for all elements e around node J_1. Hence, from the stationary condition $\delta\pi = 0$, the following conjunction conditions at node J_1 are obtained:

$$w^{(e)} = \bar{w} \quad \text{(for each element } e \text{ around node } J_1\text{)}.$$

Secondly, we consider the case of unsupported nodes J_2. The corresponding nodal terms for node J_2 in $\delta\pi$ are also composed of three terms:

(1) In the potential energy variation $\Sigma_{e_p} \delta\pi_p^{(e_p)}$ for all potential energy elements e_p around node J_2, the corresponding nodal terms are $\Sigma_{e_p} (\Delta M_{ns})^{(e_p)} \delta w^{(e_p)}$.

(2) In the complementary energy variation $-\Sigma_{e_c} \delta\pi_c^{(e_c)}$ for all complementary energy elements e_c around node J_2, the corresponding nodal terms are $-\Sigma_{e_c} w^{(e_c)} (\Delta\delta M_{ns})^{(e_c)}$.

(3) From δG_2, the corresponding nodal terms are

$$\left[\sum_e (\Delta M_{ns})^{(e)} - \bar{R}\right] \delta w^{(a)} + w^{(a)} \sum_e (\Delta\delta M_{ns})^{(e)}$$
$$- \sum_{e_p} [w^{(e_p)} (\Delta\delta M_{ns})^{(e_p)} + (\Delta M_{ns})^{(e_p)} \delta w^{(e_p)}].$$

Adding up all terms, we obtain the corresponding nodal terms for node J_2 in $\delta\pi$ as follows:

$$\left[\sum_e (\Delta M_{ns})^{(e)} - \bar{R}\right] \delta w^{(a)} - \sum_e (w^{(e)} - w^{(a)}) (\Delta\delta M_{ns})^{(e)}.$$

Hence, from the stationary condition $\delta\pi = 0$, the following conjunction conditions at node J_2 are obtained:

$$\sum_e (\Delta M_{ns})^{(e)} - \bar{R} = 0 \quad \text{(for node } J_2\text{)},$$
$$w^{(e)} = w^{(a)} \quad \text{(for each element } e \text{ around node } J_2,\ e \neq a\text{)}.$$

References

1. Chien Wei-zang, *Variational Methods and Finite Element*, Vol. 1 (in Chinese), Science Press, Beijing (1980).
2. Hu Hai-chang, *Variational Principles in Elasticity and Their Application* (in Chinese), Science Press, Beijing (1981).
3. Washizu, K., *Variational Methods in Elasticity and Plasticity*, Pergamon Press (1975).
4. Chien Wei-zang, "Non-conforming element and generalized variational principle" (in Chinese), *Symposium on Finite Element Methods*, Hefei (1981).
5. Prager, W., "Variational principles of elastic plates with relaxed continuity requirements", *Int. J. Solids Structures* **4** (1968), 837–844.
6. Long Yu-qiu, Zhi Bing-chen and Shan Jian, "Sub-region, sub-item and sub-layer generalized variational principles in elasticity", *Proceedings of the International Conference on Finite Element Methods*, Shanghai, China (2–6 August, 1982), 607–609.
7. Long Yu-qiu, Zhi Bing-chen, Kuang Wen-qi and Shan Jian, "Sub-region mixed finite element method for the calculation of stress intensity factor", *Proceedings of the International Conference on Finite Element Methods*, Shanghai, China (2–6 August, 1982), 738–740.
8. Long Yu-qiu, "Sub-region generalized variational principles for elastic thick plates" (English Edition), *Applied Mathematics and Mechanics* **4**, 2 (Apr. 1983), 175–184.

General variational principles for non-conservative problems in theory of elasticity and its application to finite element analysis

LIU DIAN-KUI and ZHANG QI-HAO

Institute of Engineering Mechanics, Harbin, PRC

Abstract. This study develops the general quasi-variational principles for non-conservative problems in the theory of elasticity such as the quasi-potential energy principle, the quasi-complementary energy principle, the generalized quasi-variational principle and the quasi-Hamilton principle. The application of these quasi-variational principles to finite element analysis is also discussed and illustrated with some examples.

In the theory of elasticity, the variational principles are very important for solving boundary problems. It is well known that variational principles are equivalent to differential equations with given boundary conditions. In this field there were many important contributions in the chinese literature [1–4]. Yet, variational principles for non-conservative systems have not been well investigated, although this problem has been brought out by Leipholz [5, 6] in the seventies.

Forces are conservative if the components are negative partial derivatives of a scalar function (potential). An elastic system is called non-conservative in cases where some of the forces acting on the system are not conservative. For example, the Beck's rod is such a non-conservative elastic system. A follower force is defined as a force which is dependent on the deformation of the elastic body. In this paper, follower forces are used instead of non-conservative forces. Hence, the problem of investigating the variational principles for non-conservative forces can be converted to that of investigating the variational principles for follower forces. If the "follower forces" are constants, the variational problems are reduced classical problems. But the "follower forces" do not always possess a potential, so the variational principles would not appear in the form of a stational value of the functional in this case. The name of quasi-variational was adopted by Oden [7]. For this reason, the following principles may be called the quasi-variational principles.

1. Quasi-potential energy principle and quasi-complementary energy principle

In this study, we are limited to static problems only. These principles are provec in three dimensional problems of elastic theory with small displacements. For convenience, the tensor notation will be used:

Yeh, K. Y. (ed) Progress in Applied Mechanics
 ISBN-13: 978-94-010-8061-3

x_i three dimensional Cartesian coordinates;
u_i displacement components;
$\mathbf{e}_{ij}$ strain tensor;
$\boldsymbol{\sigma}_{ij}$ stress tensor;
F_i, p_i follower body forces and follower surface forces respectively;
ϱ density.

The relation of the strain tensor and the displacements in the case of small displacements are as follows:

$$\mathbf{e}_{ij} = \tfrac{1}{2}(u_{i,j} + u_{j,i}) \tag{1.1}$$

and Hooke's law may be written as

$$\boldsymbol{\sigma}_{ij} = a_{ijkl}\mathbf{e}_{kl} \quad (i, j = 1, 2, 3), \tag{1.2}$$

or

$$\mathbf{e}_{ij} = b_{ijkl}\boldsymbol{\sigma}_{kl} \quad (i, j = 1, 2, 3). \tag{1.3}$$

In any body, the stresses and the follower forces must satisfy the following equation:

$$\boldsymbol{\sigma}_{ij,j} + F_i = 0. \tag{1.4}$$

To solve this elastic problem, some kind of boundary condition must be given. Two cases of boundary conditions should be considered: one is a balance type and another is continuous type, i.e., the follower surface forces $\bar{p}_i$ are given on s_p and the surface displacements $\bar{u}_i$ are given on s_u. On the overall surface $s_u + s_p$ we have

$$\boldsymbol{\sigma}_{ij}\, n_j = \bar{p}_i, \tag{1.5}$$

$$u_i = \bar{u}_i, \tag{1.6}$$

where n_i are the direction cosines of the outward normal line on the surface.

$A(\mathbf{e}_{ij})$ is the density of the potential energy, which is a function of the components of strains. So, by definition, the following formula holds:

$$\frac{\partial A(\mathbf{e}_{ij})}{\partial \mathbf{e}_{ij}} = \boldsymbol{\sigma}_{ij}. \tag{1.7}$$

Similarly, $B(\boldsymbol{\sigma}_{ij})$ is the density of the complementary energy. Observing the relationship between $A(\mathbf{e}_{ij})$ and $B(\boldsymbol{\sigma}_{ij})$, the following two equations must be satisfied:

$$A(\mathbf{e}_{ij}) + B(\boldsymbol{\sigma}_{ij}) = \boldsymbol{\sigma}_{ij}\mathbf{e}_{ij}, \tag{1.8}$$

$$\frac{\partial B(\boldsymbol{\sigma}_{ij})}{\partial \boldsymbol{\sigma}_{ij}} = \mathbf{e}_{ij}. \tag{1.9}$$

Now, the quasi-potential energy principle and the quasi-complementary energy principle are established as follows

THEOREM 1: *Quasi-potential energy principle.* For any elastic system subjected to body follower forces and surface follower forces and satisfying the strain-displacement relation with given displacement boundary conditions on the surface s_u (i.e., $u_i = \bar{u}_i$), the actual displacements and strains must satisfy the following variational equation for all permissible displacements and strains

$$\delta\Pi_{\rm I} + \delta Q + \delta P = 0 \tag{1.10}$$

in which

$$\Pi_{\rm I} = \iiint_\tau \{A(\mathbf{e}_{ij}) - F_i u_i\}\,\mathrm{d}\tau - \iint_{s_p} \bar{p}_i u_i\,\mathrm{d}s,$$

$$\delta Q = \iiint_\tau u_i \delta F_i\,\mathrm{d}\tau, \qquad \delta P = \iint_{s_p} u_i \delta\bar{p}_i\,\mathrm{d}s, \tag{1.11}$$

where F_i and $\bar{p}_i$ are nonfollower forces, i.e., when these forces are constants, it will be reduced to potential energy principle in the classical sense. Thus, it is called the quasi-potential energy principle.

PROOF: The variation of $\Pi_{\rm I}$ is given as

$$\partial\Pi_{\rm I} = \iiint_\tau \left\{\frac{\partial A(\mathbf{e}_{ij})}{\partial \mathbf{e}_{ij}}\delta\mathbf{e}_{ij} - F_i\delta u_i - u_i\delta F_i\right\}\mathrm{d}\tau - \iint_{s_p}\{\bar{p}_i\delta u_i + u_i\delta\bar{p}_i\}\,\mathrm{d}s. \tag{1.12}$$

By Green's theorem

$$\iint_\tau \boldsymbol{\sigma}_{ij}\delta\mathbf{e}_{ij}\,\mathrm{d}\tau = \iint_{s_p}\boldsymbol{\sigma}_{ij}n_j\delta u_i\,\mathrm{d}s - \iint_\tau \boldsymbol{\sigma}_{ij,j}\delta u_i\,\mathrm{d}\tau$$

will be used and the following formula

$$\delta\Pi_{\rm I} = -\iiint_\tau\{(\boldsymbol{\sigma}_{ij,j} + F_i)\,\delta u_i + u_i\delta F_i\}\,\mathrm{d}\tau - \iint_{s_p}\{(\bar{p}_i - \boldsymbol{\sigma}_{ij}n_j))\}\,\delta u_i + u_i\delta\bar{p}_i\}\,\mathrm{d}s. \tag{1.13}$$

will be obtained. By means of (1.12), (1.13), (1.4) and (1.5), Theorem 1 can be proved.

THEOREM 2. *Quasi-complementary energy principle.* For any elastic system subjected to body follower forces and surface follower forces and satisfying the equilibrium equation with the surface follower force boundary conditions on surface s_p (i.e., $\boldsymbol{\sigma}_{ij}n_j = \bar{p}_i$), the actual stresses must satisfy the following variational equation for all permissible stresses

$$\delta\Pi_{\rm II} + \delta R = 0 \tag{1.14}$$

in which

$$\Pi_{\rm II} = \iiint_\tau B(\sigma_{ij})\,\mathrm{d}\tau - \iint_{s_u}\bar{u}_i\boldsymbol{\sigma}_{ij}n_j\,\mathrm{d}s,$$

$$\delta R = -\iiint_\tau u_i\delta F_i\,\mathrm{d}\tau \tag{1.15}$$

when F_i are nonfollower forces, i.e., when these forces are constants, it will be reduced to the complementary energy principle in a classical sense. Thus, it is called the quasi-complementary energy principle.

PROOF: The variation of Π_{II} is given as

$$\delta\Pi_{II} = \iiint_\tau \left\{\boldsymbol{\sigma}_{ij} - \frac{\partial A(\mathbf{e}_{ij})}{\partial \mathbf{e}_{ij}}\right\} \delta\mathbf{e}_{ij}\, d\tau + \iiint_\tau \mathbf{e}_{ij}\delta\boldsymbol{\sigma}_{ij} - \iint_{s_u} \bar{u}_i \delta\boldsymbol{\sigma}_{ij} n_j\, ds.$$

By Green's theorem

$$\iiint_\tau u_i\, d\boldsymbol{\sigma}_{ij,j}\, d\tau = \iint_{s_u+s_p} u_i\delta\boldsymbol{\sigma}_{ij}n_j\, ds - \iint_\tau u_{i,j}\delta\boldsymbol{\sigma}_{ij}\, d\tau = - \iiint_\tau u_i\delta F_i\, d\tau$$

will be used and the following formula

$$\delta\Pi_{II} = \iiint_\tau \left\{\boldsymbol{\sigma}_{ij} - \frac{\partial A(\mathbf{e}_{ij})}{\partial \mathbf{e}_{ij}}\right\} \delta\mathbf{e}_{ij} + \iiint_\tau (\mathbf{e}_{ij} - u_{i,j})\, \delta\boldsymbol{\sigma}_{ij}\, d\tau$$
$$+ \iiint_\tau u_i\delta F_i\, d\tau + \iint_{s_u} (u_i - \bar{u}_i) n_j \delta\boldsymbol{\sigma}_{ij}\, ds \qquad (1.17)$$

will be obtained. By means of (1.17), (1.16), (1.1) and (1.6), Theorem 2 can be proved.

2. Generalized quasi-variational principle

It should be noted that not only the quasi-potential energy principles but also the quasi-complementary energy principles are variational principles with conditions. The continuous equations and the displacement boundary conditions must be satisfied for the quasi-potential energy principle. But, the equilibrium equations and force boundary conditions must be satisfied for the quasi-complementary energy principle. In the previous articles, variational principles with conditions were used. Here, quasi-variational principles without any conditions are established. Therefore, the generalized quasi-variational principles will be used. It was indicated by Prof. Chien Wei-zang [1] that the non-conservative generalized quasi-variational principles with fewer conditions may also be established. the latter phenomenon is straightforward and does not, therefore, warrant further discussion.

THEOREM 3: *Generalized quasi-potential energy principle* (the generalized quasi-variational principle with small displacements in linear elastic theory, deduced from the potential energy principle). For any elastic system subjected to the follower forces and satisfying the strain-displacement relation (1.1), the stress-strain relations (1.2), the equilibrium equations with the follower forces (1.4), the displacement boundary conditions on the surface s_u (1.6), the follower force boundary conditions on the surface s_p (1.5), and the actual displacements u_i, the strains $\mathbf{e}_{ij}$ and the stresses $\boldsymbol{\sigma}_{ij}$ must satisfy the following variational equation for all permissible displacements, strains and stresses

$$\delta\Pi_{\text{III}} + \delta Q + \delta P = 0 \tag{2.1}$$

in which

$$\Pi_{\text{III}} = \iiint_\tau \{A(\mathbf{e}_{ij}) - (\mathbf{e}_{ij} - \tfrac{1}{2}u_{i,j} - \tfrac{1}{2}u_{j,i})\boldsymbol{\sigma}_{ij} - F_i u_i\}\,\mathrm{d}\tau$$
$$- \iint_{s_p} \bar{p}_i u_i\,\mathrm{d}s - \iint_{s_u} (u_i - \bar{u}_i)\boldsymbol{\sigma}_{ij} n_j\,\mathrm{d}s \tag{2.2}$$

and

$$\delta Q = \iiint_\tau u_i \delta F_i\,\mathrm{d}\tau \qquad \delta P = \iint_{s_p} u_i \delta\bar{p}_i\,\mathrm{d}s \tag{2.3}$$

when the body and surface forces are non-follower, i.e., when they are constants as in the formula (2.3), we have $\delta Q = 0$ and $\delta P = 0$, and the expression (2.1) is reduced to the ordinary generalized potential principle. Thus, the generalized quasi-potential energy principle may be used for this theorem.

PROOF: The variation of Π_{III} is given as

$$\delta\Pi_{\text{III}} = \iiint_\tau \left\{\left(\frac{\partial A(\mathbf{e}_{ij})}{\partial \mathbf{e}_{ij}} - \boldsymbol{\sigma}_{ij}\right)\delta\mathbf{e}_{ij} - (\mathbf{e}_{ij} - \tfrac{1}{2}u_{i,j} - \tfrac{1}{2}u_{j,i})\,\delta\boldsymbol{\sigma}_{ij}\right.$$
$$\left. + \boldsymbol{\sigma}_{ij}\delta u_{i,j} - F_i\delta u_i - u_i\delta F_i\right\}\mathrm{d}\tau + \iint_{s_p} (\bar{p}_i\delta u_i + u_i\delta\bar{p}_i)\,\mathrm{d}s$$
$$- \iint_{s_u} \boldsymbol{\sigma}_{ij} n_j \delta u_i\,\mathrm{d}s - \iint_{s_u} (u_i - \bar{u}_i)\,n_j\delta\boldsymbol{\sigma}_{ij}\,\mathrm{d}s. \tag{2.4}$$

Using Green's formula

$$\iiint_\tau \boldsymbol{\sigma}_{ij}\delta u_{i,j}\,\mathrm{d}\tau = \iint_{s_p} \boldsymbol{\sigma}_{ij} n_j \delta u_i\,\mathrm{d}s - \iiint_\tau \boldsymbol{\sigma}_{ij,j}\delta u_i\,\mathrm{d}\tau \tag{2.5}$$

and substituting (2.5) into (2.4) we obtain

$$\delta\Pi_{\text{III}} = \iiint_\tau \left\{\left(\frac{\partial A(\mathbf{e}_{ij})}{\partial \mathbf{e}_{ij}} - \boldsymbol{\sigma}_{ij}\right)\delta\mathbf{e}_{ij} - (\mathbf{e}_{ij} - \tfrac{1}{2}u_{i,j} - \tfrac{1}{2}u_{j,i})\,\delta\boldsymbol{\sigma}_{ij}\right.$$
$$\left. -(\boldsymbol{\sigma}_{ij,j} + F_i)\,\delta u_i\right\}\mathrm{d}\tau + \iint_{s_p} (\boldsymbol{\sigma}_{ij} n_j - \bar{p}_i)\,\delta u_i\,\mathrm{d}s$$
$$- \iint_{s_u} (u_i - \bar{u}_i)\,n_j\delta\boldsymbol{\sigma}_{ij}\,\mathrm{d}s - \iiint_\tau u_i\delta F_i\,\mathrm{d}\tau - \iint_{s_p} u_i\delta\bar{p}_i\,\mathrm{d}s. \tag{2.6}$$

Substituting expression (2.5) into expression (2.1) and using relation (2.3), the theorem III can be proved.

THEOREM 4: *General quasi-complementary energy principle* (the generalized quasi-variational principle with small displacements in linear elastic theory deduced from the complementary energy principle). For any elastic system subjected to the follower forces and satisfying the strain displacement relations, the stress–strain relations, the equilibrium equations with the follower forces, the displacement boundary conditions on the surface s_u, the force boundary conditions on the surface s_p, and the actual displacements u_i, strains $\mathbf{e}_{ij}$ and stress $\boldsymbol{\sigma}_{ij}$ must satisfy the following variational equation for all permissible displacements, strains and stresses:

$$\delta\Pi_{\mathrm{IV}} + \delta R + \delta S = 0 \tag{2.7}$$

in which

$$\Pi_{\mathrm{IV}} = \iiint_\tau \{\boldsymbol{\sigma}_{ij}\mathbf{e}_{ij} - A(\mathbf{e}_{ij}) + (\boldsymbol{\sigma}_{ij,j} + F_i)u_i\}\,\mathrm{d}\tau - \iint_{s_p} (\boldsymbol{\sigma}_{ij}n_j - \bar{p}_i)u_i\,\mathrm{d}s - \iint_{s_u} \boldsymbol{\sigma}_{ij}\,n_j\bar{u}_i\,\mathrm{d}s, \tag{2.8}$$

$$\delta R = -\iiint_\tau u_i\delta F_i\,\mathrm{d}\tau, \qquad \delta \mathrm{S} = -\iint_{s_p} u_i\delta\bar{p}_i\,\mathrm{d}s. \tag{2.9}$$

PROOF. The variation of Π_{IV} is given as

$$\delta\Pi_{\mathrm{IV}} = \iiint_\tau \left\{\left(\boldsymbol{\sigma}_{ij} - \frac{\partial A(\mathbf{e}_{ij})}{\partial \mathbf{e}_{ij}}\right)\delta\mathbf{e}_{ij} + (\mathbf{e}_{ij}\delta\boldsymbol{\sigma}_{ij} + u_i\delta\boldsymbol{\sigma}_{ij,j}) + (\boldsymbol{\sigma}_{ij,j} + F_i)\,\delta u_i + u_i\delta F_i\right\}\mathrm{d}\tau - \iint_{s_p} (\boldsymbol{\sigma}_{ij}n_j\delta u_i - \bar{p}_i\delta u_i + u_i n_j\delta\boldsymbol{\sigma}_{ij} - u_i\delta\bar{p}_i)\,\mathrm{d}s - \iint_{s_u} \bar{u}_i n_j\delta\boldsymbol{\sigma}_{ij}\,\mathrm{d}s. \tag{2.10}$$

Using Green's formula

$$\iiint_\tau u_i\delta\boldsymbol{\sigma}_{ij,j}\,\mathrm{d}\tau = \iint_{s_p+s_u} u_i n_j\delta\boldsymbol{\sigma}_{ij}\,\mathrm{d}s - \iiint_\tau u_{i,j}\delta\boldsymbol{\sigma}_{ij}\,\mathrm{d}\tau \tag{2.11}$$

and substituting expression (2.11) into (2.10), we obtain

$$\delta\Pi_{\mathrm{IV}} = \iiint_\tau \left\{\left(\boldsymbol{\sigma}_{ij} - \frac{\partial A(\mathbf{e}_{ij})}{\partial \mathbf{e}_{ij}}\right)\delta\mathbf{e}_{ij} + (\mathbf{e}_{ij} - u_{i,j})\delta\boldsymbol{\sigma}_{ij} + (\boldsymbol{\sigma}_{ij,j} + F_i)\,\delta u_i + u_i\delta F_i\right\}\mathrm{d}\tau - \iint_{s_p}(\boldsymbol{\sigma}_{ij}n_j - \bar{p}_i)\,\delta u_i + \iint_{s_p} u_i\delta\bar{p}_i\,\mathrm{d}s + \iint_{s_u} (u_i - \bar{u}_i)\,n_j\delta\boldsymbol{\sigma}_{ij}\,\mathrm{d}s. \tag{2.12}$$

Substituting expression (2.12) into (2.7) and using relation (2.9), Theorem 4 can be proved.

According to the method of Prof. Chien Wei-zang [4] the equivalence of these generalized quasi-variational principles may be proved.

THEOREM 5: *Quasi-Hamilton principle*. For any elastic system subjected to the follower forces at the body and the surface and satisfying the strain–stress relation and the displacement boundary condition on the surface s_u, if u_i at t_1 and t_2 are prescribed so that $\delta u_i(t_1) = \delta u_i(t_2) = 0$, then the actual displacements and strains must satisfy the following variational equation for all permissible displacements and strains.

$$\delta\Pi_V + \int_{t_1}^{t_2} \delta Q \, dt + \int_{t_1}^{t_2} \delta P \, dt = 0 \tag{2.13}$$

in which

$$\Pi_V = \int_{t_1}^{t_2} \left\{ \iiint_\tau (T + \Pi_1) \, d\tau \right\} dt \tag{2.14}$$

and $T = \Sigma_{i=1}^{3} p\dot{u}_i^2/2$, δQ and δP are expressed in (1.11). When the body forces are equal to constants in the equation (2.13), we have $\delta Q = 0$ and $\delta P = 0$. In this case Theorem 5 will be reduced to the Hamilton principle in the classical sense. Thus, the quasi-Hamilton principle may be called. The proof may be omitted.

3. Application of the quasi-variational principles in the finite element method

In this study, only the case of the conforming element with continuous displacement field function is discussed. In the conservative case, the sectional variational principles of elastic theory are discussed by Prof. Chien Wei-zang [1, 4]. If the displacement is continuous on the section boundary, no energy will occur. This result is also suitable for the nonconservative case. The quasi-Hamilton principle in this section will be written as

$$\delta\Pi_{V,FE} = \delta\left\{\sum_n \frac{1}{2}\left[\int_{V_n} \varrho \dot{u}_i \dot{u}_i \, dV_n + \int_{V_n} (A(\mathbf{e}_{ij}) - F_i u_i) \, dV_n\right]\right\}$$
$$+ \sum_n \int_{V_n} u_i \delta F_i - \delta \sum_n \int_{\partial V_{nB}} p_i u_i \, ds + \sum_n \int_{\partial V_{nB}} u_i \delta p_i \, ds \tag{3.1}$$

in which, n denotes the number of elements, $\Sigma_n V_n = V$ and $V_i \cap V_j = 0$ for all V_i and V_j, $i \neq j$. ∂V_{nB} is the boundary element of V_n and $\Sigma_n \partial V_{nB} = S$. When the shape functions are sufficiently smooth functions, i.e., when the generalized displacements are continuous at the boundary between elements and the displacements of elements

$$\{u^n\} = [H^n]\{\gamma^n(t)\} \tag{3.2}$$

are assumed in which $[H^n]$ is the shape functions and $\{\gamma^n(t)\}$ is the displacements of nodes, then the strains are obtained as:

$$\{\varepsilon^n\} = [B]\{\gamma^n(t)\}. \tag{3.3}$$

Substituting (3.2), (3.3) into (3.1), we obtain

$$\delta\Pi_{V_{FE}} = \delta\left\{\sum_n \frac{1}{2}\left(\int_{V_n} \varrho\{\dot{\gamma}^n(t)\}^T[H^n]^T[H^n]\{\dot{\gamma}^n(t)\}\,\mathrm{d}V_n\right.\right.$$

$$\left.\left.+\int_{V_n}\{\gamma^n(t)\}^T[B^n]^T[D][B^n]\{\gamma^n(t)\}\,\mathrm{d}V_n\right)\right\}$$

$$-\,\delta\sum_n\int_{V_n}F([H^n]\{\gamma^n(t)\})\{\gamma^n(t)\}\,\mathrm{d}V_n$$

$$+\sum_n\int_{V_n}\{\gamma^n(t)\}^T\delta F([H^n]\{\gamma^n(t)\})\,\mathrm{d}V_n$$

$$-\,\delta\sum_n\int_{\partial V_{nB}}P([H^n]\{\gamma^n(t)\})\{\gamma^n(t)\}\,\mathrm{d}s$$

$$-\sum_n\int_{\partial V_{nB}}\{\gamma^n(t)\}\,\delta P([H^n]\{\gamma^n(t)\})\,\mathrm{d}s. \tag{3.4}$$

When the nonconservative body forces and boundary forces are proportional to the displacements, we have

$$F([H^n]\gamma^n(t)) = F\cdot f([H^n])\{\gamma^n(t)\}, \tag{3.5}$$

$$P([H^n]\gamma^n(t)) = P\cdot p([H^n])\{\gamma^n(t)\}. \tag{3.6}$$

Substituting (3.5), (3.6) into (3.4), the form of second order variation and one order quasi-variation will be obtained. The differential equations of node displacements will be

$$[M]\{\ddot{\gamma}(t)\} + ([K] + F[K_{FG}] + P[K_{PG}])\{\gamma(t)\} = 0 \tag{3.7}$$

in which $[M]$ $[K]$ $[K_{FG}]$ and $[K_{PG}]$ are the mass matrix, the stiffness matrix, the geometric stiffness matrix and the boundary geometric stiffness matrix, consisting of an element matrix. They are established in the form

$$[M^n] = \frac{1}{2}\int_{V_n}\varrho[H^n]^T[H^n]\,\mathrm{d}V_n, \tag{3.8a}$$

$$[K^n] = \frac{1}{2}\int_{V_n}[B^n]^T[D][B^n]\,\mathrm{d}V_n, \tag{3.8b}$$

$$[K^n_{FG}] = -\int_{V_n}[f([H^n])]^T[H^n]\,\mathrm{d}V_n + \int_{V_n}[H^n]^Tf([H^n])\,\mathrm{d}V_n \tag{3.8c}$$

$$[K^n_{PG}] = -\int_{\partial V_{nB}}[p([H^n])]^T[H^n]\,\mathrm{d}s + \int_{\partial V_{nB}}[H^n]^Tp([H^n])\,\mathrm{d}s, \tag{3.8d}$$

where F and P are the parameters of the body nonconservative forces and the boundary nonconservative forces. The matrix $[K_{FG}]$ and the matrix $[K_{PG}]$ are nonsymmetric, because these forces are nonconservative. When the system is stationarily varied, obviously, we have

$$\{\gamma(t)\} = \{\gamma\}\,\mathrm{e}^{\mathrm{i}\omega t}.$$

The eigenvalue equation

$$-\omega^2[M]\{\gamma\} + ([K] + F[K_{FG}] + P[K_{PG}])\{\gamma\} = 0 \tag{3.9}$$

is then established. Then, the eigenvalue problems of the non-symmetrical matrix are subjected to the non-conservative forces instead of the stability problem of the system.

4. Computational examples

Leipholz's and Beck's problems are calculated by using the finite element method from the quasi-variational principle. In Reference [10], the Hermite function

$$[H] = \begin{bmatrix} 1 & 0 & 0 & 0 \\ 0 & x & 0 & 0 \\ -\dfrac{3x^2}{l^2} & -\dfrac{2x^2}{l} & \dfrac{3x^2}{l^2} & -\dfrac{x^2}{l} \\ \dfrac{2x^3}{l^3} & \dfrac{x^3}{l^2} & -\dfrac{2x^3}{l^2} & \dfrac{x^3}{l^2} \end{bmatrix} \tag{4.1}$$

had been taken and the element mass matrix had been obtained by using the quasi-variational principles. The geometric stiffness matrix of Leipholz's rod and the boundary geometric stiffness matrix of Beck's rod had also been given. After calculating $[M]^{-1}([K] + F[K_{FG}] + P[K_{PG}])$, the eigenvalues were computed by QR method. When one of all eigenvalues is zero or complex, the critical load may be obtained. The critical loads of Beck's rod and Leipholz's rod were obtained and compared with References [6] and [8].

EXAMPLE 1: *Leipholz's rod with clamped-free conditions.* There are only nonconservative forces in this problem. The critical loads are obtained and compared with the results of Reference [8] in Table 1.

The critical loads with other boundary conditions are also obtained and compared with the results of Reference [6].

The results in Table 2 are values of FL^3/EI.

EXAMPLE 2: *Free-fixed rod subjected to Leipholz's load and Beck's load.* There is a nonconservative body force and a nonconservative boundary force in this problem. Taking eight elements, the critical loads are calculated as shown in Table 3.

The critical curve of combined Beck's and Leipholz's load is nearly a straight line.

Table 1

Number of element	2	3	4	5	6	7
This paper	40.1	38.3	37.8	37.8	37.8	37.8
Naschie [8]	25.6	40.05	51.59	53.51	54.50	55.16

Table 2

Number of elements	Free–fixed	Hinged–hinged	Fixed–fixed	Hinged–fixed
2	40.1	17.9	80.0	55.6
4	37.8	17.9	78.0	55.4
6	37.8	18.1	77.3	54.1
8	37.8	18.1	77.6	54.4
Leipholz [6] by Ritz Method	40.05	18.96	80.26	57.10

Table 3

PL^2/EI	20.03	16.04	12.03	8.02	4.01	0
FL^3/EI	0	7.42	14.90	22.43	30.14	37.80

References

1. Chien Wei-zang, *Variational Methods and Finite Element Methods* (in Chinese), Science Press (1980).
2. Quian Lind-xi, "Complementary Energy Principle", *Scientia Sinica* (in Chinese), **1** (1950) 449–465.
3. Hu Hai-chang, "On some variational principles in the theory of elasticity and the theory of plasticity", *Acta Physica Sinica* (in Chinese) **10**, 3 (1954), 259–290.
4. Chien Wei-zang, "Studies on generalized variational principles in elasticity and their applications in finite element calculations", *Mechanics and Practice* (in Chinese) **2**, 1 (1979), 14–24; **2**, 2 (1979), 18–27.
5. Leipholz, H., Variational Principles for Nonconservative Problems, A Foundation for a Finite Element Approach, *Computer Methods in Applied Mechanics and Engineering*, 17/18 Part III (1979), 609–617.
6. Leipholz, H., *Direct Variational Methods and Eigenvalue Problems in Engineering*, Noordhoff (1976).
7. Oden, J. T. and Reddy, J. N., *Variational Methods in Theoretical Mechanics*, Springer-Verlag (1976).
8. Naschie, N. S. el. and Atehl, S. al, "Remarks on the stability of flexible rods under follower forces", *Journal of Sound and Vibration* **64**, 3 (1979), 462–465.
9. Liu Dian-kui and Zhang Qi-hao, "Some general vibrational principles for non-conservative problems in theory of elasticity", *Acta Mechanica Sinica* (in Chinese), **13**, 6 (1981), 562–570.
10. Zhang Qi-hao and Shan Wen-xiu, "Finite element method applied to nonconservative stability problems", *Shanghai Mechanics* (in Chinese) **2**, 3 (1981), 40–46.

Theory of Plates and Shells

Solution in cosine series for the bending of a rectangular cantilever plate under the action of arbitrary transverse loads

SUN HUAN-CHUN, SHA DE-SONG, WANG SHOU-XIN, and TANG LI-MIN

Department of Engineering Mechanics, Dalian Institute of Technology, Dalian, PRC

Abstract. In this paper the concept of a generalized clamped edge is proposed and the bending problem of a rectangular cantilever plate under arbitrary loads is resolved in cosine series by the method of superposition.

With this approach, since the shear forces on each of the free edges and the concentrated forces on the two free corners are all zero, it is only necessary to cancel the moments on the free edges by superposing another basic system of solutions.

The results obtained are compared with the solution by the sine series method [1]. The differences between them are rather small and in some cases these results are better than those obtained by the former method [1].

Some examples have been worked under the action of certain kinds of loads distributed in the form of trigonometric functions. The boundary conditions are satisfied through the checking of global equilibrium. Accordingly the validity of this method is verified. Another advantage of this method is that only two basic systems of solutions are used, while in the sine series method, five basic systems of solutions have been adopted.

1. Introduction

Due to the importance of rectangular cantilever plates subjected to transverse loads in engineering problems, many efforts have been made to obtain the exact solutions of them. Only a few analytical solutions of particular loading have been achieved, such as in the case of Prof. Chang Fu-fan who has obtained solutions of concentrated and uniformly distributed loads [1]. An alternative approach is presented in this paper. By employing the concept of a so-called generalized clamped edge and by the use of solutions of cosine series, the difficulties of the free corner conditions are well satisfied. By the superposition of two basic solution systems and by solving simultaneous linear algebraic equations of infinite number, a general solution for arbitrarily transverse loading has been obtained for rectangular clamped plates, and this seems to be the simplest and most convenient method to date.

Yeh, K. Y. (ed) Progress in Applied Mechanics
 ISBN-13: 978-94-010-8061-3

2. The generalized clamped edge and the solution of a rectangular plate with two opposite generalized clamped edges

2.1. *Generalized clamped edge*

A generalized clamped edge is an edge on which the transverse shear force, twisting moment, and slope of the plate vanish. It may also be defined as a clamped edge free from the deflection restraint, or a free edge on which the slope is zero, and may therefore also be called a special free edge.

2.2 *The solution of the rectangular plate with two opposite generalized clamped edges*

Let OA, BC be two opposite generalized clamped edges parallel to the y axis, as shown in Fig. 1, where the boundary conditions are (according to Timoshenko's symbols):

$$\left(\frac{\partial W}{\partial x}\right)_{x=0,2a} = 0, \quad (V_x)_{x=0,2a} = 0, \quad R_0 = R_A = R_B = R_C = 0.$$

We note that the boundary conditions are automatically satisfied by choosing the deflection function in the following form

$$W = \sum_{m=0}^{\infty} Y_m \cos \alpha_m x, \tag{2.1}$$

where $\alpha_m = m\pi/2a$.

Substituting (2.1) into the governing equation $\nabla^4 W = q/D$, we obtain the following ordinary differential equation

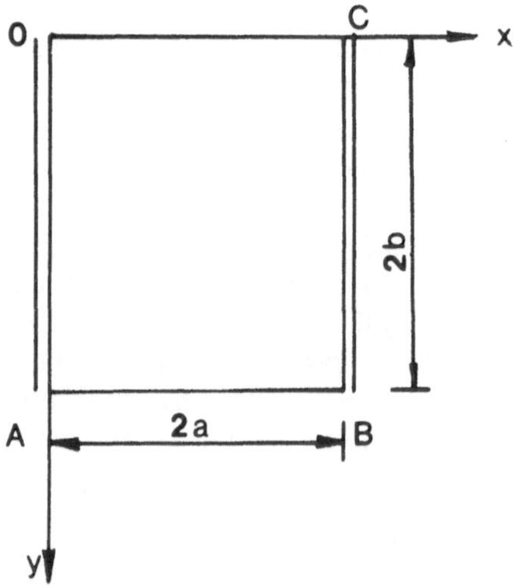

Figure 1

$$Y_m^{(4)} - 2\alpha_m^2 Y_m'' + \alpha_m^4 Y_m = \frac{q_m}{D} \quad (m = 0, 1, \ldots), \tag{2.2}$$

where q_m represents the coefficient of the known transverse load q, expanded into the form of the Fourier cosine series.

From (2.1) and (2.2), the following deflection function can be obtained

$$W = f_0(y) + A_1 y^3 + A_2 y^2 + A_3 y + A_4 + \sum_{m=1}^{\infty} [A_{1m} \operatorname{ch} \alpha_m y + A_{2m}\alpha_m y \operatorname{sh} \alpha_m y + A_{3m} \operatorname{sh} \alpha_m y + A_{4m}\alpha_m y \operatorname{ch} \alpha_m y + f_m(y)] \cos \alpha_m x \tag{2.3}$$

in which $f_0(y), f_m(y)$ are the particular solutions of (2.2) when $m = 0$ and $m \neq 0$ respectively. The coefficients A_i, A_{im} ($i = 1, 2, 3, 4$; $m = 1, 2, \ldots$) can be determined by the prescribed boundary conditions of the other two edges OC, AB.

2.3. *The solution of a rectangular plate with two opposite generalized clamped edges and the other edges having assigned boundary conditions*

As shown in Fig. 2, we assume that the edges $x = 0$, $x = 2a$ are generalized clamped edges, and on the other edges the boundary conditions are as follows

$$\left(\frac{\partial W_1}{\partial y}\right)_{y=0} = 0,$$

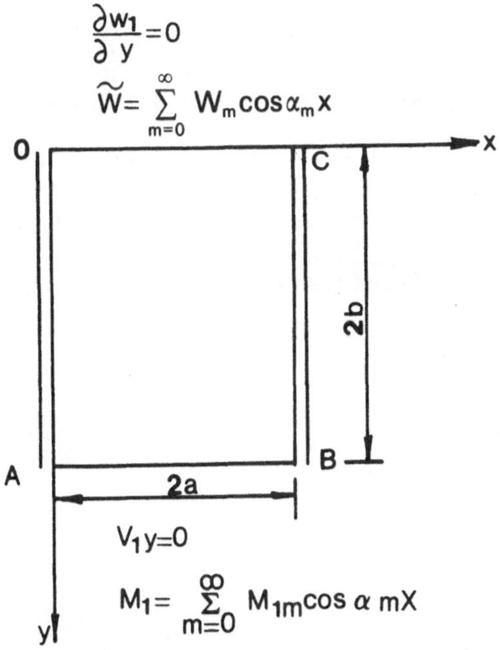

Figure 2

$$(V_{1y})_{y=2b} = 0,$$

$$(W_1)_{y=0} = \tilde{W} = \sum_{m=0}^{\infty} W_m \cos \alpha_m x,$$

$$(M_{1y})_{y=2b} = M_1 = \sum_{m=0}^{\infty} M_{1m} \cos \alpha_m x.$$

In order to maintain sufficient accuracy of the solution of the infinite linear algebraic equation system later, we take the solution as follows:

$$W_1 = f_0(y) + A_1\bar{y}^3 + A_2\bar{y}^2 + A_3\bar{y} + A_4 + \sum_{m=1}^{\infty} [A_{1m} \operatorname{ch} \alpha_m\bar{y} + A_{2m}\alpha_m\bar{y} \operatorname{sh} \alpha_m\bar{y} + A_{3m} \operatorname{sh} \alpha_m\bar{y} + A_{4m}\alpha_m\bar{y} \operatorname{ch} \alpha_m\bar{y} + f_m(y)] \cos \alpha_m x \tag{2.4}$$

in which $\bar{y} = y - b$.

$$M_{1y} = -D\Big\{f_0''(y) + 6A_1\bar{y} + 2A_2 + \sum_{m=1}^{\infty} [(1-\mu)A_{1m}\alpha_m^2 \operatorname{ch} \alpha_m\bar{y} + A_{2m}(2\alpha_m^2 \operatorname{ch} \alpha_m\bar{y} + (1-\mu)\alpha_m^3\bar{y} \operatorname{sh} \alpha_m\bar{y}) + (1-\mu)A_{3m}\alpha_m^2 \operatorname{sh} \alpha_m\bar{y} + A_{4m}(2\alpha_m^2 \operatorname{sh} \alpha_m\bar{y} + (1-\mu)\alpha_m^3\bar{y} \operatorname{ch} \alpha_m\bar{y}) + f_m''(y) - \mu\alpha_m^2 f_m(y)] \cos \alpha_m x\Big\}, \tag{2.5}$$

$$M_{1x} = -D\Big\{\mu f_0''(y) + 6\mu A_1\bar{y} + 2\mu A_2 + \sum_{m=1}^{\infty} [(\mu-1)A_{1m}\alpha_m^2 \operatorname{ch} \alpha_m\bar{y} + A_{2m}(2\mu\alpha_m^2 \operatorname{ch} \alpha_m\bar{y} + (\mu-1)\alpha_m^3\bar{y} \operatorname{sh} \alpha_m\bar{y}) + A_{3m}(\mu-1)\alpha_m^2 \operatorname{sh} \alpha_m\bar{y} + A_{4m}(2\mu\alpha_m^2 \operatorname{sh} \alpha_m\bar{y} + (\mu-1)\alpha_m^3\bar{y} \operatorname{ch} \alpha_m\bar{y}) + \mu f_m''(y) - \alpha_m^2 f_m(y)] \cos \alpha_m x\Big\}. \tag{2.6}$$

Substituting (2.4), (2.5) and (2.6) into the boundary conditions yields

$$A_1 = F_1,$$

$$A_2 = F_2 + \tfrac{1}{2}\bar{M}_{10},$$

$$A_3 = F_3 + \bar{M}_{10},$$

$$A_4 = F_4 + \bar{W}_0 + \tfrac{1}{2}\bar{M}_{10}, \tag{2.7}$$

where

$$A_i = \bar{A}_i b^i$$

$$A_{im} = \frac{\bar{A}_{im} b^4}{\operatorname{ch} \alpha_m b} \quad (i = 1, 2, 3, 4;\ m = 1, 2, \ldots)$$

$$W_m = \bar{W}_m b^4 \quad (m = 0, 1, \ldots),$$

$$M_{1m} = -\bar{M}_{1m} b^2 D \quad (m = 0, 1, \ldots),$$

For $m = 0$, let

$$F_1 = -\frac{1}{6b} f_0'''(b),$$

$$F_2 = -\frac{1}{2b^2} f_0''(b) + \frac{1}{2b} f_0'''(b),$$

$$F_3 = -\frac{1}{b^3} f_0'(0) - \frac{1}{b^2} f_0''(b) + \frac{3}{2b} f_0'''(b),$$

$$F_4 = -\frac{1}{b^4} f_0(0) - \frac{1}{b^3} f_0'(0) - \frac{1}{2b^2} f_0''(b) + \frac{5}{6b} f_0'''(b),$$

and for $m = 1, 2, \ldots$, let

$$D_{12} = \alpha_m b \text{ th } \alpha_m b,$$

$$D_{13} = -\text{th } \alpha_m b,$$

$$D_{14} = -\alpha_m b,$$

$$D_{1f} = -f_m(0)/b^4,$$

$$D_{21} = -\alpha_m b \text{ th } \alpha_m b,$$

$$D_{22} = -\alpha_m b \text{ th } \alpha_m b - \alpha_m^2 b^2,$$

$$D_{23} = \alpha_m b,$$

$$D_{24} = \alpha_m b(1 + \alpha_m b \text{ th } \alpha_m b),$$

$$D_{2f} = -f'(0)/b^3,$$

$$D_{31} = (1 - \mu)\alpha_m^2 b^2,$$

$$D_{32} = \alpha_m^2 b^2[2 + (1 - \mu)\alpha_m b \text{ th } \alpha_m b],$$

$$D_{33} = (1 - \mu)\alpha_m^2 b^2 \text{ th } \alpha_m b,$$

$$D_{34} = \alpha_m^2 b^2[2 \text{ th } \alpha_m b + (1 - \mu)\alpha_m b],$$

$$D_{3f} = -f_m''(b)/b^2 + \mu\alpha_m^2 f_m(b)/b^2$$

$$D_{41} = (1 - \mu)\alpha_m^3 b^3 \text{ th } \alpha_m b,$$

$$D_{42} = \alpha_m^3 b^3[(1 - \mu)\alpha_m b - (1 + \mu) \text{ th } \alpha_m b],$$

$$D_{43} = (1 - \mu)\alpha_m^3 b^3,$$

$$D_{44} = [(1 - \mu)\alpha_m b \text{ th } \alpha_m b - (1 + \mu)]\alpha_m^3 b^3,$$

$$D_{4f} = -(2 - \mu)\alpha_m^2 f_m'(b)/b + f_m'''(b)/b.$$

Then we obtain

$$\begin{bmatrix} 1 & D_{12} & D_{13} & D_{14} \\ D_{21} & D_{22} & D_{23} & D_{24} \\ D_{31} & D_{32} & D_{33} & D_{34} \\ D_{41} & D_{42} & D_{43} & D_{44} \end{bmatrix} \begin{Bmatrix} \bar{A}_{1m} \\ \bar{A}_{2m} \\ \bar{A}_{3m} \\ \bar{A}_{4m} \end{Bmatrix} = \begin{Bmatrix} 0 \\ 0 \\ 1 \\ 0 \end{Bmatrix} \bar{M}_{1m} + \begin{Bmatrix} 1 \\ 0 \\ 0 \\ 0 \end{Bmatrix} \bar{W}_m + \begin{Bmatrix} D_{1f} \\ D_{2f} \\ D_{3f} \\ D_{4f} \end{Bmatrix}. \tag{2.8a}$$

From (2.8a), we can obtain

$$\bar{A}_{im} = a_{0i} + a_{1i}\bar{W}_m + a_{2i}\bar{M}_{1m} \quad (i = 1, 2, 3, 4;\ m = 1, 2, \ldots), \tag{2.8}$$

where a_{ji} $(j = 0, 1, 2)$ are the solutions corresponding to the number $(3 - j)$ of vectors in the right side of (2.8a).

2.4. *The solution without transverse load of the rectangular plate with two opposite generalized clamped edges and the other edges having assigned boundary conditions*

In Fig. 3, both $y = 0$ and $y = 2b$ are generalized clamped edges; whilst on the other two edges the boundary conditions are given as follows

$$(V_{2x})_{x=0,2a} = 0,$$

$$(M_{2x})_{x=2a} = M_2 = \sum_{n=0}^{\infty} M_{2n} \cos \beta_n y,$$

$$(M_{2x})_{x=0} = M_3 = \sum_{n=0}^{\infty} M_{3n} \cos \beta_n y. \tag{2.9a}$$

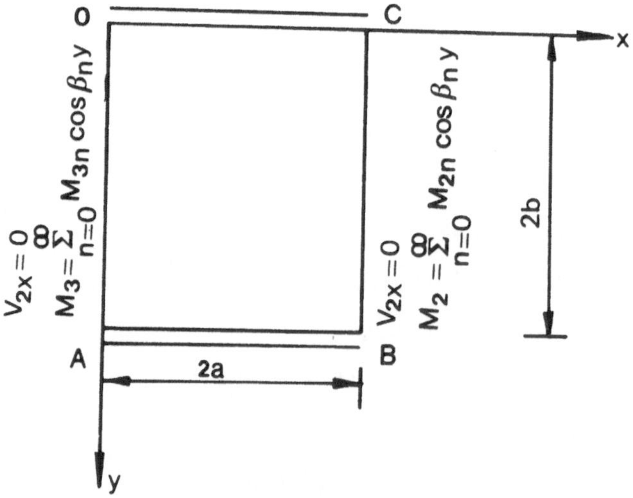

Figure 3

In addition, the transverse load $q = 0$.
Let $\bar{x} = x - a$ and take the solutions in the following forms

$$W_2 = A_5\bar{x}^3 + A_6\bar{x}^2 + A_7\bar{x} + \sum_{n=1}^{\infty} [A_{5n} \operatorname{ch} \beta_n\bar{x} + A_{6n}\beta_n\bar{x} \operatorname{sh} \beta_n\bar{x} + A_{7n} \operatorname{sh} \beta_n\bar{x} + A_{8n}\beta_n\bar{x} \operatorname{ch} \beta_n\bar{x}] \cos \beta_n y, \qquad (2.9)$$

where $\beta_n = n\pi/2b$.

$$M_{2y} = -D\Big\{6\mu A_5\bar{x} + 2\mu A_6 + \sum_{n=1}^{\infty} [(\mu - 1) A_{5n}\beta_n^2 \operatorname{ch} \beta_n\bar{x} + A_{6n}(2\mu\beta_n^2 \operatorname{ch} \beta_n\bar{x} + (\mu - 1)\beta_n^3\bar{x} \operatorname{sh} \beta_n\bar{x}) + (\mu - 1) A_{7n}\beta_n^2 \operatorname{sh} \beta_n\bar{x} + A_{8n}(2\mu\beta_n^2 \operatorname{sh} \beta_n\bar{x} + (\mu - 1)\beta_n^3\bar{x} \operatorname{ch} \beta_n\bar{x})] \cos \beta_n y\Big\}, \qquad (2.10)$$

$$M_{2x} = -D\Big\{6A_5\bar{x} + 2A_6 + \sum_{n=1}^{\infty} [(1 - \mu) A_{5n}\beta_n^2 \operatorname{ch} \beta_n\bar{x} + A_{6n}(2\beta_n^2 \operatorname{ch} \beta_n\bar{x} + (1 - \mu)\beta_n^3\bar{x} \operatorname{sh} \beta_n\bar{x}) + A_{7n}(1 - \mu)\beta_n^2 \operatorname{sh} \beta_n\bar{x} + A_{8n}(2\beta_n^2 \operatorname{sh} \beta_n\bar{x} + (1 - \mu)\beta_n^3\bar{x} \operatorname{ch} \beta_n\bar{x})] \cos \beta_n y\Big\}. \qquad (2.11)$$

Substituting (2.9), (2.10) and (2.11) into (2.9a) yields

$$\begin{bmatrix} E_{11} & E_{12} & E_{13} & E_{14} \\ E_{21} & E_{22} & E_{23} & E_{24} \\ E_{31} & E_{32} & E_{33} & E_{34} \\ E_{41} & E_{42} & E_{43} & E_{44} \end{bmatrix} \begin{Bmatrix} \bar{A}_{5n} \\ \bar{A}_{6n} \\ \bar{A}_{7n} \\ \bar{A}_{8n} \end{Bmatrix} = \begin{Bmatrix} 1 \\ 0 \\ 0 \\ 0 \end{Bmatrix} \bar{M}_{2n} + \begin{Bmatrix} 0 \\ 0 \\ 1 \\ 0 \end{Bmatrix} \bar{M}_{3n}, \qquad (2.13a)$$

in which

$$A_j = \bar{A}_j a^{j-4}$$

$$A_{jn} = \frac{\bar{A}_{jn} a^4}{\operatorname{ch} \beta_n a}, \quad (j = 5, 6, 7, 8;\ n = 1, 2, \ldots),$$

$$M_{kn} = -Da^2 \bar{M}_{kn}, \quad (k = 2, 3).$$

When $n = 0$

$$A_5 = 0$$

$$\bar{A}_6 = \tfrac{1}{2}\bar{M}_{20} = \tfrac{1}{2}\bar{M}_{30} \qquad (2.12)$$

for $n = 1, 2, \ldots$, let

$$E_{11} = E_{31} = (1-\mu)\beta_n^2 a^2,$$

$$E_{12} = E_{32} = \beta_n^2 a^2[2 + (1-\mu)\beta_n a \text{ th } \beta_n a],$$

$$E_{21} = E_{41} = (1-\mu)\beta_n^3 a^3 \text{ th } \beta_n a,$$

$$E_{22} = E_{42} = \beta_n^3 a^3[(1-\mu)\beta_n a - (1+\mu) \text{ th } \beta_n a],$$

$$E_{13} = -E_{33} = (1-\mu)\beta_n^2 a^2 \text{ th } \beta_n a,$$

$$E_{14} = -E_{34} = \beta_n^2 a^2[2 \text{ th } \beta_n a + (1-\mu)\beta_n a],$$

$$E_{23} = -E_{43} = (1-\mu)\beta_n^3 a^3,$$

$$E_{24} = -E_{44} = \beta_n^3 a^3[(1-\mu)\beta_n a \text{ th } \beta_n a - (1+\mu)].$$

The solution of (2.13a) is:

$$\bar{A}_{jn} = a_{3j}\bar{M}_{2n} + a_{4j}\bar{M}_{3n} \tag{2.13}$$

in which a_{3j} and a_{4j} are the solutions of (2.13a) corresponding to the first and second column of the right side of it respectively.

Therefore, the solutions of the rectangular cantilever plate under the action of arbitrary transverse loads can be obtained by superposing the two basic solution systems which are given in (2.3) and (2.4); while in [1], the solutions of the same problem have been obtained by superposing five basic solution systems expressed in sine series.

3. Solution of the rectangular cantilever plate bending under the action of arbitrary transverse loads

The boundary conditions of the rectangular cantilever plate are

$$(W)_{y=0} = 0, \tag{A}$$

$$\left(\frac{\partial W}{\partial y}\right)_{y=0} = 0, \tag{B}$$

$$(V_x)_{x=0,2a} = 0, \tag{C}$$

$$(V_y)_{y=2b} = 0, \tag{D}$$

$$(M_x)_{x=0,2a} = 0, \tag{E}$$

$$(M_y)_{y=2b} = 0, \tag{F}$$

$$R_A = R_B = 0. \tag{G}$$

It is obvious that conditions (B), (C), (D) and (G) are satisfied automatically when the basis solutions given in (2.3) and (2.4) are used, while conditions (A), (E) and (F) become

$$(W_2)_{y=0} + \sum_{m=0}^{\infty} W_m \cos \alpha_m x = 0, \tag{3.1}$$

$$(M_{2y})_{y=2b} + \sum_{m=0}^{\infty} M_{1m} \cos \alpha_m x = 0, \tag{3.2}$$

$$(M_{1x})_{x=2a} + \sum_{n=0}^{\infty} M_{2n} \cos \beta_n y = 0, \tag{3.3}$$

$$(M_{1x})_{x=0} + \sum_{n=1}^{\infty} M_{3n} \cos \beta_n y = 0. \tag{3.4}$$

Let us substitute (2.12), (2.13) and (2.7), (2.8) into (2.9), (2.10) and (2.6) and then further into (3.1), (3.2) and (3.3), (3.4) respectively. Then expand $(W_2)_{y=0}$, $(M_{2y})_{y=2b}$ into the Fourier cosine series with respect to x, and $(M_{1x})_{x=2a}$, $(M_{1x})_{x=a}$ into Fourier cosine series with respect to y. Comparing the coefficients of the same terms in the Fourier series, the following equations can be obtained

$$\left(\frac{b}{a}\right)^4 \bar{W}_0 + \tfrac{1}{6}\bar{M}_{20} + \sum_{n=1}^{\infty}\left[\left(\sum_{j=5}^{8} a_{3j}P_{jn0}\right)\bar{M}_{2n} + \left(\sum_{j=5}^{8} a_{4j}P_{jn0}\right)\bar{M}_{3n}\right] = 0, \tag{3.5}$$

$$\left(\frac{b}{a}\right)^4 \bar{W}_m + \tfrac{1}{2}P_{6m}\bar{M}_{20} + P_{7m}A_7 + \sum_{n=1}^{\infty}\left(\sum_{j=5}^{8} a_{3j}P_{jnm}\right)\bar{M}_{2n}$$
$$+ \left(\sum_{j=5}^{8} a_{4j}P_{jnm}\right)\bar{M}_{3n} = 0 \quad (m = 1, 2, \ldots), \tag{3.6}$$

$$\left(\frac{b}{a}\right)^2 \bar{M}_{10} + \mu\bar{M}_{20} + \sum_{n=1}^{\infty}\left[\left(\sum_{j=5}^{8} a_{3j}Q_{jn0}\right)\bar{M}_{2n} + \left(\sum_{j=5}^{8} a_{4j}Q_{jn0}\right)\bar{M}_{3n}\right](-1)^n = 0, \tag{3.7}$$

$$\left(\frac{b}{a}\right)^2 \bar{M}_{1m} + \sum_{n=1}^{\infty}\left[\left(\sum_{j=5}^{8} a_{3j}Q_{jnm}\right)\bar{M}_{2n} + \left(\sum_{j=5}^{8} a_{4j}Q_{jnm}\right)\bar{M}_{3n}\right](-1)^n = 0,$$
$$(m = 1, 2, \ldots), \tag{3.8}$$

$$\left(\frac{a}{b}\right)^2 \bar{M}_{20} + \mu\bar{M}_{10} + \sum_{m=1}^{\infty}\left[\left(\sum_{i=1}^{4} a_{1i}R_{im0}\right)\bar{W}_m + \left(\sum_{i=1}^{4} a_{2i}R_{im0}\right)\bar{M}_{1m}\right](-1)^m$$
$$= -\mu G_{200} - 2\mu F_2 - \sum_{m=1}^{\infty}\left(\sum_{i=1}^{4} a_{0i}R_{im0} + F_{m0}\right)(-1)^m, \tag{3.9}$$

$$\left(\frac{a}{b}\right)^2 \bar{M}_{2n} + \sum_{m=1}^{\infty}\left[\left(\sum_{i=1}^{4} a_{1i}R_{imn}\right)\bar{W}_m + \left(\sum_{i=1}^{4} a_{2i}R_{imn}\right)\bar{M}_{1m}\right](-1)^m$$
$$= -\mu G_{20n} - \sum_{m=1}^{\infty}\left(\sum_{i=1}^{4} a_{0i}R_{imn} + G_{mn}\right)(-1)^m, \quad (n = 1, 2, \ldots), \tag{3.10}$$

$$\left(\frac{a}{b}\right)^2 \bar{M}_{20} + \mu\bar{M}_{10} + \sum_{m=1}^{\infty}\left[\left(\sum_{i=1}^{4} a_{1i}R_{im0}\right)\bar{W}_m + \left(\sum_{i=1}^{4} a_{2i}R_{im0}\right)\bar{M}_{1m}\right]$$

$$= -\mu G_{200} - 2\mu F_2 - \sum_{m=1}^{\infty}\left(\sum_{i=1}^{4} a_{0i}R_{im0} + G_{m0}\right), \tag{3.11}$$

$$\left(\frac{a}{b}\right)^2 \bar{M}_{3n} + \sum_{m=1}^{\infty}\left[\left(\sum_{i=1}^{4} a_{1i}R_{imn}\right)\bar{W}_m + \left(\sum_{i=1}^{4} a_{2i}R_{imn}\right)\bar{M}_{1m}\right]$$

$$= -\mu G_{20n} - \sum_{m=1}^{\infty}\left[\sum_{i=1}^{4} a_{0i}R_{imn} + G_{mn}\right], \quad (n = 1, 2, \ldots), \tag{3.12}$$

in which

$$P_{5n0} = \frac{\operatorname{th}\beta_n a}{\beta_n a},$$

$$P_{6n0} = \frac{\beta_n a - \operatorname{th}\beta_n a}{\beta_n a},$$

$$P_{7n0} = P_{8n0} = 0,$$

$$P_{6m} = \frac{2}{\alpha_m^2 b^2}[(-1)^m + 1],$$

$$P_{7m} = \frac{(-1)^m - 1}{\alpha_m^2 b^2},$$

$$P_{5nm} = \frac{\beta_n a \operatorname{th}\beta_n a}{a^2(\alpha_m^2 + \beta_n^2)}[(-1)^m + 1],$$

$$P_{6nm} = \left[\frac{\beta_n a}{a^2(\alpha_m^2 + \beta_n^2)} - \frac{(\beta_n^2 - \alpha_m^2)a^2}{a^4(\alpha_m^2 + \beta_n^2)^2}\operatorname{th}\beta_n a\right]\beta_n a[(-1)^m + 1],$$

$$P_{7nm} = \frac{\beta_n a}{a^2(\alpha_m^2 + \beta_n^2)}[(-1)^m - 1],$$

$$P_{8nm} = \left[\frac{\beta_n a}{a^2(\alpha_m^2 + \beta_n^2)}\operatorname{th}\beta_n a - \frac{(\beta_n^2 - \alpha_m^2)a^2}{a^4(\alpha_m^2 + \beta_n^2)^2}\right]\beta_n a[(-1)^m - 1],$$

$$Q_{5n0} = (\mu - 1)\beta_n a \operatorname{th}\beta_n a,$$

$$Q_{6n0} = [(\mu - 1)\beta_n a + (1 + \mu)\operatorname{th}\beta_n a]\beta_n a,$$

$$Q_{7n0} = Q_{8n0} = 0,$$

$$Q_{5nm} = \frac{(\mu - 1)\beta_n^3 a^3 \operatorname{th}\beta_n a}{a^2(\alpha_m^2 + \beta_n^2)}[(-1)^m + 1],$$

$$Q_{6nm} = \left[\frac{2\mu \text{ th } \beta_n a + (\mu - 1)\beta_n a}{a^2(\alpha_m^2 + \beta_n^2)} - (\mu - 1)\frac{(\beta_n^2 - \alpha_m^2)a^2 \text{ th } \beta_n a}{a^4(\alpha_m^2 + \beta_n^2)^2}\right]\beta_n^3 a^2 [(-1)^m + 1],$$

$$Q_{7nm} = \frac{(\mu - 1)\beta_n^3 a^3[(-1)^m - 1]}{a^2(\alpha_m^2 + \beta_n^2)},$$

$$Q_{8nm} = \left[\frac{2\mu + (\mu - 1)\beta_n a \text{ th } \beta_n a}{a^2(\alpha_m^2 + \beta_n^2)} - (\mu - 1)\frac{(\beta_n^2 - \alpha_m^2)a^2}{a^4(\alpha_m^2 + \beta_n^2)^2}\right] a^3\beta_n^3 [(-1)^m - 1],$$

$$R_{1m0} = (\mu - 1)\alpha_m b \text{ th } \alpha_m b,$$

$$R_{2m0} = [(\mu - 1)\alpha_m b + (1 + \mu) \text{ th } \alpha_m b]\alpha_m b,$$

$$R_{3m0} = R_{4m0} = 0,$$

$$R_{1mn} = \frac{(\mu - 1)\alpha_m^3 b^3 \text{ th } \alpha_m b}{b^2(\alpha_m^2 + \beta_n^2)}[(-1)^n + 1],$$

$$R_{2mn} = \left[\frac{2\mu \text{ th } \alpha_m b + (\mu - 1)\alpha_m b}{b^2(\alpha_m^2 + \beta_n^2)} - (\mu - 1)\frac{(\alpha_m^2 - \beta_n^2)b^2 \text{ th } \alpha_m b}{b^4(\alpha_m^2 + \beta_n^2)^2}\right]\alpha_m^3 b^3 [(-1)^n + 1],$$

$$R_{3mn} = \frac{(\mu - 1)\alpha_m^3 b^3}{b^2(\alpha_m^2 + \beta_n^2)}[(-1)^n - 1],$$

$$R_{4mn} = \left[\frac{2\mu + (\mu - 1)\alpha_m b \text{ th } \alpha_m b}{b^2(\alpha_m^2 + \beta_n^2)} - (\mu - 1)\frac{(\alpha_m^2 - \beta_n^2)b^2}{b^4(\alpha_m^2 + \beta_n^2)^2}\right]\alpha_m^3 b^3 [(-1)^n - 1], \quad (m, n = 1, 2, \ldots)$$

G_{mn} represent the Fourier coefficients of

$$\frac{1}{b^2}[\mu f_m''(y) - \alpha_m^2 f_m(y)].$$

G_{20n} represent the Fourier coefficient of $f_0''(y)$.

$(n = 0, 1, 2, \ldots; m = 1, 2, \ldots)$.

We can work out W_0, M_{10}, M_{20}, A_7 and W_m, M_{1m}, M_{2m}, M_{3n} $(m, n = 1, 2, \ldots)$ from (3.5), (3.7), (3.9), (3.11) and four infinite simultaneous equation systems (3.6), (3.8), (3.10) and (3.12) respectively. Then A_i $(i = 1, 2, \ldots, 7)$, A_{im} $(i = 1, 2, 3, 4;$ $m = 1, 2, \ldots)$, A_{jn} $(j = 5, 6, 7, 8; n = 1, 2, \ldots)$ will be found from (2.7), (2.8), (2.12) and (2.13). Finally, the solution of this problem can be obtained by superposing as follows:

$$W = W_1 + W_2,$$
$$M_y = M_{1y} + M_{2y},$$
$$M_x = M_{1x} + M_{2x}.$$

In practical calculation, it is convenient to divide the transverse loads into two parts: symmetrical and antisymmetrical. In the symmetrical case, m are even numbers and $A_5 = A_7 = A_{7n} = A_{8n} = 0$, $M_{2n} = M_{3n}$ $(n = 1, 2, \ldots)$; and in the antisymmetrical case, m are odd numbers, $A_1 = A_2 = A_3 = A_4 = A_6 = A_{5n} = A_{6n} = f_0(y) = 0$, $M_{2n} = -M_{3n}$ $(n = 1, 2, \ldots)$. Thus the calculation becomes much more simple. It must be pointed out that there is no constant term in (2.9), because it has been merged in $\bar{W}_0$.

4. Examples

In order to make the results simple, clear and convenient so as to be comparable with the solutions in [1], we will take a, b as the length of the plates. We have calculated some solutions of the cantilever plates subjected to various distributed transverse loads as follows:

uniformly distributed load: Q,

distributed loads: $$q = Q \cos \frac{2\pi x}{a},$$

$$q = Q \cos \frac{\pi y}{b},$$

$$q = Q \cos \frac{2\pi x}{a} \cos \frac{\pi y}{b},$$

$$q = Q \cos \frac{\pi x}{a},$$

$$q = Q \cos \frac{\pi x}{a} \cos \frac{\pi y}{b}.$$

After checking the results by substituting the solution into the governing equilibrium equation and all the boundary conditions, the validity of this method is verified. Some of the results are written out in the following:

4.1. *Uniform load Q*

1. Take $b/a = 1$, $\mu = 0.3$. The results calculated up to the preceding 20 terms by the present method are listed on the following first row. The results obtained by Chang Fu-fan and Wu Liang-zhi (using the finite element method) are listed on the second and the third row respectively. Thus:

x/a	0.5	0.625	0.75	0.875	1	
$(W)_{y=b}$	0.12908	0.12895	0.12857	0.12798	0.12724	Qb^4/D
	0.13102	0.13091	0.13056	0.12998	0.12933	
	0.12905	0.12892	0.12851	0.12788	0.12708	
$(M)_{y=0}$	−0.53515	−0.53719	−0.52334	−0.51775	−0.02538	Qb^2
	−0.53560	−0.53550	−0.53353	−0.5127	0	
	−0.5309	−0.53058	−0.52760	−0.50399	−0.34571	

Checking:

The total bending moment in clamped edge (Qb^3)		Error (%)
$\int_0^a (M_y)_{y=0}\,dx = -0.50000$	(present method)	0
-0.50578	(Chang)	1.16

2. Take $b/a = 0.5$, $\mu = 0.3$. The result calculated up to the preceding 20 terms, and the results worked out by Prof. Chang are listed on the first and second row respectively as follows:

x/a	0.5	0.625	0.75	0.875	1	
$(W)_{y=b}$	0.1277	0.12764	0.12713	0.12605	0.12437	Qb^4/D
	0.12837	0.12825	0.12784	0.12671	0.12540	
$(M)_{y=0}$	−0.50778	−0.51971	−0.51031	−0.52087	−0.12520	Qb^2
	−0.51049	−0.51451	−0.51386	−0.51674	0	

Checking:

Total bending moment in clamped edge (Qb^3)	Error (%)
$\int_0^a (M_y)_{y=0}\,dx = -1.00000$	0
-1.0048	0.48

4.2. *$q = Q\cos(2\pi x/a)\cos(\pi y/b)$, $b/a = 1$, $\mu = 0.3$*

The result of the preceding 20 terms:

x/a	0.5	0.625	0.75	0.875	1	
$(W)_{y=b}$	0.0013045	0.001076	0.0004793	−0.0002879	−0.001053	Qb^4/D
$(M)_{y=0}$	0.009357	0.006279	−0.0003387	−0.0063129	−0.007413	Qb^2

The total bending moment on the boundary $y = 0$ is

$$\int_0^a (M_y)_{y=0}\,dx = -0.2664 \times 10^{-13}\, Qab^2,$$

while the exact value is zero. Thus, the error is almost zero.

y/b	0	0.5	1	
$(M_x)_{x=0.5a}$	0.002852	0.006356	0.02743	Qb^2

4.3. $q = Q \cos(\pi x/a)$, $b/a = 1$, $\mu = 0.3$

The result of the preceding 10 terms:

x/a	0.5	0.75	1	
$(W)_{y=b}$	0.33615×10^{-12}	-0.010075	-0.01769	Qb^4/D
$(M_y)_{y=0}$	-0.28337×10^{-11}	0.089234	0.27958	Qb^2

The total bending moment on the clamped edge $y = 0$ is

$$\int_0^a (M_y)_{y=0}\,dx = 0.$$

y/b	0	0.5	1	
$(Mx)_{x=0.5a}$	0.10415×10^{-12}	0.1039×10^{-11}	0.2904×10^{-11}	Qb^2

In order to find the solution of the rectangular cantilever plate subjected to arbitrary lateral loads more conveniently, we may calculate the solution due to the action of the distributed loads $\cos(m\pi x/a)$, $\cos(n\pi y/b)$ and $\cos(m\pi x/a)\cos(n\pi y/b)$ in which m, n may be any positive integer numbers, and then expand the transverse loads into double trigonometric Fourier cosine series. The solution of the problem may be obtained by superposing the solutions due to the single trigonometric term of load. Of course, the solution of this problem may still be directly obtained by the method proposed formerly in this paper.

5. Conclusion

5.1. *On the case of concentrated forces and boundary loads*

The method proposed in this paper is not only suitable for the case of distributed loads but also adaptable to arbitrary transverse loads. For instance, a concentrated force can be treated as a uniform load distributed in an appropriate small region. After expanding it into the Fourier cosine series, we may resolve this problem by the method above mentioned.

When the edges of the plate are subjected to bending moments or shear forces, certain equations should be altered. For example, if the edge BC, as shown in Fig. 3, is subjected to the action of the shear forces $P = \sum_{n=0}^{\infty} P_n \cos \beta_n y$, the boundary condition $(V_{2x})_{x=2a} = 0$ should be changed to

$$(V_{2x})_{x=2a} = \sum_{n=0}^{\infty} P_n \cos \beta_n y.$$

When the corresponding equations have been obtained by comparing the coefficients of the same term in the Fourier cosine series, the problem can be resolved in the same way.

5.2. *On the solution in cosine series for bending of rectangular plates*

Although in this paper the solution in cosine series is applied to the bending problem of the rectangular cantilever plate, it can also be applicable in the case of other boundary conditions. Since in this method the slope of the plate, shear force, and twisting moment vanish on the generalized clamped edge, it is convenient in the case of clamped edges and free edges and more especially in the case of the plate with a free corner. The conventional sine series solution is convenient in the case of simply supported edges. Hence it will be more convenient and efficient to solve the bending problem of a rectangular plate with various boundaries by selecting a proper solution in the form of either sine series or cosine series, as given in this paper.

Reference

1. Chang, Fu-fan, "Solution for the bending problem of a rectangular cantilever plate subjected to uniform loading", *Applied Mathematics and Mechanics* (English edition) **1**, 3 (1980), 371–384.

The effect of axisymmetric geometric imperfections on dynamic response to turbulent wind in a cooling tower

LU WEN-DA

Institute of Applied Mathematics and Mechanics, Shanghai University of Technology, Shanghai, PRC

LIN BAO-QING

Thermal Power Engineering Institute, The Ministry of Power, Xian, PRC

Abstract. In recent years, many investigators have studied the effect of axisymmetric meridional geometric imperfections on the static response of stress resultants in a hyperbolic cooling tower. In the present paper, using the solutions for axisymmetric shell finite elements the effect of geometric imperfections on the natural frequency and dynamic response to turbulent wind in a cooling tower has been evaluated. Through the illustrative numerical example of a typical tower with different kinds of geometric imperfections, it will be shown that the effect will increase with the decrease of imperfect wave length. Therefore, the effect of geometric imperfections on the dynamic response for determining the tolerance limits of geometric imperfection should be considered.

1. Introduction

The main loading case for hyperbolic cooling towers of reinforced concrete is produced by wind forces because they usually govern the tower design. The sudden collapse of three cooling towers, with a height of 113 m, at Ferry-Bridge Power Station during a storm led to a series of investigations over the years. Geometric imperfections are unavoidable in such huge shell structures. The hyperbolic cooling tower with a height of 91 m at the Ardeed Nylon Works of Imperial Chemical Industries Ltd. in the United Kingdom had a series meridional curvature imperfection from 50 m–60 m above the bottom linted in the shell construction. In September 1973, these induced a higher circumferential tensile force under a storm. A hoop tension failure was followed by a meridional bending failure, resulting in the total collapse of the tower. Many investigators have studied the effect of meridional geometric imperfections on the static response of cooling towers [1–5]. In these works the evaluations of specifying tolerance limits were given. However they have not considered the geometric imperfections on dynamic response to turbulent wind in cooling tower. Zerna [7] recently recommended a limit of the slope of imperfection of 15 mm/M, but this recommendation, as yet, has no foundation.

Yeh, K. Y. (ed) Progress in Applied Mechanics
 ISBN-13: 978-94-010-8061-3

In 1977 the International Association for Shell and Spatial Structures proposed interim guidelines on acceptable geometric deviations: (1) The maximum error in the slope of the meridian should not exceed $\pm 1.5\%$; (2) Within this slope, the absolute maximum radial error should not exceed 0.10 m or the value (measured in m) $\sqrt{r_i h}/47.5$, in which r_i and h are the regional values of the horizontal radius and the wall thickness of the shell, in metres. These tolerances are associated with specified minimum percentages of 0.25% high tensile steel or 0.35% mild steel in both directions. Similar recommendations are made in the new draft British Standard on the design of cooling tower, BS. 4485 (draft) Part 4): (1) the maximum error in meridional slope should not exceed 1.0%; (2) subject to a maximum error in horizontal radius of 5 cm with a specified allowance for survey inaccuracies which could increase the permitted measured error to at least 10 cm. The specified minimum reinforcement is 0.2% of the concrete cross sectional area in both directions. Neither of the above two design recommendations would seem to take adequate account of the particular shell characteristics. The specified minimum steel percentage in the two recommendations is significantly different and the British Standard adopts the lower level.

Croll [2] has given the computing formula of the tolerance limit on the meridional curvature error, and it is expressed entirely in terms of quantities that are determined at the design stage. Al-Dabbagh and Gupta [5] have also given the computing formulas of the maximum allowable slope of imperfection.

The dynamic response of cooling tower to turbulent wind has been investigated theoretically by applying the spectral method of random vibration theory. The major difficulty lies in how to describe accurately the fluctuating pressure distribution on the tower surface. Moreover, most studies of the behavior of cooling towers under turbulent wind apply the spectral method based on the results of wind-tunnel tests of cooling tower aeroelastic models; for example, Abu-Sitta *et al.* [12, 13] have suggested a simplified physical model of the cross-spectral density of fluctuating pressures were based on information from wind-tunnel tests. Singh *et al.* [14] have used directly the coherence function of tall slender building given by Davenport.

In the present paper, using the finite element solution based on the simplified physical model of fluctuating pressure cross-spectral density measured on a full-scale tower, the effects of axisymmetric geometric imperfections on natural frequency and gust wind dynamic responses in cooling towers are studied. Through the illustrative numerical example of a typical cooling tower, the appropriateness of the allowable slope of geometric imperfection is suggested. In the present paper, both the quasi-static part of the dynamic response, and the resonance response are taken into account.

2. Gust response

It is supposed that the displacement vector of a point on the middle surface of the shell may be expanded into a Fourier series in the circumferential direction. The coefficients of the Fourier series are then expanded into an infinite series of mode shapes as follows:

$$
\begin{aligned}
u &= \sum_{n=0}^{\infty}\sum_{r=1}^{\infty} q_r^n(t)\mu_{1r}^n(s)\cos n\theta + q_r^{\prime n}(t)\mu_{1r}^n(s)\sin n\theta,\\
v &= \sum_{n=0}^{\infty}\sum_{r=1}^{\infty} q_r^n(t)\mu_{2r}^n(s)\sin n\theta + q_r^{\prime n}(t)\mu_{2r}^n(s)\cos n\theta,\\
w &= \sum_{n=0}^{\infty}\sum_{r=1}^{\infty} q_r^n(t)\mu_{3r}^n(s)\cos n\theta + q_r^{\prime n}(t)\mu_{3r}^n(s)\sin n\theta,
\end{aligned} \tag{2.1}
$$

in which $q_r^n(t)$, $q_r^{\prime n}(t)$ = the generalized coordinates of symmetrical and antisymmetrical deformations for the r-th mode of the n-th harmonic; $\mu_{ir}^n(s)$ $(i = 1, 2, 3)$ = r-th normalized mode of n-th harmonic in the meridional, circumferential and normal directions; and s = arc coordinate along the meridional curve.

If we neglect the rotatory inertia and the terms of the order of h^2/r^2 compared to unity by means of Lagranges' equations of motion, the equations of forced vibration of the cooling tower shell can be obtained as follows:

$$
\begin{aligned}
\ddot{q}_r^n(t) + 2\omega_r^n\zeta_{rn}\dot{q}_r^n(t) + \omega_r^{2n}q_r^n(t) &= \frac{P_r^n}{M_r^n}\\
\ddot{q}_r^{\prime n}(t) + 2\omega_r^n\zeta_{rn}\dot{q}_r^{\prime n}(t) + \omega_r^{2n}q_r^{\prime n}(t) &= \frac{P_r^{\prime n}}{M_r^n}
\end{aligned} \tag{2.2}
$$

in which

$$
\begin{aligned}
M_r^n &= \varrho\pi\int_0^{S_0} h[\mu_{1r}^{2n}(s) + \mu_{2r}^{2n}(s) + \mu_{3r}^{2n}(s)]r(s)\,\mathrm{d}s;\\
P_r^n &= \int_0^{2\pi}\int_0^{S_0} P_w\,\mu_{3r}^n(s)\cos n\theta r(s)\,\mathrm{d}s\,\mathrm{d}\theta;\\
P_r^{\prime n} &= \int_0^{2\pi}\int_0^{S_0} P_w\,\mu_{3r}^n(s)\sin n\theta r(s)\,\mathrm{d}s\,\mathrm{d}\theta;
\end{aligned}
$$

P_w = normal component of wind pressure acting on the surface of the shell; $\omega_r^{2n} = (2\pi f_m^2)^2 = K_r^n/M_r^n$; ϱ = mass per unit volume of the tower shell; ζ_{rn} = the ratio of linear critical damping for the r-th mode of the n-th harmonic; and s_0 = are length of meridional curve of the hyperbolic cooling tower.

The natural frequency ω_r^n and mode shapes $\mu_{ir}^n(s)$ $(i = 1, 2, 3)$ in (2.2) are determined by means of the finite element method based on the Novozhilov's shell theory. The displacements are expanded into a Fourier series in the circumferential direction. The higher order Hermite interpolation functions are used in the meridional direction. In the axisymmetric rotational shell element, it is defined that the Fourier series coefficients of displacements u, v, w, their derivatives of first order with respect to arc length s, $\partial u/\partial s$, $\partial v/\partial s$, $\partial w/\partial s$, and their derivatives of second order $\partial^2 u/\partial s^2$, $\partial^2 w/\partial s^2$, are the generalized nodal circular displacements. The displacement functions of reference [9] are adopted. For the i-th element of the shell we have:

$$
\mathbf{V}^* = \mathbf{T}\sum_{k=0}^{M} \chi_{ck}\mathbf{A}\boldsymbol{\alpha}_k \tag{2.3}
$$

in which $\mathbf{V}^* = \lfloor u^*, v^*, w^* \rfloor^T$; $\mathbf{X}_{ck}\mathbf{A}$ = meridional, circumferential interpolation function matrix; $\mathbf{T}$ = transform matrix between the displacement vector $\mathbf{V}^*$ in the local system and the displacement vector $\mathbf{V}$ in the datum. The linear property of this problem is considered, then for each harmonic term, the equation of free vibration of the hyperbolic cooling tower supported by columns is obtained as follows:

$$\mathbf{M}_k \ddot{\varrho}_k + \mathbf{K}_k^* \varrho_k = 0 \tag{2.4}$$

in which $\mathbf{M}_k$, $\mathbf{K}_k^*$ = mass, stiffness and geometric stiffness matrices of shell structure in the datum system.

Let $\varrho_k = q_k \sin(\omega t + \varphi)$, then the characteristic equation of the k-th harmonic is given as:

$$(\mathbf{K}_k + \mathbf{K}_{gk} - \omega^2 \mathbf{M}_k)\mathbf{q}_k = 0. \tag{2.5}$$

This equation is solved by means of the subspace iteration method.

The normal displacement response spectrum at point (s, θ) of the cooling tower, acted on by distributed fluctuating wind pressure loads is obtained as

$$S_W(r_i; f) = \int_0^{2\pi}\int_0^{2\pi}\int_0^{S_0}\int_0^{S_0} H_{1A}^*(jf) H_{1B}(jf)\, S_p(r_A, r_B; f) \times r(S_A) r(S_B)\, \mathrm{d}S_A\, \mathrm{d}S_B\, \mathrm{d}\theta_A\, \mathrm{d}\theta_B, \tag{2.6}$$

in which r = radius vector at point (s_1, θ_1) on the shell surface; $H_{1A}^*(jf)$ and $H_{1B}(jf)$ are complex frequency response functions; $j = \sqrt{-1}$.

In the partially correlated fluctuating pressure case, according to the relation between the RMS pressure distribution coefficient $C_f(\theta)$ and the mean pressure distribution coefficient $C_P(\theta)$ by means of the two dimensions quasistationary theory given by J. C. Hunt, the simplified physical model of cross-spectral density of fluctuating pressure [10] can be obtained as:

$$S_P(r_A, r_B; f) = \frac{1}{4(\sigma_V/\bar{V}_z)^2} C_f(\theta_A) C_f(\theta_B) \left(\frac{Z(S_A) + Hd}{10}\right)^{\alpha} \left(\frac{Z(S_B) + Hd}{10}\right)^{\alpha} \times (\bar{\varrho}\bar{V}_{10})^2 S_V(r_A, r_B; f), \tag{2.7}$$

in which

$$S_V(r_A, r_B; f) = \int_{-\infty}^{\infty} [v(r_A, t) v(r_B, t + \tau)]\, \mathrm{e}^{-j2\pi f\tau}\, \mathrm{d}\tau.$$

Since the quadratic component of cross-spectral density of the fluctuating pressure is small, the quad-coherence of wind velocity may be considered to be zero. Meanwhile, it is assumed that the variation of turbulence intensity with height is small and that the power spectrum of longitudinal gust components is invariant with height. Thus, we have

$$S_V(r_A, r_B; f) = S_V(f) R_V(\Delta Z, \Delta\theta; f), \tag{2.8}$$

in which $\Delta Z = |Z(s_A) - Z(s_B)|$; $\Delta\theta = |\theta_A - \theta_B|$. In general, the coherence function $R_V(\Delta Z, \Delta\theta; f)$ is related to the tower shape and fluctuating frequency, and its values are different on the various regions on the tower surface. In cases without full-scale information of the power spectrum and cross-spectrum of fluctuating pressure, the vertical and crosswise coherence functions of gust wind suggested by Davenport can be used to determine the $S_V(r_A, r_B; f)$. Thus the RMS value of the gust response of normal displacement is given as:

$$\begin{aligned}\sigma_W(s, \theta) &= \left[2\int_0^{\infty} S_W(s, \theta; f)\,\mathrm{d}f\right]^{1/2} \\ &= \left\{\sum_{n=0}^{\infty}\sum_{m=0}^{\infty}\sum_{r=1}^{\infty}\sum_{l=1}^{\infty} 2\left[\left(\int_0^{\infty} S_{q_r^n q_l^m}(f)\,\mathrm{d}f \cos m\theta \cos n\theta\right)\right.\right. \\ &\quad \left.\left. + \left(\int_0^{\infty} S_{q_r'^n q_l'^m}(f)\,\mathrm{d}f \sin m\theta \sin n\theta\right)\right] \mu_{3r}^n(s)\mu_{3l}^m(s)\right\}^{1/2} \end{aligned} \tag{2.9}$$

in which the coupled terms between symmetrical and antisymmetrical deformation are neglected. The relations between cross-spectral values of generalized coordinates and cross-spectral values of generalized load are given as follows:

$$S_{q_r^n q_l^m}(f) = \frac{\chi_{rn}^*(jf)\chi_{lm}(jf)}{16\pi^4 f_{rn}^2 f_{lm}^2 M_r^n M_l^m} S_{P_{rn}P_{lm}}(f);$$

$$S_{q_r'^n q_l'^m}(f) = \frac{\chi_{rn}^*(jf)\chi_{lm}(jf)}{16\pi^4 f_{rn}^2 f_{lm}^2 M_r^n M_l^m} S_{P'_{rn}P'_{lm}}(f);$$

$$\chi_{rn}^*(jf) = X_{rn} + jY_{rn}; \qquad \chi_{lm}(jf) = X_{lm} + jY_{lm};$$

$$X_{rn} = \frac{1 - f^2/f_{rn}^2}{\left[\left(1 - \dfrac{f^2}{f_{rn}^2}\right)^2 + 4\dfrac{f^2}{f_{rn}^2}\zeta_{rn}^2\right]}; \qquad Y_{rn} = \frac{2\zeta_{rn}(f/f_{rn})}{\left[\left(1 - \dfrac{f^2}{f_{rn}^2}\right)^2 + 4\dfrac{f^2}{f_{rn}^2}\zeta_{rn}^2\right]}.$$

Replacing rn by lm we obtain the expressions of X_{lm} and Y_{lm} from X_{rn} and Y_{rn}.

The cross-spectral density of generalized load is defined as:

$$\begin{aligned} S_{P_{rn}P_{lm}}(f) &= \int_0^{2\pi}\int_0^{2\pi}\int_0^{S_0}\int_0^{S_0} S_P(r_A, r_B; f)\mu_{3r}^n(s_A)\mu_{3l}^m(s_B)\cos n\theta_A \\ &\quad \times \cos m\theta_B r(s_A) r(s_B)\,\mathrm{d}s_A\,\mathrm{d}s_B\,\mathrm{d}\theta_A\,\mathrm{d}\theta_B. \end{aligned}$$

Using sinusoidal function instead of cosine function in the $S_{P_{rn}P_{lm}}(f)$, the $S_{P'_{rn}P'_{lm}}(f)$ will be obtained.

The gust response factor is defined as:

$$G_W = 1 + g\frac{\sigma_W(s, \theta)}{\bar{W}(s, \theta)} \tag{2.10}$$

in which

$$g = \sqrt{2 \ln \nu^* T} + \frac{0.577}{\sqrt{2 \ln \nu T}}$$

is called the peak factor.

Let T = duration of the gust wind record:

$$\nu^* = \left[\frac{\int_0^{\infty} f^2 S_W(f)\,\mathrm{d}f}{\int_0^{\infty} S_W(f)\,\mathrm{d}f}\right]^{1/2}.$$

For a lightly damped system, $v^* \doteq$ the lowest natural frequency of the structure. $\bar{W}(s, \theta)$ is called the normal displacement response induced by mean velocity. It can be obtained by means of a static procedure and we can also use the mode-superposition procedure to obtain the static mean response of the tower, approximately as follows:

$$\bar{W}(s, \theta) = \sum_{n=0}^{\infty} \sum_{r=1}^{\infty} \frac{P_{rn} \mu_{3r}^{n}(s) \cos n\theta}{4\pi^2 f_{rn}^2 M_r^n} \tag{2.11}$$

in which

$$P_{rn} = \tfrac{1}{2} \bar{\varrho} \bar{V}_{10}^2 \int_0^{2\pi} \int_0^{S_0} \left(\frac{Z(s) + Hd}{10} \right)^{2\alpha} C_P(\theta) \mu_{3r}^{n}(s) \cos n\theta \, r(s) \, ds \, d\theta.$$

From (2.10), the normal displacement response of the hyperbolic cooling tower induced by wind gust is given as:

$$W(s, \theta) = \bar{W}(s, \theta) + g\sigma_W(s, \theta). \tag{2.12}$$

If we want to obtain the RMS of stress resultants of cooling towers, the corresponding mode shapes are used in place of $\mu_{3r}^{n}(s)$ in (2.10).

3. The effect of axisymmetric geometrical imperfections on stress resultant distributions in the cooling tower

Through the computation of a typical hyperbolic cooling tower under dead load or equivalent statical wind load, the effect of axisymmetric geometrical imperfections on stress resultant distributions in the tower shell is studied by means of the finite element solution, using a curved segment rotating around an axisymmetric axis [9]. The geometric scales of this tower are given as follows:

The height of the tower is 90 m; the height of the throat is 72 m; the height of the columns is 5.8 m; the parameters of the middle surface equation of shell

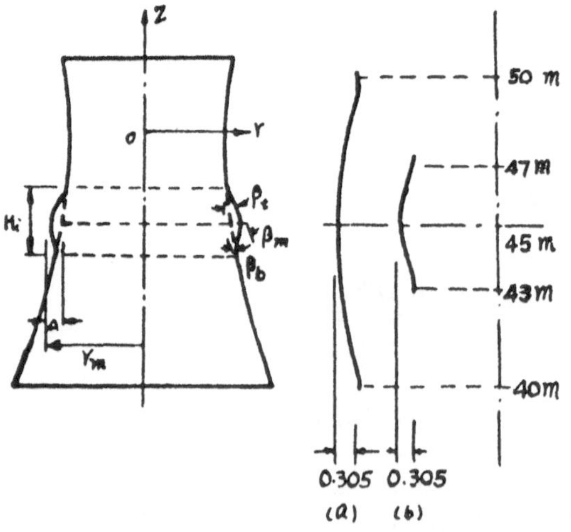

Figure 1. Assumed shapes of imperfection.

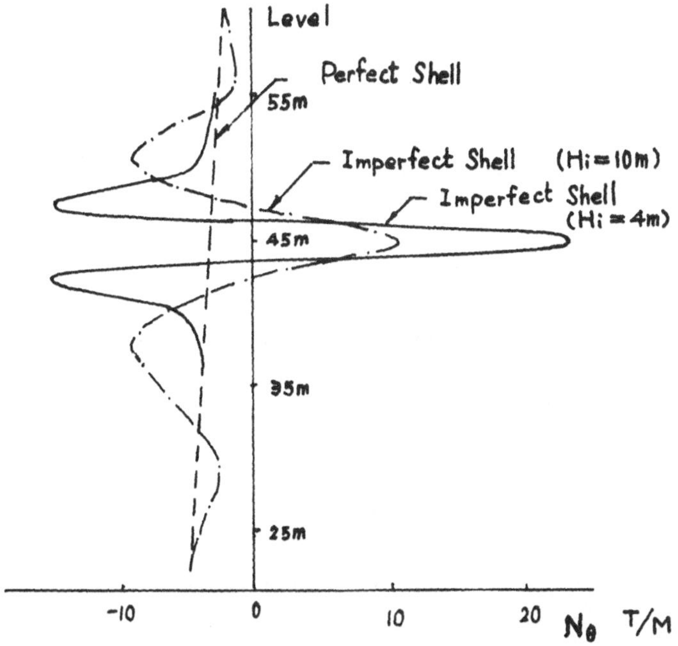

Figure 2. Meridional variation of circumferential membrane force under dead load due to geometric imperfection.

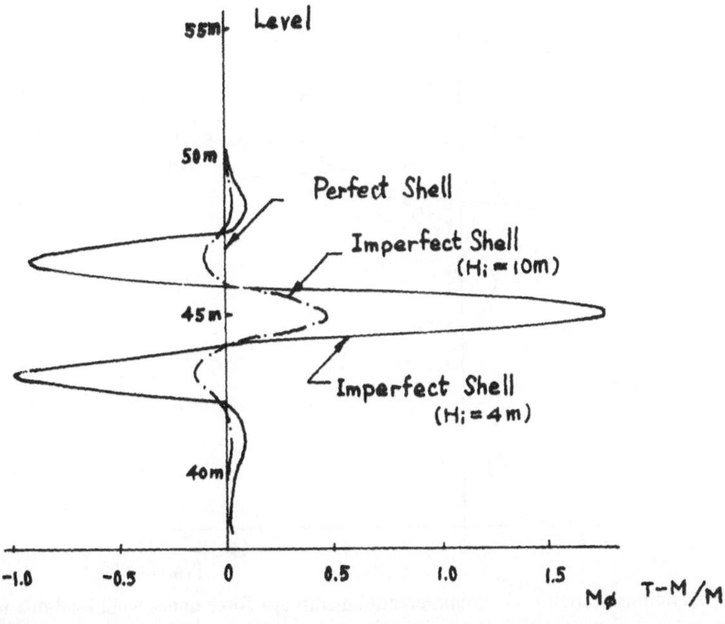

Figure 3. Meridional variation of meridional moment under dead load due to geometric imperfections.

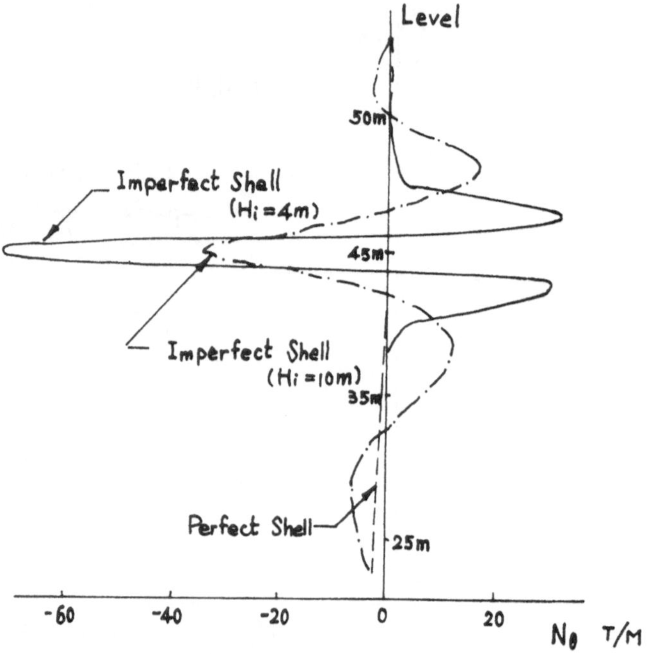

Figure 4. Meridional variation of circumferential membrane force under wind load due to geometric imperfection ($\theta = 0°$).

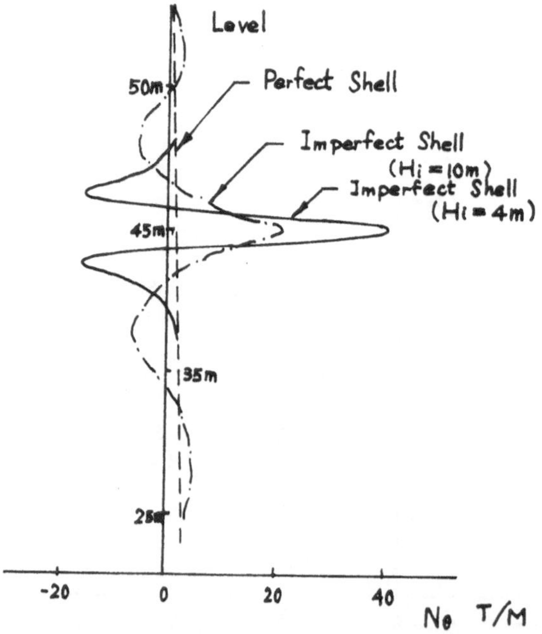

Figure 5. Meridional variation of circumferential membrane force under wind load due to geometric imperfection ($\theta = 75°$).

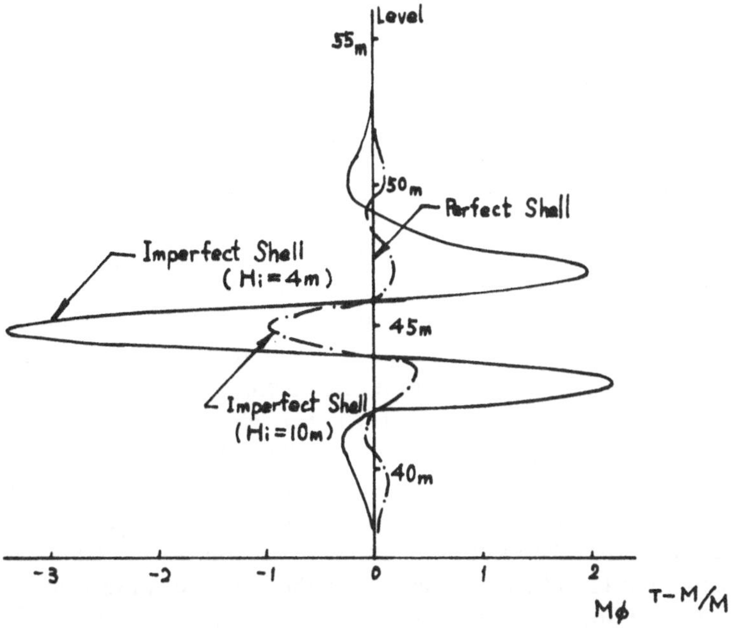

Figure 6. Meridional variation of meridional moment under wind load due to geometric imperfection ($\theta = 0°$).

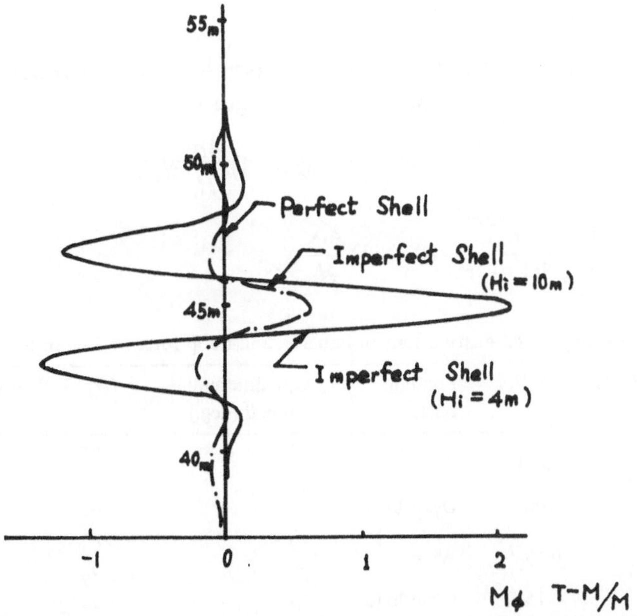

Figure 7. Meridional variation of meridional moment under wind load due to geometric imperfection ($\theta = 75°$).

($r^2 = b^2 + aZ^2$) are $a = 0.177$ m, $b = 19.4$ m; the variation of the shell thickness is of the exponential law; and their minimum, maximum thickness and exponent are $h_{min} = 0.14$ m, $h_{max} = 0.5$ m, $\eta = 2.0$.

The material constants of the cooling tower shell are:
Poisson's ratio $\nu = 0.167$, elastic modulus $E = 2{,}700{,}000\,T/M^2$, specific weight $\gamma = 2.4\,T/M^3$.

Suppose the tower shell base is fixed with supports and the top has a free boundary. Two kinds of meridional geometric imperfections are considered (Fig. 1), in which the deviation of the horizontal radius is $\Delta = 0.305$ m. In Figs. 2–7 the effect of axisymmetric geometric imperfections on circumferential membrane force N_θ and meridional bending moment M_φ are distributions under dead load or wind load, are given. It will be seen that the circumferential membrane force N_θ has a significant increase because of geometric imperfections, and the meridional force N_φ, which is primarily governed by the overall equilbrium of structure, does not change significantly. It can also be seen that the hoop force and meridional moment caused by the geometric imperfections are relatively local and that they decrease away from the location of the imperfection. In the same deviation of radius condition this effect will increase as the imperfect wave length decreases.

From Table 1 and Fig. 8, it can be seen that both $(N_\theta - N_{\theta_0})/N_{\varphi_0}$, and M_φ/N_{φ_0} are constant along the circumferential direction, and are approximately linearly related to the imperfection slope $2\Delta/H_i$. Therefore, the hoop force and meridional moment caused by the geometric imperfection can be obtained, using an axisymmetric radial force per unit length of the circumference acting on a cylinder, whose radius is the same as the horizontal radius at the middle location of the imperfect wave length.

$$p = N_{\varphi_0}\beta.$$

From [11], it is not difficult to obtain the hoop force and meridional moment at the middle location of the imperfect wave length is

$$N_\theta = N_{\theta_0} + \sqrt[4]{\frac{3(1-\nu^2)}{6}}\left(\frac{r}{h}\right)^{1/2}(\psi - 1)\frac{4\Delta}{H_i}N_{\varphi_0}, \tag{3.1}$$

$$M_\varphi = \frac{1}{4}\sqrt[4]{\frac{r^2h^2}{3(1-\nu^2)}}(\vartheta - 1)\frac{4\Delta}{H_i}N_{\varphi_0}, \tag{3.2}$$

Table 1. The circumferential distributions of imperfection hoop force and moment in cooling tower.

Z_m/Z_0	$H_i/(M)$	$2\Delta/H_i$	Loading type	Circumferential angle θ (deg.)	$(N_\theta - N_{\theta_0})/N_{\varphi_0}$	M_φ/N_{φ_0}
0.625	10	0.061	Wind	0	1.01	0.028
				75	1.01	0.032
		0.061	Dead load	0	1.02	0.035
0.625	4	0.153	Wind	0	2.03	0.10
				75	1.98	0.11
		0.153	Dead load	0	1.94	0.12

$\Delta = 0.305$ M; N_{θ_0}, N_{φ_0} = membrane forces of perfect shell at the same height; Z_m = middle height of imperfection; Z_0 = throat height.

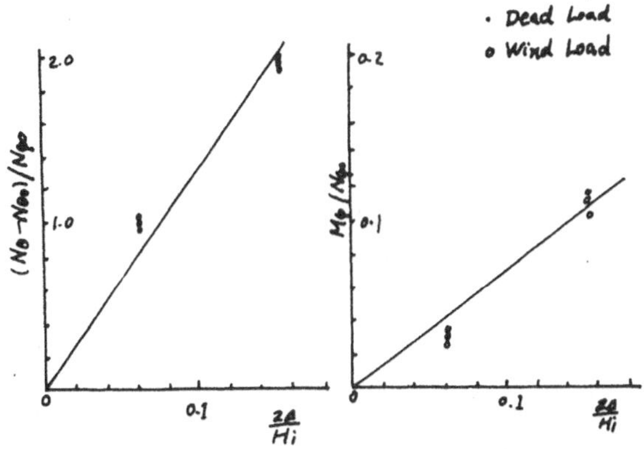

Figure 8. Variation with slope of imperfection of: (a) circumferential membrane force, (b) meridional moment.

in which

$$\begin{aligned} \psi &= \left(\sin \frac{kH_i}{2} + \cos \frac{kH_i}{2}\right) e^{-kH_i/2}, \\ \vartheta &= \left(\sin \frac{kH_i}{2} - \cos \frac{kH_i}{2}\right) e^{-kH_i/2}, \\ k &= \sqrt[4]{\frac{3(1-\nu^2)}{r^2 h^2}}. \end{aligned} \tag{3.3}$$

where h = shell thickness; and $r \doteq r_m$ horizontal radius at the middle location of the imperfect wave length.

From Table 2, it can be seen that the values of N_θ of the middle imperfection using the approximate expressions (3.1) and (3.2) agree quite well with the finite element solutions of this paper; the relative error is only about 10%. But the relative errors of M_φ are significant because its values are small near $Z_m/Z_0 = 0.625$.

Table 2. Comparison of approximate equations (3.1) and (3.2) with the finite element solution.

θ (deg.)	r_m (M)	H_i (M)	Δ (M)	Wind load N_θ (T/M)	Wind load M_φ (T)	Dead load N_θ (T/M)	Dead load M_φ (T)
0	22.79	10.0	0.305	−34.97 −35.66	−1.02 −1.52	10.80 10.99	0.50[a] 0.60[b]
75	22.79	10.0	0.305	21.26 21.99	0.65 0.85	– –	– –
0	22.79	4.0	0.305	−71.75 −64.37	−3.46 −3.06	23.92 22.18	1.65 1.21
75	22.79	4.0	0.305	41.15 37.99	2.16 1.72	– –	– –

[a] Finite element analysis results.
[b] Calculated from approximate equations (3.1) and (3.2).

4. Gust response of a shell with axisymmetric geometric imperfections

In Table 3, the natural frequencies of the imperfect shell (as Fig. 1 types *a* and *b*) by means of a finite element solution suggested in the present paper, are given. It is seen that the effect of meridional geometric imperfection on the natural frequency in the cooling tower is significant. The lowest natural frequency, which appears at the 1-st mode in 4-th harmonic, is reduced by about 7–15%. Moreover the natural frequencies at higher mode increase. Because the dominant frequency of turbulent gust wind is lower, the dynamic response induced by a wind gust will increase. From Table 4, it is seen that the dynamic response of meridional membrane force N_φ, which is primarily governed in the wind design of the tower shell, will increase as H_i decreases because of the meridional geometric imperfections. Its gust response factor will increase by about 5%. This must be considered

Table 3. The effect of axisymmetric imperfection on natural frequency in cooling tower.

Mode shapes	Frequency (Hz) / Type	Circumferential harmonic 0	1	2	3	4
1	a. Perfect shell	9.240	4.980	2.814	2.349	1.955
	b. Imperfect shell	8.838	4.640	2.673	2.403	1.827
	c. Imperfect shell	8.276	3.780	1.979	1.622	1.655
2	a.	11.836	9.875	5.239	2.841	2.444
	b. H_i = 10 m	11.270	10.309	5.373	2.701	2.861
	c. H_i = 4 m	8.965	9.646	5.195	2.959	2.439
3	a.	16.227	14.554	9.563	5.928	3.871
	b.	16.985	14.355	9.192	5.802	4.053
	c.	16.798	14.281	9.345	5.308	3.569
4	a.	17.320	16.003	12.906	9.074	6.359
	b.	19.049	16.990	13.345	9.110	5.984
	c.	18.182	16.698	13.018	9.199	6.358
5	a.	18.339	17.487	14.936	12.354	9.500
	b.	21.313	18.986	15.669	12.250	9.563
	c.	19.945	18.454	15.531	11.761	8.809
6	a.	19.458	18.810	17.344	14.160	11.613
	b.	22.972	20.868	18.224	15.177	12.272
	c.	22.124	20.141	17.206	14.766	11.512

Table 4. The comparison of gust response factors between a perfect shell and an imperfect shell.

Angle (deg.)	G_{N_θ} Perfect shell	G_{N_θ} Imperfect shell H_i = 10 m	G_{N_θ} Imperfect shell H_i = 4 m	G_{N_φ} Perfect shell	G_{N_φ} Imperfect shell H_i = 10 m	G_{N_φ} Imperfect shell H_i = 4 m
0	2.056	2.017	2.223	2.347	2.377	2.463
75	2.141	2.033	2.264	–	–	–

Z = 45 m, Computing height level of G_{N_θ}.
Z = 5.8 m, Computing height level of G_{N_φ}.

in recommending tolerance limits of geometric imperfections. On the other hand, the dynamic response of the hoop membrane force N_θ will increase by about 5% very small wavelengths because of the geometric imperfections. Therefore, in consideration of safety while we recommend the allowable imperfection for very small wave lengths of 4 m or less, the effect of imperfections on the dynamic response of hoop membrane force must also be considered.

5. Specifying tolerance limit for meridional geometric imperfections

In (3.1) the maximum possible negative value of ψ is -0.0431, then the hoop membrane force N_θ in the middle of imperfection wave length is

$$N_\theta \doteq 2.7\sqrt{\frac{r}{h}\frac{\Delta}{H_i}}N_\varphi \tag{5.1}$$

in which N_φ = the maximum compressive meridional force including wind and dead load effects.

Because the circumferential moment M_θ is very small, in general, the stress state of the circumferential section corresponds to eccentric tension. Let the tensile strength of the shell in design be R_g and the minimum reinforcement ratio be μ_{min}. Meanwhile, let the shell thickness be h, and then the allowable slope of imperfection can be obtained as:

$$\left(\frac{2\Delta}{H_i}\right)_{max} = \frac{2R_g\mu_{min}}{2.73KN_{\varphi 0}}\sqrt{\frac{h^3}{r}} - \frac{4M_\theta}{2.73(h_0 - a)N_{\varphi 0}}\sqrt{\frac{h}{r}} \tag{5.2}$$

in which h_0 = distance between tensile steel and the compressive surface; a = thickness of protective layer of steel; K = safety coefficient of structural strength. The maximum allowable slope of imperfection given by (5.2) will ensure that the hoop force will not exceed the capacity of the minimum hoop reinforcement.

In (4.2), the maximum possible negative value is -0.21, so that we have

$$M_\varphi = 0.93\sqrt{rh}\,\frac{\Delta}{H_i}(-N_{\varphi 0}).$$

The negative sign shows that the interior side of the meridional section is placed in a state of tension.

The reinforcement in the meridional direction is designed for the meridional membrane force N_φ. It is supposed that an overstress by a factor α is permitted in that reinforcement, so that we have

$$\left(\frac{2\Delta}{H_i}\right)_{max} = 1.08\,\frac{(h_0 - a)}{\sqrt{rh}}\,\alpha. \tag{5.3}$$

In Table 5 the allowable slopes of geometric imperfections indicated by (5.2) and (5.3) are given. Meanwhile, the corresponding allowable values recommended by

Table 5.1. The comparison of steel reinforcement.

Height level Z (M)	Deviation of horizontal radius Δ (M)	Direction of steel reinforcement	Imperfection wave length $H_i = 10\,\text{m}$					
			Perfect shell Ag (cm^2/M)	$\mu_{min} = 0.35\%$ imperfect shell		Perfect shell Ag (cm^2/M)	$\mu_{min} = 0.20\%$ imperfect shell	
				$\left(\frac{2\Delta}{H_i}\right)_{max}$	Ag (cm^2/M)		$\left(\frac{2\Delta}{H_i}\right)_{max}$	Ag (cm^2/M)
45	0.305	Outer layer in circumference	2.60	1.0 (BS)	1.54	1.50	1.0 (BS)	1.54
				1.5 (IASS)	1.94		1.5 (IASS)	1.94
				1.34(Eq. 5.3)	1.81		0.85 (Eq. 5.2)	1.42
				1.83 (5)	2.10		1.10 (5)	1.62
				2.90 (2)	3.05		1.90 (2)	2.25
45	0.305	Inner layer in meridian	5.88 ($\alpha = 0.2$)	1.0 (BS)	5.40	5.88 ($\alpha = 0.2$)	1.0 (BS)	5.40
				1.5 (IASS)	5.65		1.5 (IASS)	5.65
				1.34(Eq. 5.3)	5.57		0.85 (Eq. 5.2)	5.32
			6.37 ($\alpha = 0.3$)	1.83 (5)	5.82	6.37 ($\alpha = 0.3$)	1.10 (5)	5.47
				2.9 (2)	6.34		1.90 (2)	5.85

Table 5.2. The comparison of steel reinforcement.

Height level Z (M)	Deviation of horizontal radius Δ (M)	Direction of steel reinforcement	Imperfect wave length $H_i = 4\,\text{m}$				
			Perfect shell Ag (cm^2/M)	Imperfect shell $\mu_{\min} = 0.35\%$			
				$G_{N_\theta} = 2.14$			
				$\left(\frac{2\Delta}{H_i}\right)_{\max}$	Ag (cm^2/M)	$\left(\frac{2\Delta}{H_i}\right)_{\max}$	Ag (cm^2/M)
45	0.305	Outer layer in circumference	2.60	2.5 (BS) 1.5 (IASS) 1.34 (Eq. 5.3) 1.83 (5) 1.16 (2)	3.42 2.77 2.66 2.98 2.54	– – – – –	– – – – –
				$G_{N_\varphi} = 2.35$		$G_{N_\varphi} = 2.46$	
45	0.305	Inner layer in meridian	5.88 6.37	2.5 (BS) 1.5 (IASS) 1.34 (Eq. 5.3) 1.83 (5) 1.16 (2)	6.75 6.06 5.94 6.28 5.81	2.5 (BS) 1.5 (IASS) 1.34 (Eq. 5.3) 1.83 (5) 1.16 (2)	7.32 6.57 6.44 6.81 6.30
				Imperfect shell $\mu_{\min} = 0.20\%$			
				$G_{N_\theta} \mp 2.14$		$G_{N_\theta} = 2.26$	
45	0.305	Outer layer in circumference	1.50	2.5 (BS) 1.5 (IASS) 0.85 (Eq. 5.2) 1.10 (5) 0.75 (2)	3.42 2.77 2.35 2.50 2.28	2.5 (BS) 1.5 (IASS) 0.85 (Eq. 5.2) 1.10 (5) 0.75 (2)	3.51 2.85 2.40 2.57 2.33
				$G_{N_\varphi} = 2.35$		$G_{N_\varphi} = 2.46$	
45	0.305	Inner layer in meridian	5.88 6.37	2.5 (BS) 1.5 (IASS) 0.85 (Eq. 5.2) 1.10 (5) 0.75 (2)	6.75 6.06 5.60 5.77 5.52	2.5 (BS) 1.5 (IASS) 0.85 (Eq. 5.2) 1.10 (5) 0.75 (2)	7.32 6.57 6.06 6.25 5.98

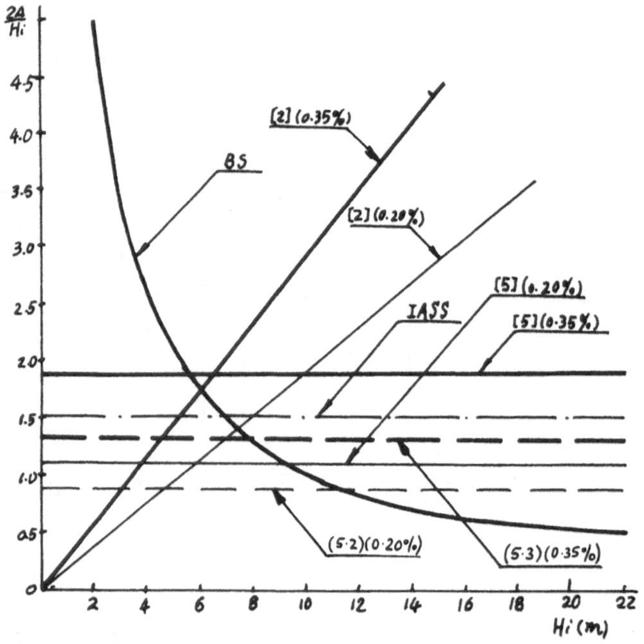

Figure 9. Variation of allowable slope of imperfection with imperfect wave length H_i.

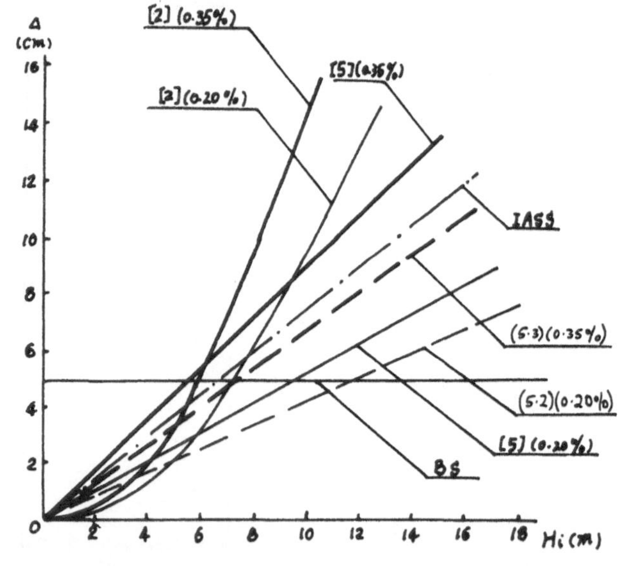

Figure 10. Variation of allowable deviation of horizontal radius with imperfect wave length H_i.

BS Standard, IASS guidelines, and equations given by references [5] and [2] are obtained in the same table. The variations of allowable imperfect slopes and allowable deviations of horizontal radius with imperfect wave length H_i are plotted in Figs. 9 and 10. It can be seen that the allowable slope of the geometric imperfection suggested by the present paper for the longer wave length would ensure the safety of the cooling tower, if an overstress by a factor $\alpha = 0.2$ were permitted in the meridional reinforcement of the perfect shell because of evaluating the effect of imperfection on gust response of the meridional membrane force N_φ. For the very small wavelength it is necessary to have an overstress by a factor $\alpha = 0.3$. It can also be seen that an allowable imperfect slope ensuring safety is recommended by the BS Standard for long wave lengths. But the allowable imperfection limit recommended by the IASS guideline agrees with the recommendation given by the present paper. however, the imperfection limits obtained in references [2] and [5] are only applied, respectively, to very small and longer wave lengths. It must be noted that all the above imperfection limits cannot ensure the safety of circumferential strength for very small imperfect wave lengths and lower levels of minimum reinforcement ratio, so that it is not applied in the design of a tower shell under wind load. From what has been said above, in specifying the tolerance limit for geometric imperfections in a cooling tower for very small wave lengths, we must consider the disadvantageous increase of gust response induced by the geometric imperfections. In order to use (5.2) and (5.3) given in the present paper to recommend the allowable geometric imperfection limit, the reinforcement in the meridional direction to be designed for a perfect shell must have an overstress by a factor $\alpha = 0.2$–0.3.

6. Conclusion

1. The effect of axisymmetrical geometric imperfections on static circumferential membrane force N_θ and meridional moment M_φ in a hyperbolic cooling tower, are significant. This effect will increase as the imperfect wave length decreases and it will damp out away from the location of imperfection.
2. In order to evaluate the effect of the imperfection on gust response in a tower shell, it is necessary to have an overstress by a factor $\alpha = 0.2$–0.3 in the meridional reinforcement of a perfect shell. Use of (5.2) and (5.3), to determine the allowable geometric imperfection limit, can ensure the safety of the tower shell.
3. Using the lower minimum reinforcement ratio in two directions, in tower design under wind load, is not recommended.
4. The allowable imperfection limit recommended by BS Standard, IASS guidelines, and reference [5], are applied to the longer wave length, H_i; but the allowable imperfection limit given by [2] is applied to small wave length, H_i.

References

1. Mircea Soare, "Cooling towers with constructional imperfections", *Concrete* (November, 1967), 367–379.
2. Croll, J. G. A. and Kemp, K. O., "Specifying tolerance limits for meridional imperfections in cooling towers", *ACI Journal* (January, 1979), 139–159.
3. Shiro Kato and Yoshitsura Yokoo, "Effect of geometric imperfections on stress distributions in the cooling tower", *Eng. Struct.* **2** (July, 1980), 150–156.
4. Croll, J. G. A., Kaleli, F. and Kemp, K. O., "A simplified approach to the analysis of geometrically imperfect cooling tower shells", *Eng. Struct.* **1** (1979), 92–98.

5. Adam Al-dabbagh and Gupta, A. K., "Meridional imperfection in cooling tower design", *Journal of the Structural Division, ASCE, Proc.* 14636, **105**, ST6 (June, 1979), 1089–1102.
6. Norton, R. L. and Weingaten, V. I., "The effect of asymmetric imperfections on the earthquake response of hyperbolic cooling towers", *Proc. of 6th I.E.A.A.* Vol. 2, 1343–1349.
7. Zerna, W., "Recommendations for the design of hyperbolic or other similar shaped cooling tower", *Proceedings of 2nd International Association of Shell Structures Colloquium* (June, 1975), 9–11.
8. Hashish, M. G. and Abu-Sitta, S. H., "Free vibration of the hyperbolic cooling tower", *Journal of the Engineering Mechanics Division, ASCE* **97**, EM2 (April, 1971), 253–269.
9. Lu Wen-da *et al.*, *Finite Element Solution of the Hyperbolic Cooling Tower Supported by Column Systems*, Research Report of Thermal Power Engineering Research Institute of the Ministry of Power (June, 1976), China.
10. Lu Wen-da *et al.*, "Response of the hyperbolic cooling tower to turbulent wind", *Journal of Electrical Engineering*, CSEE, No. 1 (1982).
11. Timoshenko, S. and Woinowsky-Krieger, S., *Theory of Plates and Shells*, McGraw-Hill, New York, Toronto and London (1959).
12. Abu-Sitta, S. H. and Hashish, M. G., "Dynamic wind stress in the hyperbolic cooling tower", *Journal of Structural Division, ASCE* **99**, St9 (Sept., 1973), 1823–1825.
13. Hashish, M. G. and Abu-Sitta, S. H., "Response of hyperbolic cooling towers to turbulent wind", *Journal of Structural Division, ASCE* **100**, ST5 (May, 1974).
14. Singh, M. P. and Gupta, A. K., "Gust factors for hyperbolic cooling towers", *Journal of Structural Division, ASCE*, Proc. Paper 11894, **102**, ST2 (Feb., 1976), 371–386.
15. Lu Wen-da *et al.*, "Effect of dead load on the dynamic properties of the hyperbolic cooling tower", *Journal of Thermal Electrical Power*, No. 5 (Feb., 1982), China.

General solution of the bending of ring shells with equations in complex form

CHEN SHAN-LIN

Department of Civil Engineering, Chongqing Institute of Architectural Engineering, Chongqing, PRC

Abstract. Using the Chernina equation, we can obtain the equation in complex form of the bending of a ring shell which takes the same form as Novozhelov's. From this, it is convenient for us to gain the general solution of this problem by making use of axial-symmetrical results, given by ours and previous works. As an example, we calculate the pure bending of a C-type corrugated tube and the result agrees well with the case subjected to an axial applied force P ($M = PR/2$). This equivalent fact was indicated by C. E. Turner [11] through experiment, and Iohelison [9] by calculation.

1. Introduction

The bending problem of a ring shell subjected to edge bending moment and horizontal force (Fig. 1) belongs to the wind load type. All the displacement components and internal force components, whose positive directions are as shown in Fig. 2 about the coordinate argument θ, vary according to the following rules:

$$\left.\begin{aligned} &[u^*, w^*, N_\varphi, N_\theta, Q_\varphi, M_\varphi, M_\theta] \\ &\quad = [u, w, n_\varphi, n_\theta, q_\varphi, m_\varphi, m_\theta] \cos\theta \\ &[v^*, N_{\varphi\theta}, N_{\theta\varphi}, Q_\theta, M_{\varphi\theta}] \\ &\quad = [v, n_{\varphi\theta}, n_{\theta\varphi}, q_\theta, m_{\varphi\theta}] \sin\theta \end{aligned}\right\} \tag{1.1}$$

The bending problem of a rotational thin shell under wind type load has been simplified successfully by V. S. Chernina [1] whose results took the form of simultaneous second order ordinary differential equations with two variables. The author obtained equations in another form with complex variables, and then reduced these results to the cases of variable thickness [2] and non-period displacement [7]. In [8], F. Y. M. Wan, obtained the Chernina equation in another form. All these results are used to treat several practical problems, such as the bending of a corrugated tube [9].

In this paper, we employ Chernina's equation with complex variables in the case of a ring shell, and obtain the equation which has the same form as Novozhelov's, whose general solution has recently been obtained by Chien Wei-zang, etc. (1980) [3]. Therefore it is convenient for us to obtain the general solution of the axial-symmetrical bending problem, to make use of the above-mentioned results, which can then be used to study problems of the pure bending and general bending of

Yeh, K. Y. (ed) Progress in Applied Mechanics
 ISBN-13: 978-94-010-8061-3

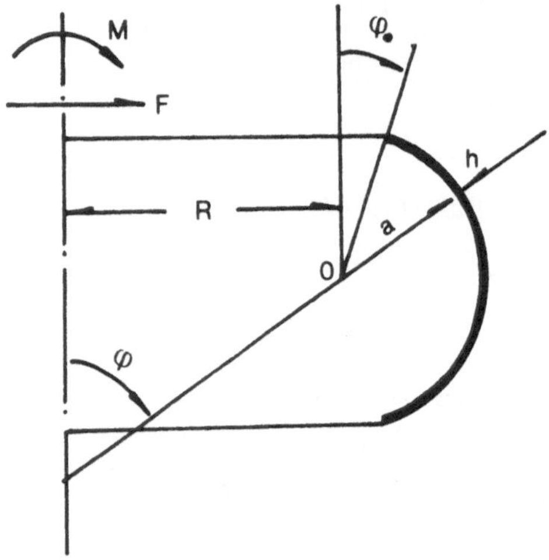

Figure 1. Load and dimension.

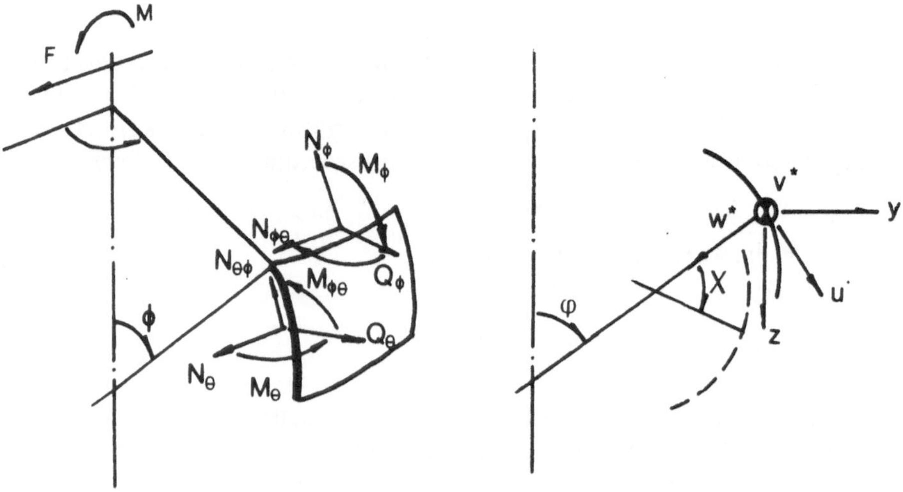

Figure 2. Internal forces and displacements.

corrugated tubes. As the application of this result, the pure bending problem of a C-type corrugated tube is investigated.

2. Equation in complex form

For the ring shell as shown in Fig. 1, we introduce

$$\left.\begin{aligned} \alpha &= \frac{a}{R}, \quad \mu = \sqrt{3(1-\nu^2)}\,\frac{a^2}{Rh}, \quad \varrho = 1 + \alpha \sin\varphi, \\ \Omega &= -Eh\psi_c, \quad \psi = -\sqrt{12(1-\nu^2)}\,\frac{\alpha}{h}\,V_c, \end{aligned}\right\} \tag{2.1}$$

where E is Young's modulus, ν = Poisson ratio, and ψ_c, V_c are functions of the Meissner type introduced in [1]. The Chernina equations are (expressions (31) and (33) in [1]):

$$\left.\begin{aligned} &(\varrho\dot{\Omega})^{\cdot} + \left[-(1+\nu)\alpha\sin\varphi - 2(1+\nu)\frac{\alpha^2\cos\varphi}{\varrho} - 2(1-\nu)\frac{\alpha^2}{\varrho}\right]\Omega \\ &\qquad -2\mu\sin\varphi\psi = f_1(\varphi), \\ &(\varrho\dot{\psi})^{\cdot} + \left[-(1-\nu)\alpha\sin\varphi - 2(1+\nu)\frac{\alpha^2}{\varrho} - 2(1-\nu)\frac{\alpha^2\cos^2\varphi}{\varrho}\right]\psi \\ &\qquad +2\mu\sin\varphi\Omega = f_2(\varphi), \end{aligned}\right\} \quad (2.2)$$

where "." = d/dφ, and terms of load are

$$\begin{aligned} f_1(\varphi) &= -\frac{4\mu^2 M}{\alpha\pi R^2}\frac{\cos\varphi}{\varrho} + \frac{4\mu^2 F}{\alpha\pi R}\left[\sin\varphi + \frac{\cos\varphi - \cos\varphi_0}{\varrho}\alpha\cos\varphi\right], \\ f_2(\varphi) &= \frac{2\mu M}{\pi R^2}\left[\frac{\nu\cos\varphi}{\varrho} + \frac{(1-2\nu)\alpha\sin\varphi\cos\varphi}{\varrho^2}\right] \\ &\quad + \frac{2\mu F}{\pi R}\left[\nu\sin\varphi + \frac{\alpha\cos^2\varphi - \alpha\nu}{\varrho}\right. \\ &\quad + \frac{\alpha\nu\cos\varphi(\cos\varphi_0 - \cos\varphi)}{\varrho} \\ &\quad \left. + \frac{(1-2\nu)\alpha^2\sin\varphi\cos\varphi}{\varrho^2}(\cos\varphi_0 - \cos\varphi)\right]. \end{aligned} \quad (2.3)$$

In deriving expressions (2.2) and (2.3), we have neglected the small quantities of order $h^2/12a^2$. Our problem belongs to the St. Venant boundary value type. Here, $\varphi = \varphi_0$ is the boundary on which the external force F and the external moment M are acted.

Denote the operator

$$L(\cdots) = [\varrho(\cdots)^{\cdot}]^{\cdot} - \left[\alpha\sin\varphi + \frac{2\alpha^2(1+\cos^2\varphi)}{\varrho}\right](\cdots), \quad (2.4)$$

therefore (2.2) can be written

$$\left.\begin{aligned} L(\Omega) - \left(\alpha\nu\sin\varphi - \frac{2\nu\alpha^2\sin^2\varphi}{\varrho}\right)\Omega - 2\mu\sin\varphi\psi &= f_1(\varphi), \\ L(\psi) + \left(\alpha\nu\sin\varphi - \frac{2\nu\alpha^2\sin^2\varphi}{\varrho}\right)\psi + 2\mu\sin\varphi\Omega &= f_2(\varphi). \end{aligned}\right\} \quad (2.5)$$

Still using the method in [5] to put the equation in complex form, let us assume we have the complex function

$$\Phi = A\Omega + B\psi, \quad (2.6)$$

where A, B are constants to be determined, and then we have

$$L(\Phi) = AL(\Omega) + BL(\psi).$$

Solving (2.5) to get $L(\Omega)$ and $L(\psi)$, then substituting to (2.6), we have

$$L(\Phi) = [(\alpha\nu A - 2\mu B)\Omega + (-\alpha\nu B + 2\mu A)\psi] \sin\varphi + \frac{2\nu\alpha^2 \sin^2\varphi}{\varrho}(-A\Omega + B\psi) + Af_1(\varphi) + Bf_2(\varphi). \tag{2.7}$$

Choose

$$\left.\begin{aligned} \alpha\nu A - 2\mu B &= -k\mathrm{i}A, \\ 2\mu A - \alpha\mu B &= -k\mathrm{i}B, \end{aligned}\right\} \tag{2.8}$$

where k is the ratio constant to be determined. From (2.8), we obtain

$$\left.\begin{aligned} B/A &= \frac{\alpha\nu}{2\mu} \pm \mathrm{i}\sqrt{1 - \left(\frac{\alpha\nu}{2\mu}\right)^2}, \\ k &= \pm 2\mu\sqrt{1 - \left(\frac{\alpha\nu}{2\mu}\right)^2}. \end{aligned}\right\} \tag{2.9}$$

Note the quantity

$$\frac{\alpha\nu}{2\mu} = \frac{1}{\sqrt{12(1-\nu^2)}}\frac{\nu h}{a} \ll 1.$$

Therefore formulae (2.9) can be simplied into

$$B/A = \pm\mathrm{i}, \qquad k = \pm 2\mu. \tag{2.10}$$

If we choose the + sign and let $A = 1$, then (2.7) can be written

$$L(\Phi) + 2\mu\mathrm{i}\sin\varphi\Phi + \frac{2\nu\alpha^2\sin^2\varphi}{\varrho}\bar{\Phi} = f_1(\varphi) + \mathrm{i}f_2(\varphi) \tag{2.11}$$

where $\Phi = \Omega + \mathrm{i}\psi$, $\bar{\Phi} = \Omega - \mathrm{i}\psi$.

Introducing the transformation $V = \varrho\Phi$, then substituting in (2.1), we obtain

$$\varrho\ddot{V} - \alpha\cos\varphi\dot{V} + \left[2\mu\mathrm{i}\sin\varphi - \frac{\alpha^2(3-\sin^2\varphi)}{\varrho}\right]V + \frac{2\nu\alpha^2\sin^2\varphi}{\varrho}\bar{V} = \varrho f_1(\varphi) + \mathrm{i}\varrho f_2(\varphi). \tag{2.12}$$

Note the quantity

$$\frac{\alpha^2}{2\mu} = \frac{1}{\sqrt{12(1-\nu^2)}}\frac{h}{a}\frac{a}{R}.$$

We can only consider the usual ring shell where $\alpha^2/2\mu \ll 1$. In (2.12), if we neglect the small quantities of order $\alpha^2/2\mu$ and $\alpha\nu/2\mu$ compared with unity in non-homogeneous terms, and obtain

$$(1+\alpha\sin\varphi)\ddot{V}-\alpha\cos\varphi\dot{V}+2\mu\left(\mathrm{i}\sin\varphi-\frac{3\alpha^2}{2\mu}\frac{1}{1+\alpha\sin\varphi}\right)V$$

$$=-\frac{4\mu^2}{\alpha}\frac{M}{\pi R^2}\cos\varphi+\frac{4\mu^2}{\alpha}\frac{F}{\pi R}[\sin\varphi+\alpha(1-\cos\varphi\cos\varphi_0)]. \quad (2.13)$$

In (2.13), neglecting the term containing the small quantity $3\alpha^2/2\mu$ in the coefficient of V, we obtain

$$(1+\alpha\sin\varphi)\ddot{V}-\alpha\cos\varphi\dot{V}+2\mu\mathrm{i}\sin\varphi V$$

$$=-\frac{4\mu^2}{\alpha}\frac{M}{\pi R}\cos\varphi+\frac{4\mu^2}{\alpha}\frac{F}{\pi R}[\sin\varphi+\alpha(1-\cos\varphi\cos\varphi_0)]. \quad (2.14)$$

Equation (2.13) is suitable in the whole region of the ring shell as the error of the Chernina equation is within the error range of the thin shell theory. The simplification from (2.13) to (2.14) is sufficiently accurate in the region $\varphi \neq 0, \pi$. In the neighbourhood of $\varphi = 0, \pi$, this simplification corresponds to the small quantity $\mathrm{i}\sin\varphi + O(3\alpha^2/2\mu)$, to be substituted for the small quantity $\mathrm{i}\sin\varphi$, which is thus in the asymptotic significance. The homogeneous equation (2.14) has the same form as Novozhelov's in the axial-symmetric case. If we do not want to study the case of the neighbourhood of two points $\varphi = 0, \pi$ especially, we can choose (2.14) as the fundamental equation of the considered problem.

3. General solution

As mentioned in the above, we can choose (2.14) as the fundamental equation of the bending problem of the ring shell and the homogeneous equation of (2.14) then has the same form as Novozhelov's. The general solution of it has been obtained by Chien Wei-zang, etc. Thus:

$$V_0 = c_1\mathrm{e}^{\lambda\varphi}\sum_{-\infty}^{\infty}c'_n\mathrm{e}^{in\varphi}+c_2\mathrm{e}^{-\lambda\varphi}\sum_{-\infty}^{\infty}(-1)^n c'_n\mathrm{e}^{-in\varphi}, \quad (3.1)$$

where c_1 and c_2 are complex constants to be determined. The calculation of the characteristic index λ and coefficient c'_n may be found in [3].

Rewriting (2.14) in the form

$$(1+\alpha\sin\varphi)\ddot{V}-\alpha\cos\varphi\dot{V}+2\mu\mathrm{i}\sin\varphi V$$

$$=-\frac{4\mu^2}{\alpha}\left(\frac{M}{\pi R^2}+\frac{F}{\pi R}\alpha\cos\varphi_0\right)\cos\varphi+\frac{4\mu^2}{\alpha}\frac{F}{\pi R}(\alpha+\sin\varphi). \quad (3.2)$$

The particular solution corresponding to the term $\cos\varphi$ of (3.2) has been obtained by Novozhelov [4]. We denote it

$$V^* = \frac{4\mu}{\alpha}\left(\frac{M}{\pi R^2}+\frac{F}{\pi R}\alpha\cos\varphi_0\right)\sum_{n=1}^{\infty}A_n\sin n\left(\frac{\pi}{2}-\varphi\right), \quad (3.3)$$

where A_n are determined by the following recurrence relations

$$\left.\begin{aligned} A_1 &= \frac{1}{\dfrac{1}{\mu} - \mathrm{i}\left(1 + \mathrm{i}\dfrac{2 \cdot 3}{2\mu}\alpha\right)\dfrac{A_2}{A_1}}, \\ \frac{A_n}{A_{n-1}} &= \frac{\mathrm{i}\left[1 + \mathrm{i}\dfrac{(n-1)(n-2)}{2\mu}\alpha\right]}{\dfrac{\eta^2}{\mu} - \mathrm{i}\left[1 + \mathrm{i}\dfrac{(n+1)(n+2)}{2\mu}\alpha\right]\dfrac{A_{n+1}}{A_n}}, \quad (n = 2, 3, \ldots). \end{aligned}\right\} \tag{3.4}$$

We now only require to find out the rest of the particular solution, i.e. that corresponding to the term $\alpha + \sin\varphi$, which is assumed to be

$$V^{**} = -\frac{4\mu}{\alpha}\frac{F}{\pi R}\sum_{n=0}^{\infty} B_n \cos n\left(\frac{\pi}{2} - \varphi\right). \tag{3.5}$$

Substituting this expression to the left side of (3.2) and making it equal to the term $\alpha + \sin\varphi$, through calculation, we can obtain

$$\left.\begin{aligned} B_0 &= \frac{\mathrm{i}}{2}\left(1 - \frac{B_1}{\mu}\right) + \frac{\mathrm{i}}{2}B_2\left(\frac{3\alpha}{\mu} - \mathrm{i}\right), \\ B_1 &= \frac{\alpha + \mu_i}{1 + (\alpha^2/\mu^2)}\frac{\alpha}{\mu} \approx \alpha\left(\frac{\alpha}{\mu} + \mathrm{i}\right), \\ \frac{B_n}{B_{n-1}} &= \frac{\mathrm{i}\left[1 + \dfrac{\alpha}{2\mu}\mathrm{i}(n-1)(n-2)\right]}{\dfrac{\eta^2}{\mu} - \mathrm{i}\left[1 + \mathrm{i}\dfrac{\alpha(n+1)(n+2)}{2\mu}\right]\dfrac{B_{n+1}}{B_n}}, \quad (n = 2, 3, \ldots). \end{aligned}\right\} \tag{3.6}$$

From this, we see that $n \geqslant 2$, B_n has the same recurrence relation as A_n.

Finally, we will obtain the general solution of the bending problem of ring shell

$$V = V_0 + V^* + V^{**}, \tag{3.7}$$

where V_0, V^*, V^{**} are given by (3.1), (3.3), (3.5) respectively. c_1, c_2 in (2.9) are determined by boundary conditions.

From (2.14), we see that in the case of pure bending ($F = 0$), if we choose $M = pR/2$ (p is axial compression), then (2.14) is the same as in the case of an axial force [3] and that therefore the solution for pure bending can be obtained directly by the axial-symmetric solution, using the substitution $M = pR/2$.

4. Internal forces

In the case of a ring shell, relations of internal forces obtained in [1] can be simplified into

$$
\begin{aligned}
m_\varphi &= \frac{\alpha a}{4\mu^2}\left[\frac{(\varrho\Omega)^{\cdot}}{\varrho} + \frac{\alpha\nu\cos\varphi}{\varrho}\Omega + O\left(\frac{\alpha^3}{2\mu}\right)\right],\\
m_\theta &= \frac{\alpha a}{4\mu^2}\left[\nu\frac{(\varrho\Omega)^{\cdot}}{\varrho} + \frac{\alpha\cos\varphi}{\varrho}\Omega + O\left(\frac{\alpha^2}{2\mu}\right)\right],\\
H &= \frac{\alpha a}{4\mu^2}\left[\alpha(1-\nu)\frac{\Omega}{\varrho} + O\left(\frac{\alpha^3}{2\mu}\right)\right],\\
n_\varphi &= \frac{\alpha}{2\mu}\left[-\frac{\cos\varphi}{\varrho}\psi + \frac{\alpha}{2\mu}\frac{\sin\varphi(\varrho\Omega)^{\cdot}}{\varrho^2} + O\left(\frac{\nu\alpha^2}{2\mu}\right)\right] - f(\varphi),\\
n_\theta &= \frac{\alpha}{2\mu}\left[-\frac{(\varrho\psi)^{\cdot}}{\alpha\varrho} + \frac{\alpha\nu}{2\mu}\frac{\sin\varphi(\varrho\Omega)^{\cdot}}{\varrho^2} + O\left(\frac{\alpha^2}{2\mu}\right)\right],\\
S &= -\frac{\alpha}{2\mu}\left[\frac{\psi}{\varrho} + O\left(\frac{\alpha^2}{2\mu}\right)\right],
\end{aligned}
\tag{4.1}
$$

where

$$
\left.
\begin{aligned}
S &= n_{\varphi\theta} - \frac{m_{\theta\varphi}}{r_2} = n_{\theta\varphi} - \frac{m_{\varphi\theta}}{r_1}, \qquad 2H = m_{\varphi\theta} + m_{\theta\varphi},\\
f(\varphi) &= \frac{M}{\pi R^2}\frac{\sin\varphi}{\varrho^2} + \frac{F}{\pi R}\frac{\cos\varphi + \alpha\sin\varphi\cos\varphi_0}{\varrho^2},
\end{aligned}
\right\}
\tag{4.2}
$$

and

$$
r_1 = a, \quad r_2 = R(1 + \alpha\sin\varphi)/\sin\varphi, \quad m_{\varphi\theta} = m_{\theta\varphi}.
$$

Neglecting the small quantity with order $\alpha^2/2\mu$ in (4.1) and using V to significant results, we obtain

$$
\begin{aligned}
m_\varphi &= \frac{\alpha a}{4\mu^2}\left[\frac{\operatorname{Re}\dot V}{\varrho} + \frac{\alpha\nu\cos\varphi}{\varrho^2}\operatorname{Re} V\right],\\
m_\theta &= \frac{\alpha a}{4\mu^2}\left[\nu\frac{\operatorname{Re}\dot V}{\varrho} + \frac{\alpha\cos\varphi}{\varrho^2}\operatorname{Re} V\right],\\
m_{\varphi\theta} &= m_{\theta\varphi} = \frac{\alpha a}{4\mu^2}\frac{\alpha(1-\nu)\operatorname{Re} V}{\varrho^2},\\
n_\varphi &= \frac{\alpha}{2\mu}\left[-\frac{\cos\varphi}{\varrho^2}\operatorname{Im} V + \frac{\alpha}{2\mu}\frac{\sin\varphi}{\varrho^2}\operatorname{Re}\dot V\right] - \frac{M}{\pi R^2}\frac{\sin\varphi}{\varrho^2}\\
&\quad + \frac{F}{\pi R}\frac{\cos\varphi + \alpha\sin\varphi\cos\varphi_0}{\varrho^2}\\
n_\theta &= \frac{\alpha}{2\mu}\left[-\frac{\operatorname{Im}\dot V}{\alpha\varrho} + \frac{\alpha\nu}{2\mu}\frac{\sin\varphi\operatorname{Re}\dot V}{\varrho^2}\right],\\
n_{\varphi\theta} &= -\frac{\alpha}{2\mu}\frac{\operatorname{Im} V}{\varrho^2},\\
n_{\theta\varphi} &= \frac{\alpha}{2\mu}\left[-\frac{\operatorname{Im} V}{\varrho^2} + \frac{\alpha}{2\mu}\frac{(1-\nu)\operatorname{Re} V}{\varrho^2}\right].
\end{aligned}
\tag{4.3}
$$

Using the results in [1], we obtain the following:

$$\left.\begin{aligned} q_\varphi &= \frac{\alpha}{2\mu}\left[\frac{\sin\varphi}{\varrho^2}\,\mathrm{Im}\,V + \frac{\alpha}{2\mu}\frac{\cos\varphi}{\varrho^2}\,\mathrm{Re}\,\dot{V}\right] - \frac{M}{\pi R^2}\frac{\cos\varphi}{\varrho^2} \\ &\quad + \frac{F}{\pi R}\frac{\sin\varphi + \alpha(1 - \cos\varphi_0\cos\varphi)}{\varrho^2}, \\ q_\theta &= -\frac{\alpha^2}{4\mu^2}\left[\frac{\mathrm{Re}\,\dot{V}}{\varrho^2} + \frac{\alpha\cos\varphi}{\varrho^2}\,\mathrm{Re}\,V\right], \end{aligned}\right\} \tag{4.4}$$

for $\alpha \ll 1$, m_φ, m_θ have the same expression as in [3] for the axial-symmetric case.

5. Displacement

The displacements u, v, w (expression (1.1) and Fig. 2) and strain components of the ring shell under wind-type load have the following relations:

$$\left.\begin{aligned} \varepsilon_\varphi &= \frac{1}{a}(\dot{u} - w), \qquad \varepsilon_\theta = \frac{1}{R\varrho}(u\cos\varphi - w\sin\varphi + v), \\ \omega &= \frac{1}{a}\left(\dot{v} - \alpha\frac{u + v\cos\varphi}{\varrho}\right) \end{aligned}\right\} \tag{5.1}$$

where

$$\left.\begin{aligned} \varepsilon_\varphi &= \frac{1}{Eh}(n_\varphi - \nu n_\theta), \qquad \varepsilon_\theta = \frac{1}{Eh}(n_\theta - \nu n_\varphi), \\ \omega &= \frac{2(1+\nu)}{Eh}S. \end{aligned}\right\} \tag{5.2}$$

Expressing u, w in (5.1) in terms of the horizontal displacement y and the vertical displacement z, and letting

$$w = z\cos\varphi - y\sin\varphi, \qquad u = z\sin\varphi + y\cos\varphi,$$

after calculation, we obtain the following equations:

$$\left.\begin{aligned} \left(\frac{z}{R\varrho}\right)^{\cdot} &= \frac{\alpha\varepsilon_\varphi}{\varrho\sin\varphi} + \mathrm{ctg}\,\varphi\left(\frac{\alpha\omega}{\varrho} - \dot{\varepsilon}_\theta\right), \\ \left(\frac{y}{R}\right)^{\cdot} &= (\varrho\varepsilon_\theta)^{\cdot} - \alpha\cos\varphi\,\varepsilon_\theta - \alpha\sin\varphi\,\frac{z}{R\varrho} - \alpha\omega, \\ \frac{v}{R} &= \varrho\varepsilon_\theta - \frac{y}{R}. \end{aligned}\right\} \tag{5.3}$$

Expressing strain components through expressions (5.2), (4.3), (4.1), and making use of (2.14) to cancel higher order derivatives, we can get the first expression of (5.3) by the following method:

$$\left(\frac{z}{R\varrho}\right)^{\cdot} = -\frac{1}{Eh}\frac{\cos\varphi}{\varrho^2}\,\mathrm{Re}\,V + O\left(\frac{\alpha\nu}{2\mu}, \frac{\alpha^2}{2\mu}\right). \tag{5.4}$$

Neglecting small quantities, we have

$$\frac{z}{R\varrho} = -\frac{1}{Eh}\int_{\varphi_0}^{\varphi}\frac{\cos\Omega\,\mathrm{Re}\,V}{\varrho^2}\,\mathrm{d}\varphi + \beta_0.$$

The above expression gives the rotational angle at the cross section $\varphi = \varphi$ of the ring shell where β_0 is the integral constant. Similarly, the second expression of (5.3) can be calculated in the following way:

$$\left(\frac{y}{R}\right)^{\cdot} = -\frac{1}{Eh}\frac{\sin\varphi}{\varrho}\,\mathrm{Re}\,V - \alpha\sin\varphi\,\frac{z}{R\varrho} + O\left(\frac{\alpha\nu}{2\mu}, \frac{\alpha^2}{2\mu}\right). \tag{5.6}$$

Neglecting small quantities, and then substituting (5.5) we obtain the following expression by integration

$$\frac{y}{R} = \frac{1}{Eh}\left(\int_{\varphi_0}^{\varphi}\frac{\sin\varphi\,\mathrm{Re}\,\dot{V}}{\varrho}\,\mathrm{d}\varphi - \cos\varphi\int_{\varphi_0}^{\varphi}\frac{\mathrm{Re}\,\dot{V}}{\varrho}\,\mathrm{d}\varphi\right) + \beta_0\alpha\cos\varphi + \gamma_0, \tag{5.7}$$

where γ_0 is the integral constant.

From the third expression of (5.3), we obtain

$$\frac{v}{R} = \varrho\varepsilon_\theta - \frac{1}{Eh}\int_{\varphi_0}^{\varphi}\frac{(\sin\xi - \cos\varphi)\,\mathrm{Re}\,\dot{V}}{1 + \alpha\sin\xi}\,\mathrm{d}\xi - \beta_0\alpha\cos\varphi - \gamma_0, \tag{5.8}$$

where

$$\varepsilon_\theta = \frac{1}{Eh}\frac{\alpha}{2\mu}\left(-\frac{1}{\alpha\varrho}\,\mathrm{Im}\,\dot{V} + \frac{\nu\cos\varphi}{\varrho^2}\,\mathrm{Im}\,V\right) + \frac{1}{Eh}\left(\frac{M}{\pi R^2}\frac{\nu\sin\varphi}{\varrho^2}\right.$$
$$\left. + \frac{F}{\pi R}\frac{\cos\varphi + \alpha\cos\varphi_0\sin\varphi}{\varrho^2}\right). \tag{5.9}$$

The integral constants in (5.5), (5.7) can be determined by the defined zero position of displacement. For example, choosing $z/R\varrho = 0$, $y/R = 0$ as $\varphi = \varphi_0$, we then have $\beta_0 = \gamma_0 = 0$.

According to the definition of ψ_c in [1], we can obtain the tangential rotational angle χ (Fig. 2)

$$\chi = \frac{\dot{w} + u}{a} = -\frac{1}{Eh\alpha}\frac{\mathrm{Re}\,V}{\varrho} + \frac{z}{R\varrho} \tag{5.10}$$

$$\chi = -\frac{1}{Eh\alpha}\frac{\mathrm{Re}\,V}{\varrho} - \frac{1}{Eh}\int_{\varphi_0}^{\varphi}\frac{\cos\varphi\,\mathrm{Re}\,V}{\varrho^2}\,\mathrm{d}\varphi + \beta_0. \tag{5.11}$$

6. Boundary condition

In general, there are four boundary conditions for the boundary $\varphi = \text{const.}$, i.e.

$$\left.\begin{array}{llll} n_\varphi \ \text{or}\ u, & m_\varphi \ \text{or}\ \chi, \\ T_{\varphi\theta} \ \text{or}\ v, & V_\varphi \ \text{Or}\ w, \end{array}\right\} \tag{6.1}$$

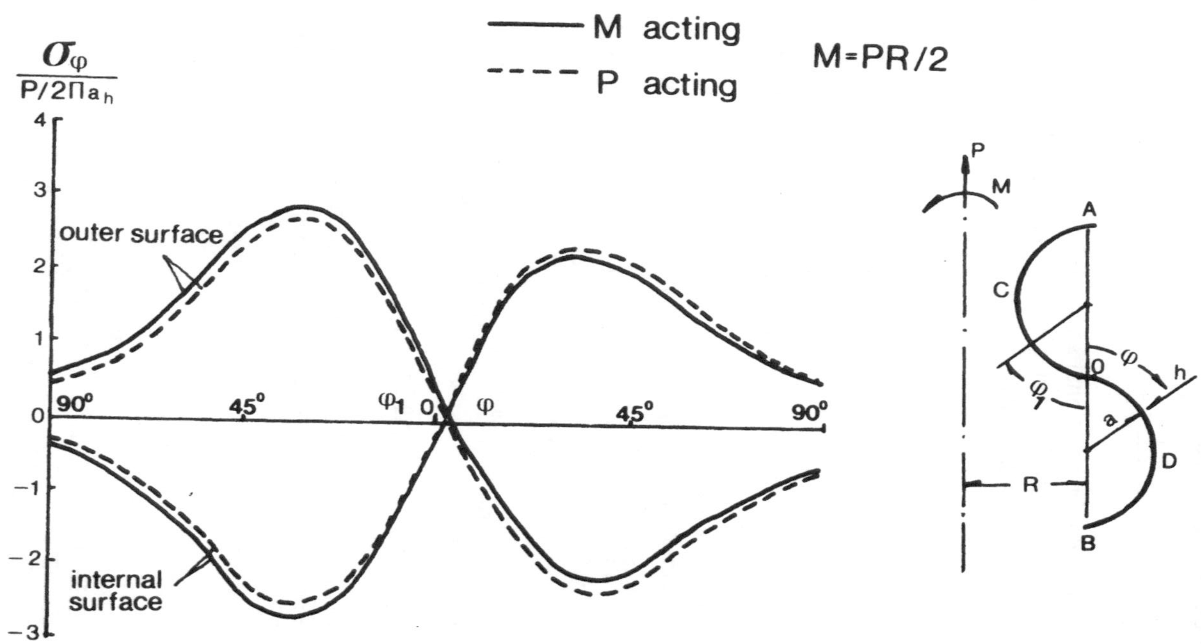

Figure 3. Pure bending of C-type corrugated tube.

where $T_{\varphi\theta}$ and V_φ are the reduced shearing force and reduced transverse force respectively, which can be represented in the form

$$T_{\varphi\theta} = n_{\varphi\theta} + \frac{m_{\theta\varphi}}{r_2}, \qquad V_\varphi = q_\varphi + \frac{m_{\varphi\theta}}{R\varrho}. \tag{6.2}$$

It is clear that terms produced by the twisting moment are small quantities with order $\alpha^2/2\mu$ by calculation, and we can therefore choose

$$T_{\varphi\theta} = n_{\varphi\theta}, \qquad V_\varphi = q_\varphi. \tag{6.3}$$

Two boundaries of the ring shell have eight boundary conditions. But during the process of deduction of (2.2), two equilibrium conditions of the whole body have been used [1]. The above mentioned expressions of displacement also contain two integral constants, and thus, the surplus independent boundary conditions are only four, which determfine two complex constants c_1 and c_2, contained in the general solution.

7. Example of calculation

For the application of results, we may consider an example of the prob'.m of pure bending in the case of C-type corrugated tubes, and compare results with the case of axial force (Fig. 3).

For arc ODB and arc OCA, writing the general solution (3.7) respectively, and making use of the symmetry about points C, D, the integral constants can be decreased by half [6]. The remaining constants can be determined by continuity conditions at point O. At point O, we have

$$\begin{aligned} m_\varphi^+ &= -m_\varphi^-, & n_{\varphi\theta}^+ &= n_{\varphi\theta}^-, \\ \chi^+ &= \chi^-, & v^+ &= v^-. \end{aligned} \tag{7.1}$$

Then, substituting the solution in (4.3), (4.4), (5.7), (5.8), we can obtain the values of internal forces and displacement.

The dimensions of the example in Fig. 3 are: $R = 217.5\,\text{mm}$, $a = 25\,\text{mm}$, $h = 2\,\text{mm}$, $E = 2.1 \times 10^4\,\text{kg/mm}^2$, $\nu = 0.266$. Through calculation, we can obtain $\alpha = 0.155$, $\mu = 2.4$. In Fig. 3, the distribution curves of stress (dimensionless) in the direction of the meridian and results under the action of axial force p, are given for $M = pR/2$. Results in the case of an axial force are calculated according to [10]. Choosing $p = 1000\,\text{kg}$, the values of axial displacement between points A and B are

M acting, $\quad z = 0.522\,\text{mm}$

p acting, $\quad z = 0.518\,\text{mm}$

We now see that if we choose $M = pR/2$, the distribution of stress and values of axial displacement are quite close in cases of pure bending and under the action of axial force. In the figure, the difference of the maximum stress in the direction of the meridian is about 6%. The substitution $M = pR/2$ was suggested by

C. E. Turner (1959) [11] according to experimental research on a corrugated tube. Subsequently, U. I. Iohelison (1971) [9] pointed out this relation by calculation. In this paper, it transpires that (2.14) of the expressions for internal forces and displacements in problems of pure bending and axial symmetry, as well as where the boundary condition are not the same, this reasonable property is only in this form in a general case. But in the definite range of parameters, especially as $\alpha \ll 1$ in the case of a slender ring-shell, the expressions for external forces and displacement are quite close for these two cases and the substitution $M = pR/2$ gives good approximate results. Therefore the study of the problem of axial symmetry can take the place of the problem of pure bending, which is useful for practical cases.

The writer wishes to express his sincere gratitude to Prof. Chien Wei-zang for his direction during the preparation of this paper.

References

1. Chernina, V. S., "On a system of differential equations of equilibrium rotationally symmetric shells under bending load", *Prokl. Math. i Mech.* **23**, 2 (1959), 258–265 (in Russian).
2. Chernina, V. S., *Statics of Rotational Thin Shells*, Nauka, Moscow (1968) (in Russian).
3. Chien Wei-zang and Zhen Se-liang, "General solutions of axial symmetrical ring shells", *Applied Mathematics and Mechanics* (English Edition) **1**, 3 (1980), 305–318.
4. Novozhelov, V. V., *Thin Shell Theory* (Chinese translation), Science Press, Beijing (1959).
5. Chien Wei-zang and Zhen Se-liang, "Equations of symmetrical ring shells in complex quantities, and their general solution for slender ring shells", *J. of Tsing-Hua University* **19**, 1 (1979), 27–47.
6. Chien Wei-zang and Zhen Se-liang, "The calculation of semi-circular corrugated tubes and an application of the general solution of ring shell", *Applied Mathematics and Mechanics* (English edition) **2**, 1 (1981), 103–116.
7. Wan, F. Y. M., "Circumferentially sinusoidal stress and strain in shells of revolution", *Int. J. Solids and Structures* **6** (1970), 959–973.
8. Wan, F. Y. M., "Laterally loaded elastic shells of revolution", *Ing. Arch.* **42**, 4 (1973), 245–258.
9. Iohelison, U. I., "Bending of bellows subjected to moment at the end", *Izv. Voz'ov, Mash.* **9–12** (1971), 16–20 (in Russian).
10. Chen Shan-lin, "The symmetrical stress and deflection of S-type corrugated tubes", *J. of Chengdu University of Science and Technology*, No. 3 (1982), 50–58.
11. Turner, C. E., "Stress and deflection studies of flat-plate and toroidal expansion bellows, subjected to axial, eccentric or internal pressure loading", *J. Mech. Engng. Sci.* **1**, 2 (1959), 130–143.

Composite materials

An analytical solution for interlaminar stresses in symmetric composite laminates

WANG REN-JIE and LOO TSU-TAO

Shanghai Jiao Tong University, PRC

Abstract. In this paper, the interlaminar stresses in a $[0/90]_s$ symmetric composite laminate under uniaxial tension are investigated, and an exact solution in series form for this class of laminates, is presented. Based on the characteristics of $[0/90]_s$ stack-sequence, the problem is reduced to a pair of plane strain problems of transversely isotropic bodies. The solution is then expressed in the form of the sum of two infinite series of Fourier-hyperbolic functions, in which the unknown coefficients are determined by the conditions of continuity at the interface as well as other related boundary conditions. As an illustration, an example is included in the paper. The numerical results are compared with those from various approximate methods. It is noted that the peak value of the interlaminar normal stress at the mid-plane of symmetry is overestimated in prevailing theories by as much as 40%. The method presented here may also be used for symmetric and antisymmetric problems of the cross-ply laminates with $2n$ layers.

1. Introduction

In the early 70's, Pipes and Pagano [1–3] studies the interlaminar stresses of composite laminates both experimentally and theoretically. They revealed, for the first time, that delamination occurs at the mid-plane of symmetry rather than at the interface of material discontinuity, and that the interlaminar stresses are indeed the edge-effect in nature. During the following decade, the problems in this respect received much attention, and many papers on approximate solutions for interlaminar stresses [4–11] have been published. The present paper treats the interlaminar stress problem of a $[0/90]_s$ symmetric laminate under the uniaxial tension rigorously with the theory of elasticity. Based on the characteristics of $[0/90]_s$ stack-sequence, it may be reduced to a pair of plane strain problems. Using the same technique as was used in [12], the displacement and thus the stress field is expressed in the form of the sum of two infinite series of Fourier-hyperbolic functions. The continuity conditions at the interface and other related boundary conditions are used to determine the arbitrary coefficients in the series. In practical computations, however, only a finite number of terms of both series is taken. As the number of truncated terms of the series increases, the numerical solution approaches its exact value. For comparison's sake, the example in [4] is chosen for numerical calculation in the present paper. It is found that the peak value of interlaminar normal stresses on the mid-plane of symmetry is much lower than

Yeh, K. Y. (ed) Progress in Applied Mechanics
 ISBN-13: 978-94-010-8061-3

those predicted by previous approximate solutions. This seems most likely to be due to lack of fulfillment of the free-edge conditions on the lateral boundaries in the previous work. Recently, Spilker [11] improved the finite element calculations by using special boundary elements, and his result did reveal some reduction of the peak value of the interlaminar normal stresses.

2. Formulation of the problem

A free-boundary problem of the $[0/90]_s$ symmetric composite laminate under uniaxial tension along the X-axis is shown in Fig. 1. The 0°-direction layer, i.e., fibre direction parallel to the X-axis, has a thickness of h_1, while the 90°-direction layer, has the thickness h_2. Let $2a$, $2b$ be the length and width of the laminate, respectively. Suppose the laminate is stretched with a tensile strain ε_0 in the OX-direction. Now we may solve this problem directly on the basis of the linear theory of elasticity.

Due to the symmetry, we need only consider the first two layers of the $[0/90]_s$ laminate. The upper surface of the first layer is perfectly free, while in the second layer, the shearing stress and normal displacement w on its lower surface both vanish. Moreover, the lateral surfaces $y = \pm b$ are free of surface tractions. To solve the problem, we may first consider a simple plane stress problem. Assume one of the layers, say the second layer, i.e., 90°-layer, is freely extended along the X-axis with a strain ε_0, so that all the other stress components vanish except $\sigma_x \neq 0$. Let the first or 0°-layer extend the same tensile strain ε_0 along the X-axis, and let the strain along the Y-axis be kept identical with that of the second layer. Moreover, let this layer be free to deform in the OZ-direction. Consequently, both layers are in states of plane stresses, i.e., $\sigma_z = \tau_{zx} = \tau_{zy} = 0$, or in other words, there exist

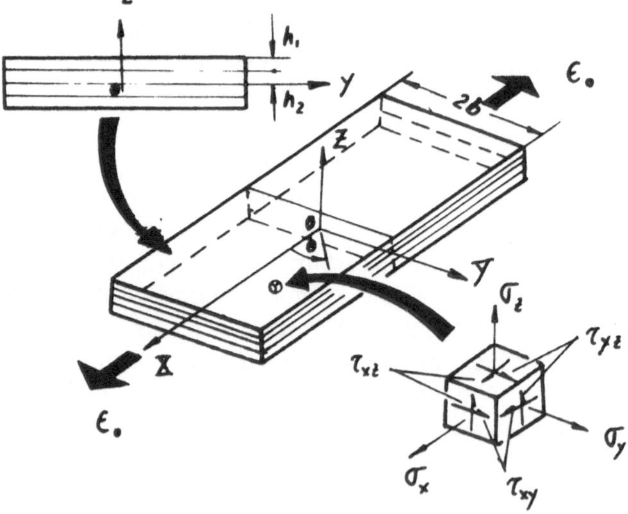

Figure 1. Free-boundary problem.

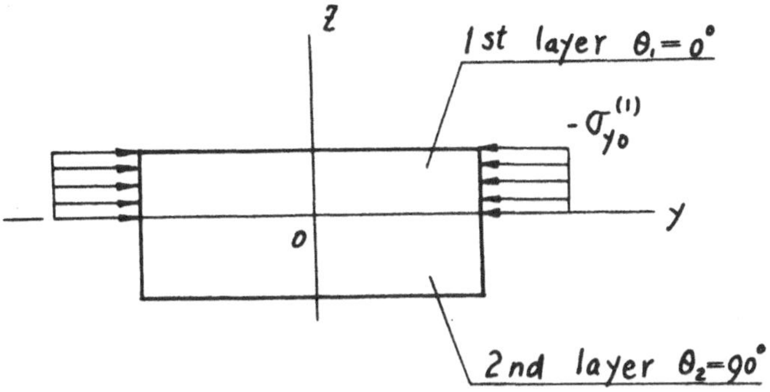

Figure 2. The boundary problem.

no interlaminar stresses between the layers. The above mentioned conditions may be realized if a uniform stress is applied on the lateral boundaries of the first layer, i.e., $(\sigma_y^{(1)})_{y=\pm b} = \sigma_{y_0}^{(1)}$ along with the tensile stress σ_x in the OX-direction. In order to comply with free boundary condition at $y = \pm b$, the next problem is to solve the residual problem as shown in Fig. 2, i.e. a plane strain problem with uniform stress $(-\sigma_{y_0}^{(1)})$ acting on the lateral boundaries of the first layer. Evidently, the interlaminar stresses of the laminate are induced in a residual problem. Hence, so far as the interlaminar stresses are concerned, we only need to solve the boundary value problem of Fig. 2.

It is clear that the present composite layers behave elastically as transversely isotropic materials. Therefore, the generalized Hooke's law for 0°-layer and 90°-layer of composite laminate are respectively as follows:

$$\begin{bmatrix} \sigma_x^{(1)} \\ \sigma_y^{(1)} \\ \sigma_z^{(1)} \end{bmatrix} = \begin{bmatrix} C_{11} & C_{12} & C_{12} \\ C_{12} & C_{22} & C_{23} \\ C_{12} & C_{23} & C_{22} \end{bmatrix} \begin{bmatrix} \varepsilon_x^{(1)} \\ \varepsilon_y^{(1)} \\ \varepsilon_z^{(1)} \end{bmatrix},$$

$$\begin{bmatrix} \tau_{yz}^{(1)} \\ \tau_{zx}^{(1)} \\ \tau_{xy}^{(1)} \end{bmatrix} = \begin{bmatrix} \frac{1}{2}(C_{22} - C_{23}) & 0 & 0 \\ 0 & C_{55} & 0 \\ 0 & 0 & C_{55} \end{bmatrix} \begin{bmatrix} \gamma_{yz}^{(1)} \\ \gamma_{zx}^{(1)} \\ \gamma_{xy}^{(1)} \end{bmatrix}, \tag{2.1}$$

$$\begin{bmatrix} \sigma_x^{(2)} \\ \sigma_y^{(2)} \\ \sigma_z^{(2)} \end{bmatrix} = \begin{bmatrix} C_{22} & C_{12} & C_{23} \\ C_{12} & C_{11} & C_{12} \\ C_{23} & C_{12} & C_{22} \end{bmatrix} \begin{bmatrix} \varepsilon_x^{(2)} \\ \varepsilon_y^{(2)} \\ \varepsilon_z^{(2)} \end{bmatrix},$$

$$\begin{bmatrix} \tau_{yz}^{(2)} \\ \tau_{zx}^{(2)} \\ \tau_{xy}^{(2)} \end{bmatrix} = \begin{bmatrix} C_{55} & 0 & 0 \\ 0 & \frac{1}{2}(C_{22} - C_{23}) & 0 \\ 0 & 0 & C_{55} \end{bmatrix} \begin{bmatrix} \gamma_{yz}^{(2)} \\ \gamma_{zx}^{(2)} \\ \gamma_{xy}^{(2)} \end{bmatrix}. \tag{2.2}$$

Since $\varepsilon_x^{(1)} = \varepsilon_x^{(2)} = \varepsilon_0$ and $\varepsilon_y^{(1)} = \varepsilon_y^{(2)}$, we find that the uniform stress $\sigma_{y_0}^{(1)}$ on the lateral boundary of the 0° layer will be

$$\sigma_{y_0}^{(1)} = C_{22}K(1 - \nu)\left(1 - \frac{1 - \nu^2}{R - K^2}\right)\varepsilon_0 \tag{2.3}$$

where

$$H = \tfrac{1}{2}(1 - \nu), \quad R = C_{11}/C_{22}, \quad K = C_{12}/C_{22}, \quad \nu = C_{23}/C_{22}. \tag{2.4}$$

In the residual problem each layer is in the state of plane strain, so the displacement field may be expressed as

$$u^{(i)} \equiv 0, \quad v^{(i)} = v^{(i)}(y, z), \quad w^{(i)} = w^{(i)}(y, z), \tag{2.5}$$

with $i = 1, 2$ for the first layer (0°-layer) and the second layer (90°-layer) respectively. It follows that

$$\tau_{zx}^{(i)} = \tau_{xy}^{(i)} = 0. \tag{2.6}$$

Thus the differential equations of equilibrium are reduced to the form

$$\left.\begin{aligned} \sigma_{y,y}^{(i)} + \tau_{yz,z}^{(i)} &= 0, \\ \tau_{yz,y}^{(i)} + \sigma_{z,z}^{(i)} &= 0, \end{aligned}\right\} \quad (i = 1, 2), \tag{2.7}$$

in which the comma represents the differentiation with respect to the coordinate variable following it.

For the first layer (i.e. $i = 1$), from the relation

$$\left.\begin{aligned} \sigma_y^{(1)} &= C_{22}v_{,y}^{(1)} + C_{23}w_{,z}^{(1)}, \\ \sigma_z^{(1)} &= C_{23}v_{,y}^{(1)} + C_{22}w_{,z}^{(1)}, \\ \tau_{yz}^{(1)} &= \tfrac{1}{2}(C_{22} - C_{23})(v_{,z}^{(1)} + w_{,y}^{(1)}), \end{aligned}\right\} \tag{2.8}$$

we have (2.7) in terms of the displacements as

$$\left(\frac{\partial^2}{\partial y^2} + H\frac{\partial^2}{\partial z^2}\right)v^{(1)} + (1 - H)\frac{\partial^2 w^{(1)}}{\partial y \partial z} = 0, \tag{2.9a}$$

$$(1 - H)\frac{\partial^2 v^{(1)}}{\partial y \partial z} + \left(H\frac{\partial^2}{\partial y^2} + \frac{\partial^2}{\partial z}\right)w^{(1)} = 0. \tag{2.9b}$$

Define a strain function $F^{(1)}(y, z)$ so that

$$\left.\begin{aligned} v^{(1)} &= -(1 - H)\frac{\partial^2}{\partial y \partial z}F^{(1)}(y, z), \\ w^{(1)} &= \left(\frac{\partial^2}{\partial y^2} + H\frac{\partial^2}{\partial z^2}\right)F^{(1)}(y, z). \end{aligned}\right\} \tag{2.10}$$

Then (2.9a) will be identically satisfied, while (2.9b) becomes the differential equation of the strain function $F^{(1)}(y, z)$ as follows

$$\nabla^2\nabla^2 F^{(1)}(y, z) = 0, \tag{2.11}$$

where

$$\nabla^2 = \frac{\partial^2}{\partial y^2} + \frac{\partial^2}{\partial z^2}.$$

Similarly, for the second layer ($i = 2$), from

$$\left.\begin{aligned} \sigma_y^{(2)} &= C_{11} v_{,y}^{(2)} + C_{12} w_{,z}^{(2)}, \\ \sigma_z^{(2)} &= C_{12} v_{,y}^{(2)} + C_{22} w_{,z}^{(2)}, \\ \tau_{yz}^{(2)} &= C_{55}(v_{,z}^{(2)} + w_{,y}^{(2)}), \end{aligned}\right\} \tag{2.12}$$

we have

$$\begin{cases} \left(\dfrac{\partial^2}{\partial y^2} + \dfrac{G}{R}\dfrac{\partial^2}{\partial z^2}\right) v^{(2)} + \dfrac{G+K}{R}\dfrac{\partial^2}{\partial y \partial z} w^{(2)} = 0, & \text{(2.13a)} \\[2ex] \dfrac{G+K}{G}\dfrac{\partial^2}{\partial y \partial z} v^{(2)} + \left(\dfrac{\partial^2}{\partial y^2} + \dfrac{1}{G}\dfrac{\partial^2}{\partial z^2}\right) w^{(2)} = 0, & \text{(2.13b)} \end{cases}$$

where

$$G = C_{55}/C_{22}. \tag{2.14}$$

Introducing $F^{(2)}(y, z)$ such that

$$\left.\begin{aligned} v^{(2)} &= -\frac{G+K}{R}\frac{\partial^2}{\partial y \partial z} F^{(2)}(y, z), \\ w^{(2)} &= \left(\frac{\partial^2}{\partial y^2} + \frac{G}{R}\frac{\partial^2}{\partial z^2}\right) F^{(2)}(y, z). \end{aligned}\right\} \tag{2.15}$$

equation (2.13a) becomes an identity and (2.13b) will be

$$\left\{\frac{\partial^4}{\partial y^4} + \left[\frac{1}{G} - \frac{K}{R}\frac{(2G+K)}{G}\right]\frac{\partial^4}{\partial y^2 \partial z^2} + \frac{1}{R}\frac{\partial^4}{\partial z^4}\right\} F^{(2)}(y, z) = 0,$$

or in more compact form

$$\left(\frac{\partial^2}{\partial y^2} + \lambda_1 \frac{\partial^2}{\partial z^2}\right)\left(\frac{\partial^2}{\partial y^2} + \lambda_2 \frac{\partial^2}{\partial z^2}\right) F^{(2)}(y, z) = 0. \tag{2.16}$$

Solving the partial differential equations (2.11) and (2.16) for strain functions $F^{(1)}(y, z)$ and $F^{(2)}(y, z)$ respectively, we may derive the displacement fields $v^{(i)}$, $w^{(i)}$ ($i = 1, 2$) with the aid of (2.10) and (2.15), and then the corresponding stress fields $\sigma_y^{(i)}$, $\sigma_z^{(i)}$ and $\tau_{yz}^{(i)}$. Finally, the boundary conditions are respectively

$$y = \pm b: \quad \tau_{yz}^{(1)}(\pm b, z) = 0, \quad \sigma_y^{(1)}(\pm b, z) = -\sigma_{y0}^{(1)}, \tag{2.17a, b}$$

$$z = h_1: \quad \tau_{yz}^{(1)}(y, h_1) = 0, \quad \sigma_z^{(1)}(y, h_1) = 0, \tag{2.17c, d}$$

for the first layer, and

$$y = \pm b: \quad \tau_{yz}^{(2)}(\pm b, z) = 0, \quad \sigma_y^{(2)}(\pm b, z) = 0, \tag{2.18a, b}$$

$$z = -h_2: \quad \tau_{yz}^{(2)}(y, -h_2) = 0, \quad w^{(2)}(y, -h_2) = 0, \tag{2.18c, d}$$

for the second layer. Furthermore, on the interface between the layers ($z = 0$), the conditions of continuity are

$$z = 0:\quad \sigma_z^{(1)}(y, 0) = \sigma_z^{(2)}(y, z),\quad \tau_{yz}^{(1)}(y, 0) = \tau_{yz}^{(2)}(y, 0), \tag{2.19a, b}$$

$$v^{(1)}(y, 0) = v^{(2)}(y, z),\quad w^{(1)}(y, 0) = w^{(2)}(y, 0). \tag{2.19c, d}$$

Under these conditions, the strain functions $F^{(1)}(y, z)$ and $F^{(2)}(y, z)$ can be uniquely determined.

3. Method of solution

Based on the above analysis, the problem is reduced to solving two linear partial differential equations (2.11) and (2.16). Let us now consider the biharmonic equation

$$\nabla^2\nabla^2 F^{(1)}(y, z) = 0. \qquad \text{see (2.11)}$$

Due to the symmetry of the problem, the displacement $v^{(1)}(y, z)$ must be an odd function of y, so that the strain function $F^{(1)}(y, z)$ may be expressed in the form of a Fourier-hyperbolic series as

$$\begin{aligned} F^{(1)}(y, z) = {} & \sum_{n=0}^{\infty} \cos \alpha_n y\{A_{1n} \operatorname{sh} \alpha_n z + A_{2n} \operatorname{ch} \alpha_n z + A_{3n} z \operatorname{sh} \alpha_n z \\ & + A_{4n} z \operatorname{ch} \alpha_n z\} + \sum_{m=0}^{\infty} \{\operatorname{ch} \beta_m y[B_{1m} \sin \beta_m z + B_{2m} \cos \beta_m z] \\ & + y \operatorname{sh} \beta_m y[B_{3m} \sin \beta_m z + B_{4m} \cos \beta_m z]\}, \end{aligned} \tag{3.1}$$

where α_n and β_m are arbitrary parameters; and A_{jn}, B_{jm} ($j = 1, 2, 3, 4$) are the undetermined constants. With the aid of (2.10) we get the displacement components of the first layer

$$\begin{aligned} \frac{v^{(1)}(y, z)}{1 - H} = {} & \sum_{n=1}^{\infty} \alpha_n^2 \sin \alpha_n y \left\{A_{1n} \operatorname{ch} \alpha_n z + A_{2n} \operatorname{sh} \alpha_n z \right. \\ & \left. + A_{3n}\left[\frac{1}{\alpha_n} \operatorname{sh} \alpha_n z + z \operatorname{ch} \alpha_n z\right] + A_{4n}\left[\frac{1}{\alpha_n} \operatorname{ch} \alpha_n z + z \operatorname{sh} \alpha_n z\right]\right\} \\ & - \sum_{m=1}^{\infty} \beta_m^2 \left\{\operatorname{sh} \beta_m y[B_{1m} \cos \beta_m z - B_{2m} \sin \beta_m z] \right. \\ & \left. + \left[\frac{1}{\beta_m} \operatorname{sh} \beta_m y + y \operatorname{ch} \beta_m y\right][B_{3m} \cos \beta_m z - B_{4m} \sin \beta_m z]\right\}, \end{aligned} \tag{3.2a}$$

$$\begin{aligned} \frac{w^{(1)}(y, z)}{1 - H} = {} & - \sum_{n=1}^{\infty} \alpha_n^2 \cos \alpha_n y \left\{A_{1n} \operatorname{sh} \alpha_n z + A_{2n} \operatorname{ch} \alpha_n z \right. \\ & + A_{3n}\left[\frac{-2H}{\alpha_n(1 - H)} \operatorname{ch} \alpha_n z + z \operatorname{sh} \alpha_n z\right] \end{aligned}$$

$$+ A_{4n}\left[\frac{-2H}{\alpha_n(1-H)}\,\mathrm{sh}\,\alpha_n z + z\,\mathrm{ch}\,\alpha_n z\right]\Bigg\}$$

$$+ \sum_{m=1}^{\infty} \beta_m^2 \Big\{\mathrm{ch}\,\beta_m y[B_{1m}\sin\beta_m z + B_{2m}\cos\beta_m z]$$

$$+ \left[\frac{2}{\beta_m(1-H)}\,\mathrm{ch}\,\beta_m y + y\,\mathrm{sh}\,\beta_m y\right][B_{3m}\sin\beta_m z$$

$$+ B_{4m}\cos\beta_m z]\Big\}, \tag{3.2b}$$

and by (2.8), the corresponding stress components

$$\frac{\sigma_y^{(1)}(y,z)}{C_{22}(1-H)(1-\nu)} = \sum_{n=1}^{\infty} \alpha_n^3 \cos\alpha_n y\Big\{A_{1n}\,\mathrm{ch}\,\alpha_n z + A_{2n}\,\mathrm{sh}\,\alpha_n z$$

$$+ A_{3n}\left[\frac{1+3\nu}{\alpha_n(1+\nu)}\,\mathrm{sh}\,\alpha_n z + z\,\mathrm{ch}\,\alpha_n z\right]$$

$$+ A_{4n}\left[\frac{1+3\nu}{\alpha_n(1+\nu)}\,\mathrm{ch}\,\alpha_n z + z\,\mathrm{sh}\,\alpha_n z\right]\Bigg\}$$

$$- \sum_{m=1}^{\infty} \beta_m^3 \Big\{\mathrm{ch}\,\beta_m y[B_{1m}\cos\beta_m z - B_{2m}\sin\beta_m z]$$

$$+ \left[\frac{2}{\beta_m(1+\nu)}\,\mathrm{ch}\,\beta_m y + y\,\mathrm{sh}\,\beta_m y\right][B_{3m}\cos\beta_m z$$

$$- B_{4m}\sin\beta_m z]\Big\}, \tag{3.3a}$$

$$\frac{\sigma_z^{(1)}(y,z)}{C_{22}(1-H)(1-\nu)} = -\sum_{n=1}^{\infty} \alpha_n^3 \cos\alpha_n y\Big\{A_{1n}\,\mathrm{ch}\,\alpha_n z + A_{2n}\,\mathrm{sh}\,\alpha_n z$$

$$+ A_{3n}\left[\frac{-(1-\nu)}{\alpha_n(1+\nu)}\,\mathrm{sh}\,\alpha_n z + z\,\mathrm{ch}\,\alpha_n z\right]$$

$$+ A_{4n}\left[\frac{-(1-\nu)}{\alpha_n(1+\nu)}\,\mathrm{ch}\,\alpha_n z + z\,\mathrm{sh}\,\alpha_n z\right]\Bigg\}$$

$$+ \sum_{m=1}^{\infty} \beta_m^3 \Big\{\mathrm{ch}\,\beta_m y[B_{1m}\cos\beta_m z - B_{2m}\sin\beta_m z]$$

$$+ \left[\frac{2(2+\nu)}{\beta_m(1+\nu)}\,\mathrm{ch}\,\beta_m y + y\,\mathrm{sh}\,\beta_m y\right][B_{3m}\cos\beta_m z$$

$$- B_{4m}\sin\beta_m z]\Big\}, \tag{3.3b}$$

$$\frac{\tau_{yz}^{(1)}(y, z)}{C_{22}(1 - H)(1 - \nu)} = \sum_{n=1}^{\infty} \alpha_n^3 \sin \alpha_n y \left\{A_{1n} \operatorname{sh} \alpha_n z + A_{2n} \operatorname{ch} \alpha_n z \right.$$
$$+ A_{3n}\left[\frac{2\nu}{\alpha_n(1 + \nu)} \operatorname{ch} \alpha_n z + z \operatorname{sh} \alpha_n z\right]$$
$$\left. + A_{4n}\left[\frac{2\nu}{\alpha_n(1 + \nu)} \operatorname{sh} \alpha_n z + z \operatorname{ch} \alpha_n z\right]\right\}$$
$$+ \sum_{m=1}^{\infty} \beta_m^3 \left\{\operatorname{sh} \beta_m y[B_{1m} \sin \beta_m z + B_{2m} \cos \beta_m z]\right.$$
$$+ \left[\frac{3 + \nu}{\beta_m(1 + \nu)} \operatorname{sh} \beta_m y + y \operatorname{ch} \beta_m y\right][B_{3m} \sin \beta_m z$$
$$\left.\left. + B_{4m} \cos \beta_m z\right]\right\}. \tag{3.3c}$$

Choosing the parameters

$$\alpha_n = \frac{n\pi}{b}, \quad (n = 1, 2, 3, \ldots) \tag{3.4}$$

and

$$B_{1/2m} = -\left[\frac{3 + \nu}{\beta_m(1 + \nu)} + b \operatorname{cth} \beta_m b\right] B_{3/4m} \tag{3.5}$$

then boundary condition (2.17a) will be satisfied, and by taking

$$\beta_m = \frac{m\pi}{h_1}, \quad (m = 1, 2, 3, \ldots) \tag{3.6}$$

boundary condition (2.17c) becomes

$$\sum_{n=1}^{\infty} \alpha_n^3 \sin \alpha_n y \left\{A_{1n} \operatorname{sh} \alpha_n h_1 + A_{2n} \operatorname{ch} \alpha_n h_1\right.$$
$$+ A_{3n}\left[\frac{2\nu}{\alpha_n(1 + \nu)} \operatorname{ch} \alpha_n h_1 + h_1 \operatorname{sh} \alpha_n h_1\right]$$
$$\left. + A_{4n}\left[\frac{2\nu}{\alpha_n(1 + \nu)} \operatorname{sh} \alpha_n h_1 + h_1 \operatorname{ch} \alpha_n h_1\right]\right\}$$
$$+ \sum_{m=1}^{\infty} (-1)^m \beta_m^3 \{ y \operatorname{ch} \beta_m y - b \operatorname{cth} \beta_m b \operatorname{sh} \beta_m y\} B_{4m} = 0. \tag{3.7}$$

Moreover, let

$$B_{1m} = B_{3m} = 0 \tag{3.8}$$

the boundary condition (2.17d) yields the relation

$$A_{1n} = -A_{2n} \operatorname{th} \alpha_n h_1 - A_{3n}\left[\frac{-(1 - \nu)}{\alpha_n(1 + \nu)} \operatorname{th} \alpha_n h_1 + h_1\right]$$
$$- A_{4n}\left[\frac{-(1 - \nu)}{\alpha_n(1 + \nu)} + h_1 \operatorname{th} \alpha_n h_1\right]. \tag{3.9}$$

Finally, from boundary conditions (2.17b) and with the aid of equations (3.4), (3.5), (3.8) and (3.9) we obtain

$$\sum_{n=1}^{\infty} (-1)^n \alpha_n^3 \left\{ [\mathrm{sh}\, \alpha_n z - \mathrm{th}\, \alpha_n h_1 \,\mathrm{ch}\, \alpha_n z] A_{2n} + \left[\frac{1+3\nu}{\alpha_n(1+\nu)} \mathrm{sh}\, \alpha_n z \right.\right.$$

$$\left. + \frac{1-\nu}{\alpha_n(1+\nu)} \mathrm{th}\, \alpha_n h_1 \,\mathrm{ch}\, \alpha_n z + (z - h_1)\,\mathrm{ch}\, \alpha_n z \right] A_{3n}$$

$$\left. + \left[\frac{2}{\alpha_n} \mathrm{ch}\, \alpha_n z + z\, \mathrm{sh}\, \alpha_n z - h_1\, \mathrm{th}\, \alpha_n h_1 \,\mathrm{ch}\, \alpha_n z \right] A_{4n} \right\}$$

$$- \sum_{m=1}^{\infty} \beta_m^3 \sin \beta_m z \left\{ \frac{b}{\mathrm{sh}\, \beta_m b} + \frac{1}{\beta_m} \mathrm{ch}\, \beta_m b \right\} B_{4m}$$

$$= - \frac{\sigma_{y0}^{(1)}}{C_{22}(1-H)(1-\nu)} = -\sigma. \tag{3.10}$$

From the differential equation

$$\left(\frac{\partial^2}{\partial y^2} + \lambda_1 \frac{\partial^2}{\partial z^2} \right) \left(\frac{\partial^2}{\partial y^2} + \lambda_2 \frac{\partial^2}{\partial z^2} \right) F^{(2)}(y, z) = 0. \tag{2.16}$$

$F^{(2)}(y, z)$ may readily be found. Similarly, it must be an even function of y, and therefore may be expressed as

$$F^{(2)}(y, z) = \sum_{n=0}^{\infty} \cos \alpha_n y \left\{ D_{1n}\, \mathrm{sh}\, \frac{\alpha_n}{\sqrt{\lambda_1}} z + D_{2n}\, \mathrm{ch}\, \frac{\alpha_n}{\sqrt{\lambda_1}} z \right.$$

$$\left. + D_{3n}\, \mathrm{sh}\, \frac{\alpha_n}{\sqrt{\lambda_2}} z + D_{4n}\, \mathrm{ch}\, \frac{\alpha_n}{\sqrt{\lambda_2}} z \right\}$$

$$+ \sum_{m=0}^{\infty} \{ \mathrm{ch}\, \sqrt{\lambda_1}\, \omega_m y [L_{1m} \sin \omega_m z + L_{2m} \cos \omega_m z]$$

$$+ \mathrm{ch}\, \sqrt{\lambda_2}\, \omega_m y [L_{3m} \sin \omega_m z + L_{4m} \cos \omega_m z] \}, \tag{3.11}$$

where α_n, ω_m; D_{jn}, L_{jm} ($j = 1, 2, 3, 4$) are arbitrary constants. The corresponding displacements and stresses will be

$$\frac{v^{(2)}(y, z)}{(G+K)/R} = \sum_{n=1}^{\infty} \alpha_n^2 \sin \alpha_n y \left\{ \frac{1}{\sqrt{\lambda_1}} D_{1n}\, \mathrm{ch}\, \frac{\alpha_n}{\sqrt{\lambda_1}} z + \frac{1}{\sqrt{\lambda_1}} D_{2n}\, \mathrm{sh}\, \frac{\alpha_n}{\sqrt{\lambda_1}} z \right.$$

$$\left. + \frac{1}{\sqrt{\lambda_2}} D_{3n}\, \mathrm{ch}\, \frac{\alpha_n}{\sqrt{\lambda_2}} z + \frac{1}{\sqrt{\lambda_2}} D_{4n}\, \mathrm{sh}\, \frac{\alpha_n}{\sqrt{\lambda_2}} z \right\}$$

$$- \sum_{m=1}^{\infty} \omega_m^2 \{ \sqrt{\lambda_1}\, \mathrm{sh}\, \sqrt{\lambda_1}\, \omega_m y [L_{1m} \cos \omega_m z - L_{2m} \sin \omega_m z]$$

$$+ \sqrt{\lambda_2}\, \mathrm{sh}\, \sqrt{\lambda_2}\, \omega_m y [L_{3m} \cos \omega_m z - L_{4m} \sin \omega_m z] \} \tag{3.12a}$$

$$\omega^{(2)}(y, z) = -\sum_{n=1}^{\infty} \alpha_n^2 \cos \alpha_n y \left\{(1 - \lambda_2 G)\left[D_{1n} \operatorname{sh} \frac{\alpha_n}{\sqrt{\lambda_1}} z\right.\right.$$
$$\left. + D_{2n} \operatorname{ch} \frac{\alpha_n}{\sqrt{\lambda_1}} z\right] + (1 - \lambda_1 G)\left[D_{3n} \operatorname{sh} \frac{\alpha_n}{\sqrt{\lambda_2}} z\right.$$
$$\left.\left. + D_{4n} \operatorname{ch} \frac{\alpha_n}{\sqrt{\lambda_2}} z\right]\right\} + \sum_{m=1}^{\infty} \omega_m^2 \{\lambda_1(1 - \lambda_2 G) \operatorname{ch} \sqrt{\lambda_1}\omega_m y$$
$$\times [L_{1m} \sin \omega_m z + L_{2m} \cos \omega_m z] + \lambda_2(1 - \lambda_1 G)$$
$$\times \operatorname{ch} \sqrt{\lambda_2}\omega_m y [L_{3m} \sin \omega_m z + L_{4m} \cos \omega_m z]\} \tag{3.12b}$$

and

$$\frac{\sigma_y^{(2)}(y, z)}{C_{22} G} = \sum_{n=1}^{\infty} \alpha_n^3 \cos \alpha_n y \left\{\frac{1}{\sqrt{\lambda_1}}(1 + \lambda_2 K)\left[D_{1n} \operatorname{ch} \frac{\alpha_n}{\sqrt{\lambda_1}} z\right.\right.$$
$$\left. + D_{2n} \operatorname{sh} \frac{\alpha_n}{\sqrt{\lambda_1}} z\right] + \frac{1}{\sqrt{\lambda_2}}(1 + \lambda_1 K)\left[D_{3n} \operatorname{ch} \frac{\alpha_n}{\sqrt{\lambda_2}} z\right.$$
$$\left.\left. + D_{4n} \operatorname{sh} \frac{\alpha_n}{\sqrt{\lambda_2}} z\right]\right\} - \sum_{m=1}^{\infty} \omega_m^3 \{\lambda_1(1 + \lambda_2 K)$$
$$\times \operatorname{ch} \sqrt{\lambda_1}\omega_m y [L_{1m} \cos \omega_m z - L_{2m} \sin \omega_m z] + \lambda_2(1 + \lambda_1 K)$$
$$\times \operatorname{ch} \sqrt{\lambda_2}\omega_m y [L_{3m} \cos \omega_m z - L_{4m} \sin \omega_m z]\}, \tag{3.13a}$$

$$\frac{\sigma_z^{(2)}(y, z)}{C_{22}} = -\sum_{n=1}^{\infty} \alpha_n^3 \cos \alpha_n y \left\{\frac{1}{\sqrt{\lambda_1}}\left(1 - \lambda_2 G - K\frac{G + K}{R}\right)\right.$$
$$\times \left[D_{1n} \operatorname{ch} \frac{\alpha_n}{\sqrt{\lambda_1}} z + D_{2n} \operatorname{sh} \frac{\alpha_n}{\sqrt{\lambda_1}} z\right]$$
$$+ \frac{1}{\sqrt{\lambda_2}}\left(1 - \lambda_1 G - K\frac{G + K}{R}\right)\left[D_{3n} \operatorname{ch} \frac{\alpha_n}{\sqrt{\lambda_2}} z\right.$$
$$\left.\left. + D_{4n} \operatorname{sh} \frac{\alpha_n}{\sqrt{\lambda_2}} z\right]\right\} + \sum_{m=1}^{\infty} \omega_m^3 \left\{\lambda_1\left(1 - \lambda_2 G - K\frac{G + K}{R}\right)\right.$$
$$\times \operatorname{ch} \sqrt{\lambda_1}\omega_m y [L_{1m} \cos \omega_m z - L_{2m} \sin \omega_m z]$$
$$+ \lambda_2\left(1 - \lambda_1 G - K\frac{G + K}{R}\right) \operatorname{ch} \sqrt{\lambda_2}\omega_m y$$
$$\left.\times [L_{3m} \cos \omega_m z - L_{4m} \sin \omega_m z]\right\}, \tag{3.13b}$$

$$\frac{\tau_{yz}^{(2)}(y, z)}{C_{22} G} = \sum_{n=1}^{\infty} \alpha_n^3 \sin \alpha_n y \left\{(1 + \lambda_2 K)\left[D_{1n} \operatorname{sh} \frac{\alpha_n}{\sqrt{\lambda_1}} z + D_{2n} \operatorname{ch} \frac{\alpha_n}{\sqrt{\lambda_1}} z\right]\right.$$
$$\left. + (1 + \lambda_1 K)\left[D_{3n} \operatorname{sh} \frac{\alpha_n}{\sqrt{\lambda_2}} z + D_{4n} \operatorname{ch} \frac{\alpha_n}{\sqrt{\lambda_2}} z\right]\right\}$$

$$+ \sum_{m=1}^{\infty} \omega_m^3 \{\lambda_1^{3/2}(1 + \lambda_2 K) \operatorname{sh} \sqrt{\lambda_1}\omega_m y$$

$$\times [L_{1m} \sin \omega_m z + L_{2m} \cos \omega_m z] + \lambda_2^{3/2}(1 + \lambda_1 K)$$

$$\times \operatorname{sh} \sqrt{\lambda_2}\omega_m y [L_{3m} \sin \omega_m z + L_{4m} \cos \omega_m z]\}. \tag{3.13c}$$

Choosing the parameter α_n, the same value as in (3.4), and applying the boundary condition (2.18a), we obtain

$$L_{1/2m} = -\frac{\lambda_2^{3/2}(1 + \lambda_1 K) \operatorname{sh} \sqrt{\lambda_2}\omega_m b}{\lambda_1^{3/2}(1 + \lambda_2 K) \operatorname{sh} \sqrt{\lambda_1}\omega_m b} L_{3/4m}. \tag{3.14}$$

Let

$$\omega_m = \frac{m\pi}{h_2}, \quad (m = 1, 2, 3, \ldots) \tag{3.15}$$

and

$$L_{2m} = L_{4m} = 0, \tag{3.16}$$

then we obtain from the boundary condition (2.18c)

$$(1 + \lambda_2 K)\left[-D_{1n} \operatorname{sh} \frac{\alpha_n}{\sqrt{\lambda_1}} h_2 + D_{2n} \operatorname{ch} \frac{\alpha_n}{\sqrt{\lambda_1}} h_2\right]$$

$$+ (1 + \lambda_1 K)\left[-D_{3n} \operatorname{sh} \frac{\alpha_n}{\sqrt{\lambda_2}} h_2 + D_{4n} \operatorname{ch} \frac{\alpha_n}{\sqrt{\lambda_2}} h_2\right] = 0. \tag{3.17}$$

According to the boundary condition (2.18d), and with the aid of (3.15), (3.16), the following relation may be derived

$$(1 - \lambda_2 G)\left[-D_{1n} \operatorname{sh} \frac{\alpha_n}{\sqrt{\lambda_1}} h_2 + D_{2n} \operatorname{ch} \frac{\alpha_n}{\sqrt{\lambda_1}} h_2\right]$$

$$+ (1 - \lambda_1 G)\left[-D_{3n} \operatorname{sh} \frac{\alpha_n}{\sqrt{\lambda_2}} h_2 + D_{4n} \operatorname{ch} \frac{\alpha_n}{\sqrt{\lambda_2}} h_2\right] = 0. \tag{3.18}$$

Solving simultaneously (3.17) and (3.18), we then obtain

$$D_{2n} = \operatorname{th} \frac{\alpha_n}{\sqrt{\lambda_1}} h_2 \cdot D_{1n}, \qquad D_{4n} = \operatorname{th} \frac{\alpha_n}{\sqrt{\lambda_2}} h_2 \cdot D_{3n}. \tag{3.19}$$

Finally, by the boundary condition (2.18b), and with the aid of (3.4), (3.14), (3.16) and (3.19), we may obtain

$$\sum_{n=1}^{\infty} (-1)^n \alpha_n^3 \left\{\frac{1}{\sqrt{\lambda_1}}(1 + \lambda_2 K)\left[\operatorname{ch} \frac{\alpha_n}{\sqrt{\lambda_1}} z + \operatorname{th} \frac{\alpha_n}{\sqrt{\lambda_1}} h_2 \operatorname{sh} \frac{\alpha_n}{\sqrt{\lambda_1}} z\right] D_{1n}\right.$$

$$\left. + \frac{1}{\sqrt{\lambda_2}}(1 + \lambda_1 K)\left[\operatorname{ch} \frac{\alpha_n}{\sqrt{\lambda_2}} z + \operatorname{th} \frac{\alpha_n}{\sqrt{\lambda_2}} h_2 \operatorname{sh} \frac{\alpha_n}{\sqrt{\lambda_2}} z\right] D_{3n}\right\}$$

$$+ \lambda_2(1 + \lambda_1 K) \sum_{m=1}^{\infty} \omega_m^3 \cos \omega_m z$$

$$\times \left\{\frac{\sqrt{\lambda_2} \operatorname{sh} \sqrt{\lambda_2}\omega_m b}{\sqrt{\lambda_1} \operatorname{sh} \sqrt{\lambda_1}\omega_m b} \operatorname{ch} \sqrt{\lambda_1}\omega_m b - \operatorname{ch} \sqrt{\lambda_2}\omega_m b\right\} L_{3m} = 0. \tag{3.20}$$

Moreover, there remain four conditions of continuity at the interface to be satisfied. The condition (2.19a), with (3.8), (3.9) and (3.14) for the coefficients $B_{1/3m}$, A_{1n} and $L_{1/2m}$ respectively, after some simplification yields

$$\sum_{n=1}^{\infty} \alpha_n^3 \cos \alpha_n y \left\{(1 - H)(1 - \nu)\left[\operatorname{th} \alpha_n h_1 A_{2n} + \left(h_1 - \frac{1 - \nu}{1 + \nu}\frac{\operatorname{th} \alpha_n h_1}{\alpha_n}\right) A_{3n} + h_1 \operatorname{th} \alpha_n h_1 A_{4n}\right] + \frac{1}{\sqrt{\lambda_1}}\left(1 - \lambda_2 G - K\frac{G + K}{R}\right) D_{1n} + \frac{1}{\sqrt{\lambda_2}}\left(1 - \lambda_1 G - K\frac{G + K}{R}\right) D_{3n}\right\}$$
$$+ \sum_{m=1}^{\infty} \omega_m^3 \lambda_2 \left\{\left(1 - \lambda_2 G - K\frac{G + K}{R}\right)\frac{\sqrt{\lambda_2}(1 + \lambda_1 K) \operatorname{sh} \sqrt{\lambda_2}\omega_m b}{\sqrt{\lambda_1}(1 + \lambda_2 K) \operatorname{sh} \sqrt{\lambda_1}\omega_m b} \times \operatorname{ch} \sqrt{\lambda_1}\omega_m y - \left(1 - \lambda_1 G - K\frac{G + K}{R}\right) \operatorname{ch} \sqrt{\lambda_2}\omega_m y\right\} L_{3m} = 0. \tag{3.21}$$

By the condition (2.19d) with (3.5), (3.16) and (3.19), we have, after simplification

$$\sum_{n=1}^{\infty} \alpha_n^2 \cos \alpha_n y \left\{-(1 - H)A_{2n} + \frac{1}{\alpha_n} 2HA_{3n} + (1 - \lambda_2 G) \operatorname{th} \frac{\alpha_n}{\sqrt{\lambda_1}} h_2 D_{1n} + (1 - \lambda_1 G) \operatorname{th} \frac{\alpha_n}{\sqrt{\lambda_2}} h_2 D_{3n}\right\} + \sum_{m=1}^{\infty} \beta_m^2 (1 - H) \left\{\left[\frac{1 - \nu}{\beta_m(1 + \nu)} - b \operatorname{cth} \beta_m b\right] \operatorname{ch} \beta_m y + y \operatorname{sh} \beta_m y\right\} B_{4m} = 0. \tag{3.22}$$

Also from the condition (2.19c), with (3.8), (3.9) and (3.14), we obtain

$$\sum_{n=1}^{\infty} \alpha_n^2 \sin \alpha_n y \left\{(1 - H) \operatorname{th} \alpha_n h_1 A_{2n} + (1 - H) \times \left[\frac{-H}{\alpha_n(1 - H)} \operatorname{th} \alpha_n h_1 + h_1\right] A_{3n} + (1 - H)\left[\frac{-1}{\alpha_n(1 - H)} + h_1 \operatorname{th} \alpha_n h_1\right] A_{4n} + \frac{G + K}{R}\frac{1}{\sqrt{\lambda_1}} D_{1n} + \frac{G + K}{R}\frac{1}{\sqrt{\lambda_2}} D_{3n}\right\}$$
$$+ \frac{G + K}{R}\sum_{m=1}^{\infty} \omega_m^2 \sqrt{\lambda_2}\left\{\frac{\lambda_2(1 + \lambda_1 K) \operatorname{sh} \sqrt{\lambda_2}\omega_m b}{\lambda_1(1 + \lambda_2 K) \operatorname{sh} \sqrt{\lambda_1}\omega_m b} \times \operatorname{sh} \sqrt{\lambda_1}\omega_m y - \operatorname{sh} \sqrt{\lambda_2}\omega_m y\right\} L_{3m} = 0, \tag{3.23}$$

and from (2.19b) with the relations of (3.5), (3.16) and (3.19), we obtain

$$\sum_{n=1}^{\infty} \alpha_n^3 \sin \alpha_n y \left\{(1 - H)(1 - \nu) A_{2n} + \frac{1}{\alpha_n} \nu(1 - \nu) A_{3n} - G(1 + \lambda_2 K) \operatorname{th} \frac{\alpha_n}{\sqrt{\lambda_1}} h_2 D_{1n} - G(1 + \lambda_1 K) \operatorname{th} \frac{\alpha_n}{\sqrt{\lambda_2}} h_2 D_{3n}\right\}$$
$$+ (1 - H)(1 - \nu) \sum_{m=1}^{\infty} \beta_m^3 \{y \operatorname{ch} \beta_m y - b \operatorname{cth} \beta_m b \operatorname{sh} \beta_m y\} B_{4m} = 0. \tag{3.24}$$

Thus, we obtain a system of algebraic equations (3.7), (3.10) and (3.20)–(3.24) in the forms of the sum of two infinite series, in which there are seven sets of arbitrary constants involved: A_{2n}, A_{3n}, A_{4n}, D_{1n}, D_{3n}, B_{4m} and L_{3m} ($n, m = 1, 2, 3, \ldots$). The orthogonality property of the trigonometric functions is applied to determine these constants. By multiplying both sides of the eqs. (3.21) and (3.22) with $\cos \alpha_n y$ and that of eqs. (3.23), (3.24) and (3.7) with $\sin \alpha_n y$, then integrating them with respect to y within the interval $[-b, b]$, we obtain, after some manipulations, the following set of equations

$$-(1 - H)A_{2n}^* + 2HA_{3n}^* + (1 - \lambda_2 G) \operatorname{th} \frac{\alpha_n}{\sqrt{\lambda_1}} h_2 D_{1n}^* + (1 - \lambda_1 G) \operatorname{th} \frac{\alpha_n}{\sqrt{\lambda_2}} h_2 D_{3n}^* + \sum_{m=1}^{\infty} 2(-1)^n (1 - H) \frac{1}{b} \frac{\beta_m}{\alpha_n^2 + \beta_m^2} \times \left\{ \frac{H}{1 - H} + \frac{\alpha_n^2 - \beta_m^2}{\alpha_n^2 + \beta_m^2} \right\} B_{4m}^* = 0, \quad (n = 1, 2, 3, \ldots), \tag{3.25a}$$

$$(1 - H)(1 - \nu)A_{2n}^* + \nu(1 - \nu)A_{3n}^* - G(1 + \lambda_2 K) \operatorname{th} \frac{\alpha_n}{\sqrt{\lambda_1}} h_2 D_{1n}^* - G(1 + \lambda_1 K) \operatorname{th} \frac{\alpha_n}{\sqrt{\lambda_2}} h_2 D_{3n}^* + \sum_{m=1}^{\infty} (-1)^n (1 - H)(1 - \nu) \frac{\beta_m}{b} \times \left(\frac{2\beta_m}{\alpha_n^2 + \beta_m^2} \right)^2 B_{4m}^* = 0, \quad (n = 1, 2, 3, \ldots), \tag{3.25b}$$

$$\frac{1}{\operatorname{ch} \alpha_n h_1} A_{2n}^* + \left[\frac{\nu}{1 - H} \operatorname{ch} \alpha_n h_1 + \frac{H}{1 - H} \operatorname{th} \alpha_n h_1 \operatorname{sh} \alpha_n h_1 \right] A_{3n}^* + \left[\operatorname{sh} \alpha_n h_1 + \frac{\alpha_n h_1}{\operatorname{ch} \alpha_n h_1} \right] A_{4n}^* + \sum_{m=1}^{\infty} (-1)^{n+m} \frac{\beta_m}{b} \times \left(\frac{2\beta_m}{\alpha_n^2 + \beta_m^2} \right)^2 B_{4m}^* = 0, \quad (n = 1, 2, 3, \ldots), \tag{3.25c}$$

$$\alpha_n (1 - H)(1 - \nu) \left[\operatorname{th} \alpha_n h_1 A_{2n}^* + \left(\alpha_n h_1 - \frac{H}{1 - H} \operatorname{th} \alpha_n h_1 \right) A_{3n}^* + \alpha_n h_1 \operatorname{th} \alpha_n h_1 A_{4n}^* \right] + \alpha_n \frac{1}{\sqrt{\lambda_1}} \left(1 - \lambda_2 G - K \frac{G + K}{R} \right) D_{1n}^* + \alpha_n \frac{1}{\sqrt{\lambda_2}} \left(1 - \lambda_1 G - K \frac{G + K}{R} \right) D_{3n}^* + \sum_{m=1}^{\infty} 2(-1)^n \omega_m^2 \frac{\lambda_2^{3/2}}{b} \times \left\{ \frac{1}{\alpha_n^2 + \lambda_1 \omega_m^2} \frac{1 + \lambda_1 K}{1 + \lambda_2 K} \left(1 - \lambda_2 G - K \frac{G + K}{R} \right) - \frac{1}{\alpha_n^2 + \lambda_2 \omega_m^2} \left(1 - \lambda_1 G - K \frac{G + K}{R} \right) \right\} L_{3m}^* = 0, \quad (n = 1, 2, 3, \ldots). \tag{3.25d}$$

$$\frac{1}{\alpha_n}(1-H)\left[\text{th}\,\alpha_n h_1 A^*_{2n} + \left(\alpha_n h_1 - \frac{H}{1-H}\text{th}\,\alpha_n h_1\right)A^*_{3n}\right.$$
$$\left.+\left(\alpha_n h_1\,\text{th}\,\alpha_n h_1 - \frac{1}{1-H}\right)A^*_{4n}\right] + \frac{1}{\alpha_n}\frac{G+K}{R}\frac{1}{\sqrt{\lambda_1}}D^*_{1n}$$
$$+\frac{1}{\alpha_n}\frac{G+K}{R}\frac{1}{\sqrt{\lambda_2}}D^*_{3n} + \sum_{m=1}^{\infty} 2(-1)^n \frac{1}{b}\frac{G+K}{R}\sqrt{\lambda_2}\left\{\frac{1}{\alpha_n^2+\lambda_2\omega_m^2}\right.$$
$$\left.-\frac{1}{\alpha_n^2+\lambda_1\omega_m^2}\frac{\lambda_2(1+\lambda_1 K)}{\lambda_1(1+\lambda_2 K)}\right\}L^*_{3m} = 0, \quad (n = 1, 2, 3, \ldots). \tag{3.25e}$$

Similarly, multiplying eq. (3.10) by $\sin \beta_m z$, eq. (3.20) by $\cos \omega_m z$, and integrating with respect to z within the interval $[0, h_1]$ and $[-h_2, 0]$ respectively, we have

$$-\sum_{n=1}^{\infty}(-1)^n\beta_m^2\frac{\alpha_n}{\alpha_n^2+\beta_m^2}\text{th}\,\alpha_n h_1 A^*_{2n} + \sum_{n=1}^{\infty}(-1)^{n+m}\beta_m^2\frac{\alpha_n}{\alpha_n^2+\beta_m^2}$$
$$\times\left\{2\left[\frac{\alpha_n^2}{\alpha_n^2+\beta_m^2}-1\right]\text{sh}\,\alpha_n h_1 + (-1)^m\left[\frac{H}{1-H}\text{th}\,\alpha_n h_1 - \alpha_n h_1\right]\right\}A^*_{3n}$$
$$+\sum_{n=1}^{\infty}(-1)^n\beta_m^2\frac{\alpha_n}{\alpha_n^2+\beta_m^2}\left\{2\left[\frac{\alpha_n^2}{\alpha_n^2+\beta_m^2}-1\right][(-1)^m\,\text{ch}\,\alpha_n h_1 - 1]\right.$$
$$\left.-\alpha_n h_1\,\text{th}\,\alpha_n h_1\right\}A^*_{4n} - \tfrac{1}{2}\beta_m^2 h_1\{\beta_m b[\text{cth}^2\beta_m b - 1]$$
$$+\,\text{cth}\,\beta_m b\}B^*_{4m} = \sigma[(-1)^m - 1], \quad (m = 1, 2, 3, \ldots), \tag{3.26a}$$

$$\sum_{n=1}^{\infty}(-1)^n(1+\lambda_2 K)\frac{\alpha_n^2}{\alpha_n^2+\lambda_1\omega_m^2}\text{th}\frac{\alpha_n}{\sqrt{\lambda_1}}h_2 D^*_{1n} + \sum_{m=1}^{\infty}(-1)^n(1+\lambda_1 K)$$
$$\times\frac{\alpha_n^2}{\alpha_n^2+\lambda_2\omega_m^2}\text{th}\frac{\alpha_n}{\sqrt{\lambda_2}}h_2 D^*_{3n} + \frac{\omega_m h_2}{2}\lambda_2(1+\lambda_1 K)$$
$$\times\left\{\frac{\sqrt{\lambda_2}}{\sqrt{\lambda_1}}\text{cth}\sqrt{\lambda_1}\omega_m b - \text{cth}\sqrt{\lambda_2}\omega_m b\right\}L^*_{3m} = 0, \quad (m = 1, 2, 3, \ldots), \tag{3.26b}$$

where

$$\left.\begin{aligned} A^*_{2n} &= \alpha_n^2 A_{2n}, \quad A^*_{3n} = \alpha_n A_{3n}, \quad A^*_{4n} = \alpha_n A_{4n},\\ D^*_{1n} &= \alpha_n^2 D_{1n}, \quad D^*_{3n} = \alpha_n^2 D_{3n},\\ B^*_{4m} &= \beta_m\,\text{sh}\,\beta_m b B_{4m}, \quad L^*_{3m} = \omega_m^2\,\text{sh}\sqrt{\lambda_2}\omega_m b L_{3m}.\end{aligned}\right\} \tag{3.27}$$

In the practical computation, we have to truncate the infinite series and take N and M terms for the series respectively. Thus, altogether we shall have $(5N + 2M)$ simultaneous linear algebraic equations, from which the unknown constants A^*_{2n}, A^*_{3n}, A^*_{4n}, D^*_{1n}, D^*_{3n} $(n = 1, 2, 3, \ldots, N)$ and B^*_{4m}, L^*_{3m} $(m = 1, 2, 3, \ldots, M)$ are to be found. Once these constants are determined, we can readily compute the displacements and stresses, and thus the interlaminar stresses of the laminate.

Evidently, the numerical solution approaches the exact value as the number of terms N and M increase.

As for the interlaminar stresses, the normal stress on the mid-plane of the $[0/90]_s$ symmetric laminate can be expressed as:

$$\begin{aligned}\frac{\sigma_z^{(2)}(y, -h_2)}{C_{22}} = &-\sum_{n=1}^{N} \alpha_n \cos \alpha_n y \left\{\frac{1}{\sqrt{\lambda_1}}\left(1 - \lambda_2 G - K\frac{G+K}{R}\right)\right.\\ &\times \frac{1}{\operatorname{ch}\frac{\alpha_n}{\sqrt{\lambda_1}} h_2} D_{1n}^* + \frac{1}{\sqrt{\lambda_2}}\left(1 - \lambda_1 G - K\frac{G+K}{R}\right)\\ &\left.\times \frac{1}{\operatorname{ch}\frac{\alpha_n}{\sqrt{\lambda_2}} h_2} D_{3n}^*\right\} + \sum_{m=1}^{M} (-1)^m \omega_m \lambda_2\\ &\times \left\{\left(1 - \lambda_1 G - K\frac{G+K}{R}\right)\frac{\operatorname{ch}\sqrt{\lambda_2}\omega_m y}{\operatorname{sh}\sqrt{\lambda_2}\omega_m b}\right.\\ &- \frac{\sqrt{\lambda_2}(1+\lambda_1 K)}{\sqrt{\lambda_1}(1+\lambda_2 K)}\left(1 - \lambda_2 G - K\frac{G+K}{R}\right)\\ &\left.\times \frac{\operatorname{ch}\sqrt{\lambda_1}\omega_m y}{\operatorname{sh}\sqrt{\lambda_1}\omega_m b}\right\} L_{3m}^*. \end{aligned} \tag{3.28}$$

The shearing stress and normal stress between the first and second layers are respectively:

$$\begin{aligned}\frac{\tau_{yz}^{(2)}(y, 0)}{C_{22}G} = &\sum_{n=1}^{N} \alpha_n \sin \alpha_n y \left\{(1+\lambda_2 K) \operatorname{th}\frac{\alpha_n}{\sqrt{\lambda_1}} h_2 D_{1n}^*\right.\\ &\left.+(1+\lambda_1 K) \operatorname{th}\frac{\alpha_n}{\sqrt{\lambda_2}} h_2 D_{3n}^*\right\}, \end{aligned} \tag{3.29}$$

$$\begin{aligned}\frac{\sigma_z^{(2)}(y, 0)}{C_{22}} = &-\sum_{n=1}^{N} \alpha_n \cos \alpha_n y \left\{\frac{1}{\sqrt{\lambda_1}}\left(1 - \lambda_2 G - K\frac{G+K}{R}\right) D_{1n}^*\right.\\ &\left.+ \frac{1}{\sqrt{\lambda_2}}\left(1 - \lambda_1 G - K\frac{G+K}{R}\right) D_{3n}^*\right\}\\ &+ \sum_{m=1}^{M} \omega_m \lambda_2 \left\{\left(1 - \lambda_1 G - K\frac{G+K}{R}\right)\frac{\operatorname{ch}\sqrt{\lambda_2}\omega_m y}{\operatorname{sh}\sqrt{\lambda_2}\omega_m b}\right.\\ &- \frac{\sqrt{\lambda_2}(1+\lambda_1 K)}{\sqrt{\lambda_1}(1+\lambda_2 K)}\left(1 - \lambda_2 G - K\frac{G+K}{R}\right)\\ &\left.\times \frac{\operatorname{ch}\sqrt{\lambda_1}\omega_m y}{\operatorname{sh}\sqrt{\lambda_1}\omega_m b}\right\} L_{3m}^*. \end{aligned} \tag{3.30}$$

4. Numerical calculation

In order to compare the present solution with the previous theories, the example in [4] is used for the numerical computation. Let the thickness $h_1 = h_2 = h$, then

$$\beta_m = \omega_m = \frac{m\pi}{h} \quad (m = 1, 2, 3, \ldots, M).$$

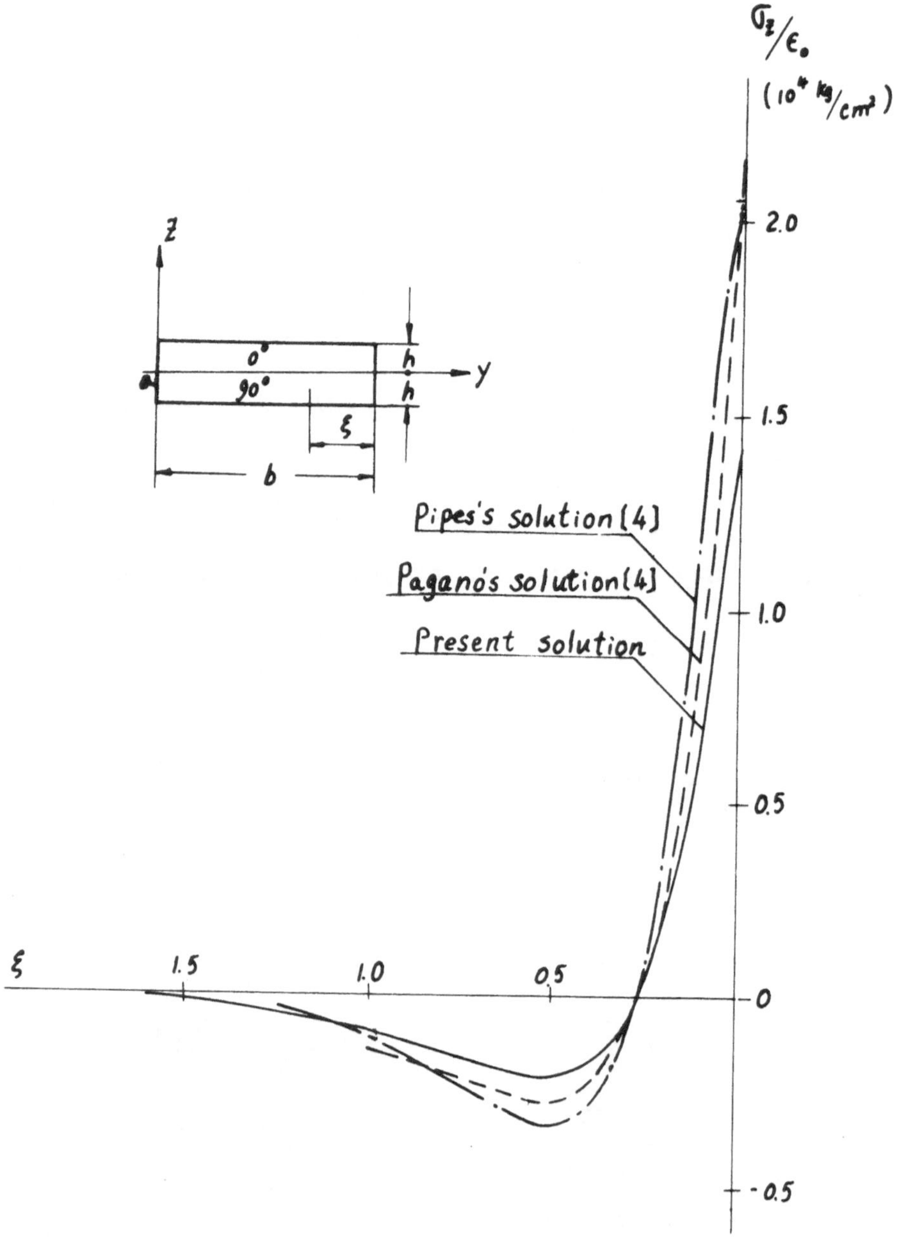

Figure 3. The distribution curve of σ_z on the mid-plane.

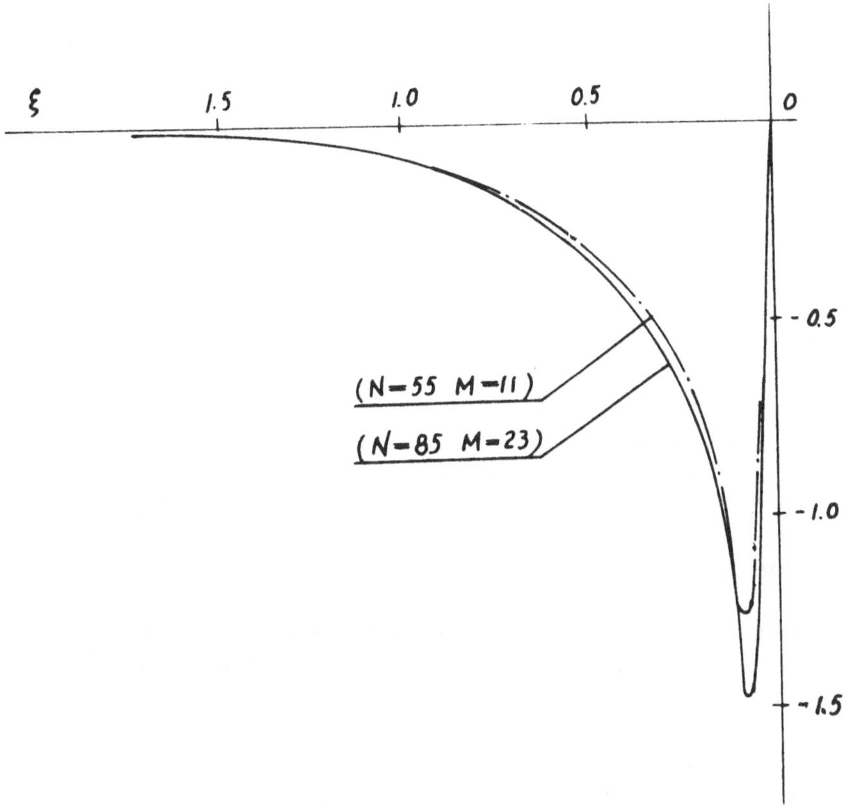

Figure 4. The distribution curve of τ_{yz} between the first and second layers.

Take ε_0 to be the unit strain, $h = 0.0625b$ and assume the material moduli (with unit of 10^6 kg/cm^2)

$$C_{11} = 1.4216, \quad C_{12} = 0.039411, \quad C_{22} = 0.15553$$

$$C_{23} = 0.033781, \quad C_{55} = 0.059820.$$

The numerical computations are performed by the '709' computer. Since the coefficient matrix of the linear algebraic equations is an asymmetric sparse one, the technique of bandwidth optimization is used to reduce it into a block diagonal matrix, and then solved by the Gaussian method of elimination. This method of calculation will greatly reduce the computer's storage required, and the computed results indicate that convergence is good. The normal stress $\sigma_z^{(2)}(y, -h)$ on the mid-plane, and the interlaminar shearing stress $\tau_{yz}^{(2)}(y, 0)$ between the first and second layers are plotted in Figs. 3 and 4, respectively, against the dimensionless length parameter $\xi = (b - y)/4h$, i.e., the relative distance with respect to the total thickness of the laminate measured from the free edge. The curve for the normal stress $\sigma_z^{(2)}(y, 0)$ between the first two layers is not given since its convergence is relatively slow, and more terms are needed in this case. Nevertheless, its overall tendency shows the close resemblance to the distribution curve of the normal stress $\sigma_z^{(2)}(y, -h)$. It can be seen from these curves that the interlaminar stress appears mainly near the free edges of the laminate. The interlaminar normal stress reaches its maximum value at the free edge, and the shearing stress, though it vanishes at the lateral surface, arrives at an apparent peak value near the edge. This may explain, from the theoretical viewpoint, the experimentally observed phenomena that the interlaminar destruction of the laminate under tension is always initiated from the free edges.

The curves are also plotted in Fig. 3 to show Pipes's finite difference solution and Pagano's approximate solution for comparison. Although the general features of both curves are very similar to those of the present study, the peak values of the interlaminar normal stress on the mid-plane are all on the high side, accounting for about 40%.

On the other hand, the shearing stress between the first and second layers of both the finite element solution of [8] and the approximate solution of [6] agree quite well with that obtained by the present theory.

Acknowledgement

The authors wish to express their gratitude to Mr. Hua Xiao-xian (Shanghai Institute of Computer Technology), for his valuable assistance in programming the computations involved.

References

1. Pipes, R. B. and Pagano, N. J., "Interlaminar stresses in composite laminates under uniform axial extension", *J. Comp. Materials* **4** (1970), 538–548.
2. Pipes, R. B., *Solution of Certain Problems in the Theory of Elasticity for Laminated Anisotropic Systems*, Ph.D. dissertation, University of Texas of Arlington, March (1972).
3. Pagano, N. J. and Pipes, R. B., "Some observations on the interlaminar strength of composite laminates", *Int. J. Mech. Sci.* **15** (1973), 679–688.
4. Pagano, N. J., "On the calculation of interlaminar normal stress in composite laminate", *J. Comp. Materials* **8** (1974), 65–81.
5. Rybicki, E. F., "Approximate three-dimensional solutions for symmetric laminates under inplane loading", *J. Comp. Materials* **5** (1971), 354–360.
6. Pagano, N. J., "Stress fields in composite laminates", *Int. J. Solids and Structures* **14** (1978), 385–400.
7. Pipes, R. B. and Pagano, N. J., "Interlaminar stresses in composite laminates – An approximate elasticity solution", *J. Appl. Mech., Trans. ASME*, Series E **41** (1974), 668–672.
8. Wang, A. S. D. and Crossman, F. N., "Some new results on edge effect in symmetric composite laminates", *J. Comp. Materials* **11** (1977), 92–106.
9. Hsu, P. W. and Herakovich, C. T., "Edge effects in angle-ply composite laminates", *J. Comp. Materials* **11** (1977), 422–428.
10. Wang, J. T. S. and Dickson, J. N., "Interlaminar stresses in symmetric composite laminates", *J. Comp. Materials* **12** (1978), 390–402.
11. Spilker, R. L. and Chou, S. C., "Edge effects in symmetric composite laminates: Importance of satisfying the traction-free edge condition", *J. Comp. Materials* **14** (1980), 2–20.
12. Pickett, G., "Application of the Fourier method to the solution of certain boundary problems in the theory of elasticity", *J. Appl. Mech.* **11** (1944), A-176–182.
13. Jones, R. M., *Mechanics of Composite Materials*, Scripta Book Company (1975), p. 35.

Finite deformation of plates and shells

Geometric theory of finite deformation of shells

CHEN ZHI-DA

Graduate School, China Institute of Mining Industry, Beijing, PRC

Abstract. The present theory is based on the classical work of plates and shells by Chien [1] and the method used by Synge [2]. By using the strain-rotation decomposition theorem in continuum mechanics, we establish a rigorous and systematic method for analysis of large deformation of shells. The accuracy of the Kirchhoff–Love hypothesis is discussed. It is shown that the hypothesis has neglected the influence of rotation at large. The present theory leads to all important classical results according to various conditions and different orders of approximation (including Love, Timoshenko, Vlasov, Novozhilov, Donnell, Koiter, Reissner, etc.)

1. Introduction

It was Kármán and Tsien [4, 5] who first pointed out that a satisfactory explanation of buckling phenomena of shells should make recourse to the non-linear large deformation theory.

The current theories of shells are mostly based on the Kirchhoff–Love hypothesis. However, due to the lack of an exact three dimensional continuum theory of shells, there has been much controversy over the accuracy of different formulations of shell theories.

Attempts at developing a rigorous theory of elasticity for shells have been made by many authors during the last few decades; notably Kirchhoff [16], Love [17], Lurie [7], Synge and Chien [1, 2], Zerna [19], and recently Ciarlet [11], etc.

The most difficult problem of the theory of shells lies in the fact that large rotations usually occur in the buckling processes, while bulk elasticity does not have such rotations. The problem can be completely solved only when we have a theory of deformation which treats rotation and strain congruently. In this paper, we shall develop a method to analyse non-linear large deformation of shells based on the co-moving coordinate method due to Chien, and the non-linear geometric field theory of the present author.

2. Parallel displacement of curve

A shell is a continuum space bounded by two surfaces, whose thickness is small compared with other dimensions. The deformation of a shell can be analysed by

Yeh, K. Y. (ed) Progress in Applied Mechanics
 ISBN-13: 978-94-010-8061-3

the spatial transformation of a co-moving coordinate system embedded in the shell. In order to determine the deformation and rotation of a shell, it is important to first set up the concept of absolutely parallel displacement of a curve. Here, we describe the concept in a slightly different way from that of Levi-Civita, in view of the geometry of deformation of shells.

Let $\mathbf{r}$ denote the position vector from a fixed point 0 to a surface G. The equation of G is:

$$\mathbf{r} = \mathbf{r}(x^1, x^2), \tag{2.1}$$

where x^1 and x^2 are curvilinear coordinates on G.

At an arbitrary point on G two basic vectors tangential to the surface, we define:

$$\mathbf{a}_\alpha = \frac{\partial \mathbf{r}}{\partial x^\alpha}, \quad (\alpha = 1, 2). \tag{2.2a}$$

The metric tensor of G is

$$a_{\alpha\beta} = \mathbf{a}_\alpha \cdot \mathbf{a}_\beta, \quad (\mathrm{d}S_0)^2 = a_{\alpha\beta}\,\mathrm{d}x^\alpha\,\mathrm{d}x^\beta, \quad (\alpha, \beta = 1, 2) \tag{2.2b}$$

where $\mathrm{d}S_0$ is arc length of the surface G.

Let $\mathbf{a}_3$ be a unit normal vector to the surface G at a point P. In the local frame $\mathbf{a}_1$, $\mathbf{a}_2$, $\mathbf{a}_3$, we have an arbitrary elemental line length $\mathrm{d}S$:

$$(\mathrm{d}S)^2 = g_{ij}\,\mathrm{d}x^i\,\mathrm{d}x^j, \quad (i, j = 1, 2, 3), \tag{2.3}$$

where g_{ij} is the metric tensor of space spanned by $\mathbf{a}_1$, $\mathbf{a}_2$, $\mathbf{a}_3$,

$$g_{33} = 1, \qquad g_{13} = g_{31} = g_{23} = g_{32} = 0. \tag{2.4}$$

On G:

$$x^3 = 0, \quad (\mathrm{d}S)^2 = (\mathrm{d}S_0)^2 = a_{\alpha\beta}\,\mathrm{d}x^\alpha\,\mathrm{d}x^\beta. \tag{2.5}$$

Later on, the Greek suffixes α, β are related to coordinates on the surface and the Latin letters i, j, k are related to the continuum space.

We consider a curve on a surface, and suppose the curve makes a parallel movement in space together with the surface in the space. Then the displacement vector $\mathbf{u}$ of all points on the curve is parallel and equal, and the condition is

$$\frac{\mathrm{d}\mathbf{u}}{\mathrm{d}S} = 0 \quad \text{on curve } x^i = f(S). \tag{2.6}$$

Because

$$\frac{\mathrm{d}\mathbf{u}}{\mathrm{d}S} = \frac{\mathrm{d}}{\mathrm{d}S}(u^i\mathbf{a}_i) = \left(\frac{\mathrm{d}u^i}{\mathrm{d}S} + \Gamma^i_{jk}u^k\,\frac{\mathrm{d}x^j}{\mathrm{d}S}\right)\mathbf{a}_i, \tag{2.7}$$

where the Christoffel symbol is

$$\Gamma^i_{jk} = \tfrac{1}{2}g^{im}\left(\frac{\partial g_{mj}}{\partial x^k} + \frac{\partial g_{mk}}{\partial x^j} - \frac{\partial g_{jk}}{\partial x^m}\right), \tag{2.8}$$

(2.7) can be simplified as

$$\frac{d\mathbf{u}}{dS} = \frac{Du^i}{DS}\mathbf{a}_i, \tag{2.9}$$

or

$$\frac{Du^i}{DS} = u^i\big|_j \frac{dx^j}{dS}. \tag{2.10}$$

Here, $u^i\big|_j$ is the covariant differential of u^i with respect to x^j, and D/DS is called the absolute differential along curve S. If curve S makes a parallel, then we have

$$\frac{Du^i}{DS} = 0, \quad (i = 1, 2, 3), \tag{2.11}$$

where u^i is the contravariant component of the displacement vector with respect to $\mathbf{a}_i$ on the surface G. For a plate, ordinary differentials should be used to replace absolute differentials. Thus, we have

$$\frac{du^i}{dS} = \frac{du_i}{dS} = 0. \tag{2.12}$$

In fact, (2.11) and (2.12) are a set of differential equations of u^i. For a plate making parallel movements, by integrating (2.12), we can obtain

$$u_1 = A, \quad u_2 = B, \quad u_3 = C,$$

where A, B, C are constants.

We shall see later, by using Levi-Civita's parallel concept, and if we substitute ordinary differentials for absolute differentials, that the expression of strain components in Cartesian coordinates can be transformed into those of curvilinear coordinates.

3. Fundamental characteristics of the deformation of surfaces

For mathematical convenience, we chose the lines of orthogonal principal curvature on the undeformed surface as embedded coordinate lines: x^1 = const. and x^2 = const. Deformation will change the orthogonal curve into a skew curve, $G_0 \to G$, $\mathbf{r} \to \bar{\mathbf{r}}$,

$$\bar{\mathbf{r}} = \mathbf{r} + \mathbf{u} \tag{3.1}$$

and $\mathbf{a}_\alpha \to \bar{\mathbf{a}}_\alpha$, $dS_0 \to dS$,

$$\begin{aligned} (dS_0)^2 &= a_{\alpha\beta}\, dx^\alpha\, dx^\beta = a_{11}(dx^1)^2 + a_{22}(dx^2)^2, \\ dS_0^1 &= \sqrt{a_{11}}\, dx^1, \qquad dS_0^2 = \sqrt{a_{22}}\, dx^2, \end{aligned} \tag{3.2}$$

where dS_0^1 and dS_0^2 are arc lengths along x^1 and x^2 respectively.

$$\begin{aligned} (dS)^2 &= \bar{a}_{\alpha\beta}\, dx^\alpha\, dx^\beta = \bar{a}_{11}(dx^1)^2 + 2\bar{a}_{12}(dx^1\, dx^2) + \bar{a}_{22}(dx^2)^2 \\ &= \bar{a}_{11}(dx^1)^2 + 2\bar{a}_{11}\bar{a}_{22}\cos\chi\,(dx^1\, dx^2) + \bar{a}_{22}(dx^2)^2, \end{aligned} \tag{3.3}$$

$$dS^1 = \sqrt{\bar{a}_{11}}\,dx^1, \qquad dS^2 = \sqrt{\bar{a}_{22}}\,dx^2,$$

where $\bar{a}_{\alpha\beta}$ is the metric tensor of surface G, χ is angle between $\bar{\mathbf{a}}_1$ and $\bar{\mathbf{a}}_2$, and dS^1 and dS^2 are arc lengths of coordinate lines on the deformed surface G. Since x^1 and x^2 are co-moving coordinates, curve x^i = constant on G_0 changes to x^i = constant on G, but their metrics are different on G_0 and G.

The unit basic vector and normal vector on G and G can be expressed as

$$\mathbf{e}_1 = \frac{1}{\sqrt{a_{11}}}\mathbf{a}_1, \quad \mathbf{e}_2 = \frac{1}{\sqrt{a_{22}}}\mathbf{a}_2, \quad \mathbf{e}_3 = \frac{\mathbf{a}_1 \times \mathbf{a}_2}{\sqrt{a}} \equiv \mathbf{n},$$

$$\bar{\mathbf{e}}_1 = \frac{1}{\sqrt{\bar{a}_{11}}}\bar{\mathbf{a}}_1, \quad \bar{\mathbf{e}}_2 = \frac{1}{\sqrt{\bar{a}_{22}}}\bar{\mathbf{a}}_2, \quad \bar{\mathbf{e}}_3 = \frac{\bar{\mathbf{a}}_1 \times \bar{\mathbf{a}}}{\sqrt{\bar{a}}} \equiv \bar{\mathbf{n}}. \tag{3.4}$$

$$\sqrt{a} = \sqrt{a_{11}a_{22}}, \qquad \sqrt{\bar{a}} = \sqrt{\bar{a}_{11}\bar{a}_{22} - (\bar{a}_{12})^2}.$$

Derivatives of $\bar{\mathbf{a}}_\alpha$, $\bar{\mathbf{a}}_\beta$, $\bar{\mathbf{n}}$, $\mathbf{n}$, with respect to x^α, may be found directly by using Gauss–Weingarten's formulae; however, for convenience of estimating errors comparable to classical theory, we derive the formulae in another form here.

Taking curve x^1 (x^2 = constant) as a space curve, we have three basic unit vectors: tangent $\bar{\mathbf{e}}_1$, normal $\bar{\mathbf{v}}_1$ and binormal $\bar{\mathbf{b}}_1$ at a point on the curve, which constitutes a right hand local frame:

$$\bar{\mathbf{b}}_1 = \bar{\mathbf{e}}_1 \times \bar{\mathbf{v}}_1. \tag{3.5}$$

As a result of deformation, $\bar{\mathbf{v}}_1$ does not generally coincide with $\bar{\mathbf{n}}$.

Frenet's formulae gives

$$\frac{\partial \bar{\mathbf{e}}_1}{\partial S^1} = \kappa_1 \bar{\mathbf{v}}_1, \quad \frac{\partial \bar{\mathbf{v}}_1}{\partial S^1} = -\kappa_1 \bar{\mathbf{e}}_1 + T_1 \bar{\mathbf{b}}_1, \quad \frac{\partial \bar{\mathbf{b}}_1}{\partial S^1} = -T_1 \bar{\mathbf{v}}_1, \tag{3.6}$$

where κ_1 is the curvature of deformed curve x^1 on G and T_1 is the torsion.

Let $\mathbf{c}_1$ be a unit vector lying in the tangent plane at P' and perpendicular to $\bar{\mathbf{e}}_1$, so that $\bar{\mathbf{e}}_1$, $\mathbf{c}_1$, $\bar{\mathbf{n}}$ form an orthogonal triad frame, $\bar{\mathbf{e}}_1 \times \mathbf{c}_1 = \bar{\mathbf{n}}$ (Fig. 1). Also let $\mathbf{c}_2$ be the unit vector, so that $\mathbf{c}_2$, $\bar{\mathbf{e}}_2$, $\bar{\mathbf{n}}$ form an orthogonal triad frame, $\mathbf{c}_2 \times \bar{\mathbf{e}}_2 = \bar{\mathbf{n}}$. Because $\mathbf{c}_1$ is perpendicular to $\bar{\mathbf{e}}_1$, and copolanar with $\mathbf{v}_1$, $\bar{\mathbf{b}}_1$, we have

$$\mathbf{c}_1 = \bar{\mathbf{v}}_1 \cos\beta + \bar{\mathbf{b}}_1 \sin\beta. \tag{3.7}$$

The angle $\beta \equiv (\bar{\mathbf{v}}_1, \mathbf{c}_1)$ is measured positive from $\bar{\mathbf{v}}_1$ toward $\mathbf{c}_1$ in counter-clockwise direct when one faces against $\bar{\mathbf{e}}_1$; therefore

$$\frac{\partial \mathbf{c}_1}{\partial S^1} = \frac{\partial}{\partial S^1}(\cos\beta\,\bar{\mathbf{v}}_1 + \sin\beta\bar{\mathbf{b}}_1) = -\kappa_1 \cos\beta\bar{\mathbf{e}}_1 + \left(T_1 + \frac{\partial\beta}{\partial S^1}\right)\bar{\mathbf{n}}. \tag{3.8}$$

We use the Frenet formulae in the derivation above, and take notice of the direction of $\bar{\mathbf{n}}$ with respect to $\bar{\mathbf{b}}_1$ and $\bar{\mathbf{v}}_1$.

On the other hand,

$$\frac{\partial \bar{\mathbf{e}}_2}{\partial S^1} = \frac{\partial}{\partial S^1}(\cos\chi\bar{\mathbf{e}}_1 + \sin\chi\mathbf{c}_1). \tag{3.9}$$

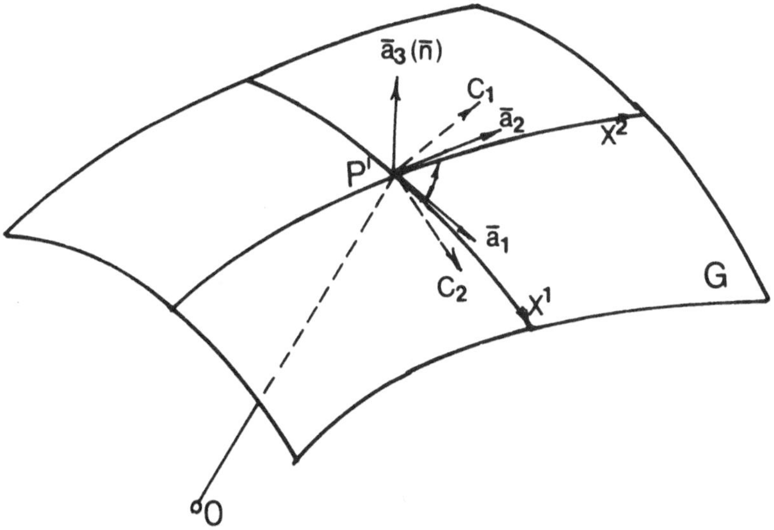

Figure 1

Applying (3.6), (3.8) to eq. (3.9), and using the relation

$$\bar{\mathbf{v}}_1 = \cos\beta \mathbf{c}_1 - \sin\beta\bar{\mathbf{n}}.$$

We find:

$$\frac{\partial\bar{\mathbf{e}}_2}{\partial S^1} = -\left(\kappa_1\cos\beta + \frac{\partial\chi}{\partial S^1}\right)\sin\chi\bar{\mathbf{e}}_1 + \left(\kappa_1\cos\beta + \frac{\partial x}{\partial S^1}\right)\cos\chi\mathbf{c}_1 + \left[\sin\chi\left(T_1 + \frac{\partial\beta}{\partial S^1}\right) - \kappa_1\sin\beta\cos\chi\right]\bar{\mathbf{n}}. \tag{3.10}$$

Similarly, we have

$$\frac{\partial\bar{\mathbf{e}}_1}{\partial S^2} = -\left(\kappa_2\cos\gamma + \frac{\partial\chi}{\partial S^2}\right)\sin\chi\bar{\mathbf{e}}_2 + \left(\kappa_2\cos\gamma + \frac{\partial\chi}{\partial S^2}\right)\cos\chi\mathbf{c}_2 + \left[\sin\chi\left(T_2 + \frac{\partial\gamma}{\partial S^2}\right) - \kappa_2\sin\gamma\cos\chi\right]\bar{\mathbf{n}}, \tag{3.11}$$

where κ_2, T_2 are the curvature and the torsion of the curve x^2, $\gamma \equiv (\bar{\mathbf{v}}_2, \mathbf{c}_2)$ and $\bar{\mathbf{v}}_2$ is the unit normal vector to curve x^2.

If a surface moves rigidly, then the angle $\chi = 90°$ will not change. Equations (3.10) and (3.11) reduce to the same formulae as $\partial\mathbf{e}_2/\partial S^1$ and $\partial\mathbf{e}_1/\partial S^1$.

From here, we will find the rate of change of the normal on a surface.

The rate of change of the normal at an arbitrary point on the deformed surface G along S^1 direction is

$$\frac{\partial\bar{\mathbf{n}}}{\partial S^1} = \frac{\partial}{\partial S^1}\left[\frac{1}{\sqrt{\bar{a}}}\bar{\mathbf{a}}_1 \times \bar{\mathbf{a}}_2\right] = \sqrt{\bar{a}}\frac{\partial}{\partial S^1}\left(\frac{1}{\sqrt{\bar{a}}}\right)\bar{\mathbf{n}} + \frac{1}{\sqrt{\bar{a}}}\frac{\partial\bar{\mathbf{a}}_1}{\partial S^1} \times \bar{\mathbf{a}}_2 + \frac{1}{\sqrt{\bar{a}}}\bar{\mathbf{a}}_1 \times \frac{\partial\bar{\mathbf{a}}_2}{\partial S^1}. \tag{3.12}$$

For

$$\frac{\partial \bar{\mathbf{a}}_1}{\partial S^1} = \left(\frac{\partial \sqrt{\bar{a}_{11}}}{\partial S^1}\right)\bar{\mathbf{e}}_1 + \sqrt{\bar{a}_{11}}\,\frac{\partial \bar{\mathbf{e}}_1}{\partial S^1},$$

$$\frac{\partial \bar{\mathbf{a}}_2}{\partial S^1} = \left(\frac{\partial \sqrt{\bar{a}_{22}}}{\partial S^1}\right)\bar{\mathbf{e}}_2 + \sqrt{\bar{a}_{22}}\,\frac{\partial \bar{\mathbf{e}}_2}{\partial S^1}. \tag{3.13}$$

Applying Frenet's formulae and (3.10) and (3.13), then substituting the result obtained into (3.12), and noting (3.9) and the following relations

$$\bar{\mathbf{v}}_1 \times \bar{\mathbf{a}}_2 = \sqrt{\bar{a}_{22}}\,(\sin\beta\mathbf{c}_2 - \cos\beta\cos\chi\mathbf{n}),$$

$$\bar{\mathbf{a}}_1 \times \mathbf{c}_1 = \sqrt{\bar{a}_{11}}\,\bar{\mathbf{n}}, \qquad \bar{\mathbf{a}}_1 \times \bar{\mathbf{n}} = -\sqrt{\bar{a}_{11}}\,\mathbf{c}_1$$

with

$$\sin\chi = \sqrt{\frac{\bar{a}}{\bar{a}_{11}\bar{a}_{22}}}.$$

We get, after some arrangements:

$$\frac{\partial \bar{\mathbf{n}}}{\partial S^1} = \sqrt{\frac{\bar{a}_{11}\bar{a}_{22}}{\bar{a}}}\left\{\left[\frac{\partial \chi}{\partial S^1}\cos\chi + \sqrt{\frac{\bar{a}}{\bar{a}_{11}\bar{a}_{22}}}\frac{\partial}{\partial S^1}\ln\sqrt{\frac{\bar{a}_{11}\bar{a}_{22}}{a}}\right]\bar{\mathbf{n}}\right.$$

$$\left. + \kappa_1\sin\beta\sin\chi\bar{\mathbf{e}}_1 - \sin\chi\left(T_1 + \frac{\partial\beta}{\partial S^1}\right)\mathbf{c}_1\right\}$$

$$= \kappa_1\sin\beta\bar{\mathbf{e}}_1 - \left(T_1 + \frac{\partial\beta}{\partial S^1}\right)\mathbf{c}_1.$$

Similarly, for $\partial\bar{\mathbf{n}}/\partial S^2$, the final results are

$$\frac{\partial \bar{\mathbf{n}}}{\partial S^1} = \kappa_1\sin\beta\bar{\mathbf{e}}_1 - \left(T_1 + \frac{\partial\beta}{\partial S^1}\right)\mathbf{c}_1,$$

$$\frac{\partial \bar{\mathbf{n}}}{\partial S^2} = \kappa_2\sin\gamma\bar{\mathbf{e}}_2 - \left(T_2 + \frac{\partial\gamma}{\partial S^2}\right)\mathbf{c}_2. \tag{3.14}$$

The geometric meaning of $\kappa_1\cos\beta$ and $\kappa_1\sin\beta$ may be explained as follows: We have at point P' on the surface G, a tangent plane and a normal plane passing $\bar{\mathbf{n}}$ and $\bar{\mathbf{a}}$. By projecting curve x^1 on the tangent plane and the normal plane, then $\kappa_1\cos\beta$ (which is often called geodesic curvature) is the curvature of the projection curve on the tangent plane at point P', while $\kappa_1\sin\beta$ is that on the normal plane. Before deformation, curve x^1 is taken as the principal line of curvature and the geodesic curvature equals zero. T_1 and $\partial\beta/\partial S^1$ denote the torsion of the deformed curve x^1 and the distortion of the surface caused by the rotation of curve x^2 about x^1 respectively.

Equations (3.10), (3.11) and (3.14) are equivalent to the well-known Gauss–Weingarten's formulae. The advantages of these equations are that they are expressed in terms of curvature, torsion and rate of change of angle of deformed coordinate lines. This will improve our comprehension of the study of shell deformations. The general orthogonal coordinate system used by Knowles and Reissner [12] is a special form of the present formulae.

If the surface coordinate lines are principal lines of curvature on an undeformed surface G, then

$$\frac{\partial \mathbf{n}}{\partial S_0^1} = \mathring{\kappa}_1 \mathbf{e}_1, \qquad \frac{\partial \mathbf{n}}{\partial S_0^2} = \mathring{\kappa}_2 \mathbf{e}_2, \tag{3.15}$$

where $\mathring{\kappa}_\theta$ is the principal curvature of the coordinate lines S_0^θ at a point.

A deformed coordinate line may extend. Let us take

$$\frac{\mathrm{d}S^1}{\mathrm{d}S_0^1} = (1 + \mathring{\varepsilon}_1), \qquad \frac{\mathrm{d}S^2}{\mathrm{d}S_0^2} = (1 + \mathring{\varepsilon}_2), \tag{3.16}$$

then we have

$$\begin{aligned} \frac{\partial \bar{\mathbf{n}}}{\partial S_0^1} &= \frac{\partial \bar{\mathbf{n}}}{\partial S^1}\frac{\mathrm{d}S^1}{\mathrm{d}S_0^1} = \frac{\mathrm{d}\bar{\mathbf{n}}}{\partial S^1}(1 + \mathring{\varepsilon}_1), \\ \frac{\partial \bar{\mathbf{n}}}{\partial S_0^2} &= \frac{\partial \bar{\mathbf{n}}}{\partial S^2}\frac{\mathrm{d}S^2}{\mathrm{d}S_0^2} = \frac{\partial \bar{\mathbf{n}}}{\partial S^2}(1 + \mathring{\varepsilon}_2). \end{aligned} \tag{3.17}$$

$\mathring{\varepsilon}_2$ is the ratio of extension of the coordinate line S_0^α.

4. Displacement of a shell continuum

Here we consider a shell as a space continuum which extends through its thickness. The midsurface $\mathbf{r}(x^1, x^2)$ of a shell is regarded as a datum surface. An arbitrary point in a shell will be denoted by coordinates $(x^1, x^2, t) \equiv (x^1, x^2, x^3)$, where t is the normal distance from the midsurface. The configuration of a shell changes after deformation, but the co-moving coordinates remain invariable. However, their metric will change generally.

The position vector of a point in an undeformed shell relative to a fixed point O is

$$\mathbf{R} = \mathbf{r}(x^1, x^2) + t\mathbf{n}, \tag{4.1}$$

while the position vector after deformation changes to

$$\bar{\mathbf{R}} = \bar{\mathbf{r}} + t\bar{\mathbf{n}} + \mathbf{w} = \mathbf{r} + \mathbf{v} + t\bar{\mathbf{n}} + \mathbf{w}. \tag{4.2}$$

Hence $\bar{\mathbf{r}}$ is the position vector of a point on the deformed midsurface, $\bar{\mathbf{n}}$ is the unit normal vector at the same point, $\mathbf{v}$ is the displacement vector of the midsurface, and $\mathbf{w}$ is the displacement vector which takes account of deviation from the normal, including the effect of thickness change (Fig. 2).

The displacement vector of a point in a shell is given by

$$\mathbf{u} = \bar{\mathbf{R}} - \mathbf{R} = \mathbf{v} + t(\bar{\mathbf{n}} - \mathbf{n}) + \mathbf{w}. \tag{4.3}$$

This expression has been used by Zerna [10], but it is convenient to estimate the degree of accuracy of classical theory.

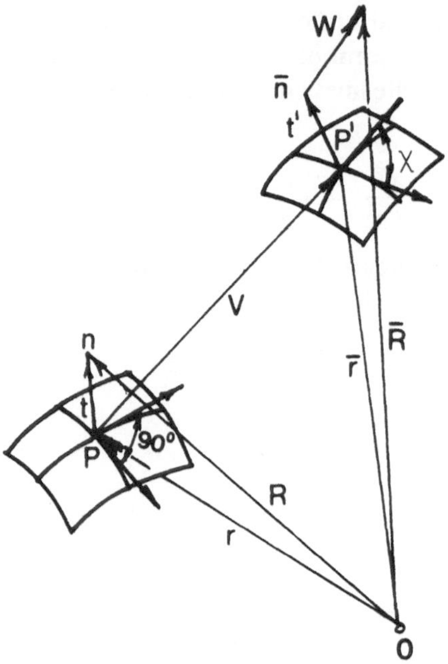

Figure 2

By (4.1), we have

$$\mathbf{dR} = \mathbf{G}_i \, dx^i = \frac{\partial \mathbf{R}}{\partial x^i} dx^i, \quad t \equiv x^3, \quad \partial \mathbf{R}/\partial x^3 = \mathbf{n}. \tag{4.4}$$

Without loss of generality, the principal lines of curvature of the midsurface will be chosen as reference coordinate lines. By using (3.15), it follows that

$$\mathbf{dR} = (1 + t\mathring{\kappa}) \, dx^1 \mathbf{a}_1 + (1 + t\mathring{\kappa}_2) \, dx^2 \mathbf{a}_2 + \mathbf{n} \, dt. \tag{4.5}$$

The arc lengths dS_0^i, along local coordinate lines before deformation in a shell, are

$$\begin{aligned} dS_0^1 &= \sqrt{G_{11}} \, dx^1, \quad dS_0^2 = \sqrt{G_{22}} \, dx^2, \quad dS_0^3 = \sqrt{G_{33}} \, dx^3, \\ \sqrt{G_{11}} &= \sqrt{a_{11}} \, (1 + t\mathring{\kappa}_1), \quad \sqrt{G_{22}} = \sqrt{a_{22}} \, (1 + t\mathring{\kappa}_2), \quad \sqrt{G_{33}} = 1, \\ G_{ij} &= 0, \quad (i \neq j). \end{aligned} \tag{4.6}$$

The displacement vector is:

$$\mathbf{U} = U^i \mathbf{G}_i. \tag{4.7}$$

If the co-moving coordinate x^2 is an angle, then the corresponding dimension of U^2 is a non-dimension. We will use $\hat{U}^i$ to denote the physical component of U^i, and $\hat{U}^i$ has the dimension of length.

$$\hat{U}^i = U^i \sqrt{G_{(ii)}}, \quad \text{no sum over } (ii). \tag{4.8}$$

5. Strain tensor of finite deformation in curvilinear coordinates

A space transformation of a shell continuum involves: translation, rotation, and deformation. It results in the change of

configuration $T_0 \rightarrow$ configuration T,

local vectors $\mathbf{G}_i \rightarrow$ local vectors $\bar{\mathbf{G}}_i$.

If we take the classical definition of Green's strain tensor as the measure of strain, we find the following disadvantages: (1) the physical meaning of second-order terms involved are not clear; (2) rotation components are not congruently defined.

The S–R decomposition theorem, which is stated in [3] and [13], defines strain, and rotation consistently remedies the defects of Green's strain tensor.

In [3], the physical components of strain tensor and mean rotation angle ϑ are derived. Using the absolute differential formula (2.10), we have

$$\begin{aligned} S_j^i &= \frac{1}{2}\left(\frac{\mathrm{D}\hat{U}^i}{\mathrm{D}S^j} + \frac{\mathrm{D}\hat{U}^j}{\mathrm{D}S^i}\right) - \hat{L}_l^i \hat{L}_j^l (1 - \cos\vartheta) \\ \vartheta &= \pm \arcsin\left\{\frac{1}{2}\left[\left(\frac{\mathrm{D}\hat{U}^2}{\mathrm{D}S^1} - \frac{\mathrm{D}\hat{U}^1}{\mathrm{D}S^3}\right)^2 + \left(\frac{\mathrm{D}\hat{U}^1}{\mathrm{D}S^2} - \frac{\mathrm{D}\hat{U}^3}{\mathrm{D}S^1}\right)^2 \right.\right. \\ &\quad \left.\left. + \left(\frac{\mathrm{D}\hat{U}^3}{\mathrm{D}S^2} - \frac{\mathrm{D}\hat{U}^2}{\mathrm{D}S^3}\right)^2\right]^{1/2}\right\} \qquad (5.1) \\ \hat{L}_2^3 &= \frac{1}{2\sin\vartheta}\left(\frac{\mathrm{D}\hat{U}^3}{\mathrm{D}S^2} - \frac{\mathrm{D}\hat{U}^2}{\mathrm{D}S^3}\right) \equiv l_1, \quad \hat{L}_3^1 = \frac{1}{2\sin\vartheta}\left(\frac{\mathrm{D}\hat{U}^1}{\mathrm{D}S^3} - \frac{\mathrm{D}\hat{U}^3}{\mathrm{D}S^1}\right) \equiv l_2, \\ \hat{L}_1^2 &= \frac{1}{2\sin\vartheta}\left(\frac{\mathrm{D}\hat{U}^2}{\mathrm{D}S^1} - \frac{\mathrm{D}U^1}{\mathrm{D}S^2}\right) \equiv l_3. \end{aligned}$$

The strain tensor S_j^i is defined in the co-moving coordinate system, and S^i is the arc length of co-moving coordinate lines. The dimension of $\mathrm{D}\hat{U}^i/\mathrm{D}S^j$ is the ratio of length, or the physical dimension of strain.

A method has been proposed by Truesdell [14] of transforming a mechanical equation in Cartesian coordinates to that of curvilinear coordinates. However, it is easier to obtain correct results if the following rules are used: (a) keep unchanged the generalized dimensions (defined by Eddington as the orders of covariant and contravariant); (b) keep unchanged the physical dimension; (c) replace ordinary differential by an absolute differential. In (5.1),

$$\frac{\mathrm{D}\hat{U}^j}{\mathrm{D}S^i} = \frac{\mathrm{D}(\sqrt{G_{(jj)}}\,U^j)}{\mathrm{D}S^i} = \frac{\mathrm{D}\sqrt{G_{(jj)}}}{\mathrm{D}S^i}U^j + \sqrt{G_{(jj)}}\frac{\mathrm{D}U^j}{\mathrm{D}S^i}.$$

Because the absolute differential of a metric tensor is zero, it follows that:

$$\frac{\mathrm{D}\hat{U}^j}{\mathrm{D}S^i} = \frac{\sqrt{G_{(jj)}}}{\sqrt{\bar{G}_{(ii)}}}U^j\Big|_i = \frac{\sqrt{G_{(jj)}}}{\sqrt{\bar{G}_{(ii)}}}\left(\frac{\partial U^j}{\partial x^i} + \Gamma_{il}^j U^l\right), \tag{5.2}$$

$$\Gamma_{il}^j = \tfrac{1}{2}G^{jm}\left(\frac{\partial G_{mi}}{\partial x^l} + \frac{\partial G_{ml}}{\partial x^i} - \frac{\partial G_{il}}{\partial x^m}\right). \tag{5.3}$$

for infinitesmal strain and rotation $\cos \vartheta \simeq 1$, the strain components reduce to

$$S_j^i \simeq \frac{1}{2}\left(\frac{D\hat{U}^i}{DS^j} + \frac{D\hat{U}^j}{DS^i}\right). \tag{5.4}$$

By substituting (5.2), (5.3) into (5.4), we obtain the expression of strain components for infinitesmal deformation in curvilinear coordinates.

If $i = j$,

$$S_{(i)}^{(i)} = \frac{\partial}{\partial x^i}\left(\frac{\hat{U}^{(i)}}{\sqrt{G_{(ii)}}}\right) + \frac{1}{2}\frac{1}{G_{(ii)}}\sum_k \frac{\partial G_{(ii)}}{\partial x^k}\frac{\hat{U}^{(k)}}{\sqrt{G_{(kk)}}} \tag{5.5}$$

(k) denotes the summation through Σ_k.

If $i \neq j$,

$$S_j^i = \frac{1}{2}\frac{1}{\sqrt{G_{(ii)}}\sqrt{G_{(jj)}}}\left[G_{(ii)}\frac{\partial}{\partial x^j}\left(\frac{\hat{U}^{(i)}}{\sqrt{G_{(ii)}}}\right) + G_{(jj)}\frac{\partial}{\partial x^i}\left(\frac{\hat{U}^{(j)}}{\sqrt{G_{(jj)}}}\right)\right]. \tag{5.6}$$

Equations (5.5), (5.6) are the same as due to Truesdell [14] and Sokolnikoff [15]. We should note that in the case of infinitesmal deformation, the error induced by choosing the initial or the deformed body as the reference state is of second order small quantity. However, it is necessary to distinguish between the different states of reference when large deformation occurs. It was Chien who first pointed out this situation in [1] and [2].

For general reference, finites strain components in cylindrical and spherical co-moving coordinates are given here.

In a cylindrical coordinate system. Let $x^1 \equiv z, x^2 \equiv \phi, x^3 \equiv r$, be the co-moving coordinates in a shell continuum, made isomorphic initially, with a space fixed coordinate system.

Before deformation,

$$(dS_0)^2 = G_{iz}\,dx^i\,dx^j, \quad G_{11} = G_{33} = 1, \quad G_{22} = r^2;$$

$$G_{ij} = 0, \quad (i \neq j).$$

After deformation,

$$(dS)^2 = \bar{G}_{ij}\,dx^i\,dx^j.$$

The physical components of the displacement vector are

$$\hat{U}^1 = U^1, \quad \hat{U}^2 = rU^2, \quad \hat{U}^3 = U^3.$$

In a cylindrical coordinate system, we have

$$\Gamma_{22}^3 = -r, \quad \Gamma_{23}^2 = \frac{1}{r}, \quad \text{others } \Gamma_{jk}^i = 0$$

Let

$$\hat{U}^1 \equiv u^z, \quad \hat{U}^2 \equiv u^\varphi, \quad \hat{U}^3 \equiv u^r$$

then

$$\| \mathrm{D}\hat{U}^i/\mathrm{D}S^j \| = \begin{bmatrix} \dfrac{1}{\sqrt{\bar{G}_{11}}}\dfrac{\partial u^z}{\partial z} & \dfrac{1}{\sqrt{\bar{G}_{22}}}\dfrac{\partial u^z}{\partial \varphi} & \dfrac{1}{\sqrt{\bar{G}_{33}}}\dfrac{\partial u^z}{\partial r} \\ \dfrac{1}{\sqrt{\bar{G}_{11}}}\dfrac{\partial u^\varphi}{\partial z} & \dfrac{1}{\sqrt{\bar{G}_{22}}}\left(\dfrac{\partial u^\varphi}{\partial \varphi} + u^r\right) & \dfrac{1}{\sqrt{\bar{G}_{33}}}\dfrac{\partial u^\varphi}{\partial r} \\ \dfrac{1}{\sqrt{\bar{G}_{11}}}\dfrac{\partial u^r}{\partial z} & \dfrac{1}{\sqrt{\bar{G}_{22}}}\left(\dfrac{\partial u^r}{\partial \varphi} - u^\varphi\right) & \dfrac{1}{\sqrt{\bar{G}_{33}}}\dfrac{\partial u^r}{\partial r} \end{bmatrix}.$$

Applying the foregoing relations to (5.1) we find that

$$\begin{aligned}
S_z^z &= \frac{1}{\sqrt{\bar{G}_{11}}}\frac{\partial u^z}{\partial z} + [1 - (l_1)^2](1 - \cos\vartheta), \\
S_\varphi^\varphi &= \frac{1}{\sqrt{\bar{G}_{22}}}\left(\frac{\partial u^\varphi}{\partial \varphi} + u^r\right) + [1 - (l_2)^2](1 - \cos\vartheta), \\
S_r^r &= \frac{1}{\sqrt{\bar{G}_{33}}}\frac{\partial u^r}{\partial r} + [1 - (l_3)^2](1 - \cos\vartheta), \\
S_r^z &= \frac{1}{2}\left(\frac{1}{\sqrt{\bar{G}_{22}}}\frac{\partial u^z}{\partial \varphi} + \frac{1}{\sqrt{\bar{G}_{11}}}\frac{\partial u^r}{\partial z}\right) - (l_1)(l_3)(1 - \cos\vartheta), \\
S_\varphi^r &= \frac{1}{2}\left[\frac{1}{\sqrt{\bar{G}_{22}}}\left(\frac{\partial u^r}{\partial \varphi} - u^\varphi\right) + \frac{1}{\sqrt{\bar{G}_{33}}}\frac{\partial u^\varphi}{\partial r}\right] - (l_2)(l_3)(1 - \cos\vartheta), \\
S_z^\varphi &= \frac{1}{2}\left[\frac{1}{\sqrt{\bar{G}_{11}}}\frac{\partial u^\varphi}{\partial z} + \frac{1}{\sqrt{\bar{G}_{22}}}\frac{\partial u^z}{\partial \varphi}\right] - (l_1)(l_2)(1 - \cos\vartheta),
\end{aligned} \tag{5.7a}$$

$$\sin\vartheta = \pm\frac{1}{2} \times \sqrt{\left(\frac{1}{\sqrt{\bar{G}_{11}}}\frac{\partial u^r}{\partial z} - \frac{1}{\sqrt{\bar{G}_{33}}}\frac{\partial u^z}{\partial r}\right)^2 + \left[\frac{1}{\sqrt{\bar{G}_{33}}}\frac{\partial u^\varphi}{\partial r} - \frac{1}{\sqrt{\bar{G}_{22}}}\left(\frac{\partial u^r}{\partial \varphi} - u^\varphi\right)\right]^2 + \left(\frac{1}{\sqrt{\bar{G}_{22}}}\frac{\partial u^z}{\partial \varphi} - \frac{1}{\sqrt{\bar{G}_{11}}}\frac{\partial u^\varphi}{\partial z}\right)^2}, \tag{5.7b}$$

$$\left.\begin{aligned}
l_1 &= \frac{1}{2\sin\vartheta}\left[\frac{1}{\sqrt{\bar{G}_{33}}}\frac{\partial u^\varphi}{\partial r} - \frac{1}{\sqrt{\bar{G}_{22}}}\left(\frac{\partial u^r}{\partial \varphi} - u^\varphi\right)\right], \\
l_2 &= \frac{1}{2\sin\vartheta}\left[\frac{1}{\sqrt{\bar{G}_{11}}}\frac{\partial u^r}{\partial z} - \frac{1}{\sqrt{\bar{G}_{33}}}\frac{\partial u^z}{\partial r}\right], \\
l_3 &= \frac{1}{2\sin\vartheta}\left[\frac{1}{\sqrt{\bar{G}_{22}}}\frac{\partial u^z}{\partial \varphi} - \frac{1}{\sqrt{\bar{G}_{11}}}\frac{\partial u^\varphi}{\partial z}\right],
\end{aligned}\right\} \tag{5.7c}$$

In a spherical coordinate system. Let $x^1 \equiv \varphi$, $x^2 \equiv \beta$, $x^3 \equiv r$, be the co-moving coordinates in a shell continuum, made isomorphic initially with a space fixed coordinate system.

Before deformation,

$$(\mathrm{d}S_0)^2 = G_{ij}\,\mathrm{d}x^i\,\mathrm{d}x^j,$$

$$G_{11} = r^2, \quad G_{22} = r^2\sin^2\varphi, \quad G_{33} = 1; \quad G_{ij} = 0, \quad (i \neq j).$$

After deformation,

$$(\mathrm{d}S)^2 = \bar{G}_{ij}\,\mathrm{d}x^i\,\mathrm{d}x^j.$$

The physical components of the displacement vector are

$$\hat{U}^1 = rU^1, \quad \hat{U}^2 = r\sin\varphi\, U^2, \quad \hat{U}^3 = U^3.$$

Let

$$\hat{U}^1 \equiv u^\varphi, \quad \hat{U}^2 \equiv u^\beta, \quad \hat{U}^3 \equiv u^r.$$

then we have

$$\Gamma^3_{22} = -r\sin^2\phi, \quad \Gamma^1_{22} = -\cos\phi\sin\phi, \quad \Gamma^3_1 = -r,$$

$$\Gamma^2_{32} = \Gamma^1_{31} = \frac{1}{r}, \quad \Gamma^2_{21} = \mathrm{ctg}\,\phi, \quad \text{others } \Gamma^i_{jk} = 0,$$

$$\|\mathrm{D}\hat{U}^i/\mathrm{D}S^j\| =$$

$$\begin{bmatrix} \frac{1}{\sqrt{\bar{G}_{11}}}\left(\frac{\partial u^\varphi}{\partial\varphi} + u^r\right) & \frac{1}{\sqrt{\bar{G}_{22}}}\left(\frac{\partial u^\varphi}{\partial\beta} - \cos\phi u^\beta\right) & \frac{1}{\sqrt{\bar{G}_{33}}}\left(\frac{\partial u^\varphi}{\partial r} - \frac{1}{r}\frac{\partial u^\varphi}{\partial r} + \frac{u^\varphi}{r}\right) \\ \frac{1}{\sqrt{\bar{G}_{11}}}\frac{\partial u^\beta}{\partial\varphi} & \frac{1}{\sqrt{\bar{G}_{22}}}\left(\frac{\partial u^\beta}{\partial\beta} + \cos\phi u^\varphi + \sin\varphi u^r\right) & \frac{1}{\sqrt{\bar{G}_{33}}}\frac{\partial u^\beta}{\partial r} \\ \frac{1}{\sqrt{\bar{G}_{11}}}\left(\frac{\partial u^r}{\partial\varphi} - u^\varphi\right) & \frac{1}{\sqrt{\bar{G}_{22}}}\left(\frac{\partial u^r}{\partial\beta} - \sin\varphi u^\beta\right) & \frac{1}{\sqrt{\bar{G}_{33}}}\frac{\partial u^r}{\partial r} \end{bmatrix}$$

$$S^\varphi_\varphi = \frac{1}{\sqrt{\bar{G}_{11}}}\left(\frac{\partial u^\varphi}{\partial\varphi} + u^r\right) + [1 - (l_1)^2](1 - \cos\vartheta),$$

$$S^\beta_\beta = \frac{1}{\sqrt{\bar{G}_{22}}}\left(\frac{\partial u^\beta}{\partial\beta} + \cos\phi u^\varphi + \sin\varphi u^r\right) + [1 - (l_2)^2](1 - \cos\vartheta),$$

$$S^r_r = \frac{1}{\sqrt{\bar{G}_{33}}}\frac{\partial u^r}{\partial r} + [1 - (l_3)^2](1 - \cos\vartheta),$$

$$S^r_\beta = \frac{1}{2}\left[\frac{1}{\sqrt{\bar{G}_{22}}}\left(\frac{\partial u^r}{\partial\beta} - \sin\varphi u^\beta\right) + \frac{1}{\sqrt{\bar{G}_{33}}}\frac{\partial u^\beta}{\partial r}\right] - (l_3)(l_2)(1 - \cos\vartheta),$$

$$S^\beta_\varphi = \frac{1}{2}\left[\frac{1}{\sqrt{\bar{G}_{11}}}\frac{\partial u^\beta}{\partial\varphi} + \frac{1}{\sqrt{\bar{G}_{22}}}\left(\frac{\partial u^\varphi}{\partial\beta} - \cos\varphi u^\beta\right)\right] - (l_2)(l_1)(1 - \cos\vartheta),$$

$$S^\varphi_r = \frac{1}{2}\left[\frac{1}{\sqrt{\bar{G}_{33}}}\left(\frac{\partial u^\varphi}{\partial r} - \frac{1}{r}\frac{\partial u^\varphi}{\partial r} + \frac{u^\varphi}{r}\right) + \frac{1}{\sqrt{\bar{G}_{11}}}\left(\frac{\partial u^r}{\partial\varphi} - u^\varphi\right)\right]$$
$$- (l_3)(l_1)(1 - \cos\vartheta),$$

(5.8a)

$$\sin\vartheta = \pm\frac{1}{2}\left\{\left[\frac{1}{\sqrt{\bar{G}_{22}}}\left(\frac{\partial u^r}{\partial\beta} - \sin\varphi u^\beta\right) - \frac{1}{\sqrt{\bar{G}_{33}}}\frac{\partial u^\beta}{\partial r}\right]^2 + \left[\frac{1}{\sqrt{\bar{G}_{11}}}\frac{\partial u^\beta}{\partial\varphi} - \frac{1}{\sqrt{\bar{G}_{22}}}\left(\frac{\partial u^\varphi}{\partial\beta} - \cos\varphi u^\beta\right)\right]^2 + \left[\frac{1}{\sqrt{\bar{G}_{33}}}\left(\frac{\partial u^\varphi}{\partial r} - \frac{1}{r}\frac{\partial u^\varphi}{\partial r} + \frac{u^\varphi}{\dot{r}}\right) - \frac{1}{\sqrt{\bar{G}_{11}}}\left(\frac{\partial u^r}{\partial\varphi} - u^\varphi\right)\right]^2\right\}^{1/2}, \tag{5.8b}$$

$$\left.\begin{aligned} l_1 &= \frac{1}{2\sin\vartheta}\left[\frac{1}{\sqrt{\bar{G}_{22}}}\left(\frac{\partial u^r}{\partial\beta} - \sin\varphi u^\beta\right) - \frac{1}{\sqrt{\bar{G}_{33}}}\frac{\partial u^\beta}{\partial r}\right],\\ l_2 &= \frac{1}{2\sin\vartheta}\left[\frac{1}{\sqrt{\bar{G}_{33}}}\left(\frac{\partial u^\varphi}{\partial r} - \frac{1}{r}\frac{\partial u^\varphi}{\partial r} + \frac{u^\varphi}{r}\right) - \frac{1}{\sqrt{\bar{G}_{11}}}\left(\frac{\partial u^r}{\partial\varphi} - u^\varphi\right)\right],\\ l_3 &= \frac{1}{2\sin\vartheta}\left[\frac{1}{\sqrt{\bar{G}_{11}}}\frac{\partial u^\beta}{\partial\varphi} - \frac{1}{\sqrt{\bar{G}_{22}}}\left(\frac{\partial u^\varphi}{\partial\beta} - \cos\varphi u^\beta\right)\right]. \end{aligned}\right\} \tag{5.8c}$$

If a shell rotates rigidly, then all its strain components must be zero; equation (5.8) meet this sufficient and necessary condition. Now we will prove the statement.

Suppose a shell rotates about a fixed axis ξ^3 as shown in Fig. 3, the displacement vector of a point P is perpendicular to axis ξ^3:

$$U = |\mathbf{U}| = 2R\sin\vartheta/2.$$

R is the distance from point p to the ξ_3, ϑ is the rotation angle. Let us examine an example of rotation in a spherical co-moving coordinate system. (φ, β, r) are measured as shown in Fig. 3. The physical components of the displacement vector $\mathbf{U}$ in a local frame $\mathbf{G}_i$ are

$$u^\varphi = -U\sin(\vartheta_0/2)\cos\varphi, \quad u^\beta = U\cos(\vartheta_0/2),$$

$$u^r = -U\sin(\vartheta_0/2)\sin\varphi$$

or

$$u^\varphi = -r\sin\varphi\cos\varphi\,(1 - \cos\vartheta_0), \quad u^\beta = r\sin\varphi\sin\vartheta_0,$$

$$u^r = -r\sin^2\varphi\,(1 - \cos\vartheta_0).$$

Then we find from (5.8a), (5.8b), (5.8c) that:

$$S_j^i = 0, \quad (i, j = 1, 2, 3),$$

$$\vartheta = \vartheta_0$$

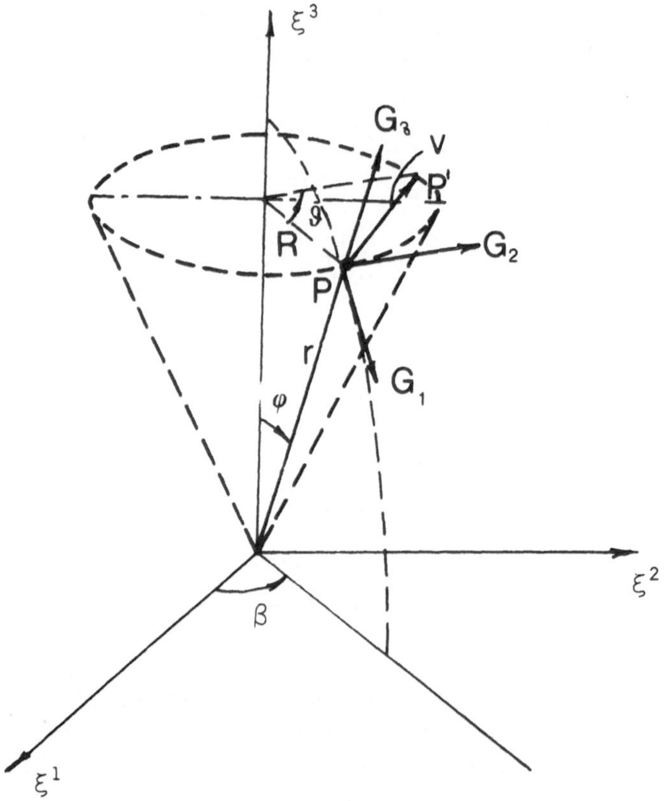

Figure 3

$$l_1 \;=\; -\sin\varphi, \quad l_2 \;=\; 0, \quad l_3 \;=\; \cos\varphi.$$

This proves that the shell is in a state of rigid rotation without strain. The rotation angle is ϑ, and the direction cosines of the axis show that it is parallel to axis ξ^3, where l_i relate to a local frame.

6. The finite strain of a shell

The displacement vector function of a shell is expressed by (4.3):

$$\mathbf{U} \;=\; U^i\mathbf{G}_i \;=\; [V^i + t(\bar{n}^i - n^i) + W^i]\,\mathbf{G}_i. \tag{6.1}$$

By (4.6) and (4.8) the physical components of the displacement vector are

$$\left.\begin{aligned} \hat{u}^1 &= \sqrt{G_{11}}\,U^1 = \sqrt{a_{11}}\,(1 + t\mathring{\kappa}_1)\,[V^1 + t(\bar{n}^1 + n^1) + W^1] \\ \hat{u}^2 &= \sqrt{G_{22}}\,U^2 = \sqrt{a_{22}}\,(1 + t\mathring{\kappa}_2)\,[V^2 + t(\bar{n}^2 - n^2) + W^2] \\ \hat{u}^3 &= \sqrt{G_{33}}\,U^3 = [V^3 + t(\bar{n}^3 - n^3) + W^3]. \end{aligned}\right\} \tag{6.2}$$

Substituting (6.2) into (5.1) to get the finite strain components of a shell, we find:

$$S_1^1 = \sqrt{G_{11}}\left[\frac{DV^1}{DS^1} + \left(\frac{D\bar{n}^1}{DS^1} - \frac{Dn^1}{DS^1}\right)t + \frac{DW^1}{DS^1}\right]$$

$$+ [1 - (l_1)^2](1 - \cos \vartheta),$$

$$S_2^1 = \tfrac{1}{2}\sqrt{G_{11}}\left[\frac{DV^1}{DS^2} + \left(\frac{D\bar{n}^1}{DS^2} - \frac{Dn^1}{DS^2}\right)t + \frac{DW^1}{DS^2}\right] \tag{6.3}$$

$$+ \tfrac{1}{2}\sqrt{G_{22}}\left[\frac{DV^2}{DS^1} + \left(\frac{D\bar{n}^2}{DS^1} - \frac{Dn^2}{DS^1}\right)t + \frac{DW^2}{DS^1}\right]$$

$$- (l_1)(l_2)(1 - \cos \vartheta).$$

To clarify the geometrical meaning of the above equations, it is necessary to use characteristic parameters which relate to $D\bar{n}^i/DS^j$ and Dn^i/DS^j in the deformed and undeformed surfaces.

By (3.15), we have

$$\frac{\partial \mathbf{n}}{\partial S^1} = \frac{1}{\sqrt{G_{11}}}\frac{\partial \mathbf{n}}{\partial x^1} = \frac{\sqrt{a_{11}}}{\sqrt{G_{11}}}\frac{\partial \mathbf{n}}{\partial S_0^1} = \frac{\sqrt{a_{11}}}{\sqrt{G_{11}}}\mathring{\kappa}_1 \mathbf{e}_1 \tag{6.4}$$

Similarly,

$$\frac{\partial \mathbf{n}}{\partial S^2} = \frac{\sqrt{a_{22}}}{\sqrt{G_{22}}}\mathring{\kappa}_2 \mathbf{e}_2, \tag{6.5}$$

$$\frac{\partial \mathbf{n}}{\partial S^3} = 0. \tag{6.6}$$

Equation (3.14) gives the following relations:

$$\frac{\partial \mathbf{n}}{\partial S^1} = \frac{1}{\sqrt{\bar{G}_{11}}}\frac{\partial \mathbf{n}}{\partial x^1} = \frac{\sqrt{\bar{a}_{11}}}{\sqrt{\bar{G}_{11}}}\frac{\partial \mathbf{n}}{\partial S^1}$$

$$= \frac{\sqrt{\bar{a}_{11}}}{\sqrt{\bar{G}_{11}}}\left[\kappa_1 \sin \beta \bar{\mathbf{e}}_1 - \left(T_1 + \frac{\partial \beta}{\partial S^1}\right)\mathbf{c}_1\right], \tag{6.7}$$

$$\frac{\partial \mathbf{n}}{\partial S^2} = \frac{\sqrt{\bar{a}_{22}}}{\sqrt{\bar{G}_{22}}}\left[\kappa_2 \sin \gamma \bar{\mathbf{e}}_2 - \left(T_2 + \frac{\partial \gamma}{\partial S^2}\right)\mathbf{c}_2\right] \tag{6.8}$$

$$\frac{\partial \mathbf{n}}{\partial S^3} = \frac{\partial \mathbf{n}}{\partial t} = 0. \tag{6.9}$$

The four unit vectors $\bar{\mathbf{e}}_1$, $\bar{\mathbf{e}}_2$, $\mathbf{c}_1$, $\mathbf{c}_2$ can be projected to the local frame at the same point on the undeformed midsurface. Let

$$
\left.\begin{aligned}
\bar{\mathbf{e}}_1 &= p_1^i \mathbf{e}_i = \sum_i p_1^i \frac{\mathbf{G}_i}{\sqrt{G_{(ii)}}},\\
\bar{\mathbf{e}}_2 &= p_2^i \mathbf{e}_i = \sum_i p_2^i \frac{\mathbf{G}_i}{\sqrt{G_{(ii)}}},\\
\mathbf{c}_1 &= p_3^i \mathbf{e}_i = \sum_i p_3^i \frac{\mathbf{G}_i}{\sqrt{G_{(ii)}}},\\
\mathbf{c}_2 &= p_4^i \mathbf{e}_i = \sum_i p_4^i \frac{\mathbf{G}_i}{\sqrt{G_{(ii)}}}.
\end{aligned}\right. \tag{6.10}
$$

Notice

$$
\frac{\partial \mathbf{n}}{\partial S^i} = \frac{Dn^j}{DS^i}\mathbf{G}_j, \qquad \frac{\partial \bar{\mathbf{n}}}{\partial S^i} = \frac{D\bar{n}^j}{DS^i}\mathbf{G}_j. \tag{6.11}
$$

By substituting (6.10) into (6.7)–(6.9), and then comparing the coefficients of $\mathbf{G}_i$, we get

$$
\left.\begin{aligned}
\frac{Dn^1}{DS^1} &= \frac{1}{\sqrt{\bar{G}_{11}}}\mathring{\kappa}_1, \qquad \frac{Dn^2}{DS^2} = \frac{1}{\sqrt{\bar{G}_{22}}}\mathring{\kappa}_2,\\
\frac{Dn^i}{DS^j} &= 0, \quad (i \neq j),\\
\frac{D\bar{n}^1}{DS^1} &= \frac{\sqrt{\bar{a}_{11}}}{\sqrt{\bar{G}_{11}\, G_{11}}}(\kappa_1^* p_1^1 - T_1^* p_3^1),\\
\frac{D\bar{n}^2}{DS^1} &= \frac{\sqrt{\bar{a}_{11}}}{\sqrt{\bar{G}_{11}\, G_{22}}}(\kappa_1^* p_1^2 - T_1^* p_3^2),\\
\frac{D\bar{n}^3}{DS^1} &= \frac{\sqrt{\bar{a}_{11}}}{\sqrt{\bar{G}_{11}\, G_{33}}}(\kappa_1^* p_1^3 - T_1^* p_3^3),\\
\frac{D\bar{n}^1}{DS^2} &= \frac{\sqrt{\bar{a}_{22}}}{\sqrt{\bar{G}_{22}\, G_{11}}}(\kappa_2^* p_2^1 - T_2^* p_4^1),\\
\frac{D\bar{n}^2}{DS^2} &= \frac{\sqrt{\bar{a}_{22}}}{\sqrt{\bar{G}_{22}\, G_{22}}}(\kappa_2^* p_2^2 - T_2^* p_4^2),\\
\frac{D\bar{n}^3}{DS^2} &= \frac{\sqrt{\bar{a}_{22}}}{\sqrt{\bar{G}_{22}\, G_{33}}}(\kappa_2^* p_2^3 - T_2^* p_4^3),\\
\frac{D\bar{n}^1}{DS^3} &= \frac{D\bar{n}^2}{DS^3} = \frac{D\bar{n}^3}{DS^3} = 0,
\end{aligned}\right\} \tag{6.12}
$$

where

$$\left.\begin{aligned} \kappa_1^* &= \kappa_1 \sin\beta, & T_1^* &= T_1 + \frac{\partial\beta}{\partial S^1}, \\ \kappa_2^* &= \kappa_2 \sin\gamma, & T_2^* &= T_2 + \frac{\partial\gamma}{\partial S^2}. \end{aligned}\right\} \tag{6.13}$$

The finite strain components of a shell are finally found by substituting (6.12) into (6.3):

$$\begin{aligned}
S_1^1 &= \frac{\mathrm{D}v^1}{\mathrm{D}S^1} + \frac{\mathrm{D}w^1}{\mathrm{D}S^1} + \frac{1}{\sqrt{\bar{G}_{11}}}[\sqrt{\bar{a}_{11}}\,(\kappa_1^* p_1^1 - T_1^* p_3^1) - \mathring{\kappa}_1]\,t \\
&\quad + [1 - (l_1)^2](1 - \cos\vartheta), \\
S_2^2 &= \frac{\mathrm{D}v^2}{\mathrm{D}S^2} + \frac{\mathrm{D}w^2}{\mathrm{D}S^2} + \frac{1}{\sqrt{\bar{G}_{22}}}[\sqrt{\bar{a}_{22}}\,(\kappa_2^* p_2^2 - T_2^* p_4^2) - \mathring{\kappa}_2]\,t \\
&\quad + [1 - (l_2)^2](1 - \cos\vartheta), \\
S_3^3 &= \frac{\mathrm{D}v^3}{\mathrm{D}S^3} + \frac{\mathrm{D}w^3}{\mathrm{D}S^3} + (\bar{n}^3 - n^3) + [1 - (l_3)^2](1 - \cos\vartheta), \\
S_2^1 &= \frac{1}{2}\left(\frac{\mathrm{D}v^1}{\mathrm{D}S^2} + \frac{\mathrm{D}v^2}{\mathrm{D}S^1}\right) + \frac{1}{2}\left(\frac{\mathrm{D}w^1}{\mathrm{D}S^2} + \frac{\mathrm{D}w^2}{\mathrm{D}S^1}\right) \\
&\quad + \frac{1}{2}\left[\sqrt{\frac{\bar{a}_{22}}{\bar{G}_{22}}}\,(\kappa_2^* p_2^1 - T_2^* p_4^1) + \sqrt{\frac{\bar{a}_{11}}{\bar{G}_{11}}}\,(\kappa_1^* p_1^2 - T_1^* p_3^2)\right] t \\
&\quad - (l_1)(l_2)(1 - \cos\vartheta), \\
S_3^2 &= \frac{1}{2}\frac{\mathrm{D}v^3}{\mathrm{D}S^2} + \frac{1}{2}\left(\frac{\mathrm{D}w^2}{\mathrm{D}S^2} + \frac{\mathrm{D}w^3}{\mathrm{D}S^2}\right) + \tfrac{1}{2}\sqrt{G_{22}}\,(\bar{n}^2 - n^2) \\
&\quad + \frac{1}{2}\sqrt{\frac{\bar{a}_{22}}{\bar{G}_{22}}}\,(\kappa_2^* p_2^3 - T_2^* p_4^3)\,t - (l_2)(l_3)(1 - \cos\vartheta), \\
S_1^3 &= \frac{1}{2}\frac{\mathrm{D}v^3}{\mathrm{D}S^1} + \frac{1}{2}\left(\frac{\mathrm{D}w^3}{\mathrm{D}S^1} + \frac{\mathrm{D}w^1}{\mathrm{D}S^3}\right) + \tfrac{1}{2}\sqrt{G_{11}}\,(\bar{n}^1 - n^1) \\
&\quad + \frac{1}{2}\sqrt{\frac{\bar{a}_{11}}{\bar{G}_{11}}}\,(\kappa_1^* p_1^3 - T_1^* p_3^3)\,t - (l_3)(l_1)(1 - \cos\vartheta).
\end{aligned} \tag{6.14}$$

Here v^i and w^i are the physical components of displacement of a point on the midsurface and the deviation at the vertex of a normal.

2. Discussion of the Kirchoff–Love hypothesis

Since we have the exact formulation of finite strain for a shell, we can estimate the degree of accuracy of the Kirchoff–Love hypothesis which was developed especially for a thin shell. In the following discussion, we shall see that the error induced by Kirchoff–Love hypothesis was caused chiefly by ignoring the influence of finite rotation.

According to the results given previously, the statements of the hypothesis may be expressed as follows:

(1) Both strain and rotation are small quantities of the same order, and the square of the rotation angle is negligible with respect to unity. This condition occurs when the change of direction of local vectors and the normal on a shell is small, so that

$$1 - \cos \vartheta \simeq 0$$

and $p^i_{(m)}$ of (6.10) may be approximated as

	$\bar{e}_1$	$\bar{e}_2$	c_1	c_2
e_1	1	$O(\lambda)$	$O(\lambda)$	1
e_2	$O(\lambda)$	1	1	$O(\lambda)$
e_3	$O(\theta)$	$O(\theta)$	$O(\theta)$	$O(\theta)$

$\lambda = \pi/2 - \chi$, χ is the angle between $\bar{\mathbf{e}}_1$ and $\bar{\mathbf{e}}_2$ after deformation. $O(\lambda)$ and $O(\theta)$ represent quantities that are of the same order as λ and θ or smaller than λ or θ.

(2) Assume that the normal of an undeformed midsurface is still perpendicular to the midsurface after deformation.

(3) Assume that the distance of a point in a shell to the midsurface is unchanged after deformation.

Assumptions (2) and (3) are equivalent to $W = 0$, or $w^1 = w^2 = w^3 = 0$.

If we are not limited to the consideration of a thin shell, the assumptions stated above simplify (6.14) as follows:

$$S^1_1 = \frac{Dv^1}{DS^1} + \frac{1}{\sqrt{\bar{G}_{11}}}[\sqrt{\bar{a}_{11}}\,(\kappa^*_2 - T^*_1 p^1_3) - \mathring{\kappa}_1]\,t,$$

$$S^2_2 = \frac{Dv^2}{DS^2} + \frac{1}{\sqrt{\bar{G}_{22}}}[\sqrt{\bar{a}_{22}}\,(\kappa^*_2 - T^*_2 p^2_4) - \mathring{\kappa}_2]\,t,$$

$$S^3_3 = \frac{Dv^3}{DS^3} + (\bar{n}^3 - n^3) = (\bar{n}^3 - n^3),$$

$$S^1_2 = \frac{1}{2}\left(\frac{Dv^1}{DS^2} + \frac{Dv^2}{DS^1}\right) + \frac{1}{2}\left[\sqrt{\frac{\bar{a}_{22}}{\bar{G}_{22}}}\,(\kappa^*_2 p^1_4 - T^*_2) + \sqrt{\frac{\bar{a}_{11}}{\bar{G}_{11}}}\,(\kappa^*_1 p^2_1 - T_1{}^*)\right]$$

t,

$$S^2_3 = \frac{1}{2}\frac{Dv^3}{DS^2} + \tfrac{1}{2}\sqrt{G_{22}}\,(\bar{n}^2 - n^2) + \frac{1}{2}\sqrt{\frac{\bar{a}_{22}}{\bar{G}_{22}}}\,(\kappa^*_2 p^3_2 - T^*_2 p^3_4)\,t,$$

$$S^3_1 = \frac{1}{2}\frac{Dv^3}{DS^1} + \tfrac{1}{2}\sqrt{G_{11}}\,(\bar{n}^1 - n^1) + \frac{1}{2}\sqrt{\frac{\bar{a}_{11}}{\bar{G}_{11}}}\,(\kappa^*_1 p^3_1 - T^*_1 p^3_3)\,t.$$

(7.1)

Let us examine the meaning of S_3^3. Suppose a volume element at a point P in the midsurface of the shell makes a rigid rotation, the rotation axis passes through the point P and is perpendicular to the normal line as shown in Fig. 4,

$$l_3 = 0, \qquad \bar{n}^3 - n^3 = \cos\theta - 1.$$

Therefore

$$(\bar{n}^3 - n^3) + [1 - (l_3)^2](1 - \cos\vartheta) = 0$$

and we see that $(\bar{n}^3 - n^3)$ is a small quantity of the same order as $(1 - \cos\vartheta)$. If $(1 - \cos\vartheta)$ is neglected, then $(\bar{n}^3 - n^3)$ must be neglected at the same time in the expression of S_3^3, so $S_3^3 = 0$. Otherwise, there is a contradiction in variation of thickness during rigid motion. If we keep the term $(\bar{n}^3 - n^3)$, then the error induced under this condition is $S_3^3 \simeq O(\theta^2)$.

Next, let us examine the term S_3^2. On the midsurface, $t = 0$, S_3^2 of (7.1) becomes

$$S_{3[0]}^2 = \frac{1}{2}\frac{\mathrm{D}v^3}{\mathrm{D}S^2} + \tfrac{1}{2}(\bar{n}^2 - n^2)\sqrt{G_{22}}.$$

Consider the tangent plane at a point on a surface. If the tangent plane rotates with an angle θ about axis x^1, then

$$S_{3[0]}^2 = \tfrac{1}{2}\sin\theta - \tfrac{1}{2}\sin\theta = 0.$$

Hence, if we neglect the influence of rotation, the error is $S_3^2 \simeq O(\theta^2)$.

The arguments above remain although the strain is small, but with large rotation, the terms involving θ in the strain components cannot be neglected, if large errors are to be avoided (Fig. 4).

Now, we will introduce the further assumptions:

(1) The thickness of a shell is very small. Assume

$$t\kappa_1,\ t\kappa_2 \leqslant 1/100.$$

(2) The extension S_α^α and angular strain $\lambda = \pi/2 - \chi$ of the midsurface are small quantities of the same order as $t\kappa_\theta$, so that

$$\sqrt{\frac{\bar{a}_{11}}{\bar{G}_{11}}} \simeq \sqrt{\frac{(1 + \varepsilon_1)a_{11}}{(1 + t\kappa_1)a_{11}}} \simeq 1, \qquad \sqrt{\frac{\bar{a}_{22}}{\bar{G}_{22}}} \simeq 2.$$

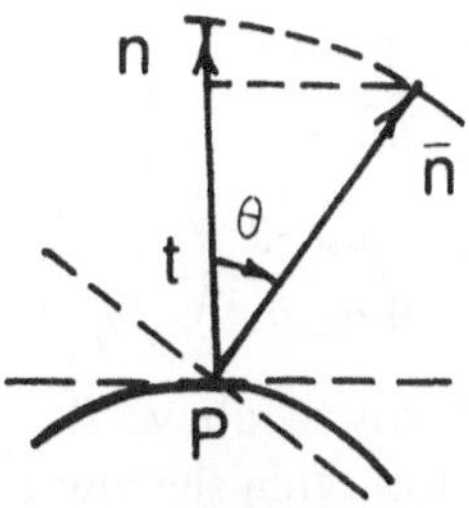

Figure 4

Moreover, $(\kappa_2^* t)p_2^3$ and $(\kappa_1^* t)p_1^3$ are of a negligible higher order small quantity. In addition, the terms $T_1^* p_3^1$ and $T_2^* p_4^2$ are negligible if the change of torsion of coordinate lines is very small.

The foregoing approximation results in:

$$\begin{aligned} S_1^1 &= \frac{Dv^1}{DS^1} + [\kappa_1^* - \mathring{\kappa}_1]t, \\ S_2^2 &= \frac{Dv^2}{DS^2} + [\kappa_2^* - \mathring{\kappa}_2]t, \\ S_2^1 &= \frac{1}{2}\left(\frac{Dv^1}{DS^2} - \frac{Dv^2}{DS^1}\right) - \tfrac{1}{2}(T_2^* + T_1^*)t, \\ S_3^3 &= S_3^2 = S_1^3 = 0, \end{aligned} \tag{7.2}$$

where

$$\begin{aligned} \frac{Dv^1}{DS^1} &= \frac{D(\sqrt{G_{11}}\, V^1}{\sqrt{\bar{G}_{11}}\, Dx^1} = \frac{\sqrt{G_{11}}}{\sqrt{\bar{G}_{11}}}\frac{DV^1}{Dx^1} \\ &= \frac{1}{\sqrt{\bar{G}_{11}}}\frac{\partial v^1}{\partial x^1} + \frac{v^2}{\sqrt{\bar{G}_{11}\, G_{22}}}\frac{\partial\sqrt{G_{11}}}{\partial x^2} + \frac{1}{2}\frac{v^3}{\sqrt{\bar{G}_{11}}}\frac{\partial G_{11}}{\partial x^3}. \end{aligned} \tag{7.3}$$

For the shallow thin shell,

$$t\kappa_1,\ t\kappa_2 \leqslant 1/100,$$

we approximate the metric tensor in (7.3) by that of the undeformed midsurface. Put

$$(\bar{G}_{11})^{1/2} \simeq (G_{11})^{1/2}(1 + \kappa_1 t)^{1/2} \simeq (G_{11})^{1/2} \simeq (a_{11})^{1/2}$$

and for similar terms. Then (7.3) appears as

$$\frac{Dv^1}{DS^1} \simeq \frac{1}{\sqrt{a_{11}}}\frac{\partial v^1}{\partial x^1} + \frac{v^2}{\sqrt{a_{11}a_{22}}}\frac{\partial\sqrt{a_{11}}}{\partial x^2} + v^3\mathring{\kappa}_1. \tag{7.4}$$

This leads to the same formula as Love's [17], where v^3 is taken as positive in the convex direction). Similarly, we have ([24]),

$$\frac{Dv^2}{DS^2} \simeq \frac{1}{\sqrt{a_{22}}}\frac{\partial v^2}{\partial x^2} + \frac{v^1}{\sqrt{a_{11}a_{22}}}\frac{\partial\sqrt{a_{22}}}{\partial x^1} + v^3\mathring{\kappa}_2, \tag{7.5}$$

$$\begin{aligned} \frac{1}{2}\left(\frac{Dv^1}{DS^2} + \frac{Dv^2}{DS^1}\right) &\simeq \frac{1}{2}\left[\frac{1}{\sqrt{a_{22}}}\frac{\partial v^1}{\partial x^2} + \frac{1}{\sqrt{a_{11}}}\frac{\partial v^2}{\partial x^1} - \frac{v^1}{\sqrt{a_{11}a_{22}}}\frac{\partial\sqrt{a_{11}}}{\partial x^2}\right. \\ &\left. - \frac{v^2}{\sqrt{a_{11}a_{22}}}\frac{\partial\sqrt{a_{22}}}{\partial x^1}\right) = \sqrt{\frac{a_{11}}{a_{22}}}\frac{\partial}{\partial x^2}\left(\frac{v^1}{\sqrt{a_{11}}}\right) + \sqrt{\frac{a_{22}}{a_{11}}}\frac{\partial}{\partial x^1}\left(\frac{v^2}{\sqrt{a_{22}}}\right). \end{aligned} \tag{7.6}$$

The derivations we have followed above show that the approximate strain formulae of Love are suitable for a thin shallow shell, $t\kappa \leqslant 1/100$. The estimated error will be smaller than 2%.

Another implication of (7.2) is the neglect of the difference between the neutral surface and the midsurface. The distance between the neutral surface and the midsurface is estimated to be of the order of magnitude $t^2\mathring{\kappa}$. Therefore, neglect of terms smaller than $t^2\mathring{\kappa}$ is equivalent to the assumption of the coincidence of the two surfaces.

A plate is a special case of a shell, and we can therefore derive formulae for finite strains for plates of a moderate degree of deflection by (7.2).

8. Reissner–Vlasov's approximate formulae for the displacement of a shell

Suppose we accept the Kirchhoff–Love's hypothesis; let $W^1 = W^2 = W^3 = 0$, and the displacement vector in a shell continuum by (6.1) will then be:

$$\mathbf{U} = U^i\mathbf{G}_i = [V^i + t(\bar{n}^i - n^i)]\mathbf{G}_i. \tag{8.1}$$

Since

$$\bar{\mathbf{a}}_i = \frac{\partial\bar{\mathbf{r}}}{\partial x^i} = \frac{\partial\mathbf{r}}{\partial x^i} + \frac{\partial\mathbf{V}}{\partial x^i} = \mathbf{a}_i + \frac{\mathrm{D}V^j}{\mathrm{D}x^i}\mathbf{G}_j \tag{8.2}$$

and

$$\bar{\mathbf{n}} = \frac{1}{\sqrt{\bar{\mathbf{a}}}}\bar{\mathbf{a}}_1 \times \bar{\mathbf{a}}_2, \qquad \mathbf{n} = \frac{1}{\sqrt{a}}\mathbf{a}_1 \times \mathbf{a}_2, \tag{8.3}$$

after neglecting small quantities of the second order, we then have

$$\bar{\mathbf{a}}_1 \times \bar{\mathbf{a}}_2 = \left[1 + \frac{\mathrm{D}V^1}{\mathrm{D}x^1} + \frac{\mathrm{D}V^2}{\mathrm{D}x^2}\right]\sqrt{a}\,\mathbf{n} - \sqrt{a_{22}}\frac{\mathrm{D}V^3}{\mathrm{D}x^1}\mathbf{E}_1 - \sqrt{a_{11}}\frac{\mathrm{D}V^3}{\mathrm{D}x^2}\mathbf{E}_2,$$

where $\mathbf{E}_i = \mathbf{G}_i/|\mathbf{G}_i|$. To the first approximation, $\sqrt{\bar{a}} \simeq \sqrt{a}$, we find

$$\bar{\mathbf{n}} - \mathbf{n} = (\bar{n}^i - n^i)\mathbf{G}_i = -\frac{1}{a_{11}(1 + t\mathring{\kappa}_1)}\frac{\mathrm{D}V^3}{\mathrm{D}x^1}\mathbf{G}_1$$

$$-\frac{1}{a_{22}(1 + t\mathring{\kappa}_1)}\frac{\mathrm{D}V^3}{\mathrm{D}x^2}\mathbf{G}_2 + \left(\frac{\mathrm{D}V^1}{\mathrm{D}x^1} + \frac{\mathrm{D}V^2}{\mathrm{D}x^2}\right)\mathbf{n}$$

Comparing the coefficients of $\mathbf{G}_i$, we find the expression $(\bar{n}^i - n^i)$, then substitute it into (8.1), so that

$$\begin{aligned} U^1 &= V^1 - \frac{t}{a_{11}(1 + t\mathring{\kappa}_1)}\frac{\mathrm{D}V^3}{\mathrm{D}x^1}, \\ U^2 &= V^2 - \frac{t}{a_{22}(1 + t\mathring{\kappa}_2)}\frac{\mathrm{D}V^3}{\mathrm{D}x^2}, \\ U^3 &= V^3 + t\left(\frac{\mathrm{D}V^1}{\mathrm{D}x^1} + \frac{\mathrm{D}V^2}{\mathrm{D}x^2}\right), \end{aligned} \tag{8.4}$$

we note

$$\frac{\mathrm{D}V^3}{\mathrm{D}x^1} = \frac{\partial V^3}{\partial x^1} - a_{11}\mathring{\kappa}_1(1 + t\mathring{\kappa}_1)V^1,$$
$$\frac{\mathrm{D}V^3}{\mathrm{D}x^2} = \frac{\partial V^3}{\partial x^2} - a_{22}\mathring{\kappa}_2(1 + t\mathring{\kappa}_2)V^2. \tag{8.5}$$

Inserting (8.5) into (8.4), we obtain

$$\hat{u}^1 = U^1\sqrt{G_{11}} = (1 + t\mathring{\kappa}_1)v^1 - \frac{t}{\sqrt{a_{11}}}\frac{\partial v^3}{\partial x^1},$$
$$\hat{u}^2 = U^2\sqrt{G_{22}} = (1 + t\mathring{\kappa}_2)v^2 - \frac{t}{\sqrt{a_{22}}}\frac{\partial v^3}{\partial x^2}, \tag{8.6}$$
$$\hat{u}^3 = U^3\sqrt{G_{33}} = v^3 + t\left[\frac{\mathrm{D}V^1}{\mathrm{D}x^1} + \frac{\mathrm{D}V^2}{\mathrm{D}x^2}\right],$$

where $\hat{u}^i$ is the physical component of **U**, and v^i is the physical component of **V**. The formulae above show that displacement along normal direction is affected by the extension of midsurface. If we ignore the effect, let

$$\hat{u}^3 = v^3. \tag{8.6a}$$

The first two equations of (8.6) and (8.6a) are the well known formulae for displacements of a shell by Reissner [18] and Vlasov [19]. They obtained the formulae by integrating Cauchy's strain components for infinitesmal deformation under the conditions of $S_3^1 = S_3^2 = S_3^3 = 0$.

Formulae (8.6) enable us to calculate the approximate strain components for shells with accuracy of up to second-order small quantities.

9. Finite strain components of thin cylindrical shells

When a thin shell is under large deformation, experiments prove that the Kirchhoff–Love hypothesis still holds reliably. However, when the strain of the shell is small, rotation may be large. If Cauchy's infinitesmal strain tensor is used, large errors may result. In this section, we shall discuss the influence of rotation on the strain tensor of a cylindrical shell when deformation of moderate degree occurs.

Let a_0 be the radius of the midsurface of a cylindrical shell. We chose cylindrical coordinates (z, r, φ) as co-movings coordinates:

$$x^1 \equiv z, \quad x^2 \equiv \varphi, \quad x^3 \equiv r = a_0 + t.$$

The displacement of a point on the midsurface of the shell is

$$v^1 \equiv v^z(z, \varphi), \quad v^2 \equiv v^\varphi(z, \varphi), \quad v^3 \equiv v^r(z, \vartheta).$$

The displacement of a point in the shell continuum is

$$u^1 \equiv u^z(z, \varphi, t), \quad u^2 \equiv u^\varphi(z, \varphi, t), \quad u^3 \equiv u^r(z, \varphi, t).$$

Because $\mathring{\kappa}_1 = 0$, $\mathring{\kappa}_2 = 1/a_0$, the metric measure of the shell is given by

$$(\mathrm{d}S_0)^2 = (\mathrm{d}z)^2 + r^2(\mathrm{d}\varphi)^2 + (\mathrm{d}r)^2,$$

$$a_{11} = 1, \quad a_{22} = r^2, \quad a_{33} = 1.$$

The approximate equation of displacement is from (8.6) and (8.6a),

$$
\begin{aligned}
u^z &= v^3 - t\frac{\partial v^r}{\partial z}, \\
u^\varphi &= \left(1 + \frac{t}{a_0}\right) v^\varphi - \frac{t}{a_0}\frac{\partial v^r}{\partial \varphi}, \\
u^r &= v^r.
\end{aligned}
\tag{9.1}
$$

Substituting (9.1) into (5.7), let $\bar{G}_{11} \simeq a_{11}$, $\bar{G}_{22} \simeq a_{22}$, $\bar{G}_{33} \simeq a_{33}$, $r = a_0 + t$, $\partial/\partial r = \partial/\partial t$, and then we find:

$$
\begin{aligned}
\sin\vartheta = &\left\{\left(\frac{\partial v^r}{\partial z}\right)^2 + \left(\frac{\partial v^r}{a_0 \partial\varphi} - \frac{v^\varphi}{a_0}\right)^2 + \frac{1}{4}\left[\left(1 + \frac{t}{a_0}\right)\frac{\partial v^\varphi}{\partial z}\right.\right. \\
&\left.\left. - \frac{1}{a_0 + t}\frac{\partial v^3}{\partial\varphi}\frac{t^2}{a_0(a_0 + t)}\frac{\partial^2 v^r}{\partial\varphi\partial z}\right\}^{1/2},
\end{aligned}
\tag{9.2}
$$

$$
\begin{aligned}
l_1 &= \frac{1}{\sin\vartheta}\left(\frac{\partial v^r}{a_0\partial\varphi} - \frac{v^\varphi}{a_0}\right), \\
l_2 &= -\frac{1}{\sin\vartheta}\frac{\partial v^r}{\partial z}, \\
l_3 &= \frac{1}{2\sin\vartheta}\left[\left(1 + \frac{t}{a_0}\right)\frac{\partial v^\varphi}{\partial z} - \frac{1}{a_0 + t}\frac{\partial v^3}{\partial\varphi} - \frac{t^2}{a_0(a_0 + t)}\frac{\partial^2 v^r}{\partial\varphi\partial z}\right].
\end{aligned}
\tag{9.3}
$$

Finite strain components are approximately calculated by the following formulae. First, we consider

$$
\begin{aligned}
S_z^z &= \frac{\partial u^z}{\partial z} + [1 - (l_1)^2](1 - \cos\vartheta) \\
&= \frac{\partial v^z}{\partial z} - t\frac{\partial^2 v^r}{\partial z^2} + \left[1 - \frac{1}{\sin^2\vartheta}\left(\frac{\partial v^r}{a_0\partial\varphi} - \frac{v^\varphi}{a_0}\right)^2\right](1 - \cos\vartheta) \\
&= \frac{\partial v^z}{\partial z} - t\frac{\partial^2 v^r}{\partial z^2} + \frac{1}{2\cos^2\vartheta/2}\left\{\left(\frac{\partial v^r}{\partial z}\right)^2 + \frac{1}{4}\left[\left(1 + \frac{t}{a_0}\right)\frac{\partial v^\varphi}{\partial z}\right.\right. \\
&\qquad \left.\left. - \frac{1}{a_0 + t}\frac{\partial v^z}{\partial\varphi} - \frac{t^2}{a_0(a_0 + t)}\frac{\partial^2 v^r}{\partial\varphi\partial z}\right]^2\right\}.
\end{aligned}
$$

Suppose the rotation angle in $\theta \leqslant 10°$, then $\cos^2\theta/2 \simeq 1$ (a more rough approximation would be $\theta < 20°$). Therefore in the case of a thin shell, $t/a_0 \ll 1$, terms smaller than t^2/a_0 may be neglected according to this degree of approximation,

$$S_z^z = \frac{\partial v^z}{\partial z} + \frac{1}{2}\left(\frac{\partial v^r}{\partial z}\right)^2 + \frac{1}{8}\left(\frac{\partial v^\varphi}{\partial z} - \frac{\partial v^z}{a_0 \partial \varphi}\right)^2 - t\left[\frac{\partial^2 v^r}{\partial z^2} - \frac{1}{4a_0}\left(\frac{\partial v^\varphi}{\partial z}\right)^2 + \frac{1}{4a_0^3}\left(\frac{\partial v^z}{\partial \varphi}\right)^2\right].$$

Similarly,

$$\begin{aligned}
S_\varphi^\varphi &= \frac{1}{a_0}\frac{\partial v^\varphi}{\partial \varphi} + \frac{v^r}{a_0} + \frac{1}{2}\left(\frac{1}{a_0}\frac{\partial v^r}{\partial \varphi} - \frac{v^\varphi}{a_0}\right)^2 + \frac{1}{8}\left(\frac{\partial v^\varphi}{\partial z} - \frac{1}{a_0}\frac{\partial v^z}{\partial \varphi}\right)^2 \\
&\quad - t\left[\frac{1}{a_0^2}\frac{\partial^2 v^r}{\partial \varphi^2} + \frac{v^r}{a_0^2} + \frac{1}{4a_0^3}\left(\frac{\partial v^z}{\partial \varphi}\right) - \frac{1}{4a_0}\left(\frac{\partial v^\varphi}{\partial z}\right)^2\right], \\
S_r^r &= \frac{1}{2}\left[\left(\frac{\partial v^r}{\partial z}\right)^2 + \left(\frac{1}{a_0}\frac{\partial v^r}{\partial \varphi} - \frac{v^\varphi}{a_0}\right)^2\right], \\
S_z^\varphi &= \frac{1}{2}\left[\frac{\partial v^\varphi}{\partial z} + \frac{1}{a_0}\frac{\partial v^z}{\partial \varphi} + \frac{\partial v^r}{\partial z}\left(\frac{1}{a_0}\frac{\partial v^r}{\partial \varphi} - \frac{v^\varphi}{a_0}\right)\right] \\
&\quad - t\left[\frac{1}{a_0}\frac{\partial^2 v^r}{\partial \varphi \partial z} + \frac{1}{2a_0^2}\frac{\partial v^z}{\partial \varphi} - \frac{1}{2a_0}\frac{\partial v^\varphi}{\partial z}\right], \\
S_r^z &= \frac{1}{4}\left(\frac{1}{a_0}\frac{\partial v^r}{\partial \varphi} - \frac{v^\varphi}{a_0}\right)\left(\frac{1}{a_0}\frac{\partial v^z}{\partial \varphi} - \frac{\partial v^\varphi}{\partial z}\right) \\
&\quad - \frac{t}{4}\left(\frac{1}{a_0}\frac{\partial v^r}{\partial \varphi} - \frac{v^\varphi}{a_0}\right)\left(\frac{1}{a_0}\frac{\partial v^\varphi}{\partial z} + \frac{1}{a_0^2}\frac{\partial v^z}{\partial \varphi}\right), \\
S_\varphi^r &= \frac{1}{4}\frac{\partial v^r}{\partial z}\left(\frac{\partial v^\varphi}{\partial z} - \frac{1}{a_0}\frac{\partial v^z}{\partial \varphi}\right) + \frac{t}{4}\frac{\partial v^r}{\partial z}\left(\frac{1}{a_0}\frac{\partial v^\varphi}{\partial z} + \frac{1}{a_0^2}\frac{\partial v^z}{\partial \varphi}\right).
\end{aligned} \tag{9.4}$$

First order approximation of (9.4) gives Vlasov's results:

$$\begin{aligned}
S_z^z &= \frac{\partial v^z}{\partial z} - t\frac{\partial^2 v^r}{\partial z^2} \\
S_\varphi^\varphi &= \frac{1}{a_0}\frac{\partial v^\varphi}{\partial \varphi} + \frac{v^r}{a_0} - t\left(\frac{1}{a_0^2}\frac{\partial^2 v^r}{\partial \varphi^2} + \frac{v^r}{a_0^2}\right), \\
S_z^\varphi &= \frac{1}{2}\left(\frac{\partial v^\varphi}{\partial z} + \frac{1}{a_0}\frac{\partial v^z}{\partial \varphi}\right) - \frac{t}{2}\left[2\frac{1}{a_0}\frac{\partial^2 v^r}{\partial \varphi \partial z} - \frac{1}{a_0^2}\frac{\partial v^z}{\partial \varphi} - \frac{1}{a_0}\frac{\partial v^\varphi}{\partial z}\right]. \\
S_r^r &= S_r^z = S_\varphi^r = 0
\end{aligned} \tag{9.5}$$

The coefficients of t are always considered as the rate change of curvature. They are

$$\kappa'_z = -\frac{\partial^2 v^r}{\partial z^2}, \quad \kappa'_\varphi = -\frac{1}{a_0^2}\frac{\partial^2 v^r}{\partial \varphi^2} + \frac{v^r}{a_0^2},$$
$$\kappa'_{z\varphi} = -2\frac{1}{a_0}\frac{\partial^2 v^r}{\partial \varphi \partial z} - \frac{1}{a_0^2}\frac{\partial v^z}{\partial \varphi} - \frac{1}{a_0}\frac{\partial v^\varphi}{\partial z}. \tag{9.6}$$

A rigorous analysis of the relations show that these do not only include the curvature change (see (6.14)), but also, the effects of torsion and the extension of the midsurface. Many investigators have overlooked it.

If in the calculation of κ'_z, κ'_φ, $\kappa'_{z\varphi}$, we neglect the effect of the extension of the midsurface and the angular deformation on the change of curvature, and introduce the following relations into (9.6):

$$\frac{1}{a_0}\frac{\partial v^\varphi}{\partial \varphi} + \frac{v^r}{a_0} = 0, \quad \frac{\partial v^\varphi}{\partial z} + \frac{1}{a_0}\frac{\partial v^z}{\partial \varphi} = 0$$

we have

$$\kappa''_z = -\frac{\partial^2 v^r}{\partial z^2}, \quad \kappa''_\varphi = -\frac{1}{a_0}\frac{\partial^2 v^r}{\partial \varphi^2} + \frac{1}{a_0}\frac{\partial v^\varphi}{\partial \varphi},$$
$$\kappa''_{z\varphi} = -2\frac{\partial^2 v^r}{a_0 \partial \varphi \partial z} + \frac{2}{a_0}\frac{\partial v^\varphi}{\partial z}. \tag{9.7}$$

These equations derived from Love and Timoshenko. As Vlasov corrected some of the approximations in the equations of Love and Timoshenko, they were regarded as one of his contributions. Nevertheless, the derivation here proves that the result of Love and Timoshenko is mathematically of the same order of approximation as that of Vlasov.

Donnell [20] had already generalized Korchhoff–Karman's strain components of large deflection of a plate to the cylindrical shell, but no rigorous proof exists. Let us now examine the domain of applicability of Donnell's formulae.

We choose local coordinates $(x^1, x^2, x^3) \equiv (x, y, z)$ on the midsurface of the cylindrical shell, as shown in Fig. 5. Use the notations

$$v^r \equiv w, \quad v^\varrho \equiv v, \quad v^z \equiv u; \quad a_0\,\mathrm{d}\varphi = \mathrm{d}y, \quad \mathrm{d}z = \mathrm{d}z, \quad \mathrm{d}t = \mathrm{d}x.$$

Then S^z_z, S^φ_φ, S^φ_z reduces to

$$S^z_z = \frac{\partial u}{\partial z} + \frac{1}{2}\left(\frac{\partial w}{\partial z}\right)^2 + \frac{1}{8}\left(\frac{\partial v}{\partial z} - \frac{\partial u}{\partial y}\right)^2$$
$$- t\left[\frac{\partial^2 w}{\partial z^2} - \frac{1}{4a_0}\left(\frac{\partial v}{\partial z}\right)^2 + \frac{1}{4a_0}\left(\frac{\partial u}{\partial y}\right)^2\right],$$

$$S^y_y = \frac{\partial v}{\partial y} + \frac{w}{a_0} + \frac{1}{2}\left(\frac{\partial w}{\partial y} - \frac{v}{a_0}\right)^2 + \frac{1}{8}\left(\frac{\partial v}{\partial z} - \frac{\partial u}{\partial y}\right)^2$$
$$- t\left[\frac{\partial^2 w}{\partial y^2} + \frac{w}{a_0^2} + \frac{1}{4a_0}\left(\frac{\partial u}{\partial y}\right)^2 - \frac{1}{4a_0}\left(\frac{\partial v}{\partial z}\right)^2\right],$$

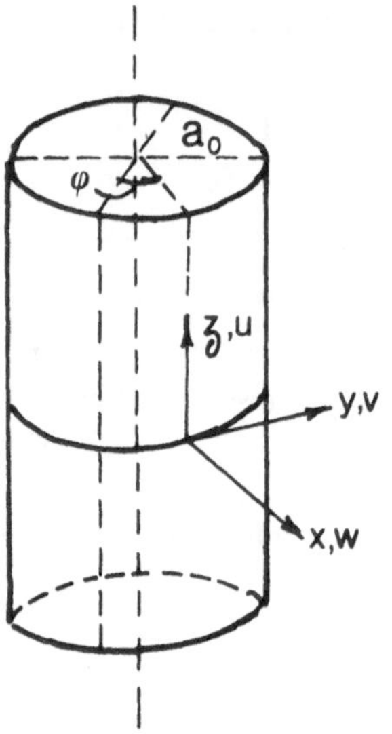

Figure 5

$$S_y^x = \frac{1}{2}\left[\frac{\partial v}{\partial z} + \frac{\partial u}{\partial y} + \frac{\partial w}{\partial z}\left(\frac{\partial w}{\partial y} - \frac{v}{a_0}\right)\right] - t\left[\frac{\partial^2 w}{\partial y \partial z} + \frac{1}{2a_0}\frac{\partial u}{\partial y} - \frac{1}{2a_0}\frac{\partial v}{\partial x}\right]. \tag{9.8}$$

Assume: (1) When buckling occurs in a cylindrical shell, w is the principal quantity. $(\partial v/\partial z)^2$, $(\partial u/\partial y)^2$ and other small quantities of the same order are negligible in comparison with $(\partial w/\partial z)^2$, $(\partial w/\partial y)^2$.

(2) From Timoshenko's work [21], the terms

$$\frac{\partial w}{\partial y} - \frac{v}{a}, \qquad \frac{\partial v}{\partial z} - \frac{\partial u}{\partial y},$$

represent the rotation angle of an elemental area on the midsurface about a generatrix and a normal of the shell, respectively. They can be found by taking principal terms of the numerator in (9.3), which are the direction cosines of the rotation axis. If rotation is small, and its influence on strain components is negligible, then the term $(\partial v/\partial z - \partial u/\partial y)^2$ is omitted. Moreover, v/a_0 is smaller than $\partial w/\partial y$, and is also omitted. t/a_0^2 is also negligible. Following the approximations given, we simplify (9.7) in order to get Donnell's formulae:

$$\begin{aligned} S_z^z &= \frac{\partial u}{\partial z} + \frac{1}{2}\left(\frac{\partial w}{\partial z}\right)^2 - t\frac{\partial^2 w}{\partial z^2}, \\ S_y^y &= \frac{\partial v}{\partial y} + \frac{w}{a_0} + \frac{1}{2}\left(\frac{\partial w}{\partial y}\right)^2 - t\frac{\partial^2 w}{\partial y^2}, \\ S_y^x &= \frac{1}{2}\left(\frac{\partial v}{\partial z} + \frac{\partial u}{\partial y} + \frac{\partial w}{\partial z}\frac{\partial w}{\partial y}\right) - t\frac{\partial^2 w}{\partial y \partial z}. \end{aligned} \tag{9.9}$$

The reasoning above shows that Donnell's formulae are suitable for calculating critical loads in stability problems of cylindrical shells. But if the shell suffers from large rotation, the post buckling behaviours calculated by Donnell's formulae may bring about large errors.

10. Finite strain components of thin spherical shells

Let a_0 be the radius of the midsurface of a spherical shell $x^1 \equiv \varphi$, and $x^2 \equiv \beta$, $x^3 \equiv r$ be the co-moving coordinates of a point in the shell continuum, where ϕ is the angle between radius and axis, β is the longitude, and $r = a_0 + t$, where t denotes the perpendicular distance from a point to the midsurface.

The physical components of the displacement vector of the midsurface are denoted by

$$v^1 \equiv v^\varphi(\varphi, \beta), \quad v^2 \equiv v^\beta(\varphi, \beta), \quad v^3 \equiv v^\gamma(\varphi, \beta).$$

The physical components of the displacement vector of the shell continuum are denoted by

$$u^1 \equiv u^\varphi(\varphi, \beta, t), \quad u^2 \equiv u^\beta(\varphi, \beta, t), \quad u^3 \equiv u^r(\varphi, \beta, t).$$

We find the approximate formulae for the displacement of a point in a thin spherical shell to be:

$$\begin{aligned} u^\varphi &= v^\varphi(1 + t/a_0) - \frac{t}{a_0}\frac{\partial v^r}{\partial \varphi} \\ u^\beta &= v^\beta(1 + t/a_0) - \frac{t}{a_0 \sin\varphi}\frac{\partial v^r}{\partial \beta}. \\ u^r &= v^r. \end{aligned} \tag{10.1}$$

Substituting (10.1) into (5.8), and noting that $r = a_0 + t$, $\partial/\partial r = \partial/\partial t$, we get

$$\begin{aligned} \sin\vartheta &= \left\{\left(\frac{1}{a_0 \sin\varphi}\frac{\partial v^r}{\partial \beta} - \frac{v^\beta}{a_0}\right)^2 + \left(\frac{v^\varphi}{a_0} - \frac{1}{a_0}\frac{\partial v^r}{\partial \varphi}\right)^2 \right. \\ &\quad \left. + \frac{1}{4}\left(\frac{1}{a_0}\operatorname{ctg}\varphi v^\beta + \frac{1}{a_0}\frac{\partial v^\beta}{\partial \varphi} - \frac{1}{a_0 \sin\varphi}\frac{\partial v^\varphi}{\partial \beta}\right)^2\right\}^{1/2}, \\ l_1 &= \frac{1}{\sin\vartheta}\left[\frac{1}{a_0 \sin\varphi}\frac{\partial v^r}{\partial \beta} - \frac{v^\beta}{a_0}\right], \\ l_2 &= \frac{1}{\sin\vartheta}\left[\frac{v^\varphi}{a_0} - \frac{1}{a_0}\frac{\partial v^r}{\partial \varphi}\right], \\ l_3 &= \frac{1}{\sin\vartheta}\left[\frac{1}{a_0}\operatorname{ctg}\varphi v^\beta + \frac{1}{a_0}\frac{\partial v^\beta}{\partial \varphi} - \frac{1}{a_0 \sin\varphi}\frac{\partial v^\varphi}{\partial \beta}\right]. \end{aligned}$$

The mean rotation angle is actually a function of the location of points. Since the shell is thin and deformation is small, we may use the mean rotation angle of the midsurface to approximate this in the shell continuum.

A second order approximation of finite strain components for a thin spherical shell is:

$$
\begin{aligned}
S_\varphi^\varphi &= \frac{1}{a_0}\frac{\partial v^\varphi}{\partial \varphi} + \frac{v^r}{a_0} + \frac{1}{2}\left(\frac{v^\varphi}{a_0} - \frac{1}{a_0}\frac{\partial v^r}{\partial \varphi}\right)^2 \\
&\quad + \frac{1}{8}\left[\frac{1}{a_0}\operatorname{ctg}\varphi v^\beta + \frac{1}{a_0}\frac{\partial v^\beta}{\partial \varphi} - \frac{1}{a_0 \sin\varphi}\frac{\partial v^\varphi}{\partial \beta}\right]^2 - t\left(\frac{1}{a_0^2}\frac{\partial^2 v^r}{\partial \varphi^2} + \frac{v^r}{a_0^2}\right), \\
S_\beta^\beta &= \frac{1}{a_0 \sin\beta}\frac{\partial v^\beta}{\partial \beta} + \frac{v^r}{a_0} + \frac{1}{a_0}\operatorname{ctg}\varphi v^\varphi \\
&\quad + \frac{1}{8}\left[\frac{1}{a_0}\operatorname{ctg}\varphi v^\beta + \frac{1}{a_0}\frac{\partial v^\beta}{\partial \varphi} - \frac{1}{a_0 \sin\varphi}\frac{\partial v^\varphi}{\partial \beta}\right]^2 \\
&\quad + \frac{1}{2}\left[\frac{1}{a_0 \sin\varphi}\frac{\partial v^r}{\partial \beta} - \frac{v^\beta}{a_0}\right]^2 - t\left(\frac{1}{a_0^2 \sin^2\varphi}\frac{\partial^2 v^r}{\partial \beta^2} + \frac{v^r}{a_0^2} + \frac{1}{a_0^2}\operatorname{ctg}\varphi\frac{\partial v^r}{\partial \varphi}\right), \\
S_r^r &= \frac{1}{2}\left[\left(\frac{1}{a_0}\frac{\partial v^r}{\partial \varphi} - \frac{v^\varphi}{a_0}\right)^2 + \left(\frac{1}{a_0 \sin\varphi}\frac{\partial v^r}{\partial \beta} - \frac{v^\beta}{a_0}\right)^2\right], \\
S_\beta^\varphi &= \frac{1}{2}\left[\frac{1}{a_0}\frac{\partial v^\beta}{\partial \varphi} - \frac{1}{a_0}\operatorname{ctg}\varphi v^\beta + \frac{1}{a_0 \sin\varphi}\frac{\partial v^\varphi}{\partial \beta}\right. \\
&\quad \left. - \left(\frac{v^\varphi}{a_0} - \frac{1}{a_0}\frac{\partial v^r}{\partial \varphi}\right)\left(\frac{1}{a_0 \sin\varphi}\frac{\partial v^r}{\partial \beta} - \frac{v^\beta}{a_0}\right)\right] - t\frac{2}{a_0^2}\frac{\partial}{\partial \varphi}\left(\frac{1}{\sin\varphi}\frac{\partial v^r}{\partial \beta}\right), \\
S_\beta^r &= \frac{1}{4}\left(\frac{1}{a_0}\frac{\partial v^r}{\partial \varphi} - \frac{v^\varphi}{a_0}\right)\left(\frac{1}{a_0 \sin\varphi}\frac{\partial v^\varphi}{\partial \beta} - \frac{\operatorname{ctg}\varphi}{a_0}v^\beta - \frac{1}{a_0}\frac{\partial v^\beta}{\partial \varphi}\right), \\
S_r^\varphi &= \frac{1}{2}\left[\frac{v^\beta}{a_0} - \frac{1}{a_0 \sin\varphi}\frac{\partial v^r}{\partial \beta}\right]\left[\frac{\operatorname{ctg}\varphi}{a_0}v^\beta + \frac{1}{a_0}\frac{\partial v^\beta}{\partial \varphi} - \frac{1}{a_0 \sin\varphi}\frac{\partial v^\varphi}{\partial \beta}\right].
\end{aligned}
\tag{10.2}
$$

A first order approximation of the above equations gives Vlasov's results.

For the case of symmetrical deformation, we derive the formulae of Friedrichs [22] from (10.2). This can be done as follows: Let $v^\beta = 0$, $v^\varphi = v^\varphi(\varphi)$, $v^r = v^r(\varphi)$ from (10.2), and we obtain

$$
\begin{aligned}
S_\varphi^\varphi &= \frac{1}{a_0}\frac{\mathrm{d}v^\varphi}{\mathrm{d}\varphi} + \frac{v^r}{a_0} + \frac{1}{2}\left(\frac{v^\varphi}{a_0} - \frac{1}{a_0}\frac{\mathrm{d}v^r}{\mathrm{d}\varphi}\right)^2 - t\left(\frac{1}{a_0^2}\frac{\mathrm{d}^2 v^r}{\mathrm{d}\varphi^2} + \frac{v^r}{a_0^2}\right), \\
S_\beta^\beta &= \frac{v^r}{a_0} + \frac{\operatorname{ctg}\varphi}{a_0}v^\varphi - t\left(\frac{v^r}{a_0^2} + \frac{\operatorname{ctg}\varphi}{a_0^2}\frac{\mathrm{d}v^r}{\mathrm{d}\varphi}\right), \\
S_\varphi^\varphi &= 0, \quad S_\beta^r = 0, \quad S_r^\varphi = 0.
\end{aligned}
\tag{10.3}
$$

S_r^r is neglected. In Friedrich's result, the term $t/2a_0(v^\varphi/a_0 - \mathrm{d}v^r/a_0\,\mathrm{d}\varphi)^2$ appears in the equation of S_φ^φ, but this can be neglected in a second order approximation.

As a direct application of (10.2) to the extreme case of a shallow shell, the circular plate, we can derive from (10.3) the finite strain components of non-symmetrical deflection for the plate.

11. Concluding remark

Using the method of co-moving coordinates first proposed by Synge and Chien in their study of finite deformation of plates and shells, and the author's non-linear geometric field theory; we develop in this paper, an analytical method to find the effects of rotation and strain congruently. The systematically mathematical results lead to important classical formulae so far as we know.

This theory enables us to find mathematically practical formulae for the finite strain and rotation of shells under various conditions and with different degrees of accuracy for engineering usages.

I am grateful to Prof. Chien Wei-zang for his comments on the paper, and would also thank Mr. Zeng Zong for his translation of it into English.

References

1. Chien Wei-zang, "The intrinsic theory of thin shells and plates, I, II, III", *Quart. Appl. Math.* **1** (1943), 297–327; **2** (1944), 120–135.
2. Synge, J. L., and Chien Wei-zang, "The intrinsic theory of shells and plates", *Kàrmàn Anniversary Volume* (1947), 111.
3. Chen Zhi-da, "Geometric field theory of finite deformation for continuum mechanics", *Acta Mechanica Sinica*, No. 2 (1979), 107–117.
4. Th. von Kàrmàn, and Tsien Hsue-shen, "The buckling of spherical shells by external pressure", *J. Aero. Sci.* **7** (1939), 43–50.
5. Th. von Kàrmàn, and Tsien Hsue-shen, "The buckling of thin cylindrical shells under axial compression", *J. Aero. Sci.* **8** (1941), 303–312.
6. Oniashvili, O. J., "Computation of shells and other thin walled space structures", *Acta Mechanica Sinica* **3** (1959), 133–176.
7. Lurie, A. I., "General theory of elastic thin shells", *PMM*, No. 4 (1940), 7–34.
8. Koiter, W. T., *A Consistent First Approximation in the General Theory of Thin Elastic Shells*, Tech. Univ., Delft (1959).
9. Donnell, L. H., "General thin shell displacement–strain relations", *Proc. of the 4th U.S. Ccngr. of Appl. Mech.* (1962), 529.
10. Zerna, W., "Mathematical strenge Theorie elastisher Schalen", *ZAMM* **42** (1962), 333–341.
11. Ciarlet, P. G., "A justification of the von Kàrmàn equation", *Arch. Rational Mech. Analy.* **73** (1980), 351.
12. Knowles, J. K., and Reissner, E., "A new derivation of the equation of shell theory for general orthogonal coordinates", *J. Math. Phys.* **35** (1956), 351–358.
13. Chien Zhi-da, *Lectures on Rational Mechanics* (in Chinese), Graduate School, China Institute of Mining.
14. Truesdell, C., "The physical components of vectors and tensors", *ZAMM* **33** (1953), 345–356.
15. Sokolnikoff, I. S., *Mathematical Theory of Elasticity*, McGraw-Hill, (1946), 48.
16. Kirchhoff, G. R., *Vorlesungen über mathematische Physik, Mechanik*, Leipzig, (1874), Vorlesungen XXX.
17. Love, A. E. H., *The Mathematical theory of Elasticity*, 4th ed., Cambridge, (1927), Ch. 24.

18. Reissner, E., "A new derivation of the equations for the deformation of elastic shells", *Am. J. Math.* **63** (1941), 177–184.
19. Vlasov, V. Z., *General Theory of Shells and its Applications in Technik*, Gostechizdat (1949) (in Russian). Available in English as NASA translation.
20. Donnel, L. H., "A new theory for the buckling of thin cylinders under axial compression and bending", *Trans. ASME* **56** (1934), 795.
21. Timoshenko, S., *Theory of Elastic Stability* (Chinese translation), Science Press, Beijing (1958), 79.
22. Friedrichs, "On the minimum buckling load for spherical shells", *Kàrmàn Anniversary Volume, CIT* (1951), 258–272.
23. Green, A. E. and Zerna, W., *Theoretical Elasticity*, Oxford, (1954), Ch. XI.
24. Novozhilov, V. V., *Theory of Thin Shells* (1951) (in Russian).
25. Chen Zhi-da, *Theory of Large Deformations of Rods, Plates and Shells* (in Chinese), Graduate School, China Institute of Mining (1982).
26. Flügge, W., *Stresses in Shells*, Springer-Verlag (1973).

Perturbation solutions for uniformly clamped rectangular plates with large deflections

YEH KAI-YUAN

Lanzhou University, Lanzhou, Gansu 730001, PRC

FANG JU-XIAN and WANG YUN

Sixth Design Institute of the Ministry of Chemical Engineering, PRC

Abstract. In this paper we use the perturbation method to solve the large deflection problems of uniformly loaded clamped rectangular plates. Using the maximum deflection as a perturbation parameter, we solve twelve cases with the ratio of width against breadth ($= \lambda$): 1; 1.1; 1.2; 1.3; 1.4; 1.5; 1.6; 1.7; 1.8; 1.9; 2.0; ∞, and give the numerical calculation formulae for stresses at the central point of plates and the central points on the boundaries, as well as graphs of maximum deflection and stress, both as functions of loads. As $\lambda = 1$; 1.5; 2.0; ∞, present results are compared with those in the works of S. Timoshenko *et al.* [2], and A. S. Volmir [3].

1. Introduction

The large deflection problems of a rectangular plate with four edges clamped under uniform normal load have been studied by S. Way [5] using the Ritz method, and S. Levy [4] using the double Fourier series, giving only approximate solutions and the ratio of width to breadth ($= \lambda$) in only four cases: 1; 1.5; 2.0; and ∞. Here we use the maximum deflection, as the parameter of systematically approximate solutions, to solve twelve cases: $\lambda = 1$; 1.1; 1.2; 1.3; 1.4; 1.5; 1.6; 1.7; 1.8; 1.9; 2.0; ∞ and give approximate analytic expressions of displacement, deflection and stress as well as graphs of the maximum deflection and maximum stress, both against loads. For $\lambda = 1$; 1.5; 2.0; ∞, the present results are compared with those in S. Timoshenko, S. Woinowsky-Krieger [2] and A. S. Volmir [3]. When $f = w_0/h \leqslant 1$, these three are very close, and when $f < 2$, the maximum relative error is not less than 6%, where w_0 is the central deflection of the plate, i.e. the maximum deflection, and h is the thickness of the plate.

2. Fundamental equations and boundary conditions

As shown in Fig. 1, the neutral plane of the rectangular plate has width $2a$, breadth $2b$, normal load $q =$ const., thickness h, and the displacement functions in directions X, Y, Z are $u(x, y)$, $v(x, y)$, $w(x, y)$.

Yeh, K. Y. (ed) Progress in Applied Mechanics
 ISBN-13: 978-94-010-8061-3

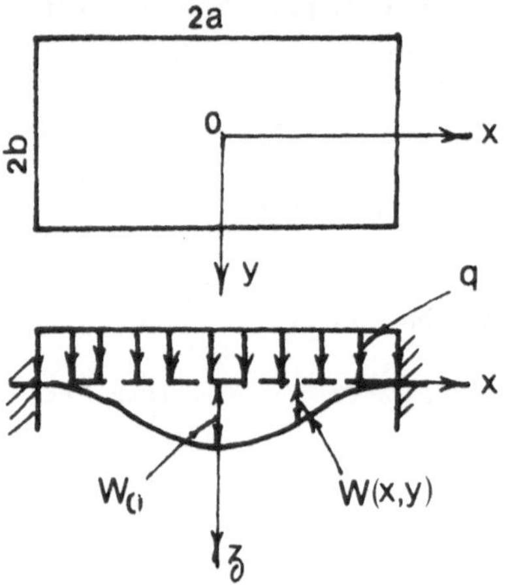

Figure 1

In the case of four edges clamped, the boundary conditions are:

$$u = v = w = \frac{\partial w}{\partial x} = 0, \qquad \text{when } x = \pm a,$$

$$u = v = w = \frac{\partial w}{\partial y} = 0, \qquad \text{when } y = \pm b.$$

The relations between stress and strain are

$$\left.\begin{aligned}
\sigma_x &= \frac{E}{1-\mu^2}(\varepsilon_x + \mu\varepsilon_y) \\
&= \frac{E}{1-\mu^2}\left\{\frac{\partial u}{\partial x} + \frac{1}{2}\left(\frac{\partial w}{\partial x}\right)^2 + \mu\left[\frac{\partial v}{\partial y} + \frac{1}{2}\left(\frac{\partial w}{\partial y}\right)^2\right]\right\}, \\
\sigma_y &= \frac{E}{1-\mu^2}(\varepsilon_y + \mu\varepsilon_x) \\
&= \frac{E}{1-\mu^2}\left\{\frac{\partial v}{\partial y} + \frac{1}{2}\left(\frac{\partial w}{\partial y}\right)^2 + \mu\left[\frac{\partial u}{\partial x} + \frac{1}{2}\left(\frac{\partial w}{\partial x}\right)^2\right]\right\}, \\
\tau_{xy} &= \frac{E}{2(1+\mu)}\gamma_{xy} = \frac{E}{2(1+\mu)}\left(\frac{\partial u}{\partial y} + \frac{\partial v}{\partial x} + \frac{\partial w}{\partial x}\frac{\partial w}{\partial y}\right),
\end{aligned}\right\} \qquad (2.1)$$

where μ is Poisson's ratio.

The equilibrium equations are

$$\left.\begin{aligned}
\frac{\partial \sigma_x}{\partial x} + \frac{\partial \tau_{xy}}{\partial y} &= 0, \\
\frac{\partial \tau_{xy}}{\partial x} + \frac{\partial \sigma_y}{\partial y} &= 0,
\end{aligned}\right\} \qquad (2.2)$$

$$D\nabla^2\nabla^2 w = q + h\left(\sigma x \frac{\partial^2 w}{\partial x^2} + \sigma_y \frac{\partial^2 w}{\partial y^2} + 2\tau_{xy}\frac{\partial^2 w}{\partial x \partial y}\right),$$

where $D = Eh^3/12(1 - \mu^2)$ is the flexural rigidity,

$$\nabla^2\nabla^2(\ ) = \frac{\partial^4(\ )}{\partial x^4} + 2\frac{\partial^4(\ )}{\partial x^2 \partial y^2} + \frac{\partial^4(\ )}{\partial y^4}.$$

Substituting (2.1) in (2.2), we obtain the Kàrmàn equations of large deflection in displacement form for a rectangular plate:

$$2\frac{\partial^2 u}{\partial x^2} + (1-\mu)\frac{\partial^2 u}{\partial y^2} + (1+\mu)\frac{\partial^2 v}{\partial x \partial y} = -(1-\mu)\frac{\partial w}{\partial x}\nabla^2 w - \frac{1+\mu}{2}\frac{\partial}{\partial x}\left[\left(\frac{\partial w}{\partial x}\right)^2 + \left(\frac{\partial w}{\partial y}\right)^2\right], \tag{2.3a}$$

$$(1+\mu)\frac{\partial^2 u}{\partial x \partial y} + (1-\mu)\frac{\partial^2 v}{\partial x^2} + 2\frac{\partial^2 v}{\partial y^2} = -(1-\mu)\frac{\partial w}{\partial y}\nabla^2 w - \frac{1+\mu}{2}\frac{\partial}{\partial y}\left[\left(\frac{\partial w}{\partial x}\right)^2 + \left(\frac{\partial w}{\partial y}\right)^2\right], \tag{2.3b}$$

$$\begin{aligned} D\nabla^2\nabla^2 w = {} & q + \frac{Eh}{1-\mu^2}\left\{\frac{\partial^2 w}{\partial x^2}\left(\frac{\partial u}{\partial x} + \mu\frac{\partial v}{\partial y}\right) + \frac{\partial^2 w}{\partial y^2}\left(\frac{\partial v}{\partial y} + \mu\frac{\partial u}{\partial x}\right)\right. \\ & \left. + (1-\mu)\frac{\partial^2 w}{\partial x \partial y}\left(\frac{\partial u}{\partial y} + \mu\frac{\partial v}{\partial x}\right)\right\} \\ & + \frac{Eh}{2(1-\mu^2)}\left\{\left[\left(\frac{\partial w}{\partial x}\right)^2 + \mu\left(\frac{\partial w}{\partial y}\right)^2\right]\frac{\partial^2 w}{\partial x^2}\right. \\ & \left. + \left[\left(\frac{\partial w}{\partial y}\right)^2 + \mu\left(\frac{\partial w}{\partial x}\right)^2\right]\right\}\frac{\partial^2 w}{\partial y^2} + 2(1-\mu)\frac{\partial w}{\partial x}\frac{\partial w}{\partial y}\frac{\partial^2 w}{\partial x^2 \partial y^2}. \end{aligned} \tag{2.3c}$$

3. Perturbation method

Firstly, we introduce the following dimensionless quantities:

$$\left.\begin{aligned} \lambda &= \frac{a}{b}, \quad \xi = \frac{x}{a}, \quad \eta = \frac{y}{b}, \quad U = \frac{12au}{h^2}, \\ V &= \frac{12av}{h^2}, \quad W = \frac{2\sqrt{3}w}{h}, \quad Q = \frac{24\sqrt{3}(1-\mu^2)a^4 q}{Eh^4}. \end{aligned}\right\} \tag{3.1}$$

Substituting (3.1) in (2.3), we obtain the following equations in dimensionless form:

$$\begin{aligned} & 2\frac{\partial^2 U}{\partial \xi^2} + (1-\mu)\lambda^2\frac{\partial^2 U}{\partial \eta^2} + (1+\mu)\lambda\frac{\partial^2 V}{\partial \xi \partial \eta} \\ & = -(1-\mu)\frac{\partial W}{\partial \xi}\left(\frac{\partial^2 W}{\partial \xi^2} + \lambda^2\frac{\partial^2 W}{\partial \eta^2}\right) - \frac{1+\mu}{2}\frac{\partial W}{\partial \xi}\left[\left(\frac{\partial W}{\partial \xi}\right)^2 + \lambda^2\left(\frac{\partial W}{\partial \eta}\right)^2\right], \end{aligned} \tag{3.2a}$$

$$(1+\mu)\lambda\frac{\partial^2 U}{\partial\xi\partial\eta}+(1-\mu)\frac{\partial^2 V}{\partial\xi^2}+2\lambda^2\frac{\partial^2 V}{\partial\eta^2}$$
$$=-(1-\mu)\lambda\frac{\partial W}{\partial\eta}\left(\frac{\partial^2 W}{\partial\xi^2}+\lambda^2\frac{\partial^2 W}{\partial\eta^2}\right)$$
$$-\frac{1+\mu}{2}\lambda\frac{\partial}{\partial\eta}\left[\left(\frac{\partial W}{\partial\xi}\right)^2+\lambda^2\left(\frac{\partial W}{\partial\eta}\right)^2\right], \tag{3.2b}$$

$$\frac{\partial^4 W}{\partial\xi^4}+2\lambda^2\frac{\partial^4 W}{\partial\xi^2\partial\eta^2}+\lambda^4\frac{\partial^4 W}{\partial\eta^4}=Q+\frac{\partial^2 W}{\partial\xi^2}\left(\frac{\partial U}{\partial\xi}+\lambda\mu\frac{\partial V}{\partial\eta}\right)$$
$$+\lambda^2\frac{\partial^2 W}{\partial\eta^2}\left(\lambda\frac{\partial V}{\partial\eta}+\mu\frac{\partial U}{\partial\xi}\right)+\lambda(1-\mu)\left(\lambda\frac{\partial U}{\partial\eta}+\frac{\partial V}{\partial\xi}\right)\frac{\partial^2 W}{\partial\xi\partial\eta}$$
$$+\frac{1}{2}\frac{\partial^2 W}{\partial\xi^2}\left[\left(\frac{\partial W}{\partial\xi}\right)^2+\mu\lambda\left(\frac{\partial W}{\partial\eta}\right)^2\right]$$
$$+\frac{1}{2}\frac{\partial^2 W}{\partial\eta^2}\left[\lambda^2\left(\frac{\partial W}{\partial\eta}\right)^2+\left(\frac{\partial W}{\partial\xi}\right)^2\right]+\lambda^2(1-\mu)\frac{\partial W}{\partial\xi}\frac{\partial W}{\partial\eta}\frac{\partial^2 W}{\partial\xi\partial\eta}. \tag{3.2c}$$

The boundary conditions are

$$\left.\begin{aligned}&W=U=V=0,\quad \text{at } \xi=\pm1 \text{ and } \eta=\pm1,\\&\frac{\partial W}{\partial\xi}=0,\quad \text{at } \xi=\pm1,\\&\frac{\partial W}{\partial\eta}=0,\quad \text{at } \eta=\pm1.\end{aligned}\right\} \tag{3.3}$$

Let the reduced central deflection of the plate be $W(0,0)=W_0=2\sqrt{3}w_0/h$, where w_0 is the central deflection of the plate. Under the possible condition of convergence, we expand Q, U, V, W into a power series of W_0 as the following:

$$\left.\begin{aligned}Q&=\alpha_1 W_0+\alpha_3 W_0^3+\alpha_5 W_0^5+\cdots,\\W&=W_1(\xi,\eta)W_0+W_3(\xi,\eta)W_0^3+\cdots,\\U&=s_2(\xi,\eta)W_0^2+s_4(\xi,\eta)W_0^4+\cdots,\\V&=t_2(\xi,\eta)W_0^2+t_4(\xi,\eta)W_0^4+\cdots,\end{aligned}\right\} \tag{3.4a, b, c, d}$$

and let quantities in (3.4) satisfy $W_1(0,0)=1$, $W_3(0,0)=W_5(0,0)=\cdots=0$. Substituting (3.4) to (3.2) and (3.3), then comparing the coefficients with the same power index term on both sides of the equations, we obtain the simultaneous differential equations ($\mu=1/3$):

$$\left.\begin{aligned}&\frac{\partial^4 W_1}{\partial\xi^4}+2\lambda^2\frac{\partial^4 W_1}{\partial\xi^2\partial\eta^2}+\lambda^4\frac{\partial^4 W_1}{\partial\eta^4}=\alpha_1,\\&W_1=\frac{\partial W_1}{\partial\xi}=0,\quad \text{at } \xi=\pm1,\\&W_1=\frac{\partial W_1}{\partial\eta}=0,\quad \text{at } \eta=\pm1 \quad (\text{coefficients of } W_0),\end{aligned}\right\} \tag{3.5}$$

$$\left.\begin{aligned}
\frac{\partial^2 s_2}{\partial \xi^2} + \frac{\lambda^2}{3}\frac{\partial^2 s_2}{\partial \eta^2} + \frac{2}{3}\lambda\frac{\partial^2 t_2}{\partial \xi \partial \eta} &= -L_1(W_1, W_1), \\
\frac{2}{3}\lambda\frac{\partial^2 s_2}{\partial \xi \partial \eta} + \frac{1}{3}\frac{\partial^2 t_2}{\partial \xi^2} + \lambda^2\frac{\partial^2 t_2}{\partial \eta^2} &= -L_2(W_1, W_1), \\
s_2 = t_2 = 0, \quad \text{at } \xi = \pm 1, \quad \eta &= \pm 1 \quad (\text{coefficients of } W_0^2),
\end{aligned}\right\} \tag{3.6}$$

where

$$\left.\begin{aligned}
L_1(\ ,\ ) &= \frac{\partial(\ )}{\partial \xi}\frac{\partial^2(\ )}{\partial \xi^2} + \frac{\lambda^2}{3}\frac{\partial(\ )}{\partial \xi}\frac{\partial^2(\ )}{\partial \eta^2} + \frac{2}{3}\lambda^2\frac{\partial(\ )}{\partial \eta}\frac{\partial^2(\ )}{\partial \xi \partial \eta} \\
L_2(\ ,\ ) &= \frac{\lambda}{3}\frac{\partial(\ )}{\partial \eta}\frac{\partial^2(\ )}{\partial \xi^2} + \lambda^3\frac{\partial(\ )}{\partial \eta}\frac{\partial^2(\ )}{\partial \eta^2} + \frac{2}{3}\lambda\frac{\partial(\ )}{\partial \xi}\frac{\partial^2(\ )}{\partial \xi \partial \eta}
\end{aligned}\right\} \tag{3.6$'$}$$

$$\frac{\partial^4 W_3}{\partial \xi^4} + 2\lambda^2\frac{\partial^4 W_3}{\partial \xi^2 \partial \eta^2} + \lambda^4\frac{\partial^4 W_3}{\partial \eta^4} = \alpha_3 + L_1(s_2, W_1)$$

$$+ L_2(t_2, W_1) + L_3(W_1, W_1, W_1),$$

$$W_3 = \frac{\partial W_3}{\partial \xi} = 0, \quad \text{at } \xi = \pm 1,$$

$$W_3 = \frac{\partial W_3}{\partial \eta} = 0, \quad \text{at } \eta = \pm 1 \quad (\text{coefficients of } W_0^3),$$

where

$$L_3(\ ,\ ,\ ) = \frac{1}{2}\frac{\partial(\ )}{\partial \xi}\frac{\partial(\ )}{\partial \xi}\frac{\partial^2(\ )}{\partial \xi^2} + \frac{1}{6}\lambda^2\frac{\partial(\ )}{\partial \eta}\frac{\partial(\ )}{\partial \eta}\frac{\partial^2(\ )}{\partial \xi^2}$$

$$+ \frac{1}{2}\lambda^4\frac{\partial(\ )}{\partial \eta}\frac{\partial(\ )}{\partial \eta}\frac{\partial^2(\ )}{\partial \eta^2} + \frac{1}{6}\lambda^2\frac{\partial(\ )}{\partial \xi}\frac{\partial(\ )}{\partial \xi}\frac{\partial^2(\ )}{\partial \eta^2}$$

$$+ \frac{2}{3}\lambda^2\frac{\partial(\ )}{\partial \xi}\frac{\partial(\ )}{\partial \eta}\frac{\partial^2(\ )}{\partial \xi \partial \eta}.$$

Similarly, we can obtain the fourth, fifth, . . . simultaneous equations. For (3.5), (3.6) and (3.7), we may choose the following polynomial series which is made to satisfy boundary conditions:

$$W_1(\xi, \eta) = (1 - \xi^2)^2(1 - \eta^2)^2(1 + B_1\xi^2 + C_1\eta^2 + D_1\xi^4 + E_1\eta^4$$

$$+ F_1\xi^2\eta^2 + G_1\xi^2\eta^4 + H_1\xi^4\eta^2 + I_1\xi^6 + J_1\eta^6). \tag{3.8}$$

Obviously, (3.8) satisfies the boundary conditions in (3.5) and $W_1(0, 0) = 1$. Substituting (3.8) to (3.5), and making coefficients of the same power index equal on both sides of the equation, we obtain simultaneous linear algebraic equations with unknowns $B_1, C_1, \ldots, I_1, J_1, \alpha_1$. It is not difficult to solve them by computer.

Similarly, for (3.6), we may choose the following polynomial expressions:

$$\begin{aligned} s_2 &= (1-\xi^2)(1-\eta^2)\xi(A_2 + B_2\xi^2 + C_2\eta^2 + D_2\xi^4 + E_2\eta^4 \\ &\quad + F_2\xi^2\eta^2 + G_2\xi^2\eta^4 + H_2\xi^4\eta^2 + I_2\xi^6 + J_2\eta^6), \\ t_2 &= (1-\xi^2)(1-\eta^2)\eta(A_3 + B_3\xi^2 + C_3\eta^2 + D_3\xi^4 + E_3\eta^4 \\ &\quad + F_3\xi^2\eta^2 + G_3\xi^2\eta^4 + H_3\xi^4\eta^2 + I_3\xi^4 + J_3\eta^4). \end{aligned} \tag{3.9}$$

Obviously, s_2, t_2 satisfy the boundary conditions and $s_2(0, 0) = t_2(0, 0) = 0$ (symmetry to the centre of the plate). Substituting s_2, t_2 in (3.6), using the same method, we obtain simultaneous linear algebraic equations with unknowns $A_2, B_2, \ldots, I_2, J_2$; $A_3, B_3, \ldots, I_3, J_3$, and solve them.

For (3.7), we choose W_3 as the following expression:

$$\begin{aligned} W_3(\xi, \eta) &= (1-\xi^2)^2(1-\eta^2)^2(B_4\xi^2 + C_4\eta^2 + D_4\xi^4 + E_4\eta^4 \\ &\quad + F_4\xi^2\eta^2 + G_4\xi^2\eta^4 + H_4\xi^4\eta^2 + I_4\xi^6 + J_4\eta^6). \end{aligned} \tag{3.10}$$

By the same method, we obtain solutions of simultaneous linear algebraic equation with unknowns $B_4, C_4, \ldots, I_4, J_4$. Here we only find results to the third approximation.

4. Relations between the maximum deflection and load

From (3.1) and (3.4a), we obtain

$$24\sqrt{3}(1-\mu^2)a^4q/Eh^4 = \alpha_1 W_0 + \alpha_3 W_0^3.$$

Let $w_0/h = f$, we have

$$\alpha_1 f + 12\alpha_3 f^3 = \lambda^4 b^4 q/Dh. \tag{4.1}$$

Formula (4.1) is the relation between f (ratio of maximum deflection against thickness) and load as λ is finite. For various λ, the values of α_1, α_3 are indicated in Table 1.

5. Formulae for stress

Bending stresses with a distance z from the neutral plane of the plate are:

$$\left.\begin{aligned} \sigma_x'' &= -\frac{Ez}{1-\mu^2}\left(\frac{\partial^2 w}{\partial x^2} + \mu\frac{\partial^2 w}{\partial y^2}\right), \\ \sigma_y'' &= -\frac{Ez}{1-\mu^2}\left(\frac{\partial^2 w}{\partial y^2} + \mu\frac{\partial^2 w}{\partial x^2}\right), \\ \tau_{xy}'' &= -\frac{Ez}{1+\mu}\frac{\partial^2 w}{\partial x \partial y}. \end{aligned}\right\} \tag{5.1}$$

Putting $h = -h/2$ from (5.1), we obtain:

$$\left.\begin{aligned} \sigma''_x &= \frac{Eh}{2(1-\mu^2)}\left(\frac{\partial^2 w}{\partial x^2} + \mu\frac{\partial^2 w}{\partial y^2}\right), \\ \sigma''_y &= \frac{Eh}{2(1-\mu^2)}\left(\frac{\partial^2 w}{\partial y^2} + \mu\frac{\partial^2 w}{\partial x^2}\right), \\ \tau''_{xy} &= \frac{Eh}{2(1+\mu)}\frac{\partial^2 w}{\partial x\partial y}. \end{aligned}\right\} \tag{5.2}$$

$$\left.\begin{aligned} \Sigma'_x(\xi, \eta) &= \frac{12(1-\mu^2)a^2}{Eh^2}\sigma'_x, \\ \Sigma'_y(\xi, \eta) &= \frac{12(1-\mu^2)a^2}{Eh^2}\sigma'_y, \\ \Sigma'_{xy}(\xi, \eta) &= \frac{4\sqrt{3}(1+\mu)ab}{Eh^2}\tau'_{xy}, \end{aligned}\right\} \tag{5.3}$$

$$\left.\begin{aligned} \Sigma''_x(\xi, \eta) &= \frac{4\sqrt{3}(1-\mu^2)a^2}{Eh^2}\sigma''_x, \\ \Sigma''_y(\xi, \eta) &= \frac{4\sqrt{3}(1-\mu^2)a^2}{Eh^2}\sigma''_y, \\ \Sigma''_{xy}(\xi, \eta) &= \frac{4\sqrt{3}(1+\mu^2)ab}{Eh^2}\tau''_{xy}. \end{aligned}\right\} \tag{5.4}$$

Formulae (5.3) are dimensionless expressions for membrane stresses in the neutral plane, and formulae (5.4) are dimensionless expressions for bending stress on the surfaces of the plate. Substituting (3.1) in (2.1) and (5.2), then substituting (3.4a, b, c) in them, (5.3) and (5.4) can have the following relations:

$$\left.\begin{aligned} \Sigma'_x &= \frac{\partial U}{\partial \xi} + \mu\lambda\frac{\partial V}{\partial \eta} + \frac{1}{2}\left[\left(\frac{\partial W}{\partial \xi}\right)^2 + \lambda^2\mu\left(\frac{\partial W}{\partial \eta}\right)^2\right], \\ \Sigma'_y &= \lambda\frac{\partial V}{\partial \eta} + \mu\frac{\partial U}{\partial \xi} + \frac{1}{2}\left[\mu\left(\frac{\partial W}{\partial \xi}\right)^2 + \lambda^2\left(\frac{\partial W}{\partial \eta}\right)^2\right], \\ \Sigma'_{xy} &= \lambda\frac{\partial U}{\partial \eta} + \frac{\partial V}{\partial \xi} + \lambda\frac{\partial W}{\partial \xi}\frac{\partial W}{\partial \eta}, \end{aligned}\right\} \tag{5.5}$$

$$\left.\begin{aligned} \Sigma''_x &= \frac{\partial^2 W}{\partial \xi^2} + \mu\lambda^2\frac{\partial^2 W}{\partial \eta^2}, \\ \Sigma''_y &= \lambda^2\frac{\partial^2 W}{\partial \eta^2} + \mu\frac{\partial^2 W}{\partial \xi^2}, \\ \Sigma''_{xy} &= \lambda\frac{\partial^2 W}{\partial \xi\partial \eta}. \end{aligned}\right\} \tag{5.6}$$

Substituting the third approximation in (5.5), (5.6), we obtain the appropriate expressions for membrane stress as in the neutral plane, and bending stresses on the surface of the plate:

$$\begin{aligned}
\Sigma'_x &= W_0^2\left(\frac{\partial s_2}{\partial \xi} + \mu\lambda\frac{\partial t_2}{\partial \eta}\right) + \tfrac{1}{2}W_0^2\left\{\left(\frac{\partial W_1}{\partial \xi}\right)^2 + \mu\lambda^2\left(\frac{\partial W_1}{\partial \eta}\right)^2\right. \\
&\quad + 2W_0^2\left(\frac{\partial W_1}{\partial \xi}\frac{\partial W_3}{\partial \xi} + \mu\lambda^2\frac{\partial W_1}{\partial \eta}\frac{\partial W_3}{\partial \eta}\right) \\
&\quad \left. + W_0^4\left[\left(\frac{\partial W_3}{\partial \xi}\right)^2 + \mu\lambda^2\left(\frac{\partial W_3}{\partial \eta}\right)^2\right]\right\}, \\
\Sigma'_y &= W_0^2\left(\mu\frac{\partial s_2}{\partial \xi} + \lambda\frac{\partial t_2}{\partial \eta}\right) + \tfrac{1}{2}W_0^2\left\{\mu\left(\frac{\partial W_1}{\partial \xi}\right)^2 + \lambda^2\left(\frac{\partial W_1}{\partial \eta}\right)^2\right. \\
&\quad + 2W_0^2\left(\mu\frac{\partial W_1}{\partial \xi}\frac{\partial W_3}{\partial \xi} + \lambda^2\frac{\partial W_1}{\partial \eta}\frac{\partial W_3}{\partial \eta}\right) \\
&\quad \left. + W_0^4\left[\mu\left(\frac{\partial W_3}{\partial \xi}\right)^2 + \lambda^2\left(\frac{\partial W_3}{\partial \eta}\right)^2\right]\right\}, \\
\Sigma'_{xy} &= W_0^2\left(\lambda\frac{\partial s_2}{\partial \eta} + \frac{\partial t_2}{\partial \xi}\right) + \lambda\, W_0^2\left[\frac{\partial W_1}{\partial \xi}\frac{\partial W_1}{\partial \eta}\right. \\
&\quad \left. + W_0^2\left(\frac{\partial W_1}{\partial \xi}\frac{\partial W_3}{\partial \eta} + \frac{\delta W_1}{\partial \eta}\frac{\partial W_3}{\partial \xi}\right) + W_0^4\frac{\partial W_3}{\partial \xi}\frac{\partial W_3}{\partial \eta}\right],
\end{aligned} \qquad (5.7)$$

$$\left.\begin{aligned}
\Sigma''_x &= W_0\left(\frac{\partial^2 W_1}{\partial \xi^2} + \mu\lambda^2\frac{\partial^2 W_1}{\partial \eta^2}\right) + W_0^3\left(\frac{\partial^2 W_3}{\partial \xi^2} + \mu\lambda^2\frac{\partial^2 W_3}{\partial \eta^2}\right), \\
\Sigma''_y &= W_0\left(\mu\frac{\partial^2 W_1}{\partial \xi^2} + \lambda^2\frac{\partial^2 W_1}{\partial \eta^2}\right) + W_0^3\left(\mu\frac{\partial^2 W_3}{\partial \xi^2} + \lambda^2\frac{\partial^2 W_3}{\partial \eta^2}\right), \\
\Sigma''_{xy} &= \lambda W_0\frac{\partial^2 W_1}{\partial \xi\partial \eta} + \lambda W_0^3\frac{\partial^2 W_3}{\partial \xi\partial \eta}.
\end{aligned}\right\} \qquad (5.8)$$

The total stresses on any point of the surface of the plate are:

$$\left.\begin{aligned}
S_x(\xi, \eta) &= \Sigma'_x + \Sigma''_x, \\
S_y(\xi, \eta) &= \Sigma'_y + \Sigma''_y, \\
S_{xy}(\xi, \eta) &= \Sigma'_{xy} + \Sigma''_{xy}.
\end{aligned}\right\} \qquad (5.9)$$

Substituting the third approximation of W_1, W_3, s_2, t_2 in (5.7), (5.8) and (5.9), we obtain the total stresses at any point of the surface of the plate. Choosing $\mu = 1/3$, we find the expressions for total stresses at (0, 0), (0, 1), (1, 0) to be the following:

$$\left.\begin{aligned}\Sigma_x'(0,0) &= 12f^2(\beta_1 + \tfrac{1}{3}\lambda\beta_2),\\ \Sigma_y'(0,0) &= 12f^2(\tfrac{1}{3}\beta_1 + \lambda\beta_2),\\ \Sigma_x''(0,0) &= 4\sqrt{3}f[\beta_3 + \tfrac{1}{3}\lambda^2\beta_4 + 12f^2(\beta_5 + \tfrac{1}{3}\lambda^2\beta_6],\\ \Sigma_y''(0,0) &= 4\sqrt{3}f[\tfrac{1}{3}\beta_2 + \lambda^2\beta_4 + 12f^2(\tfrac{1}{3}\beta_5 + \lambda^2\beta_6),\end{aligned}\right\} \tag{5.10}$$

$$\left.\begin{aligned}\Sigma_x'(0,1) &= -8\lambda\beta_7 f^2,\\ \Sigma_y'(0,1) &= -24\beta_7 f^2,\\ \Sigma_x''(0,1) &= \frac{16\sqrt{3}}{3}\lambda^2(\beta_8 f + 12\beta_9 f^3),\\ \Sigma_y''(0,1) &= 16\sqrt{3}\lambda^2(\beta_8 f + 12\beta_9 f^3),\end{aligned}\right\} \tag{5.11}$$

$$\left.\begin{aligned}\Sigma_x'(1,0) &= -24\beta_{10} f^2,\\ \Sigma_y'(1,0) &= -8\beta_{10} f^2,\\ \Sigma_x''(1,0) &= 16\sqrt{3}(\beta_{11} f + 12\beta_{12} f^3),\\ \Sigma_y''(1,0) &= \frac{16\sqrt{3}}{3}(\beta_{11} f + 12\beta_{12} f^3).\end{aligned}\right\} \tag{5.12}$$

All the β_i ($i = 1, 2, \ldots, 12$) for various λ are shown in Table 1. The maximum stress is at the middle point of the long edge, i.e. (0, 1). The total stress at this point is

$$S_y(0,1) = \Sigma_y'(0,1) + \Sigma_y''(0,1).$$

From (5.11) and Table 1, we can find the numerical values of maximum stress.

6. Equations and formulae for stress at $\lambda = \infty$

We know that since $\lambda = a/b$, as $\lambda = 0$, it is signified that one edge of the rectangular plate $a = 0$, and similarly, as $\lambda = \infty$, it is signified that $b = 0$. We also know that $\xi = x/a$, $\eta = y/b$, and therefore, we only require to have differential equations and calculation formulae of stress as $\lambda = 0$; then exchanging $\xi \to \eta$, $s_2 \leftrightarrow t_2$, we can obtain the required equations and formulae for stress.

From (3.5), as $\lambda = 0$, we have

$$\frac{\partial^4 W_1}{\partial \xi} = \alpha_1.$$

Then making the exchange $\xi \to \eta$, we obtain the first approximate equation and

Table 1

$\lambda = \frac{a}{b}$	α_1	α_3	Present w_0 (b^4q/D)	Timoshenko, S. w_0 (b^4q/D)	β_1	β_2	β_3	β_4	β_5	β_6
1.0	49.6114	2.00167	0.00126	0.00126	0.522667	0.533160	−1.743792	−1.743792	0.022183	0.022256
1.1	60.9514	2.46311	0.00150	0.00150	0.488868	0.607776	−1.673231	−1.800758	0.024087	0.020502
1.2	75.4987	3.08427	0.00172	0.00172	0.448668	0.679381	−1.594615	−1.843034	0.026494	0.019239
1.3	93.8683	3.89687	0.00190	0.00191	0.404051	0.748478	−1.508529	−1.875180	0.029697	0.018309
1.4	116.753	4.94013	0.00206	0.00207	0.358500	0.815270	−1.415932	−1.900124	0.003986	0.017618
1.5	114.924	6.26539	0.00218	0.00220	0.310965	0.879752	−1.318122	−1.919791	0.039525	0.017115
1.6	179.231	7.93943	0.00229	0.00230	0.285058	0.941832	−1.21669	−1.935480	0.046244	0.016773
1.7	220.603	10.0448	0.00237	0.00238	0.268665	1.00146	−1.113322	−1.948101	0.053798	0.016579
1.8	270.039	12.6760	0.00243	0.00245	0.272148	1.05877	−1.009900	−1.958304	0.061616	0.016524
1.9	328.614	15.9330	0.00248	0.00249	0.297506	1.11409	−0.90819	−1.966572	0.069018	0.016594
2.0	397.463	19.9130	0.00252	0.00254	0.343618	1.16804	−0.80982	−1.973274	0.074373	0.016800
∞	24.0	1.46281	0.00260	0.00260	0	0.60952	0	−2	−0.0000039	0.020319

Table 1. Continued

$\lambda = \frac{a}{b}$	β_7	β_8	β_9	β_{10}	β_{11}	β_{12}
1.0	−0.73057	1.265202	0.023268	−0.769109	1.265202	0.023131
1.1	−0.62263	1.202255	0.022673	−1.060624	1.345728	0.022478
1.2	−0.55458	1.156729	0.021701	−1.427148	1.438937	0.021224
1.3	−0.51400	1.122824	0.020650	−1.854689	1.545557	0.020276
1.4	−0.49302	1.096976	0.019674	−2.312588	1.666087	0.021027
1.5	−0.48673	1.076914	0.018848	−2.752925	1.800741	0.025141
1.6	−0.49203	1.061140	0.018205	−3.113990	1.949399	0.034211
1.7	−0.50681	1.048620	0.017759	−3.327717	2.111593	0.049428
1.8	−0.52934	1.038628	0.017507	−3.329982	2.286489	0.071328
1.9	−0.55811	1.030627	0.017430	−3.071592	2.472894	0.099681
2.0	−0.59154	1.024217	0.017530	−2.52710	2.669319	0.133597
∞	−0.30472	1	0.020320	0	9.000000	−0.200657

boundary condition in the case of $\lambda = \infty$:

$$\left.\begin{aligned} \frac{\partial^4 W_1}{\partial \eta^4} &= \alpha_1, \\ W_1 &= \frac{\partial W_1}{\partial \eta} = 0, \quad \text{at } \eta = \pm 1. \end{aligned}\right\} \tag{6.1}$$

Similarly, we can obtain the second and third approximate equations as the following:

$$\left.\begin{aligned} \frac{\partial^2 t_2}{\partial \eta^2} &= -\frac{\partial W_1}{\partial \eta}\frac{\partial^2 W_1}{\partial \eta^2}, \quad \frac{\partial^2 s_2}{\partial \eta^2} = 0 \\ t_2 &= s_2 = 0, \quad \text{at } \eta = \pm 1, \end{aligned}\right\} \tag{6.2}$$

$$\left.\begin{aligned} \frac{\partial^4 W_3}{\partial \eta^4} &= \alpha_3 + \frac{\partial t_2}{\partial \eta}\frac{\partial^2 W_1}{\partial \eta^2} + \frac{1}{2}\left(\frac{\partial W_1}{\partial \eta}\right)^2 \frac{\partial^2 W_1}{\partial \eta^2}, \\ W_3 &= \frac{\partial W_3}{\partial \eta} = 0, \quad \text{at } \eta = \pm 1. \end{aligned}\right\} \tag{6.3}$$

For (6.1), (6.2) and (6.3), choosing polynomial functions similar to (3.8), (3.9) and (3.10), the method of solving the equations is the same as the previous and the results are shown in Fig. 1. Similarly for treating formulae for stress, we can obtain the following at points (0, 0) and (0, 1):

$$\left.\begin{aligned} \Sigma_x'(0, 0) &= \tfrac{1}{3}\beta_2 W_0^2, \\ \Sigma_y'(0, 0) &= \beta_2 W_0^2, \\ \Sigma_x''(0, 0) &= \tfrac{2}{3}\beta_4 W_0 + \tfrac{2}{3}\beta_6 W_0^3, \\ \Sigma_y''(0, 0) &= 2\beta_4 W_0 + 2\beta_6 W_0^3, \end{aligned}\right\} \tag{6.4}$$

$$\left.\begin{aligned} \Sigma_x'(0, 1) &= -\tfrac{2}{3}\beta_7 W_0^2, \\ \Sigma_y'(0, 1) &= -2\beta_7 W_0^2, \\ \Sigma_x''(0, 1) &= \tfrac{8}{3}\beta_8 W_0 + \tfrac{8}{3}\beta_8 W_0^3, \\ \Sigma_y''(0, 1) &= 8\beta_9 W_0 + 8\beta_9 W_0^3, \end{aligned}\right\} \tag{6.5}$$

The maximum stress is at point (0, 1)

$$S_y(0, 1) = \Sigma_y'(0, 1) + \Sigma_y''(0, 1). \tag{6.6}$$

7. Discussion of results

In this paper, we put $\mu = \frac{1}{3}$, but use $\mu = 0.3$ in the other writings.

(1) The first set of approximate results of maximum deflection in this paper, compared with S. Timoshenko and Woinowsky-Krieger, S. [2], are shown in Table 1. The two sets of results are quite close.

(2) The third set of approximate results are compared with the graphs of the above mentioned book (pp. 451–453) in the caes of $\lambda = 1$; 1.5; 2.0; ∞. The maximum relative error of these two sets of results is less than 6%.

Compared with formula (2.155) in p. 101 of Volmir's work [3] in the cases $\lambda = 1.5$; 1.6, the results of numerical calculation show that the present perturbation solution is relatively greater than (2.155) by 2–3%, and that Levy's series solution [4] is relatively greater than (2.155) by 3%, which means that the present solution is closer to Levy's.

As $\lambda = 1, \infty$, changing the coordinates in Volmir's book [3] into those shown in Fig. 1, the formulae of these two sets of results can be compared in the following:

$\lambda = 1$:

$$26.9f^3 + 53.5f = b^4q/Dh \quad \text{(Volmir)},$$

$$24f^3 + 49.5f = b^4q/Dh \quad \text{(present)},$$

$\lambda = \infty$:

$$24f + 16.8f^3 = b^4q/Dh \quad \text{(Volmir)},$$

$$24f + 17.5f^3 = b^4q/Dh \quad \text{(present)},$$

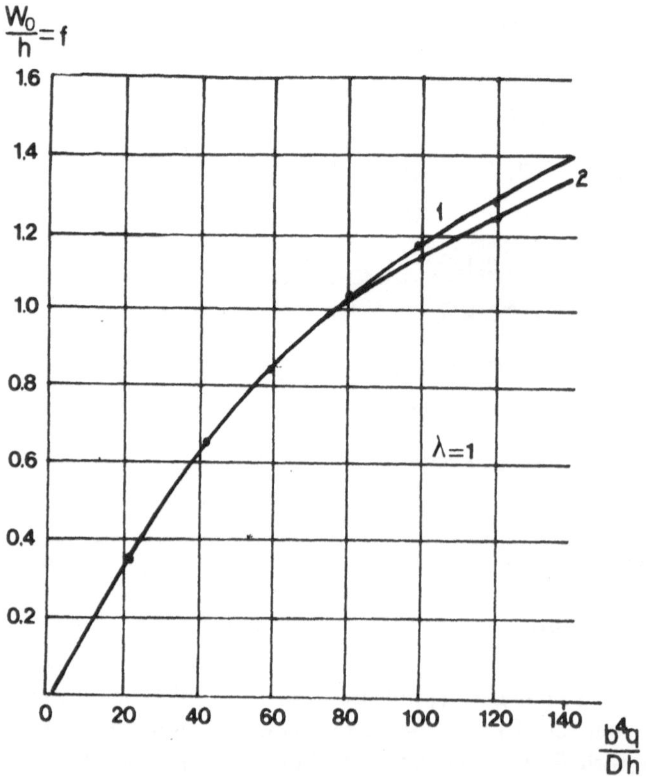

Figure 2

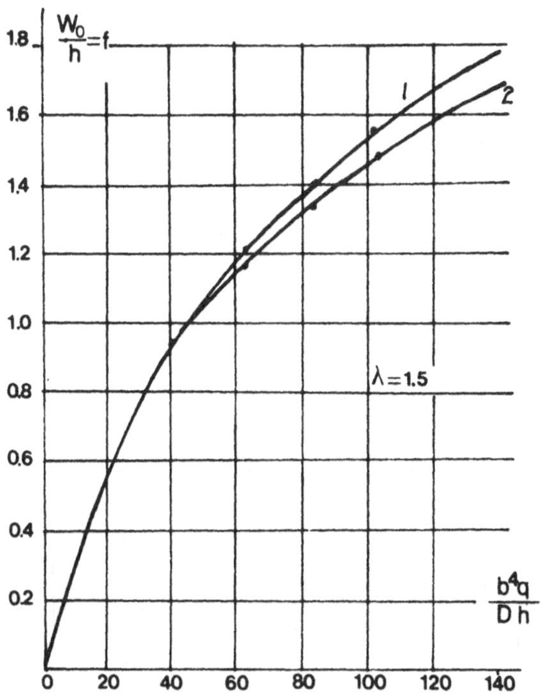

Figure 3

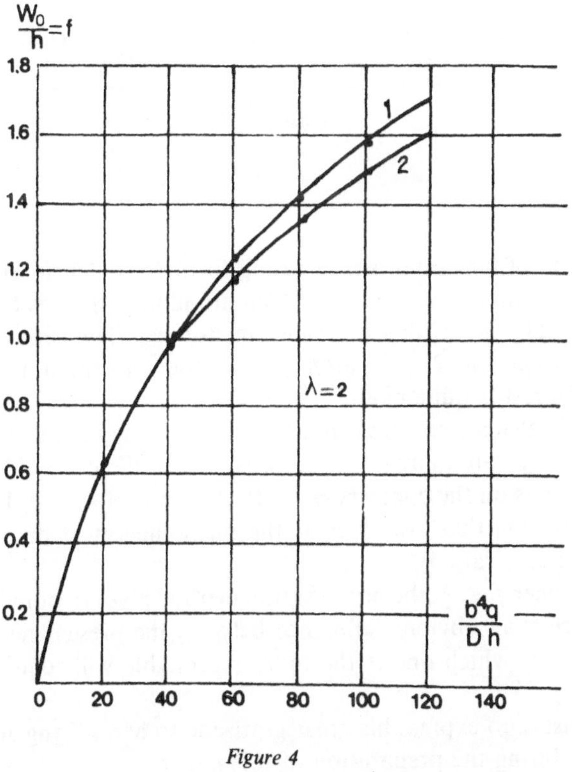

Figure 4

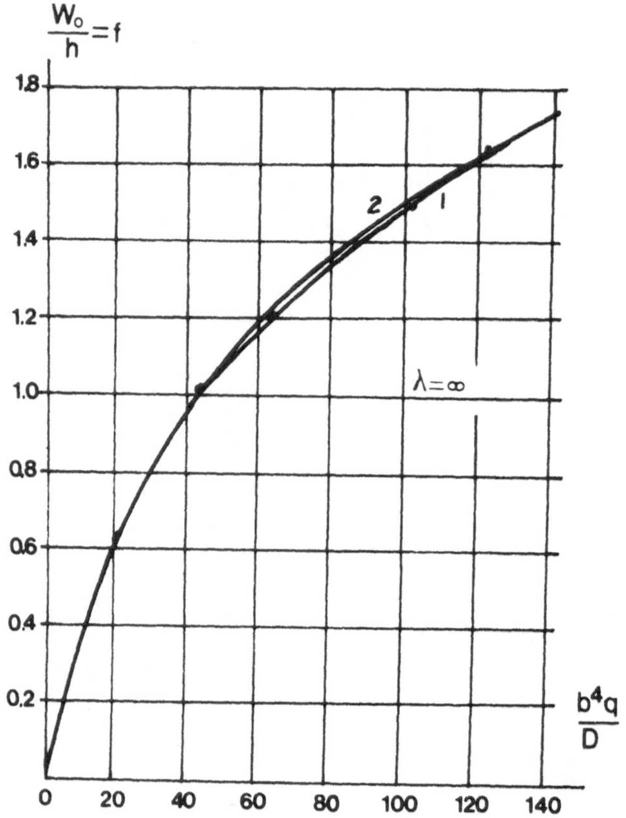

Figure 5

(3) Comparison of maximum stress: Using formulae (5.11, (6.5) and (6.6) and comparing the maximum stress curves which are plotted in Figs 6–9 in the cases of $\lambda = 1$; 1.5; 2.0; ∞ with the above-mentioned Timoshenko's work [2, pp. 452–453]; except for $\lambda = 1$, after $f > 1$, the maximum relative error is 10–20%, and the rest is quite close.

(4) Numerical calculation shows that when $f > 1$, the relative error increases; and when $f \leqslant 1$, they are more or less consistent. On the other side, the curve in the case of $\lambda = \infty$ is on the contrary beneath the cases of $\lambda = 2$; 1.5 (see Figs 10 and 11 which indicate that after $f > 1$, the phenomenon of fluctuation due to poorer convergence is raised).

In conclusion when $f \leqslant 1$, the perturbation method gives reasonable results, but after $f > 1$, there is an obvious difference between the present results and those of others. However, which one is the more reasonable will require still further investigation.

The author wishes to express his great gratitude to Mr. Chang Jun-yan for his laborious effort during the preparation of this paper.

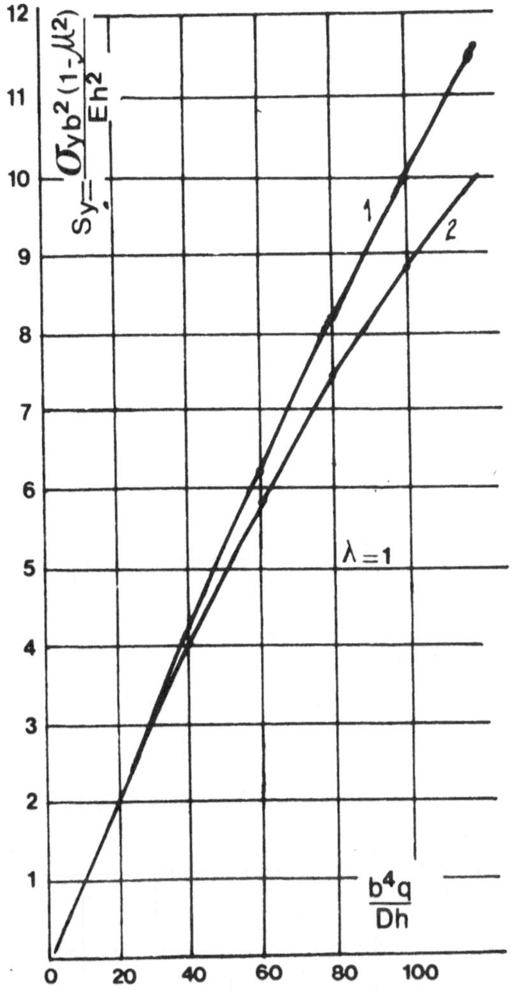

Figure 6

References

1. Chien Wei-zang, Li Hung-sun, Hu Hai-chang, and Yeh Kai-yuan, *Large Deflection of Thin Circular Plates*, Science Press, Beijing (1954).
2. Timoshenko, S. and Woinowsky-Krieger, S., *Theory of Plates and Shells* (Chinese Translation), Science Press, Beijing (1977), 213; 452–453.
3. Volmir, A. S., *Flexible Plates and Flexible Shells* (Chinese Translation), Science Press, Beijing (1963), 73; 101.
4. Levy, S., "Bending of rectangular plates with large deflections", *NACA Rep.*, No. 737 (1942).
5. Way, S., "Uniformly loaded clamped rectangular plates with large deflections", *Proc. Fifth Intern. Congr. Appl. Mech.*, Cambridge, Mass. (1983), 123.

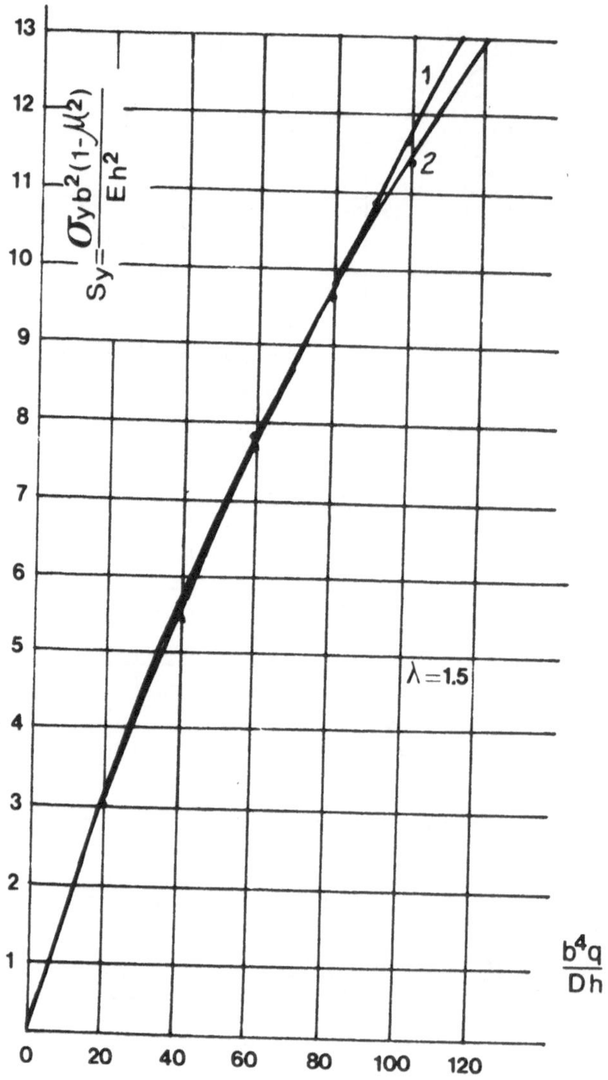
13
12
11
10
9
8
7
6
5
4
3
2
1
0
20
40
60
80
100
120
$S_y = \frac{\sigma_y b^2 (1-\mu^2)}{E h^2}$
1
2
λ=1.5
$\frac{b^4 q}{D h}$

Figure 7

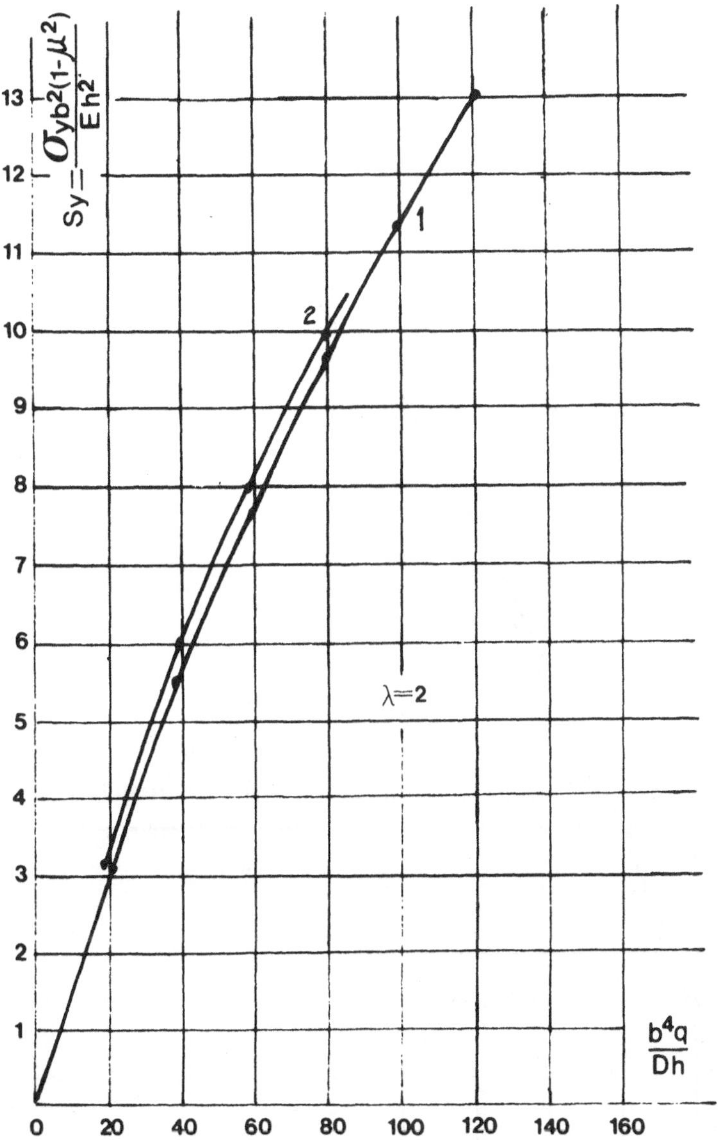

$S_y = \frac{\sigma_y b^2(1-\mu^2)}{Eh^2}$
13
12
11
10
9
8
7
6
5
4
3
2
1
0
20
40
60
80
100
120
140
160
1
2
λ=2
$\frac{b^4q}{Dh}$

Figure 8

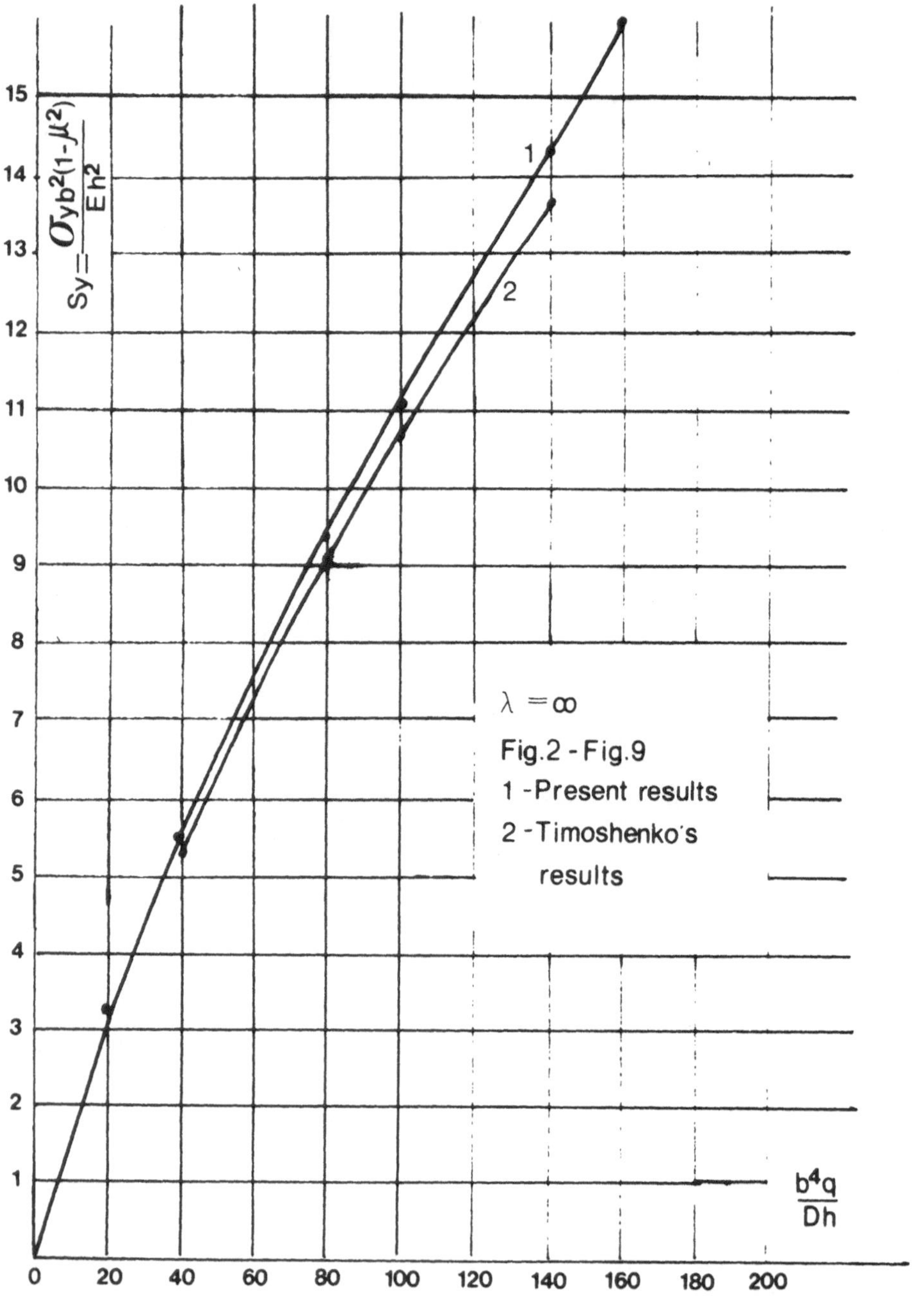

$S_y = \frac{\sigma_y b^2 (1-\mu^2)}{E h^2}$
15
14
13
12
11
10
9
8
7
6
5
4
3
2
1
1
2
λ =∞
Fig.2 - Fig.9
1 - Present results
2 - Timoshenko's results
$\frac{b^4 q}{Dh}$
0
20
40
60
80
100
120
140
160
180
200

Figure 9

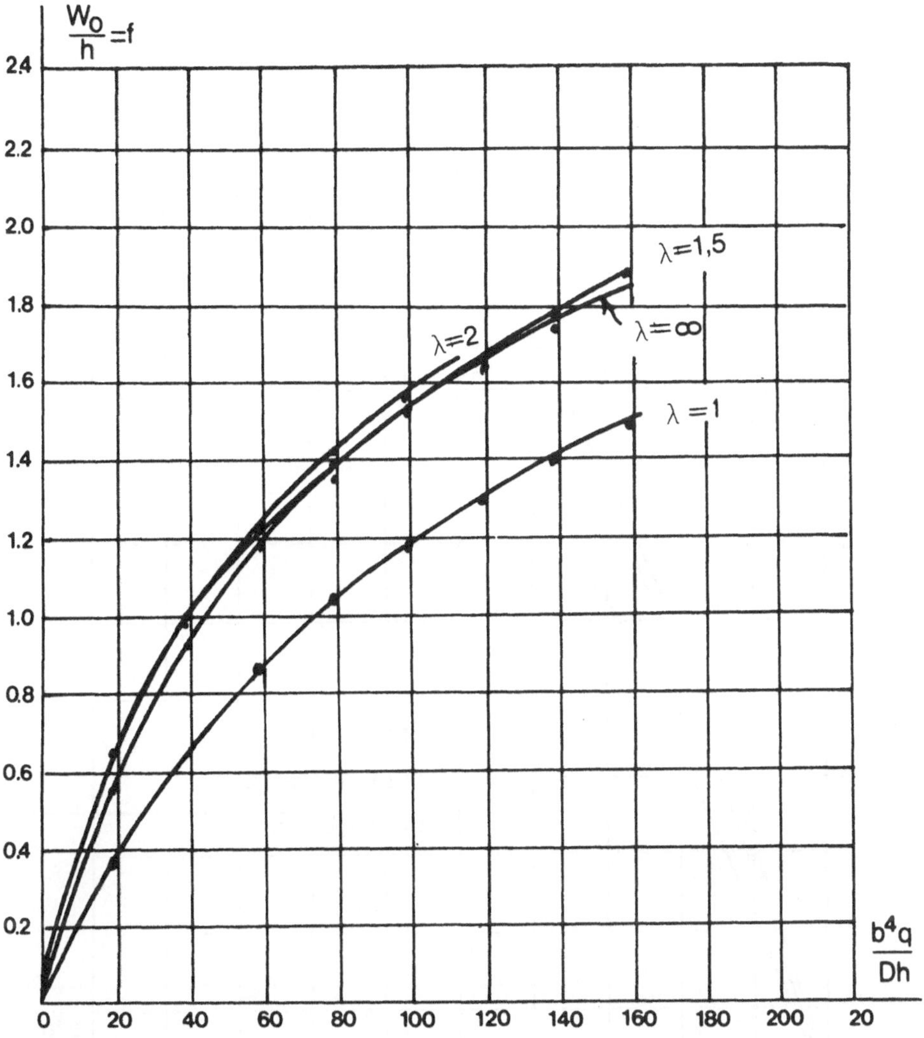

$\frac{W_0}{h}=f$
2.4
2.2
2.0
1.8
1.6
1.4
1.2
1.0
0.8
0.6
0.4
0.2
λ=1,5
λ=2
λ=∞
λ =1
$\frac{b^4q}{Dh}$
0
20
40
60
80
100
120
140
160
180
200
20

Figure 10

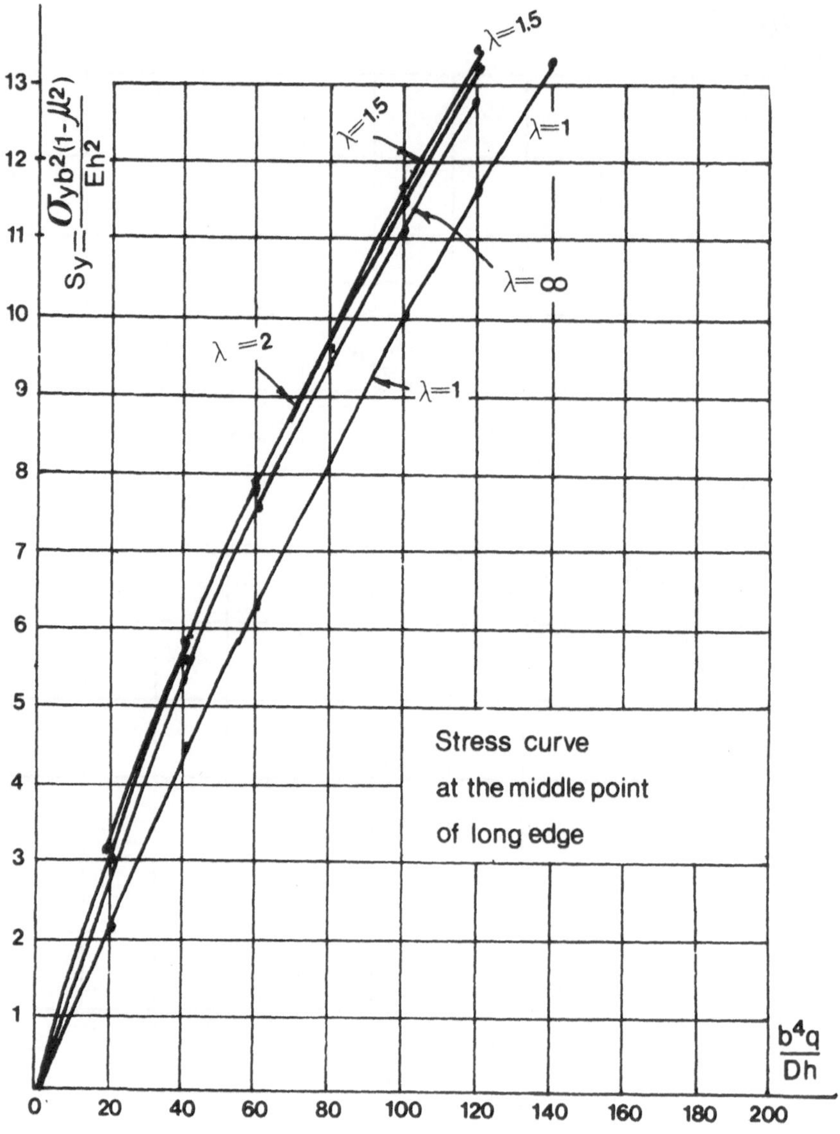

$Sy=\frac{\sigma_y b^2(1-\mu^2)}{Eh^2}$
13
12
11
10
9
8
7
6
5
4
3
2
1
0
20
40
60
80
100
120
140
160
180
200
$\frac{b^4q}{Dh}$
λ=1.5
λ=1.5
λ=1
λ=∞
λ=2
λ=1
Stress curve
at the middle point
of long edge

Figure 11

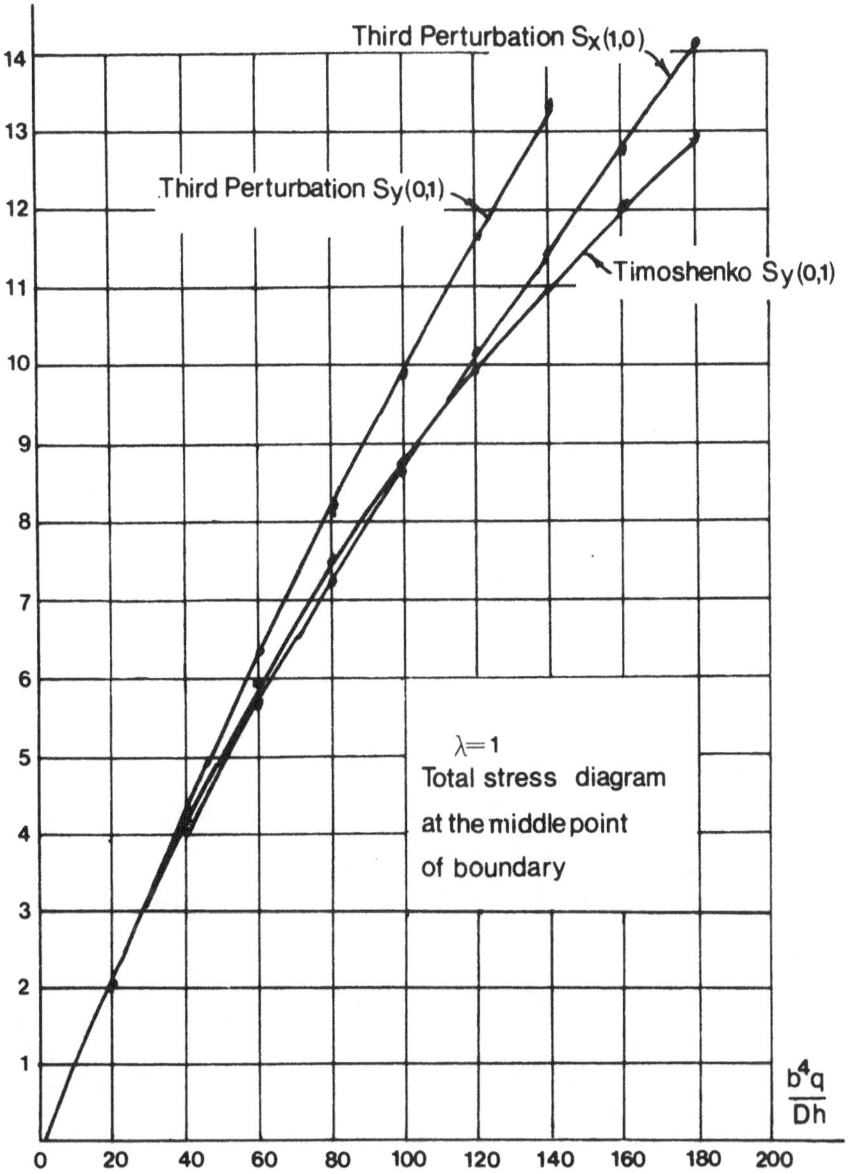

Third Perturbation Sx(1,0)
Third Perturbation Sy(0,1)
Timoshenko Sy(0,1)
λ=1
Total stress diagram
at the middle point
of boundary
14
13
12
11
10
9
8
7
6
5
4
3
2
1
0
20
40
60
80
100
120
140
160
180
200
b⁴q / Dh

Figure 12

The large deflection method in the analysis of elasto-plastic instability of thin shells with initial imperfections

ZHOU CHENG-TI

Dalian University, Dalian, PRC

Abstract. The theory of the instability of thin elasto-plastic shells with initial imperfections is wrought with great difficulties in respect of the complexity of geometrical and physical nonlinearities. This paper proposes a general method for analyzing the instability of stiffened thin elasto-plastic shells, with initial imperfections, under hydrostatic pressure. The nonlinearity of the inelastic shell buckling problems is linearized by a series of anisotropic elastic shell buckling problems; the process of linearization, the computation of the bifurcation point and the limit load of the buckled shell are completed by a general computer programme. This method has been applied to various shell structure problems such as the general instability of ring-reinforced cylindrical shells; the local buckling of shell plates between stiffeners; the buckling of stiffened conical shells and stiffened cylindrical panels; and the buckling of cylindrical shells with nonhomogeneity in the ring-stiffeners and the shell-plates. Many experiments are made, and the analytical results are confirmed with the experimental data. This proposed computer programme has been used in various engineering structural designs, such as the design of pressure hulls of submarines and the design of deep living torpedo shells. The general idea of this theory and the computational method can also be used in the design of some aeronautical structures. Recently, this method has been successfully used to solve the nonlinear buckling problems of cylindrical shells of composite materials with initial imperfections. With these buckling problems, not only the geometrical nonlinearities of shells, but also the physical nonlinearities of the composite materials are considered. This theory and the large deflection method have been introduced and discussed in detail in the recently published book *Theory of stability of elasto-plastic thin shells* [1].

1. Introduction

As early as the 1950's, Professor Qian Wei-chan proposed the perturbation method of large deflection theory for the analysis of instability of thin shells. A number of difficult shell buckling problems were solved by this method. According to the research direction oriented by Prof. Qian, the author successfully used the large deflection method to solve the elasto-plastic instability of thin shells with initial imperfections. Many engineering problems were investigated by this method and all showed good results. Therefore, up to now, the large deflection theory in the investigation of the instability of thin shells is still an important fundamental research field.

The research of nonlinear instability of thin shells has been concentrated on the nature of large deflection post-buckling equilibrium configurations over the last

Yeh, K. Y. (ed) Progress in Applied Mechanics
 ISBN-13: 978-94-010-8061-3

few decades. Now, it is clear that there may exist a kind of equilibrium configuration corresponding to a minimum load, but this minimum load is not a critical load of buckling. These investigations, however, are still not sufficient for the explanation of the differences between the results of classical theory and the experimental data. After the 1960's, Stein suggested a new theory, called "Nonlinear prebuckling consistent theory" [2, 3]. The theoretical results of Stein's theory approach the value of experimental data obtained from the so called "perfect model" tests. This explains the nature and causes of the differences between the results of the linear classical theory and the test data prior to the 1960's. But Stein's theory is still a stability theory founded on ideal perfect shells. Moreover, Koiter proposed a theory of sensitivity based on initial imperfections [4]. He concentrated his investigations on the nature of initial postbuckling behaviour direct near a critical point, and he used this point of view to explain the influence of initial imperfections on the buckling behaviour of shells. Hutchinson [5] extended Koiter's theory to the plastic buckling problems, by discussing a general bifurcation criterion of the DMV theory, and solved several numerical classic examples in doing so. But, nowadays, obtaining a general solution of the elasto-plastic instability of stiffened thin shells with initial imperfections is still a difficult problem.

From the above discussion, though different theories of instability give different concepts and analyses about the nature of prebuckling and postbuckling behaviour, they all provide approvable ideas about the value of the small deflection linear solution of buckling problems. Therefore, in this paper we first find a small deflection solution of the elasto-plastic shell buckling problem in an analytical form, without considering the influence of initial imperfections of the shell. If the material of the shell is ideal elastic, the solution will degenerate and conform with the classical linear elastic solution. When the shell material is a kind of real elasto-plastic material, this solution will represent the influence of the physical nonlinearity of this shell material during elasto-plastic buckling. Then, in the second stage, the initial imperfections are considered, and we use the methods of inelastic nonlinear large deflection theory to compute a series of nonlinear load-displacement curves during the elasto-plastic buckling of the shell. This set of nonlinear curves gives the description of the interactions between the influences of geometrical nonlinearities and the physical nonlinearities. From this set of nonlinear displacement curves, the maximum limit load can be determined. This limit load is the critical buckling load of the elasto-plastic shell with specified initial imperfections.

2. Elasto-plastic constitutive relations

Suppose the shell is made of an isotropic hardening material and is compressible; we will set up the fundamental stability equations of elasto-plastic shell buckling, according to Shanley's continuous loading criterion of instability. According to our previous works [1, 6, 7], the elasto-plastic constitutive relations can be formed from the Prandtl–Reuss equations of incremental theory or by the Hencky-Nadai relations of deformation theory, as follows:

2.1. *Incremental theory*

Prandtl–Reuss equations:

$$d\varepsilon_{ij} = [(ds_{ij}/2G) + (1 - 2\nu)\delta_{ij}(d\sigma_0/E)] + h(J_2)\, dJ_2 S_{ij}. \tag{2.1}$$

Here, G, E, ν, are elastic moduli, δ_{ij} is a Kronecker δ, $d\sigma_0$ is the average stress increment, S_{ij} is the stress deviator, and ds_{ij} is its increment. Also dJ_2 is:

$$dJ_2 = [S_x\, d\sigma_x + S_y\, d\sigma_y + S_z\, d\sigma_z + 2(\tau_{yz}\, d\tau_{yz} + \tau_{zx}\, d\tau_{zx} + \tau_{xy}\, d\tau_{xy})].$$

The function $h(J_2)$ is determined by uniaxial tension or compression tests, and it is:

$$h(J_2) = \frac{3}{4}\left(\frac{1}{E_t^0} - \frac{1}{E}\right)\frac{1}{J_2} = (9/4\sigma_i^2)\left(\frac{1}{E_t^0} - \frac{1}{E}\right), \tag{2.2}$$

where σ_i is the stress intensity (or effective stress) and $E_t^0 = d\sigma_x/d\varepsilon_x$ is the tangent modulus of the shell material determined by simple tension or compression tests. If we suppose the tangent modulus $E_t = d\sigma_i/d\varepsilon_i$, and if the influence of the compressibility of the shell material can be ignored, then we obtain $E_t^0 = E_t = d\sigma_i/d\varepsilon_i$. Substituting equation (2.2) into (2.1), we get the constitutive relations between the increments of stress and strain during the instant of instability, i.e.

$$d\varepsilon_{ij} = [a_{ij}]\, d\sigma_{ij}. \tag{2.3}$$

In physical meaning, these relations represent the nature of anisotropy of the stress field and strain field during the instant of elasto-plastic buckling. In the case of thin shells, the coefficients $[a_{ij}]$ are:

$$\begin{aligned}
a_{11} &= E^{-1} + \tilde{E}_t^{-1} + \tilde{E}_t^{-1}(\sigma_x - 0.5\sigma_y), & a_{22} &= E^{-1} + \tilde{E}_t^{-1}(\sigma_y - 0.5\sigma_x),\\
a_{12} &= -\nu\tilde{E}^{-1} + \tilde{E}_t^{-1}(\sigma_x - 0.5\sigma_y) & a_{26} &= 3\tilde{E}_t^{-1}(\sigma_y - 0.5\sigma_x)\tau_{xy},\\
&\quad \times(\sigma_y - 0.5\sigma_x), & &\\
a_{16} &= 3\tilde{E}_t^{-1}(\sigma_x - 0.5\sigma_y)\tau_{xy}, & a_{66} &= G^{-1} + 9\tilde{E}_t^{-1}\tau_{xy}^2.
\end{aligned} \tag{2.4}$$

Here $\tilde{E}_t^{-1} = \sigma_i^{-2}[(E_t^0)^{-1} - E^{-1}]$. For the three dimensional cases, the coefficients a_{ij} can also be derived similarly (see [1] in detail).

The inverse form of the relation (2.3) is

$$d\sigma_{ij} = [b_{ij}]\, d\varepsilon_{ij}. \tag{2.5}$$

For the case of thin shells, let Δ be the determinant of a_{ij}, then the coefficients b_{ij} are

$$\begin{aligned}
b_{11} &= (a_{22}a_{66} - a_{26}^2)\Delta^{-1}, & b_{22} &= (a_{11}a_{66} - a_{16}^2)\Delta^{-1},\\
b_{12} &= (a_{26}a_{61} - a_{21}a_{66})\Delta^{-1}, & b_{26} &= (a_{61}a_{12} - a_{11}a_{62})\Delta^{-1},\\
b_{16} &= (a_{21}a_{62} - a_{61}a_{22})\Delta^{-1}, & b_{66} &= (a_{11}a_{22} - a_{12}^2)\Delta^{-1}.
\end{aligned} \tag{2.6}$$

2.2. *Deformation theory*

According to the Hencky–Nadai equations:

$$\varepsilon_{ij} = [S_{ij}(2G)^{-1} + (1 - 2\nu)\delta_{ij}\sigma_0 E^{-1}] + g(\sigma_i)S_{ij}. \tag{2.7}$$

Here, the function $g(\sigma_i)$ is determined by simple uniaxial tension or compression tests. It is

$$g(\sigma_i) = (3/2)[(E_s^0)^{-1} - E^{-1}]. \tag{2.8}$$

When considering the compressibility of the material, the relations between the secant moduli $E_s = \sigma_i/\varepsilon_i$ and $E_s^0 = \sigma_x/\varepsilon_x$ are:

$$E_s^{-1} = (E_s^0)^{-1} - (1 - 2\nu)(3E)^{-1}. \tag{2.9}$$

It has been proved in [6] that the above equation (2.9) is equivalent to the following equations:

$$\mu = 0.5 - (0.5 - \nu)E_s^0 E^{-1}. \tag{2.10}$$

This is the Gerard–Wildhorn formula, in which μ is a variable Poisson's ratio during plastic deformation. From plastic buckling, the relations are:

$$\mathrm{d}\varepsilon_{ij} = \left(\frac{\partial \varepsilon_{ij}}{\partial \sigma_{ij}}\right) \mathrm{d}\sigma_{ij}. \tag{2.11}$$

Here, the coefficients $[a_{ij}]$ that belong to the deformation theory are

$$[a_{ij}] = \left(\frac{\partial \varepsilon_{ij}}{\partial \sigma_{ij}}\right). \tag{2.12}$$

In the case of thin shells, the coefficients a_{ij} are

$$\begin{aligned}
a_{11} &= \partial\varepsilon_x/\partial\sigma_x = \tilde{E}_s^{-1}(\sigma_x - 0.5\sigma_y)^2 + (E_s^0)^{-1},\\
a_{12} &= \partial\varepsilon_x/\partial\sigma_y = -\nu E^{-1} + \tilde{E}^{-1}(\sigma_x - 0.5\sigma_y)(\sigma_y - 0.5\sigma_x)\\
&\qquad -0.5[(E_s^0)^{-1} - E^{-1}],\\
a_{16} &= \partial\varepsilon_x/\partial\tau_{xy} = 3\tilde{E}_s^{-1}(\sigma_x - 0.5\sigma_y)\tau_{xy},\\
a_{22} &= \partial\varepsilon_y/\partial\sigma_y = \tilde{E}_s^{-1}(\sigma_y - 0.5\sigma_x)^2 + (E_s^0)^{-1},\\
a_{26} &= \partial\varepsilon_y/\partial\tau_{xy} = 3\tilde{E}_s^{-1}(\sigma_y - 0.5\sigma_x)\tau_{xy},\\
a_{66} &= \partial\gamma_{xy}/\partial\tau_{xy} = 2(1 + \nu)E^{-1} + 9\tilde{E}_s^{-1}\tau_{xy}^2 + 3[(E_s^0)^{-1} - E^{-1}],
\end{aligned} \tag{2.13}$$

where $\tilde{E}_s^{-1} = (E_t^{-1} - E_s^{-1})\sigma_i^{-2}$. For three dimensional cases, the a_{ij} have been derived in [1].

3. Elasto-plastic buckling in small deflection theory

Consider a ring-stiffened cylindrical shell under external hydrostatic pressure. Assume there is no bending moment during the prebuckling state. Then, the internal forces are

$$N_x = 0.5qR, \quad \sigma_x = 0.5qR/h, \quad N_y = qR, \quad \sigma_y = qR/h' = \gamma qR/h. \tag{3.1}$$

Here R is the radius of the cylindrical shell, q is the hydrostatic pressure, h' is an imaginary thickness of the shell when considering the stiffness, and γ is the reduction factor of the shell thickness:

$$\gamma = [1 + (NA/Lh)]^{-1},$$

where N is the number of stiffeners, and A is the cross sectional area of the ring stiffener. The stress intensity is

$$\sigma_x^2 = (1 - 2\gamma + 4\gamma^2)(qR/2h)^2. \tag{3.2}$$

Suppose the buckling wave form of the cylindrical shell is

$$u = f_u \sin \beta y \cos \xi\alpha x, \quad v = f_v \cos \beta y \cos \xi\alpha x, \quad w = f \sin \beta y \sin \xi\alpha x, \tag{3.3}$$

where $\alpha = \pi/L$, $\beta = n/R$, ξ is the number of longitudinal half-waves, and n is the number of circumferential waves. When buckling occurs, the induced additional increments of strain are

$$\begin{aligned} d\varepsilon_x &= \partial u/\partial x, & d\varepsilon_y &= (\partial v/\partial y) - (w/R), \\ d\gamma_{xy} &= (\partial v/\partial x) + (\partial u/\partial y), & d\chi_x &= \partial^2 w/\partial x^2, \\ d\chi_y &= (\delta^2 w/\partial y^2) + (w/R^2), & d\chi_{xy} &= (\partial^2 w/\partial x\partial y) + (\partial v/R\partial x). \end{aligned} \tag{3.4}$$

Introducing $E_1 = E(1 - \nu^2)^{-1}$, $B_{ij} = [b_{ij}]/E_l$, $\alpha_1 = \alpha R = \pi R/L$, $D = Eh^3[12(1 - \nu^2)]^{-1}$, where the expressions for strain energy of the shell U_1, the strain energy of the ring stiffeners U_2, and the work done by the external hydrostatic pressure W, can also be derived, and then the total energy functional is

$$\Pi = W - U_1 - U_2. \tag{3.5}$$

The conditions for extreme values of the energy are:

$$(\partial\Pi/\partial f_u) = 0, \quad (\partial\Pi/\partial f_v) = 0, \quad (\partial\Pi/\partial f) = 0.$$

From the above three equations, the non-trivial solution referred to f_u, f_v, f can be obtained, and this is the solution for the bifurcation point. We will omit the derivation and just give the final result, i.e., the critical pressure for the elasto-plastic buckling of the reinforced shell q_{cr} as follows:

$$q_{cr} = \left[1 + \frac{\alpha_1^2}{2(n^2 - 1)}\right]^{-1} \left[\frac{Eh^3}{12(1 - \nu^2)R^3} F_1(B) + \frac{Eh}{R} F_2(B) + \frac{E_t I(n^2 - 1)}{1R^3}\right]. \tag{3.6}$$

When deriving the above formula, we only consider the bending moment terms for the ring-stiffener, and under hydrostatic pressure, take $\xi = 1$ in the formula (3.5), where E_t is the tangent modulus of the ring-stiffeners, and I is the moment of inertia of the ring-stiffeners. The functions $F_1(B)$ and $F_2(B)$ are:

$$F_1(B) = (n^2 - 1)^{-1}[B_{11}\alpha_1^4 + 2B_{12}\alpha_1^2(n^2 - 1) + B_{22}(n^2 - 1)^2 + 4B_{66}n^2\alpha_1^2],$$

$$F_2(B) = [(1 - \nu^2)(n^2 - 1)]^{-1}\Bigg[B_{22}$$

$$+ \frac{[2B_{11}B_{22}n^2\alpha_1^2(B_{12} + B_{66}) - n^2B_{22}^2(B_{11}\alpha_1^2 + B_{66}n^2) - B_{12}^2\alpha_1^2(B_{22}n^2 + B_{66}\alpha_1^2)]}{[(B_{11}\alpha_1^2 + B_{66}n^2)(B_{22}n^2 + B_{66}\alpha_1^2) - (B_{12} + B_{66})^2n^2\alpha_1^2]}\Bigg].$$

The fundamental formula (3.6) is a nonlinear equation. It mainly represents the physical nonlinearity of the materials of the shell and its ring-stiffeners. So, in the computer programme, the critical pressure of the bifurcation point is computed by the method of iteration.

When the shell is in a state of elastic buckling, the coefficients $B_{11} = 1$, $B_{22} = 1$, $B_{12} = \nu$, $B_{66} = 0.5(1 - \nu)$. Then the formula (3.6) degenerates to the classical linear solution:

$$q_{cr} = \left[1 + \frac{\alpha_1^2}{2(n^2 - 1)}\right]^{-1}\left[\frac{Eh^3}{12(1 - \nu^2)R^3}\frac{(\alpha_1^2 + n^2 - 1)^2}{(n^2 - 1)}\right.$$

$$\left. + \frac{Eh}{R}\frac{\alpha_1^4}{(\alpha_1^2 + n^2)^2(n^2 - 1)} + \frac{EI(n^2 - 1)}{1R^3}\right]. \tag{3.7}$$

In the computer programme, we use the method of iteration to approach the critical pressure q_{cr} of the elasto-plastic buckling of the shell. In the iterative process, it is only when the fundamental formula (3.6), together with the pre-buckling stress state equation (3.2) and the nonlinear stress strain curve of the shell material are all satisfied simultaneously, that the bifurcation point of the instability can be found, and we can obtain the small deflection solution of the elasto-plastic buckling of a reinforced shell.

4. Analysis of the geometrical nonlinearity of ring-stiffened shells in elasto-plastic buckling

The deformation compatibility equation of the ring-stiffened shell in elasto-plastic buckling is

$$a_{22}\varphi_{,xxxx} + (a_{66} + 2a_{12})\varphi_{,xxyy} + a_{11}\varphi_{,yyyy}$$

$$= \lambda[(w_{,xy})^2 - (w_{,xx})(w_{,yy})] - (1/R)w_{,xx}. \tag{4.1}$$

Here φ is the stress function:

$$d\sigma_x = \varphi_{,yy}, \quad d\sigma_y = \varphi_{,xx}, \quad d\tau_{xy} = \varphi_{,xy}$$

and λ is the initial imperfection factor:

$$\lambda = 1 + 2(w_0/w),$$

where w_0 is the initial imperfection deflection. The initial imperfection is a random factor, and it may be represented by harmonic analysis. In most experiments, it can be observed that if the wave form of the initial imperfection is represented by a

Fourier Series, the term of the series which is similar to the buckling wave form during the instant of the instability will always contribute the most important influence to the buckling of the shell. So we can take this main term of the series as w_0, and suppose that the wave form of w_0 is the same as the buckling wave w. Therefore, we take the deflection function w and w_0 as follows:

$$
\begin{aligned}
w &= f(\sin \alpha x \sin \beta y + \delta \sin^2 \alpha x + k), \\
w_0 &= f_0(\sin \alpha x \sin \beta y + \delta \sin^2 \alpha x + k).
\end{aligned} \tag{4.2}
$$

Here f and f_0 are coefficients of the deflections, and

$$\alpha = \pi/L, \quad \beta = n/R, \quad m = \alpha/\beta = \pi R/nL.$$

We solve the differential equation (4.1) for the stress function φ, and get

$$
\begin{aligned}
\varphi = {} & \psi_1 \cos 2\alpha x + \psi_2 \cos 2\beta y + \psi_3 \sin \alpha x \sin \beta y \\
& + \psi_4 \sin \beta y \sin 3\alpha x - (p_1 y^2/2h) - (p_2 x^2/2h),
\end{aligned}
$$

where $p_1 = Rq/2$, $p_2 = Rq$ for hydrostatic pressure, and the coefficients ψ_1 ($i = 1$–4) are as follows:

$$\psi_1 = \left(\frac{f}{a_{22} m^2}\right)\left(\frac{\lambda f}{32} - \frac{\delta}{8R^2}\right), \qquad \psi_2 = \left(\frac{\lambda f^2}{32 a_{11}}\right) m^2,$$

$$\psi_3 = \left(\frac{f}{a_{11} m_1}\right)\left(\frac{1}{R\beta^2} - \lambda f \delta\right) m^2, \qquad \psi_4 = \left(\frac{\lambda f^2 \delta}{a_{11} m_2}\right) m^2,$$

$$m_1 = \left(1 + \frac{a_{66} + 2a_{12}}{a_{11}} m^2 + \frac{a_{22}}{a_{11}} m^4\right),$$

$$m_2 = \left[1 + \frac{9(a_{66} + 2a_{12})}{a_{11}} m^2 + \frac{81 a_{22}}{a_{11}} m^4\right].$$

When, in elasto-plastic buckling, the relations between increments of stress and strain are:

$$
\begin{aligned}
d\varepsilon_x &= a_{11}\, d\sigma_x + a_{12}\, d\sigma_y + a_{16}\, d\sigma_{xy} = a_{11}\varphi_{,yy} + a_{12}\varphi_{,xx} - a_{16}\varphi_{,xy}, \\
d\varepsilon_y &= a_{12}\, d\sigma_x + a_{22}\, d\sigma_y + a_{26}\, d\sigma_{xy} = a_{21}\varphi_{,yy} + a_{22}\varphi_{,xx} - a_{26}\varphi_{,xy}, \\
d\gamma_{xy} &= a_{61}\, d\sigma_x + a_{62}\, d\sigma_y + a_{66}\, d\sigma_{xy} = a_{61}\varphi_{,yy} + a_{62}\varphi_{,xx} - a_{66}\varphi_{,xy},
\end{aligned} \tag{4.3}
$$

The strain equations in the middle surface of the shell will be:

$$
\begin{aligned}
\varepsilon_x^0 &= u_{0,x} + (\lambda/2)(w_{,x})^2 - k_x w, & \chi_x &= -w_{,xx}, \\
\varepsilon_y^0 &= v_{0,y} + (\lambda/2)(w_{,y})^2 - k_y w, & \chi_y &= -w_{,yy}, \\
\gamma_{xy}^0 &= u_{0,y} + v_{0,x} + \lambda(w_{,x})(w_{,y}), & \chi_{xy} &= -w_{,xy}.
\end{aligned} \tag{4.4}
$$

Now, the equation of total energy Π can be written and we can derive the two conditions: $\Pi_{,f} = 0$, and $\Pi_{,\delta} = 0$ referred to the two parameters f and δ, here $f = f/h$, and $\delta = f\ /h$. From this, two nonlinear equations are obtained:

$$\bar{q} = 2\lambda(C_1/C)\bar{f}^2 + [2\lambda/(1+\lambda)](C_2/C)\bar{\delta}^2$$
$$-2(C_3/C)\bar{\delta} + [2/(1+\lambda)](C_4/C), \tag{4.5}$$

$$\bar{\delta}^3 - [1 + (2/\lambda)](C_3/C_2)\bar{\delta}^2 + [((1+\lambda)/\lambda^2)(C_1 C_7/C_2^2)\bar{q}$$
$$-((1+\lambda)/2\lambda)(C/C_2^2)\bar{q} + ((1+\lambda)/\lambda^2)(C_3^2/C_2^2) + \lambda^{-1}C_4/C_2$$
$$-((1+\lambda)/\lambda^2)(C_1 C_5/C_2^2)]\bar{\delta} + [((1+\lambda)/2\lambda^2)(CC_3/C_2^2)\bar{q}$$
$$-(1/\lambda^2)(C_3 C_4/C_2^2)] = 0, \tag{4.6}$$

where the coefficients C_i are:

$$C_1 = (1/64)[E^{-1}(a_{22}^{-1} + a_{11}^{-1}m^4) + \omega((1 + (a_{21}^2/a_{11}^2)m^4)]\eta^2,$$
$$C_2 = (1/2)[(1/Ea_{11})(m_1^{-1} + m_2^{-1})m^4$$
$$+\omega(a_{21}^2/a_{11}^2)((m_3/m_1^2) + (m_4/m_2^2))m^4]\eta^2,$$
$$C_3 = (1/16)[E^{-1}(a_{22}^{-1} + (8/a_{11})(m^4/m_1)) + (1 + 8(a_{21}^2/a_{11}^2)(m_3/m_1^2)m^4)]\eta,$$
$$C_4 = (1/2)[(1/Ea_{11})(m^4/m_1) + (1/12)(\eta^2/E)(b_{11}m^4 + 2(b_{12} + 2b_{66})m^2$$
$$+b_{22}) + \omega((a_{21}^2/a_{11}^2)(m_3/m_1^2)m^4 + (\eta^2 I/Ah^2))],$$
$$C_5 = (1/4)[(1/Ea_{22}) + (4/3)\eta^2(b_{11}/E)m^4 + \omega],$$
$$C_6 = (1/4)\eta, \quad C_7 = (1/2)\zeta,$$
$$C_8 = [-E((a_{11}/2) + 2a_{12} + 2a_{22}) + \omega E^2((a_{12}^2/2) + 2a_{12}a_{22} + 2a_{22}^2)],$$
$$C = (1/2)\eta + (1/4)\zeta,$$
$$\eta = \beta^2 Rh = n^2 h/R, \quad \zeta = \alpha^2 Rh, \quad \omega = E_t A/(Elh), \quad \bar{q} = qR^2/(Eh^2),$$
$$m_3 = [1 + (a_{22}/a_{11})m^2]^2, \qquad m_4 = [1 + 9(a_{22}/a_{21})m^2]^2.$$

In Fig. 1, a set of large deflection curves for the buckling of a ring-stiffened cylindrical steel shell is shown.

5. Elasto-plastic buckling of shell plates between ring-stiffeners

There may be three different kinds of failure of ring-stiffened cylindrical shells under external pressure. (1) The whole reinforced cylindrical shell or the shell plate between ring-stiffeners loses its load carrying capacity. (2) There may occur elasto-plastic buckling with symmetric wave form or with asymmetric wave form depending on the parameters of shell size. (3) There also may occur elastic buckling with asymmetric wave form.

To find when the shell loses its load carrying capacity is a problem of limit analysis (or limit design), and in such cases, small imperfection will not produce any effect on the limit load or the carrying capacity of the shell. However, in the experiments, it can be observed that during the theoretical limit state, the shell may still carry some increase in load and with only small deformation; or the applied

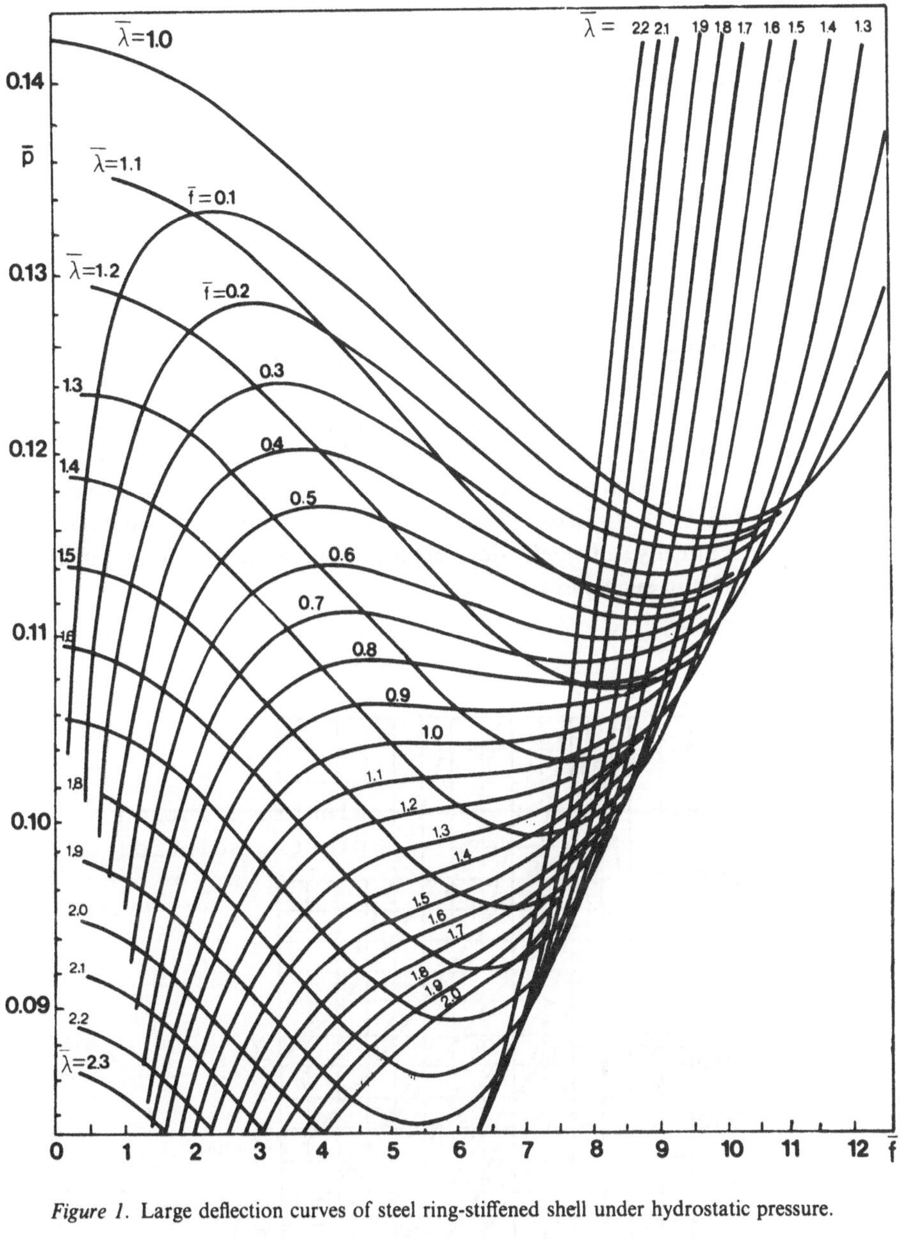

Figure 1. Large deflection curves of steel ring-stiffened shell under hydrostatic pressure.

load may decrease, and at the same time may be accompanied with rapidly increased deformation. In the latter case, the phenomena means that to maintain the original equilibrium state at that instant becomes impossible, and is really a different kind of unstable phenomena. Such kinds of instability problem must be analyzed by a more exact method than that of the ordinary method of limit equilibrium analysis [8]. In this paper we will only consider the elastic or elasto-plastic buckling of shell plates between ring-stiffeners, and basic equation (3.5) can be used. We have computed the following parameters of cylindrical shells:

$$R/h = 90\text{–}150, \quad u = 0.6425L/(Rh)^{1/2} = 0.8\text{–}3.0.$$

The material of the shell plates is nickel-chromium alloy steel. The computational results are shown in Fig. 2. From the results, it can be seen that when $u \geqslant 2.0$, the buckling is elastic or small elasto-plastic; when $u \leqslant 1.2$, the prebuckling stress state already approaches or even reaches the yielding limit state, and in general it is a problem of limit analysis. When the value of u is between the above numbers, the buckling will be elasto-plastic. For the cases of $u \leqslant 1.6$, the prebuckling plastic deformation is already sufficiently developed, and therefore the initial imperfection is not so sensible in such cases.

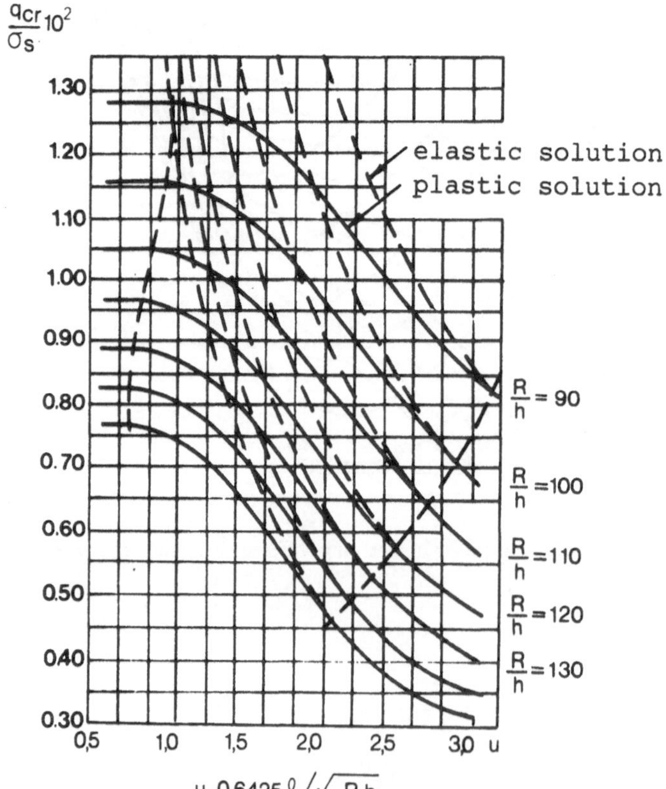

Figure 2. Critical pressure of plastic buckling for different values of u.

6. The general instability of ring-stiffened conical shells in elasto-plastic buckling

Consider a ring-stiffened conical shell as shown in Fig. 3. Introduce the assumption: $\bar{M}_s = \bar{Q}_s = \bar{M}_{s\theta} = 0$, and the equilibrium equations are

$$\bar{N}_{s,s} + (s \sin \alpha)^{-1}\bar{N}_{s\theta,\theta} + (\bar{N}_s/S)(\bar{N}_\theta/S) - \bar{q}_s = 0,$$
$$(s \sin \alpha)^{-1}\bar{N}_{\theta,\theta} + \bar{N}_{s\theta,s} + (2/s)\bar{N}_{s\theta} - (\bar{Q}_\theta/s \operatorname{tg} \alpha) + \bar{q}_\theta = 0, \tag{6.1}$$
$$(s \sin \alpha)^{-1}\bar{Q}_{\theta,\theta} + (\bar{N}_\theta/s \operatorname{tg} \alpha) + \bar{q}_z = 0, \qquad (s \sin \alpha)^{-1}\bar{M}_{\theta,\theta} - \bar{Q}_\theta = 0.$$

The above equations are partial differential equations with variable coefficients. It is difficult to solve them, so we introduce the assumption of an "equivalent cylindrical shell" to simplify these equations. Thus, we use an average value $S_c = (S_1 + S_2)/2$, instead of variable S, and under hydrostatic pressure $\bar{q}_s = \bar{q}_\theta = 0$, we have

$$\bar{N}_{s,s} + R_c^{-1}\bar{N}_{s\theta,\theta} = 0, \qquad \bar{N}_{s\theta,\theta} + R_c^{-1}\bar{N}_{\theta,\theta} + (R_c R_*)^{-1}\bar{M}_{\theta,\theta} = 0,$$
$$R_c^{-2}\bar{M}_{\theta,\theta\theta} + (\bar{N}_\theta/R_*) + \bar{q}_z = 0, \tag{6.2}$$

where $R_c = S_c \sin \alpha$, $R_* = S_c \operatorname{tg} \alpha$. Suppose the conical shell is buckling, and there are n waves in the circumferential direction. Introducing the transform matrix $[\theta]$, we get

$$\{\bar{N}\} = [\theta]\{N\},$$

where the matrices $\{\bar{N}\}$ and $\{N\}$ are:

$$\{\bar{N}\} = \{\bar{N}_s(S,\theta) \;\; \bar{N}_{s\theta}(S,\theta) \;\; \bar{N}_\theta(S,\theta) \;\; \bar{M}_\theta(S,\theta) \;\; \bar{q}_z(S,\theta)\}^T,$$
$$\{N\} = \{N_s(S) \;\; N_{s\theta}(S) \;\; N_\theta(S) \;\; M_\theta(S) \;\; q_z(S)\}^T$$

$$[\theta] = \begin{bmatrix} \cos n\theta & & & & \\ & \sin n\theta & & 0 & \\ & & \cos n\theta & & \\ & 0 & & \cos n\theta & \\ & & & & \cos n\theta \end{bmatrix}. \tag{6.3}$$

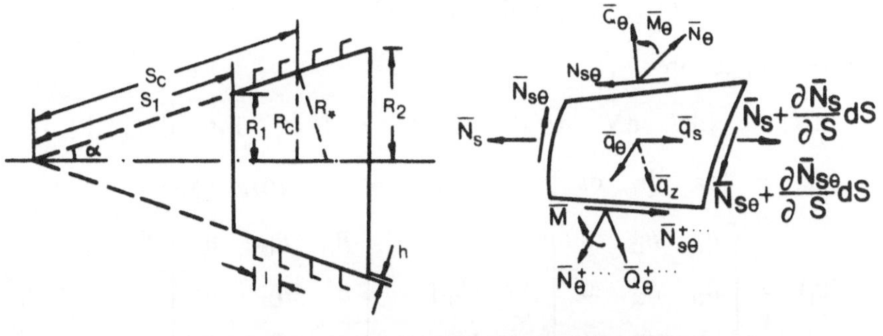

Figure 3. A truncated ring-stiffened conical shell.

The fundamental equations after transformation are

$$N_{s,s} + (n/R_c)N_{s\theta} = 0, \qquad N_{s\theta,s} - (n/R_c)N_\theta + \cos\alpha(n/R_c^2)M_\theta = 0,$$
$$-[n^2/(R_c R_* \cos\alpha)]M_\theta + (N_\theta/R_*) + q_z = 0. \tag{6.4}$$

Let the prebuckling stress state in the middle surface of the shell σ_{x_0} and σ_{θ_0} be:

$$\sigma_{s_0} = N_{s_0}/h = qR_*(S)/(2h) = qs \operatorname{tg}\alpha/(2h),$$

$$\sigma_{\theta_0} = qR_*(S)/h' = rqs \operatorname{tg}\alpha/h.$$

The stress intensity σ_i is:

$$\sigma_i = (1 + 2\gamma + 4\gamma^2)^{1/2} qs \operatorname{tg}\alpha/(2h).$$

When buckling, the coefficients a_{ij} and b_{ij} are the same as stated in Section 2. For the sake of simplicity, we write the increments of stress and strain during buckling not in incremental form, but in the form of the entire stress and strain.

The deformation equations in the middle surface of the shell are:

$$\bar{\varepsilon}_s = \bar{u}_{,s}, \quad \bar{\varepsilon}_\theta = (s\sin\alpha)^{-1}\bar{v}_{,\theta} - \bar{w}/(s \operatorname{tg}\alpha) \approx R_c^{-1}\bar{v}_{,\theta} - (\bar{w}/R_*),$$

$$\bar{\gamma}_{s\theta} = (s\sin\alpha)^{-1}\bar{u}_{,\theta} + \bar{v}_{,\theta} \approx R_c^{-1}\bar{u}_{,\theta} + \bar{v}_{,s},$$

$$\bar{\chi}_s = \bar{w}_{,ss}, \quad \chi_\theta = (s^2\sin^2\alpha)^{-1}\bar{w}_{,\theta\theta} + (\bar{w}/(s^2 \operatorname{tg}^2\alpha)) \approx R_c^{-2}\bar{w}_{,\theta\theta} + (\bar{w}/R_*^2).$$

When buckling, suppose $\bar{u} = u\cos n\theta$, $\bar{v} = v\sin n\theta$, $\bar{w} = w\cos n\theta$, then take the wave form of the displacement (satisfying the boundary conditions of simple support on both ends of the conical shell) to be:

$$u = f_u \cos\pi(s - s_1)/L, \quad v = f_v \sin\pi(s - s_1)/L, \quad w = f_w \sin\pi(s - s_1)/L.$$

Using the constitutive relations of stress and strain

$$\{\mathrm{d}\varepsilon_{ij}\} = [a_{ij}]\{\mathrm{d}\sigma_{ij}\}, \qquad \{\mathrm{d}\sigma_{ij}\} = [b_{ij}]\{\mathrm{d}\varepsilon_{ij}\}$$

the following equations can be derived

$$\{\mathrm{d}N\} = Eh(1 - v^2)^{-1}[B_{\mathrm{I}}]\{\mathrm{d}\varepsilon\},$$

$$\{\mathrm{d}M\} = -Eh^3[12(1 - v^2)]^{-1}[B_{\mathrm{II}}]\{\mathrm{d}x\}.$$

Here the matrices are

$$\{\mathrm{d}N\} = \{\mathrm{d}N_s \ \ \mathrm{d}N_\theta \ \ \mathrm{d}N_{s\theta}\}^T, \quad \{\mathrm{d}M\} = \{\mathrm{d}M_s \ \ \mathrm{d}M_\theta \ \ \mathrm{d}M_{s\theta}\}^T,$$

$$\{\mathrm{d}\varepsilon\} = \{\mathrm{d}\varepsilon_s \ \ \mathrm{d}\varepsilon_\theta \ \ \mathrm{d}\varepsilon_{s\theta}\}^T, \qquad \{\mathrm{d}\chi\} = \{\mathrm{d}\chi_s \ \ \mathrm{d}\chi_\theta \ \ \mathrm{d}\chi_{s\theta}\}^T,$$

$$[B_{\mathrm{I}}] = \begin{bmatrix} B_{11} & B_{12} & 0 \\ B_{21} & B_{22} & 0 \\ 0 & 0 & B_{66} \end{bmatrix}, \qquad [B_{\mathrm{II}}] = \begin{bmatrix} B_{11} & B_{12} & 0 \\ B_{21} & B_{22} & 0 \\ 0 & 0 & 2B_{66} \end{bmatrix}.$$

Substituting the above relations into the fundamental equilibrium equations (6.3), we can get a system of homogeneous equations:

$$\begin{bmatrix} C_{11} & C_{12} & C_{13} \\ C_{21} & C_{22} & C_{23} \\ C_{31} & C_{32} & C_{33} \end{bmatrix} \begin{Bmatrix} f_u \\ f_v \\ f_w \end{Bmatrix} = 0$$

From the above equations, set the determinant of the coefficients equal to zero, then the critical pressure q_{cr} of the conical shell during buckling can be found and we get the following final result:

$$q_{cr} = \left[1 + \frac{\alpha_1^2}{2(n^2 - \cos^2\alpha)}\right]^{-1} \left[\frac{Eh}{R_c} \frac{\alpha_c^4 \cos^3\alpha}{(n^2 - \cos^2\alpha)} F_1(B) + \frac{K_s(n^2 - \cos^2\alpha)}{R_*^2 \cos^2\alpha} F_2(B)\right], \tag{6.5}$$

where $\alpha_c = \pi R_c/L$, $K_s = EI_s/l$, I_s is the moment of inertia of the stiffeners referred to as the principal curvature line S. The functions $F_1(B)$ and $F_2(B)$ are

$$F_1(B) = \frac{B_{66}(B_{11}B_{22} - B_{12}^2)}{[(B_{11}B_{22} - B_{12}^2)n^2\alpha_c^2 + B_{66}(B_{22}n^4 - 2B_{12}n^2\alpha_c^4 + B_{11}\alpha_c^4)](1 - v^2)}$$

$$F_2(B) = \frac{B_{22}[(B_{11}B_{22} - B_{12}^2)n^2\alpha_c^2 + B_{66}(B_{22}n^4 - B_{12}\alpha_c^2 n^2) + B_c]}{[(B_{11}B_{22} - B_{12}^2)n^2\alpha_c^2 + B_{66}(B_{22}n^4 - 2B_{12}n^2\alpha_c^2 + B_{11}\alpha_c^4)]},$$

where B_c is

$$B_c = [B_{66}(B_{11}\alpha_c^2 - B_{12}n^2)n^2\alpha_c^2]/(n^2 - \cos^2\alpha).$$

When in elastic buckling, $B_{11} = B_{22} = 1$, $B_{12} = v$, $B_{66} = 0.5(1 - v)$, then we get

$$F_1(B) = [(n^2 + \alpha_c^2)^{-2}], \quad F_2(B) = \left[1 - \frac{\alpha_c^2(n^2 - \alpha_c^2)\cos^2\alpha}{(n^2 + \alpha_c^2)^2(n^2 - \cos^2\alpha)}\right] \approx 1,$$

$$q_{cr} = \left[1 + \frac{\alpha_c^2}{2(n^2 - \cos^2\alpha_c)}\right]^{-1} \left[\frac{Eh}{R_c} \frac{\alpha_c^4 \cos^3\alpha}{(n^2 - \cos^2\alpha)(n^2 + \alpha_c^2)^2} + \frac{K_s(n^2 - \cos^2\alpha)}{R_*^2 \cos^2\alpha}\right]. \tag{6.6}$$

The above equations (4.1) and (6.6) are the fundamental formulae for the computation of the critical pressure q_{cr} of ring-stiffened conical shells during elasto-plastic buckling.

If the axis of the ring-stiffener is inclined with the line of principal curvature of the shell, this influence on the critical pressure q_{cr} can be considered through the coefficient K_s in the fundamental formulae (see [1] in detail).

7. Elasto-plastic buckling of spherical shells

The instability theory of inelastic spherical shells was developed by Lunchick, Gerard and others. All of their solutions were made to modify the classical linear solution by a factor of modification. But Krenzke's [9] tests did not support their results. Therefore, the nonlinear buckling theory of inelastic spherical shells is still a problem to be investigated.

Now we are going to give the general equations for the nonlinear large deflection theory of the instability of spherical shells. We introduce an initial imperfection factor $\lambda = 1 + (w_0/w)$; for the spherical shell $K_x = K_y = 1/R$, and thus obtain the fundamental equilibrium equation and the equation of compatibility:

$$\begin{aligned} b_{11} w_{,xxxx} &+ 2(b_{12} + 2b_{66}) w_{,xxyy} + b_{22} w_{,yyyy} \\ &= (12/h^2)[(1/R)\nabla^2 \varphi - \sigma \nabla^2 w + (\varphi_{,yy} w_{,xx} + \varphi_{,xx} w_{,yy})], \end{aligned} \tag{7.1}$$

$$\begin{aligned} a_{22} \varphi_{,xxxx} &+ (a_{66} + 2a_{12}) \varphi_{,xxyy} + a_{11} \varphi_{,yyyy} \\ &= \lambda[(w_{,xy})^2 - (w_{,xx})(w_{,yy})] - (1/R)(\nabla^2 w). \end{aligned} \tag{7.2}$$

The coefficients a_{ij} and b_{ij} are:

(1) For incremental theory

$$\begin{aligned} a_{11} &= (1/E) + (1/4)[(E_t^0)^{-1} - E^{-1}], \\ a_{12} &= -(\nu/E) + (1/4)[(E_t^0)^{-1} - E^{-1}], \\ a_{22} &= (1/E) + (1/4)[(E_t^0)^{-1} - E^{-1}], \quad a_{66} = 2(1 + \nu)/E, \\ a_{16} &= a_{26} = 0, \quad a_{11} = a_{22}. \end{aligned}$$

(2) For deformation theory

$$\begin{aligned} a_{11} &= (E_s^0)^{-1} + (1/4)[(E_t)^{-1} - E_s^{-1}], \\ a_{12} &= -(\nu/E) + (1/4)[(E_t)^{-1} - E_s^{-1}] \\ &\quad -(1/2)[(E_s^0)^{-1} - E^{-1}], \\ a_{22} &= (E_s^0)^{-1} + (1/4)[(E_t)^{-1} - E_s^{-1}], \\ a_{66} &= 2(1 + \nu)/E + 3[(E_s^0)^{-1} - E^{-1}], \\ a_{16} &= a_{26} = 0, \quad a_{11} = a_{22}. \end{aligned}$$

The relations between b_{ij} and a_{ij} are:

$$\begin{aligned} b_{11} &= a_{22}/(a_{11}a_{22} - a_{12}^2), \quad b_{12} = -a_{12}/(a_{11}a_{22} - a_{12}^2), \\ b_{22} &= a_{11}/(a_{11}a_{12} - a_{12}^2), \quad b_{66} = 1/a_{66}. \end{aligned}$$

When in the elastic state:

$$\begin{aligned} a_{11} &= 1/E, \quad a_{12} = -\nu/E, \quad a_{66} = 2(1 + \nu)/E, \\ b_{11} &= b_{22} = E/(1 - \nu^2), \quad b_{12} = E\nu/(1 - \nu^2), \\ b_{66} &= E/2(1 + \nu); \quad (b_{12} + 2b_{66})/b_{11} = 1, \quad (a_{66} + 2a_{12})/2a_{22} = 1. \end{aligned}$$

In the symmetrical case, the fundamental equations (7.1) and (7.2) can be transformed to spherical coordinates:

$$b_{11}w_{,rrrr} + 2(b_{12} + 2b_{66})[(1/r)w_{,r}]_{,rr} + b_{22}(1/r)[(1/r)w_{,r}]_{,r}$$
$$= (12/h^2)[(1/R)\nabla^2\varphi - \sigma\nabla^2 w + [(1/r)\varphi_{,r}w_{,rr} + \varphi_{,rr}(1/r)w_{,r}]], \tag{7.3}$$

$$a_{22}\varphi_{,rrrr} + (a_{66} + 2a_{12})[(1/r)\varphi_{,r}]_{,rr} + a_{11}(1/r)r[(1/r)\varphi_{,r}]_{,r}$$
$$= [-w_{,rr}(1/r)w_{,r}] - (1/R)\nabla^2 w. \tag{7.4}$$

The Laplace operator is

$$\nabla^2 = (1/r)\frac{\mathrm{d}}{\mathrm{d}r}\left(r\frac{\mathrm{d}}{\mathrm{d}r}\right).$$

If the elastic part and the plastic part are separated, the fundamental equations will be

$$\nabla^4 w + 2\left(\frac{b_{12} + 2b_{66}}{b_{11}} - 1\right)[(1/r)w_{,r}]_{,rr}$$
$$= (12/b_{11}h^2)[(1/R)\nabla^2\varphi - \sigma\nabla^2 w + (r^{-1}\varphi_{,r}w_{,rr} + r^{-1}w_{,r}\varphi_{,rr})], \tag{7.5}$$

$$\nabla^4\varphi + 2\left(\frac{a_{66} + 2a_{12}}{2a_{22}} - 1\right)(r^{-1}\varphi_{,r})_{,rr}$$
$$= -(1/a_{22})(r^{-1}w_{,r}w_{,rr}\lambda + r^{-1}\nabla^2 w). \tag{7.6}$$

In the elastic case:

$$12/(b_{11}h^2) = h/D, \quad (b_{12} + 2b_{66})/b_{11} = 1, \quad (a_{66} + 2a_{12})/2a_{22} = 1,$$

then we get the equations for the elastic spherical shells:

$$\nabla^4 w = (h/D)[r^{-1}\nabla^2\varphi - \sigma\nabla^2 w + (r^{-1}\varphi_{,r}w_{,rr} + r^{-1}w_{,r}\varphi_{,rr})],$$
$$\nabla^4\varphi = -E(r^{-1}w_{,r}w_{,rr}\lambda + r^{-1}\nabla^2 w). \tag{7.7}$$

If we neglect the nonlinear terms, we get the following fundamental equations for the elastic theory of small deflection:

$$\nabla^2\nabla^2 w = (h/d)(R^{-1}\nabla^2\varphi - \sigma\nabla^2 w),$$
$$\nabla^2\nabla^2\varphi = -ER^{-1}\nabla^2 w. \tag{7.8}$$

Here, the stress function φ is:

$$\sigma_r = (1/r)\varphi_{,r}, \qquad \sigma_\varphi = \varphi_{,rr} = (\sigma_r r)_{,r}.$$

For the elasto-plastic theory of small deflection, the fundamental equations are:

$$\nabla^4 w + 2\left(\frac{b_{12} + 2b_{66}}{b_{11}} - 1\right)(r^{-1}w_{,r})_{,rr} = (12/b_{11}h^2)(R^{-1}\nabla^2\varphi - \nabla^2\omega),$$
$$\nabla^4\varphi + 2\left(\frac{a_{66} + 2a_{12}}{2a_{22}} - 1\right)(r^{-1}\varphi_{,r})_{,rr} = -[1/(a_{22}R)]\nabla^2 w. \tag{7.9}$$

Introducing the coefficients α and β:

$$\alpha = \left(\frac{a_{66} + 2a_{12}}{2a_{22}} - 1\right), \qquad \beta = \left(\frac{b_{12} + 2b_{66}}{b_{11}} - 1\right),$$

and we get:

$$\begin{aligned} \nabla^4 w + 2\beta(r^{-1}w_{,r})_{,rr} &= (12/b_{11}h^2)(R^{-1}\nabla^2\varphi - \sigma\nabla^2 w), \\ \nabla^4\varphi + 2\alpha[(1/r)_{,r}]_{,rr} &= -(1/(Ra_{22}))\nabla^2 w. \end{aligned} \tag{7.10}$$

These are the fundamental equations for plastic buckling of spherical shells in the small deflection theory.

References

1. Zhou Cheng-ti, *Theory of Stability of Elasto-Plastic Thin Shells*, National Defence Publishing Company, PRC (1979).
2. Stein, M., "The effect on the buckling of perfect cylinders of prebuckling deformation and stresses induced by edge support", *NASA* TN D1510, Dec. (1962), 217.
3. Stein, M., "The influence of prebuckling deformations and stresses in the buckling of perfect cylinders", *NASA* TR-190, (1964).
4. Koiter, W. T., "Elastic stability and postbuckling behaviour", *Nonlinear Problems* (edited by R. E. Langer) (1963), 257–275.
5. Hutchinson, J. W., "Plastic buckling", *Advances in Applied Mechanics*, Vol. 14 (1974).
6. Zhou Cheng-ti "Elasto-plastic instability of ring-stiffened cylindrical shells under hydrostatic pressure", *Journal of Dalian Institute of Technology*, No. 1, Mar. (1973) (in Chinese).
7. Zhou Cheng-ti, "Theory of plastic buckling of stiffened thin shells with initial imperfections", *Journal of Dalian Institute of Technology*, No. 2, June (1978) (in Chinese).
8. Zhou Cheng-ti, "The influence of variation of geometrical shape to the load carrying capacity of structures", *Journal of Dalian Institute of Technology*, No. 1 (1964) (in Chinese).
9. Krenzke, M. A., "Tests on machined deep spherical shells under external hydrostatic pressure", AD 278075, *DTMB Rept.* 1601 (1962).

Composite expansion method applied to large deflection problems of spherical shells

CHOU HUAN-WEN

Department of Mathematics, Wuhan University, Wuhan, PRC

Abstract. In this paper, the method of composite expansions which was proposed by W. Z. Chien [1] is presented to obtain asymptotic solutions for large deflection problems of shallow spherical shells. For shallow spherical shells with simple or clamped support under a uniformly distributed load, the problems, which are to solve the corresponding asymptotic differential equations, may all be reduced to solving the minimal value of a certain functional.

1. Introduction

In many problems in mechanics, the highest order derivative term of the governing equations often contains a small parameter. These problems are called "singular perturbation" types. The composite expansion method is one of the better ways of solving these problems. As early as 1948, Chien Wei-zang proposed this method in investigating the problem of large deflection of circular plates [1]. Recently the composite expansion method has seen new developments. The small parameter in the highest order derivative term may be non-fixed, but can be chosen properly by means of the method of undetermined small parameter (see W. Z. Chien [2]). Here we will use W. Z. Chien's composite expansion method of the undetermined parameter method [3, 10] to obtain the asymptotic solutions of large deflection problems of shallow spherical shells.

2. Simplification of basic equations

Now let us discuss the axisymmetric nonlinear problem of a shallow spherical shell with a thickness of h, a span of $2a$ and a rise of f. The governing equations are [4, 5]

$$Dr\frac{\mathrm{d}}{\mathrm{d}r}\frac{1}{r}\frac{\mathrm{d}}{\mathrm{d}r}r\frac{\mathrm{d}\omega}{\mathrm{d}r} - \frac{\mathrm{d}\Phi}{\mathrm{d}r}\frac{\mathrm{d}\omega}{\mathrm{d}r} - \frac{1}{R}r\frac{\mathrm{d}\Phi}{\mathrm{d}r} = \int_0^r q(r)r\,\mathrm{d}r, \tag{2.1a}$$

$$r\frac{\mathrm{d}}{\mathrm{d}r}\frac{1}{r}\frac{\mathrm{d}}{\mathrm{d}r}r\frac{\mathrm{d}\Phi}{\mathrm{d}r} + \frac{Eh}{2}\left(\frac{\mathrm{d}\omega}{\mathrm{d}r}\right)^2 + \frac{Eh}{R}r\frac{\mathrm{d}\omega}{\mathrm{d}r} = 0. \tag{2.1b}$$

Yeh, K. Y. (ed) Progress in Applied Mechanics
 ISBN-13: 978-94-010-8061-3

If we let

$$N_r = \frac{1}{r}\frac{d\Phi}{dr}, \tag{2.2}$$

then the above are

$$Dr\frac{d}{dr}\frac{1}{r}\frac{d}{dr}r\frac{d\omega}{dr} - rN_r\frac{d\omega}{dr} - \frac{2f}{a^2}r^2N_r - \int_0^r rq\,dr = 0, \tag{2.3a}$$

$$r\frac{d}{dr}\frac{1}{r}\frac{d}{dr}r^2N_r + \frac{Eh}{2}\left(\frac{d\omega}{dr}\right)^2 + \frac{2f}{a^2}Ehr\frac{d\omega}{dr} = 0, \tag{2.3b}$$

where D is the rigidity of bending, and μ is Poisson's ratio.

In order to use the singular perturbation method, we introduce the following transformation

$$x = r^2/a^2, \quad \beta = \sqrt{12(1-\mu^2)}\,\varepsilon(\omega/h), \quad F = \frac{a^2}{D}\varepsilon^2 N_r \tag{2.4}$$

and

$$k^2 = \frac{\sqrt{12(1-\mu^2)}}{4}\frac{f}{h}, \qquad p(x) = \frac{[12(1-\mu^2)]^{3/2}\,a^4}{16Eh^4}\frac{1}{x}\int_0^x q\,dx, \tag{2.5}$$

then the equations (2.3a and b) becomes

$$\varepsilon^2\frac{d^2}{dx^2}\left(x\frac{d\beta}{dx}\right) - \tfrac{1}{4}F\frac{d\beta}{dx} - \varepsilon k^2F - \varepsilon^3p(x) = 0, \tag{2.6a}$$

$$\frac{d^2}{dx^2}(xF) + \frac{1}{2}\left(\frac{d\beta}{dx}\right)^2 + 4\varepsilon k^2\frac{d\beta}{dx} = 0, \tag{2.6b}$$

where ε may be arbitrary. E.g., let

$$\varepsilon^2 = \frac{h^2}{12(1-\mu^2)a^2}.$$

This was discussed by L. S. Srubshchik [7].

If the singular perturbation method is not used, we may let $\varepsilon = 1$, and the equations (2.6a and b) become

$$\frac{d^2}{dx^2}\left(x\frac{d\beta}{dx}\right) - \tfrac{1}{4}F\frac{d\beta}{dx} - k^2F - p(x) = 0, \tag{2.7a}$$

$$\frac{d^2}{dx^2}(xF) + \frac{1}{2}\left(\frac{d\beta}{dx}\right)^2 + 4k^2\frac{d\beta}{dx} = 0. \tag{2.7b}$$

Moreover, some authors, such as H. J. Weinitchke [8], E. L. Reiss [9], etc., have derived formulae such as (2.7a) and (2.7b). All have used Chien Wei-zang's results [4].

Now we transform the related boundary conditions that were discussed in [6].

When $r = a$,

$$\omega = 0, \tag{2.8a}$$

$$D\left(\frac{d^2\omega}{dr^2} + \frac{\mu}{r}\frac{d\omega}{dr}\right) = -k_b\frac{d\omega}{dr}, \tag{2.8b}$$

$$N_r = -\frac{k_a}{Eh}\left[r\frac{dN_r}{dr} + (1-\mu)N_r\right]. \tag{2.8c}$$

When $r = 0$

$$N_r,\ d\omega/dr \text{ are finite,} \tag{2.9}$$

and k_a and k_b are given constants, they take on various values for various boundary conditions.

Using the transformation (2.4), the boundary conditions (2.8) and (2.9) become as follows:

When $x = 1$

$$\beta = 0, \tag{2.10a}$$

$$\frac{d\beta}{dx} + k_\beta\frac{d^2\beta}{dx^2} = 0, \tag{2.10b}$$

$$F + k_F\frac{dF}{dx} = 0. \tag{2.10c}$$

Where

$$k_\beta = \frac{2}{1 + \mu + ak_b/D}, \tag{2.11a}$$

$$k_F = \frac{2}{1 - \mu + Eh/k_a}. \tag{2.11b}$$

When $x = 0$

$$F,\ \frac{d\beta}{dx} \text{ are finite.} \tag{2.12}$$

It is worth noticing that after transformation, the boundary conditions in [6] become the conditions (2.10), (2.12), which are independent of ε. Therefore, the boundary conditions have not been perturbed.

3. Expressions of asymptotic expansion

Here we discuss the asymptotic expansion of the system (2.6). In the system there are a given parameter k and a parameter function $p(x)$ besides the undetermined parameter ε. We suppose that k and $p(x)$ can be all asymptotically expanded:

$$\varepsilon k^2 = k_0 + \varepsilon k_1 + \varepsilon^2 k^2 + \cdots, \tag{3.1}$$

$$\varepsilon^3 p(x) = p_0(x) + \varepsilon p_1(x) + \varepsilon^2 p_2(x) + \cdots, \tag{3.2}$$

$$\frac{d\beta}{dx} = \omega(x, \varepsilon) + \Omega(\zeta, \varepsilon), \tag{3.3a}$$

$$F = \varphi(x, \varepsilon) + \psi(\zeta, \varepsilon), \tag{3.3b}$$

where

$$\omega(x, \varepsilon) = \omega_0(x) + \omega_1(x)\varepsilon + \omega_2(x)\varepsilon^2 + \cdots, \tag{3.4a}$$

$$\varphi(x, \varepsilon) = \varphi_0(x) + \varphi_1(x)\varepsilon + \varphi_2(x)\varepsilon^2 + \cdots, \tag{3.4b}$$

$$\Omega(\zeta, \varepsilon) = \Omega_0(\zeta) + \Omega_1(\zeta)\varepsilon + \Omega_2(\zeta)\varepsilon^2 + \cdots, \tag{3.4c}$$

$$\psi(\zeta, \varepsilon) = \psi_0(\zeta) + \psi_1(\zeta)\varepsilon + \psi_2(\zeta)\varepsilon^2 + \cdots, \tag{3.4d}$$

and

$$\zeta = 1 - x/\varepsilon. \tag{3.5}$$

Substituting (3.3) into (2.6), we obtain

$$\varepsilon^2 \frac{d^2}{dx^2} x\omega - \tfrac{1}{4}\varphi\omega - \varepsilon k^2 \varphi - \varepsilon^3 p(x) = 0, \tag{3.6a}$$

$$\frac{d^2}{dx^2} x\varphi + \tfrac{1}{2}\omega^2 + 4k^2\varepsilon\omega = 0, \tag{3.6b}$$

and

$$\frac{d^2}{d\zeta^2}(1 - \varepsilon\zeta)\Omega - \tfrac{1}{4}(\varphi\Omega + \psi\omega + \psi\Omega) - \varepsilon k^2\psi = 0, \tag{3.7a}$$

$$\varepsilon^{-2}\frac{d}{d\zeta^2}(1 - \varepsilon\zeta)\psi + \tfrac{1}{2}(2\omega\Omega + \Omega^2) + 4\varepsilon k^2\Omega = 0. \tag{3.7b}$$

Substituting (3.1), (3.2), (3.4a) and (3.4b) into (3.6), the asymptotic expansion formulas for ω and φ are obtained in the following form

$$\tfrac{1}{4}\varphi_0\omega_0 + k_0\varphi_0 + p_0 = 0, \tag{3.8a}$$

$$\frac{d^2}{dx^2} x\varphi_0 + \tfrac{1}{2}\omega_0^2 + 4k_0\omega_0 = 0, \tag{3.8b}$$

$$\tfrac{1}{4}(\varphi_0\omega_1 + \varphi_1\omega_0) + k_0\varphi_1 + k_1\varphi_0 + p_1 = 0, \tag{3.9a}$$

$$\frac{d^2}{dx^2}(x\varphi_1) + \omega_0\omega_1 + 4(k_0\omega_1 + k_1\omega_0) = 0, \tag{3.9b}$$

$$\varphi_0\omega_n + \varphi_n\omega_0 + 4k_0\varphi_n = \frac{d^2}{dx^2} x\omega_{n-2} - \sum_{j=1}^{n-1} \varphi_j\omega_{n-j} - 4\sum_{j=1}^{n} k_j\varphi_{n-j} - 4p_n, \tag{3.10a}$$

$$\frac{d^2}{dx^2}x\varphi_n + \omega_0\omega_n + 4k_0\omega_n = -\frac{1}{2}\sum_{j=1}^{n-1}\omega_j\omega_{n-j} - 4\sum_{j=1}^{n}k_j\omega_{n-j},$$

$$(n = 2, 3, 4, \ldots) \tag{3.10b}$$

At the same time substituting (3.1), (3.4c) and (3.4d) into (3.7), we have

$$\frac{d^2}{d\zeta^2}\psi_0 = 0, \tag{3.11a}$$

$$\frac{d^2\psi_1}{d\zeta^2} - \frac{d^2}{d\zeta^2}(\zeta\psi_0) = 0, \tag{3.11b}$$

$$\frac{d^2\psi_n}{d\zeta^2} - \frac{d^2}{d\zeta^2}(\zeta\psi_{n-1}) + \sum_{j=0}^{n-2}\sum_{k=0}^{j}\zeta^{j-k}\omega_{k,j-k}\Omega_{n-2-j} + \frac{1}{2}\sum_{j=0}^{n-2}\Omega_j\Omega_{n-2-j}$$

$$+ 4\sum_{j=0}^{n-2}k_j\Omega_{n-2-j} = 0, \quad n = 2, 3, 4, \ldots), \tag{3.12}$$

$$\frac{d^2}{d\zeta^2}\Omega_n - \frac{d^2}{d\zeta^2}(\zeta\Omega_{n-1}) - \sum_{j=0}^{n}k_j\psi_{n-j} - \frac{1}{4}\sum_{j=0}^{n}\sum_{k=0}^{j}\zeta^{j-k}\varphi_{k,j-k}\Omega_{n-j}$$

$$-\frac{1}{4}\sum_{j=0}^{n}\sum_{k=0}^{j}\zeta^{j-k}\omega_{k,j-k}\psi_{n-j} - \frac{1}{4}\sum_{j=0}^{n}\psi_j\Omega_{n-j} = 0, \quad (n = 0, 1, 2, \ldots),$$

$$\tag{3.13}$$

where

$$\Omega_{-1} \equiv 0, \tag{3.14}$$

$$\omega_{ij} = (-1)^j\frac{\omega_i^{(j)}(1)}{j!}, \tag{3.15a}$$

$$\varphi_{ij} = (-1)^j\frac{\varphi_i^{(j)}(1)}{j!}. \tag{3.15b}$$

As above, the derivation of the equations are completed. Their relating conditions of determined solutions are based on the supposition that:

$$\psi_i(\infty) = 0, \tag{3.16a}$$

$$\Omega_i(\infty) = 0, \quad (i = 0, 1, 2, 3, \ldots) \tag{3.16b}$$

and supposing that:

$$\varphi_i(0) \text{ is finite} \quad (i = 0, 1, 2, 3, \ldots), \tag{3.17}$$

and

$$\omega_i(1) + \Omega_i(0) + k_\beta\frac{d\omega_i(1)}{dx} - k_\beta\frac{d\Omega_{i+1}(0)}{d\zeta} = 0* \tag{3.18a}$$

*$d\omega_i(1)/dx$ means $(d\omega_i/dx)|_{x=1}$, $d\Omega_{i+1}(0)/d\zeta$ means $(d\Omega_{i+1}/d\zeta)|_{\zeta=0}$, the following are the same as this.

$$\varphi_i(1) + \psi_i(0) + k_F \frac{d\varphi_i(1)}{dx} - k_F \frac{d\psi_{i+1}(0)}{d\zeta} = 0, \quad (i = -1, 0, 1, 2, 3, \ldots), \tag{3.18b}$$

where ω_{-1}, Ω_{-1}, φ_{-1} and ψ_{-1} are all supposed to be zero.

As above, this is a general discussion. It must be pointed out that because the membrane equations are nonlinear, the proof of the existence and uniqueness of the solutions of these equations will be specially discussed in my other paper [10], one also may refer to L. S. Srubschchik's work [7].

4. The case of uniformly distributed load

We here discuss the cases of clamped or simple support. Now

$$k_F = \frac{2}{1-\mu},$$

$$k_\beta = \begin{cases} 0 & \text{for clamped support,} \\ \dfrac{2}{1+\mu} & \text{for simple support.} \end{cases}$$

From (3.11) and (3.16) we know that ψ_0 and ψ_1 are zero functions, i.e.,

$$\psi_0 \equiv 0, \qquad \psi_1 \equiv 0. \tag{4.1}$$

Now let us find the solutions of (3.8). After obtaining ω_0 from (3.8a) and substituting it into (3.8b), we obtain

$$\frac{d^2}{dx^2} x\varphi_0 - \frac{8p_0^2}{\varphi_0^2} - 8k_0^2 = 0 \tag{4.2}$$

for (3.17) and (3.18b) the boundary conditions are

$$\varphi_0(1) + \frac{2}{1-\mu} \varphi_0'(1) = 0, \tag{4.3a}$$

$$\varphi_0(0) \text{ is finite.} \tag{4.3b}$$

The problem of solving (4.2), satisfying boundary conditions (4.3), can become equivalent to one of finding the minimal value of $I(\varphi_0)$:

$$I(\varphi_0) = \frac{1}{2}\int_0^1 \left[x^2\varphi_0'^2 + \frac{1-\mu}{2}\varphi_0(\varphi_0 + 2x\varphi_0') + \frac{16p_0^2 x}{\varphi_0} + 16k_0^2 x\varphi_0 \right] dx. \tag{4.4}$$

After following the procedure, the existence and uniqueness of the solutions can be proved.

Substituting φ_0 obtained from the above into (3.8a), we get

$$\omega_0 = -\frac{4(p_0 + k_0\varphi_0)}{\varphi_0}. \tag{4.5}$$

Now let us find Ω_0. From (3.13) we obtain

$$\frac{d^2\Omega_0}{d\zeta^2} - \tfrac{1}{4}\varphi_{00}\Omega_0 = 0. \tag{4.6}$$

By (3.15b), we know

$$\varphi_{00} = \varphi_0(1). \tag{4.7}$$

By the conditions of determining solutions (3.16) and (3.18a), we get

$$\Omega_0 = \begin{cases} 0, & \text{for simple support,} \\ \dfrac{4(p_0 + k_0\varphi_{00})}{\varphi_{00}} e^{-(\sqrt{\varphi_{00}}/2)\zeta}, & \text{for clamped support.} \end{cases} \tag{4.8}$$

Substituting Ω_0 into (3.12), we obtain

$$\frac{d^2\psi_2}{d\zeta^2} = -\omega_0(1)\Omega_0 - \tfrac{1}{2}\Omega_0^2 - 4k_0\Omega_0. \tag{4.9}$$

After integration, we have

$$\psi_2 = \frac{64(p_0 + k_0\varphi_{00})^2}{\varphi_{00}\zeta}[e^{-(\sqrt{\varphi_{00}}/2)\zeta} - \tfrac{1}{8}e^{-\sqrt{\varphi_{00}}\zeta}] - \frac{64k_0(p_0 + k_0\varphi_{00})}{\varphi_{00}^2}e^{-(\sqrt{\varphi_{00}}/2)\zeta}. \tag{4.10a}$$

For the clamped support; and

$$\psi_2 = 0 \tag{4.10b}$$

for the simple support.

In the following we will find φ_1 and ω_1. From (3.9) we obtain

$$(x\varphi_1)'' - \frac{16p_0^2}{\varphi_0^3}\varphi_1 = -\left[4k_1\omega_0 + \frac{16p_0p_1}{\varphi_0^2} + \frac{16k_1p_0}{\varphi_0}\right], \tag{4.11}$$

the boundary conditions are

$$\varphi_1(0) < \infty, \tag{4.12a}$$

$$\varphi_1(1) + \frac{2}{1-\mu}\varphi_1'(1) - \frac{2}{1-\mu}\psi_2'(0) = 0. \tag{4.12b}$$

The problem of finding solutions of the equation (4.11) with conditions (4.12) can equivalently become the problem of finding the minimal value of functional $I_1(\varphi_1)$,

$$I_1(\varphi_1) = \frac{1}{2}\int_0^1 \left[x^2\varphi_1'^2 + \frac{16xp_0^2}{\varphi_0^3}\varphi_1^2 + \frac{1-\mu}{2}\varphi_1(\varphi_1 + 2x\varphi_1') - 2\psi_2'(0)(\varphi_1 + x\varphi_1') - 2x\lambda_0(x)\varphi_1\right]dx, \tag{4.13}$$

where

$$\lambda_0(\chi) = \frac{16p_0p_1}{\varphi_0^2} - 16k_0k_1. \tag{4.14}$$

After finding φ_1 we substitute it into (3.9a) and obtain ω_1:

$$\omega_1 = \frac{4p_0\varphi_1}{\varphi_0^2} - \frac{4p_1}{\varphi_0} - 4k_1. \tag{4.15}$$

Using the boundary condition (3.18), we can find Ω_1 from (3.13), for the boundary condition of clamped support:

$$\Omega_1 = -\omega_1(1)\,\mathrm{e}^{-(\varphi_{00}/2)\zeta} - \left[\frac{\varphi_{01}}{4\varphi_{00}} + \frac{\varphi_{10}}{4\sqrt{\varphi_{00}}} - \frac{3}{4}\right]\zeta\Omega_0$$
$$- \frac{1}{8\sqrt{\varphi_{00}}}(\varphi_{01} + \varphi_{00})\zeta^2\Omega. \tag{4.15a}$$

For the boundary condition of simple support

$$\Omega_1 = -\left[\frac{2}{\sqrt{\varphi_{00}}}\omega_0'(1) + \frac{1+\mu}{\sqrt{\varphi_{00}}}\omega_0(1)\right]\mathrm{e}^{-(\sqrt{\varphi_{00}}/2)\zeta} \tag{4.15b}$$

from (3.12) ψ_3 can be found.

According to the above method, we can find the asymptotic approximation of a higher order:

$$(x\varphi_i)'' - \frac{16p_0^2}{\varphi_0^3}\varphi_i = -\lambda_{i-1}(x), \quad (i = 2, 3, 4, \ldots), \tag{4.16}$$

where

$$\lambda_{i-1}(x) = \frac{1}{2}\sum_{j=1}^{i-1}\left(\omega_j + \frac{8p_0}{\varphi_0^2}\varphi_j\right)\omega_{i-j} + 4\sum_{j=1}^{i}k_j\left(\omega_{i-j} + \frac{4p_0}{\varphi_0^2}\varphi_{i-j}\right)$$
$$+ \frac{16p_0p_i}{\varphi_0^2} - \frac{16p_0}{\varphi_0^2}(x\omega_{i-2}). \tag{4.17}$$

Obviously, $\lambda_{i-1}(x)$ is a known function.

The boundary conditions for (4.16) are

$$\varphi_i(0) < \infty, \tag{4.18a}$$

$$\varphi_i(1) + \psi_i(0) + \frac{2}{1-\mu}[\varphi_i'(1) - \psi_{i+1}'(0)] = 0. \tag{4.18b}$$

The problem of solving (4.16) with boundary conditions (4.18) can be equivalently changed to the problem of finding a minimal value of the functional $I_i(\varphi_i)$.

$$I_i(\varphi_i) = \frac{1}{2}\int_0^1\left\{x^2\varphi_i'^2 + \left[\frac{1-\mu}{2} + \frac{16xp_0^2}{\varphi_0^3}\right]\varphi_i^2\right.$$
$$+ (1-\mu)x\varphi_i\varphi_i' + [(1-\mu)\psi_i(0) - 2\psi_{i+1}(0)]x\varphi_i$$
$$\left. + [(1-\mu)\psi_i(0) - 2\psi_{i+1}'(0) - 2\lambda_{i-1}x]\varphi_i\right\}\mathrm{d}x. \tag{4.19}$$

After finding $\varphi_i(x)$, according to the preceding method, we can find ω_i, Ω_i and ψ_{i+2} $(i \geqslant 2)$.

In the above asymptotic functions there are arbitrary constants k_j and p_j $(j = 1, 2, 3, \ldots)$. These constants may be determined through applications of mathematics and mechanics. For example, supposing

$$\varepsilon^3 p = 1 - \varepsilon, \tag{4.20}$$

or

$$\varepsilon = \left(\frac{1}{2p}\right)^{1/3} \left\{\left[\left(1 + \frac{4}{27p}\right)^{1/2} + 1\right]^{1/3} - \left[\left(1 + \frac{4}{27p}\right)^{1/2} - 1\right]^{1/3}\right\}. \tag{4.21}$$

Obviously, from (4.20) or (4.21) we know that for any values of p, we can ensure that ε is larger than zero and smaller than one, i.e.,

$$0 < \varepsilon < 1. \tag{4.22}$$

Let the load be of uniform distribution. Hence p and p_j are all constants. Comparing (4.20) and (3.2), obviously $p_0 = 1$, $p_1 = -1$, and $p_j \equiv 0$. For determining k_j, we may let $\omega_j(0) = 0$, for example $\omega_1(0) = 0$, then k_1 is determined as follows:

$$k_1 = \frac{\varphi_1(0)}{\varphi_0^2(0)} + \frac{1}{\varphi_0(0)}, \tag{4.23}$$

etc.

It is very suitable to determine the parameter ε by the method of (4.20) in the large load case, because ε decreases with the increase of γ shown in the following table:

p	500	750	1000	1500	1997.66
γ	2000	3000	4000	6000	7990.64
ε	0.1207	0.106	0.09665	0.0848	0.0773

the symbol γ in the table has the same meaning as in [8].

We use (4.23) to determine k_1, and use the methods of determining other k_j in order that the central area for "the membrane solutions" does not become concave.

References

1. Chien Wei-zang, "Asymptotic behaviour of a thin clamped circular plate under uniform normal pressure at very large deflection", *Science Report of Tsing Hua University* **5**, 1 (1948), 71–86.
2. Chien Wei-zang, "Large deflection of a circular clamped plate under uniform pressure", *Chinese Journal of Physics* **7**, 2 (1947), 102–113.
3. Chien Wei-zang, *Singular Perturbation Theory* (in Chinese), to be published in Szechuan People's Publishing House, Chengtu.

4. Chien Wei-zang, "The intrinsic theory of thin shells and plates. Part 1, General theory", *Quart. Appl. Math.* **1** (1944), 43–59; "Part 2, Application to thin plates", *ibid.* **2** (1944), 43–59; "Part 3, Application to thin shells", *ibid.* **2** (1944), 120–135.
5. Hu Hai-chang, "On the snapping of a thin spherical cap", *Acta Scientia Sinica* **3**, 4 (1954), 437–461.
6. Chien Wei-zang and Yeh Kai-yuan, "On the large deflection of circular plates" (in Chinese), *Acta Physica Sinica* **10**, 3 (1954), 209–236; (in English), *Acta Scientia Sinica* **III**, 4 (1954), 415–436.
7. Srubshchik, L. S., "Asymptotic integration of a system of the large deflection equation of symmetrically loaded shells of revolution" (in Russian) *PMM* **26**, 5 (1962).
8. Weinitchke, H. J., "On the stability problem for shallow spherical shells", *J. Math. and Phys.* **38**, 4 (1960), 209–231.
9. Reiss, E. L., "Bifurcation buckling of spherical caps", *Communs. Pure and Appl. Math.* **18**, 1/2 (1965), 65–82.
10. Chou Huan-wen, "The singular perturbation method applied to the problems of large deflection of circular plates", in *Singular Perturbation Theory and its Application in Mechanics* (edited by Chien Wei-zang) (1981).

*Nonlinear bending of circular sandwich plates under the action of axisymmetric uniformly distributed line loads**

LIU REN-HUAI

University of Science and Technology of China, Hefei, Anhui, PRC

Abstract. In the nonlinear bending theory of circular sandwich plates, nonlinear bending of the plates under the action of axisymmetric uniformly distributed line loads is a difficult problem. As far as we know, the problem has not yet been studied. In this paper, applying the perturbation method, a solution of this problem has been obtained.

1. Introduction

In recent years, many investigators have paid great attention to sandwich plates because of the importance of such plates in engineering. But few studies have considered the nonlinear bending problems of circular sandwich plates. This paper is an extension of the author's previous papers [1–3]. Here, we will consider a more difficult problem, i.e., the nonlinear bending problems of the plates with rigidly clamped edges under the action of axisymmetric uniformly distributed line loads. As we know, the problem has not yet been studied because of serious difficulties. It is apparent that the difficulties are due to the nonlinear differential equations and the discontinuity of the loads. To conquer these difficulties the Heaviside function will be introduced, and the perturbation method proposed by Chien Wei-zang [4] will be used. The obtained solution of this problem may be applied directly to engineering design.

2. Fundamental equations

Let us consider a rigidly clamped circular sandwich plate in which the load p is uniformly distributed along a circle of radius b as shown in Fig. 1. In order to simplify the following relation, the system of notation for the displacements and the stresses in this paper is the same as that used in the discussion of [2]. Proceeding

*This paper was completed in Ruhr-University, Bochum, Federal Republic of Germany. The author wishes to express his thanks to the Alexander von Humboldt Foundation, Ruhr-University, Bochum, and especially to W. Zerna for his warm hospitality.

Yeh, K. Y. (ed) Progress in Applied Mechanics
 ISBN-13: 978-94-010-8061-3

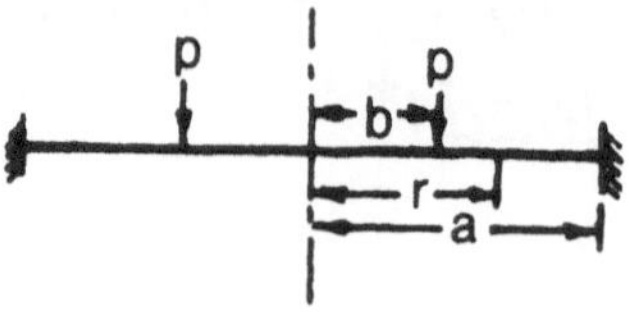

Figure 1

by the simplified equations in [2], we can easily obtain the fundamental equations of nonlinear bending for the plate. The equations satisfied by the deflection w and the radial stress σ_{r0} of the middle plane of the plate are

at $0 \leqslant r < b$,

$$Dr\frac{\mathrm{d}}{\mathrm{d}r}\frac{1}{r}\frac{\mathrm{d}}{\mathrm{d}r}r\frac{\mathrm{d}w}{\mathrm{d}r} + 2t\left(\frac{D}{G_2 h_0}r\frac{\mathrm{d}}{\mathrm{d}r}\frac{1}{r}\frac{\mathrm{d}}{\mathrm{d}r} - 1\right)\left(r\sigma_{r0}\frac{\mathrm{d}w}{\mathrm{d}r}\right) = 0, \tag{2.1a}$$

at $b < r \leqslant a$,

$$Dr\frac{\mathrm{d}}{\mathrm{d}r}\frac{1}{r}\frac{\mathrm{d}}{\mathrm{d}r}r\frac{\mathrm{d}w}{\mathrm{d}r} + 2t\left(\frac{D}{G_2 h_0}r\frac{\mathrm{d}}{\mathrm{d}r}\frac{1}{r}\frac{\mathrm{d}}{\mathrm{d}r} - 1\right)\left(r\sigma_{r0}\frac{\mathrm{d}w}{\mathrm{d}r}\right) - pb = 0, \tag{2.1b}$$

and

$$r\frac{\mathrm{d}}{\mathrm{d}r}\frac{1}{r}\frac{\mathrm{d}}{\mathrm{d}r}(r^2\sigma_{r0}) = -\frac{E}{2}\left(\frac{\mathrm{d}w}{\mathrm{d}r}\right)^2. \tag{2.2}$$

To simplify the calculations we now make use of the Heaviside function

$$H(r-b) = \begin{cases} 1 & \text{at } r \geqslant b \\ 0 & \text{at } r < b \end{cases} \tag{2.3}$$

thus (2.1a) and (2.1b) can be reduced to one:

$$Dr\frac{\mathrm{d}}{\mathrm{d}r}\frac{1}{r}\frac{\mathrm{d}}{\mathrm{d}r}r\frac{\mathrm{d}w}{\mathrm{d}r} + 2t\left(\frac{D}{G_2 h_0}\frac{\mathrm{d}}{\mathrm{d}r}\frac{1}{r}\frac{\mathrm{d}}{\mathrm{d}r} - 1\right)\left(r\sigma_{r0}\frac{\mathrm{d}w}{\mathrm{d}r}\right) - H(r-b)pb = 0. \tag{2.4}$$

In this way the solution of (2.2) and (2.4) can automatically satisfy jump condition for the shearing force and continuity conditions at the place of the action of the load.

The boundary conditions for a rigidly clamped edge are

$$w = 0, \quad \psi = 0, \quad u = 0 \qquad \text{for } r = a;$$

$$\psi = 0, \quad \sigma_{r0} \text{ finite} \qquad \text{for } r = 0. \tag{2.5}$$

Using expressions for the rotation ψ of a normal to the middle plane of the plate in the diametral plane and the radial displacement u of the middle plane of the plate, we can represent the boundary conditions in the following form:

$$w = 0, \quad \left(\frac{2t}{G_2 h_0}\sigma_{r0} + 1\right)\frac{\mathrm{d}w}{\mathrm{d}r} + \frac{pb}{G_2 h_0 a} = 0,$$

$$\frac{\mathrm{d}}{\mathrm{d}r}(r\sigma_{r0}) - \nu\sigma_{r0} = 0 \quad \text{for } r = a;$$

$$\left(\frac{2t}{G_2 h_0}\sigma_{r0} + 1\right)\frac{\mathrm{d}w}{\mathrm{d}r} = 0, \quad \sigma_{r0} \text{ finite} \quad \text{for } r = 0. \tag{2.6}$$

Equations (2.4), (2.2), and the boundary conditions (2.6) containing the two unknown functions w and σ_{r0}, will now be used in solving the problem. Having the solution for the nonlinear boundary value problem, we can find the tangential stress $\sigma_{\theta 0}$ of the middle plane of the plate:

$$\sigma_{\theta 0} = \frac{\mathrm{d}(r\sigma_{r0})}{\mathrm{d}r}. \tag{2.7}$$

To represent these equations (2.4), (2.2) and the conditions (2.6) in a simpler dimensionless form let us introduce the following notations:

$$y = \frac{r}{a}, \quad \beta = \frac{b}{a}, \quad W = \frac{\sqrt{2(1-\nu^2)}}{h_0}w, \quad S_r = \frac{2ta^2}{D}\sigma_{r0},$$
$$S_\theta = \frac{2ta^2}{D}\sigma_{\theta 0}, \quad k = \frac{D}{G_2 h_0 a^2}, \quad P = \frac{a^3\sqrt{2(1-\nu^2)}}{h_0 D}p. \tag{2.8}$$

With these notations, (2.4), (2.2) and the conditions (2.6) become, respectively,

$$L\left(y\frac{\mathrm{d}W}{\mathrm{d}y}\right) = -(kL - 1)\left(yS_r\frac{\mathrm{d}W}{\mathrm{d}y}\right) + H(y-\beta)\beta P,$$
$$L(y^2 S_r) = -\left(\frac{\mathrm{d}W}{\mathrm{d}y}\right)^2, \tag{2.9}$$

$$W = 0, \quad (kS_r + 1)\frac{\mathrm{d}W}{\mathrm{d}y} + k\beta P = 0, \quad \frac{\mathrm{d}S_r}{\mathrm{d}y} + \lambda_1 S_r = 0 \quad \text{for } y = 1;$$
$$(kS_r + 1)\frac{\mathrm{d}W}{\mathrm{d}y} = 0, \quad S_r \text{ finite} \quad \text{for } y = 0 \tag{2.10}$$

where

$$L(\ldots) = y\frac{\mathrm{d}}{\mathrm{d}y}\frac{1}{y}\frac{\mathrm{d}}{\mathrm{d}y}(\ldots), \tag{2.11}$$

$$\lambda_1 = 1 - \nu, \tag{2.12}$$

and (2.7) becomes

$$S_\theta = \frac{\mathrm{d}(yS_r)}{\mathrm{d}y}. \tag{2.13}$$

Equations (2.9) and (2.10) are the dimensionless nonlinear boundary value problem for discussing a rigidly clamped circular sandwich plate under the action of axisymmetric uniformly distributed line load. It should be noted that the difficulty in solving this problem is great. It is only just possible to find the solution by the Heaviside function. However, even in this case, results are still very complicated.

3. To solve the nonlinear boundary value problem by the perturbation method

To obtain a satisfactory solution of the above nonlinear boundary value problem, it is necessary to use the perturbation method. Assuming that the perturbation parameter is the dimensionless centre deflection W_0 for the circular sandwich plate, we represent dimensionless deflection W, dimensionless radial stress S_r, and dimensionless axisymmetric uniformly distributed line load P by the following power series for W_0:

$$\begin{aligned} W &= w_1(y)W_0 + w_3(y)W_0^3 + \dots, \\ S_r &= S_2(y)W_0^2 + S_4(y)W_0^4 + \dots, \\ P &= p_1W_0 + p_3W_0^3 + \dots, \end{aligned} \tag{3.1}$$

in which $w_1, w_3, \dots, S_2, S_4, \dots$ and $p_1, p_3, \dots$ are functions and constants to be determined later, respectively. Substituting these series in the problem (2.9) and (2.10), and observing that these equations must be satisfied for any power of W_0, we find a system of linear boundary value problems for $w_1, S_2, p_1, w_3, S_4, p_3, \dots$. For w_1 we have the linear boundary value problem as follows

$$L\left(y\frac{dw_1}{dy}\right) = H(y-\beta)\beta p_1, \tag{3.2}$$

$$\begin{aligned} w_1 &= 0, \quad \frac{dw_1}{dy} = -k\beta p_1 \quad \text{for } y = 1; \\ w_1 &= 1, \quad \frac{dw_1}{dy} = 0 \quad \text{for } y = 0. \end{aligned} \tag{3.3}$$

The solution of this problem is

$$\begin{aligned} w_1 = &-\frac{\beta p_1}{4}\{\alpha y^2 - (\beta^2\ln\beta - \tfrac{1}{2}\beta^2 + 2k + \tfrac{1}{2}) - H(y-\beta)[y^2\ln y \\ &- (\ln\beta + 1)y^2 + \beta^2\ln y - \beta^2(\ln\beta - 1)]\}, \end{aligned} \tag{3.4}$$

where

$$p_1 = \frac{8}{\beta(2\beta^2\ln\beta - \beta^2 + 4k + 1)}, \tag{3.5}$$

$$\alpha = \tfrac{1}{2}\beta^2 - \ln\beta + 2k - \tfrac{1}{2}. \tag{3.6}$$

For S_2 we have the linear boundary value problem as follows

$$L(y^2S_2) = -\left(\frac{dw_1}{dy}\right)^2, \tag{3.7}$$

$$\frac{dS_2}{dy} + \lambda_1 S_2 = 0 \quad \text{for } y = 1; \tag{3.8}$$

$$S_2 \text{ finite} \qquad \text{for } y = 0.$$

The solution of this problem is

$$S_2 = -\frac{\beta^2 p_1^2}{32}\left\{\alpha^2 y^2 - \frac{1}{\lambda_1}\left[\frac{1}{4}\lambda_2\beta^6 - \frac{1-7\nu}{2}\beta^4 \ln\beta\right.\right.$$
$$+ \left(\lambda_2 k - \frac{13-19\nu}{8}\right)\beta^4 - 2\lambda_1\beta^2\ln^2\beta + 4\lambda_1(2k+1)\beta^2\ln\beta$$
$$\left. - \left(2(3-\nu)k - \frac{1-7\nu}{4}\right)\beta^2 + 4(3-\nu)k^2 + (5-3\nu)k + \frac{9-7\nu}{8}\right]$$
$$+ H(y-\beta)\,[y^2\ln^2 y - (\beta^2 + 4k + \tfrac{3}{2})y^2\ln y + (\beta^2\ln\beta + \tfrac{5}{4}\beta^2$$
$$- \ln^2\beta + (4k-1)\ln\beta + 5k + \tfrac{5}{8})y^2 + 2\beta^2\ln^2 y$$
$$- 2\beta^2(\beta^2 + 4k + 1)\ln y + \beta^2(2\beta^2\ln\beta - \beta^2$$
$$- 2\ln^2\beta + 4(2k+1)\ln\beta - 4k + 1) - \beta^4 y^{-2}\ln y$$
$$\left. - \beta^4(\tfrac{1}{4}\beta^2 - \tfrac{3}{2}\ln\beta + k + \tfrac{3}{8})y^{-2}]\right\}, \tag{3.9}$$

where

$$\lambda_2 = 1 + \nu. \tag{3.10}$$

For w_3, we have the linear boundary value problem as follows

$$L\left(y\frac{dw_3}{dy}\right) = -(kL-1)\left(yS_2\frac{dw_1}{dy}\right) + H(y-\beta)\,\beta p_3, \tag{3.11}$$

$$w_3 = 0, \quad \frac{dw_3}{dy} = -k\left(S_2\frac{dw_1}{dy} + \beta p_3\right) \quad \text{for } y = 1; \tag{3.12}$$

$$w_3 = 0, \quad \frac{dw_3}{dy} = -kS_2\frac{dw_1}{dy} \quad \text{for } y = 0.$$

The solution of this problem is

$$w_3 = \frac{\beta^3 p_1^3}{512}\left[\sum_{i=1}^{3} f_{2i}y^{2i} + H(y-\beta)\sum_{i=-1}^{3}\sum_{j=0}^{3} g_{2i,j}\,y^{2i}\ln^j y\right] - \frac{\beta p_3}{4}\{\alpha y^2$$
$$- H(y-\beta)[y^2\ln y - (\ln\beta + 1)y^2 + \beta^2\ln y - \beta^2(\ln\beta - 1)]\}, \tag{3.13}$$

where

$$p_3 = -\frac{\beta^3 p_1^4}{1024\lambda_1}\left\{\frac{1+5\nu}{24}\beta^{10}\ln\beta - \frac{5+31\nu}{144}\beta^{10} + \frac{13-31\nu}{12}\beta^8\ln^2\beta\right.$$
$$- \left(\frac{1+17\nu}{6}k + \frac{139-229\nu}{48}\right)\beta^8\ln\beta$$

$$- \left(\frac{91 - 127\nu}{36} k - \frac{79}{36} \lambda_1 \right) \beta^8 - 3\lambda_1 \beta^6 \ln^3 \beta$$

$$+ \left(\frac{4 + 68\nu}{3} k + \frac{86}{9} \lambda_1 \right) \beta^6 \ln^2 \beta + \left(\frac{20 - 68\nu}{3} k^2 \right.$$

$$\left. + \frac{92 - 152\nu}{3} k - \frac{601 - 493\nu}{54} \right) \beta^6 \ln \beta - \left(\frac{221}{9} \lambda_1 k^2 \right.$$

$$\left. + \frac{391 - 463\nu}{18} k - \frac{5213 - 2135\nu}{2592} \right) \beta^6 + 12\lambda_1 k \beta^4 \ln^3 \beta$$

$$- 24\lambda_1 k (3k + 2) \beta^4 \ln^2 \beta + \left[32\lambda_1 k^3 + (206 - 122\nu) k^2 \right.$$

$$\left. + (79 - 51\nu) k + \frac{63 - 49\nu}{16} \right] \beta^4 \ln \beta - \left[(60 - 28\nu) k^3 \right.$$

$$\left. + (65 - 23\nu) k^2 + \frac{73 + 269\nu}{24} k + \frac{1607 - 1193\nu}{288} \right] \beta^4$$

$$- k [(96 - 32\nu) k^2 + (40 - 24\nu) k + 9 - 7\nu] \beta^2 \ln \beta$$

$$+ \left[(152 - 56\nu) k^3 + \frac{352 - 178\nu}{3} k^2 + \frac{280 - 181\nu}{9} k \right.$$

$$\left. + \frac{445 - 283\nu}{288} \right] \beta^2 - 32(5 - \nu) k^4 - \frac{920 - 272\nu}{9} k^3$$

$$\left. - \frac{250 - 106\nu}{9} k^2 - \frac{275 - 149\nu}{72} k - \frac{353 - 191\nu}{2592} \right\}, \tag{3.14}$$

$$f_2 = -\frac{1}{\lambda_1} \left\{ \frac{1 + 5\nu}{96} \beta^{10} + \frac{19 - 61\nu}{48} \beta^8 \ln \beta - \left(\frac{13 + 5\nu}{24} k \right. \right.$$

$$\left. + \frac{491 - 617\nu}{576} \right) \beta^8 - \frac{25 - 61\nu}{12} \beta^6 \ln^2 \beta + \left(\frac{10 + 8\nu}{3} k \right.$$

$$\left. + \frac{146 - 173\nu}{18} \right) \beta^6 \ln \beta - \left(\frac{13 + 11\nu}{3} k^2 + \frac{13 + 5\nu}{18} k \right.$$

$$\left. + \frac{3463 - 3301\nu}{864} \right) \beta^6 + \tfrac{3}{2}\lambda_1 \beta^4 \ln^3 \beta - [(3 + 9\nu) k + 6\lambda_1] \beta^4 \ln^2 \beta$$

$$- \left[(12 - 44\nu) k^2 - (7 - 3\nu) k - \frac{123 - 81\nu}{16} \right] \beta^4 \ln \beta$$

$$- \left[8\lambda_2 k^3 - (24 - 19\nu) k^2 + \frac{49 - 67\nu}{8} k - \frac{263 - 323\nu}{96} \right] \beta^4$$

$$- 8\lambda_1 k \beta^2 \ln^3 \beta + 12\lambda_1 k (4k + 1) \beta^2 \ln^2 \beta$$

$$- \left[64\lambda_1 k^3 + (52 - 28\nu) k^2 - (4 - 12\nu) k + \frac{9 - 7\nu}{8} \right] \beta^2 \ln \beta$$

$$+ \left[(24 - 8v)k^3 - (13 - 19v)k^2 + \frac{43}{6}\lambda_1 k + \frac{341 - 251v}{144}\right]\beta^2$$

$$+ k\left[(48 - 16v)k^2 + (20 - 12v)k + \frac{9 - 7v}{2}\right]\ln\beta$$

$$- 32(3 - v)k^4 - \frac{80 - 56v}{3}k^3 - \frac{20 - 11v}{3}k^2$$

$$+ \frac{4 - 13v}{18}k - \frac{445 - 283v}{1728}\Bigg\},$$

$$f_4 = -\frac{1}{\lambda_1}\Bigg\{\frac{1}{32}\lambda_2\beta^8 - \frac{v}{2}\beta^6\ln\beta + \left(\frac{1}{2}k - \frac{15 - 17v}{64}\right)\beta^6 - \frac{3 - 9v}{8}\beta^4\ln^2\beta$$

$$- \left(\frac{1 + 3v}{2}k - \frac{27 - 21v}{32}\right)\beta^4\ln\beta + \left(\frac{7 - 5v}{2}k^2 - \frac{39 - 33v}{16}k\right.$$

$$\left.+ \frac{15 - 33v}{64}\right)\beta^4 + \tfrac{1}{2}\lambda_1\beta^2\ln^3\beta - \tfrac{3}{4}\lambda_1\beta^2\ln^2\beta - \left(8\lambda_1 k^2 - \frac{11 - 9v}{2}k\right.$$

$$\left.+ \frac{9 - 15v}{16}\right)\beta^2\ln\beta + \left(12\lambda_1 k^3 - \frac{15 - 13v}{2}k^2 + \frac{9}{4}\lambda_1 k + \frac{7}{64}\lambda_2\right)\beta^2$$

$$- 2\lambda_1 k\ln^3\beta + 3\lambda_1 k(4k - 1)\ln^2\beta - \left[24\lambda_1 k^3 - (9 - 11v)k^2\right.$$

$$\left.+ \frac{11 - 9v}{4}k + \frac{9 - 7v}{32}\right]\ln\beta + 16\lambda_1 k^4 - 2(3 - 5v)k^3$$

$$+ 4\lambda_1 k^2 - \frac{5 - 3v}{16}k - \frac{9 - 7v}{64}\Bigg\},$$

$$f_6 = \tfrac{1}{18}\alpha^3,$$

$$g_{-2,0} = k\beta^6(\tfrac{1}{2}\beta^2 - 3\ln\beta + 2k + \tfrac{17}{4}),$$

$$g_{-2,1} = 2k\beta^6,$$

$$g_{-2,2} = g_{-2,3} = 0,$$

$$g_{0,0} = -\frac{\beta^2}{\lambda_1}\Bigg\{\frac{1 + 5v}{48}\beta^8\ln\beta - \frac{4 + 23v}{144}\beta^8 + \frac{13 - 31v}{24}\beta^6\ln^2\beta$$

$$- \left(\frac{1 + 17v}{12}k + \frac{59 - 101v}{32}\right)\beta^6\ln\beta - \left(\frac{8 - 53v}{36}k\right.$$

$$\left.- \frac{1051 - 1213v}{576}\right)\beta^6 - \tfrac{3}{2}\lambda_1\beta^4\ln^3\beta + \left(\frac{2 + 34v}{3}k + \frac{55}{9}\lambda_1\right)\beta^4\ln^2\beta$$

$$+ \left[\frac{10 - 34v}{3}k^2 + (7 - 23v)k - \frac{4045 - 3883v}{432}\right]\beta^4\ln\beta$$

$$- \left(\frac{107 - 251v}{18}k^2 + \frac{203 - 365v}{36}k - \frac{14107 - 12649v}{2592}\right)\beta^4$$

$$+ 6\lambda_1 k\beta^2 \ln^3\beta - 12\lambda_1 k(3k+2)\beta^2\ln^2\beta + \left[16\lambda_1 k^3 + (91 - 73\nu)k^2 + \frac{131 - 105\nu}{4}k + \frac{9 - 7\nu}{32}\right]\beta^2 \ln\beta - \left(22\lambda_1 k^3 + \frac{107 - 69\nu}{2}k^2 + \frac{45 + 9\nu}{8}k + \frac{27 - 21\nu}{64}\right)\beta^2 - k\left[(48 - 16\nu)k^2 + (20 - 12\nu)k + \frac{9 - 7\nu}{2}\right]\ln\beta + k\left[(48 - 16\nu)k^2 + (20 - 12\nu)k + \frac{9 - 7\nu}{2}\right]\Bigg\},$$

$$g_{0,1} = \frac{\beta^2}{\lambda_1}\left\{\frac{1 + 5\nu}{48}\beta^8 + \frac{19 - 37\nu}{24}\beta^6\ln\beta - \left(\frac{1 + 17\nu}{12}k + \frac{401 - 455\nu}{288}\right)\beta^6 - \tfrac{8}{3}\lambda_1\beta^4\ln^2\beta + \left(\frac{14 + 22\nu}{3}k + \frac{65}{9}\lambda_1\right)\beta^4\ln\beta + \left(\frac{10 - 34\nu}{3}k^2 + \frac{32 - 68\nu}{9}k - \frac{2347 - 2185\nu}{432}\right)\beta^4 - 24\lambda_1 k^2\beta^2\ln\beta + \left[16\lambda_1 k^3 + (37 - 19\nu)k^2 + \frac{17 + 9\nu}{4}k + \frac{9 - 7\nu}{32}\right]\beta^2 - k\left[(48 - 16\nu)k^2 + (20 - 12\nu)k + \frac{9 - 7\nu}{2}\right]\right\},$$

$$g_{0,2} = -\beta^4\left[\tfrac{1}{4}\beta^4 - \tfrac{3}{2}\beta^2\ln\beta + (4k + \tfrac{17}{8})\beta^2 - 6k\ln\beta + k(12k + \tfrac{21}{2})\right],$$

$$g_{0,3} = -\tfrac{1}{3}\beta^6,$$

$$g_{2,0} = -\frac{1}{\lambda_1}\left\{\frac{\nu}{2}\beta^8\ln\beta + \frac{3}{32}\lambda_1\beta^8 + \frac{3 - 15\nu}{4}\beta^6\ln^2\beta + \left(2\lambda_1 k - \frac{33 - 39\nu}{8}\right)\beta^6\ln\beta - \left(\frac{3 + 5\nu}{4}k + \frac{3}{16}\lambda_1\right)\beta^6 - \tfrac{3}{2}\lambda_1\beta^4\ln^3\beta + [(3 + 9\nu)k + 6\lambda_1]\beta^4\ln^2\beta + \left[(24 - 32\nu)k^2 + \frac{1 + 9\nu}{2}k - \frac{45 - 33\nu}{8}\right]\beta^4\ln\beta - \left(\frac{13 + 3\nu}{2}k^2 - \frac{45 - 57\nu}{4}k + \frac{15}{4}\lambda_1\right)\beta^4 + 8\lambda_1 k\beta^2\ln^3\beta - 12\lambda_1 k(4k + 1)\beta^2\ln^2\beta + \left[64\lambda_1 k^3 + (52 - 28\nu)k^2 - (4 - 12\nu)k + \frac{9 - 7\nu}{8}\right]\beta^2\ln\beta - k\left[8\lambda_1 k^2 - (46 - 30\nu)k - \frac{27 - 3\nu}{4}\right]\beta^2 - k\left[(48 - 16\nu)k^2 + (20 - 12\nu)k + \frac{9 - 7\nu}{2}\right]\ln\beta - k\left[(48 - 16\nu)k^2 + (20 - 12\nu)k + \frac{9 - 7\nu}{2}\right]\right\},$$

$$g_{2,1} = \frac{1}{\lambda_1}\left\{\frac{v}{2}\beta^8 - 3v\beta^6\ln\beta + \left(2\lambda_1 k - \frac{15-18v}{4}\right)\beta^6\right.$$

$$+ (12k + 3\lambda_1)\beta^4\ln\beta + \left[(24-32v)k^2 + (2-9v)k\right.$$

$$\left. - \frac{12-9v}{2}\right]\beta^4 + \left[64\lambda_1 k^3 + (76-52v)k^2 + (23-15v)k\right.$$

$$\left.\left. + \frac{9-7v}{8}\right]\beta^2 - k\left[(48-16v)k^2 + (20-12v)k + \frac{9-7v}{2}\right]\right\},$$

$$g_{2,2} = 3\beta^2[\tfrac{1}{4}\beta^4 - \tfrac{1}{2}\beta^2\ln\beta - (3k - \tfrac{7}{8})\beta^2 - 2k(8k+3)],$$

$$g_{2,3} = 8k\beta^2,$$

$$g_{4,0} = \frac{1}{\lambda_1}\left\{\frac{3-5v}{16}\beta^6\ln\beta - \frac{3v}{16}\beta^6 - \frac{3-9v}{8}\beta^4\ln^2\beta\right.$$

$$+ \left(\frac{1-3v}{4}k + \frac{39-9v}{32}\right)\beta^4\ln\beta - \left(\frac{6-3v}{4}k - \frac{99-117v}{64}\right)\beta^4$$

$$+ \tfrac{1}{2}\lambda_1\beta^2\ln^3\beta - \tfrac{3}{4}\lambda_1\beta^2\ln^2\beta - \left(8\lambda_1 k^2 - \frac{11-9v}{2}k\right.$$

$$\left. + \frac{9-15v}{16}\right)\beta^2\ln\beta - \left(\frac{27}{2}\lambda_1 k^2 - \frac{19-13v}{4}k - \frac{51-33v}{32}\right)\beta^2$$

$$- 2\lambda_1 k\ln^3\beta + 3\lambda_1 k(4k-1)\ln^2\beta - \left[24\lambda_1 k^3 - (9-11v)k^2\right.$$

$$\left. + \frac{11-9v}{4}k + \frac{9-7v}{32}\right]\ln\beta - 30\lambda_1 k^3 - \frac{19-13v}{2}k^2$$

$$\left. - \frac{25-19v}{8}k - \frac{27-21v}{64}\right\},$$

$$g_{4,1} = -\frac{1}{\lambda_1}\left[\frac{3-5v}{16}\beta^6 + \frac{3}{8}\lambda_2\beta^4\ln\beta + \left(\frac{1-3v}{4}k + \frac{57-63v}{32}\right)\beta^4\right.$$

$$- \left(8\lambda_1 k^2 - \frac{5-3v}{2}k - \frac{45-39v}{16}\right)\beta^2 - 24\lambda_1 k^3 - (21-19v)k^2$$

$$\left. - \frac{21-19v}{4}k - \frac{9-7v}{32}\right]$$

$$g_{4,2} = \tfrac{3}{4}\beta^4 + \tfrac{15}{4}\beta^2 - 12k^2 - \tfrac{9}{2}k,$$

$$g_{4,3} = -\tfrac{1}{2}\beta^2 + 2k,$$

$$g_{6,0} = \tfrac{1}{6}[\tfrac{1}{4}\beta^4\ln\beta + \tfrac{19}{48}\beta^4 - \tfrac{1}{2}\beta^2\ln^2\beta + (2k - \tfrac{1}{2})\beta^2\ln\beta + (\tfrac{19}{6}k + \tfrac{13}{24})\beta^2$$

$$+ \tfrac{1}{3}\ln^3\beta - (2k - \tfrac{1}{2})\ln^2\beta + (4k^2 - 2k + \tfrac{1}{4})\ln\beta$$

$$+ \tfrac{19}{3}k^2 + \tfrac{13}{6}k + \tfrac{16}{27}],$$

$$
\begin{aligned}
g_{6,1} &= -\tfrac{1}{3}\,[\tfrac{1}{8}\beta^4 + (k + \tfrac{13}{24})\beta^2 + 2k^2 + \tfrac{13}{6}k + \tfrac{2}{3}],\\
g_{6,2} &= \tfrac{1}{3}(\tfrac{1}{4}\beta^2 + k + \tfrac{13}{24}),\\
g_{6,3} &= -\tfrac{1}{18}.
\end{aligned}
\tag{3.15}
$$

For S_4 we have the linear boundary value problem as follows

$$
L(y^2 S_4) = -2\,\frac{\mathrm{d}w_1}{\mathrm{d}y}\,\frac{\mathrm{d}w_3}{\mathrm{d}y},
\tag{3.16}
$$

$$
\begin{aligned}
\frac{\mathrm{d}S_4}{\mathrm{d}y} + \lambda_1 S_4 &= 0 \quad \text{for } y = 1;\\
S_4 \text{ finite} & \quad\quad\ \, \text{for } y = 0.
\end{aligned}
\tag{3.17}
$$

The solution of this problem is

$$
S_4 = \frac{\beta^4 p_1^4}{512}\left[\sum_{i=0}^{3} m_{2i}y^{2i} - H(y-\beta)\sum_{i=-2}^{3}\sum_{j=0}^{4} n_{2i,j}\,y^{2i}\ln^j y\right],
\tag{3.18}
$$

where

$$
\begin{aligned}
m_0 &= \frac{1}{\lambda_1}\left\{\sum_{i=-2}^{3}[n_{2i,1} + (2i+1-\nu)n_{2i,0}] - \sum_{i=1}^{3}(2i+1-\nu)m_{2i}\right\},\\
m_2 &= \frac{\alpha}{4}\left(f_2 - \frac{128}{\beta^2 p_1^3}\,p_3\alpha\right),\\
m_4 &= \tfrac{1}{6}\alpha f_4,\\
m_6 &= \tfrac{1}{144}\alpha^4,\\
n_{-4,0} &= -\frac{k\beta^8}{8}\,(\tfrac{1}{2}\beta^2 - 3\ln\beta + 2k + \tfrac{19}{4}),\\
n_{-4,1} &= -\tfrac{1}{4}k\beta^8,\\
n_{-4,2} &= n_{-4,3} = n_{-4,4} = 0,\\
n_{-2,0} &= -\frac{\beta^5 p_1}{16\lambda_1}\left\{\frac{1-2\nu}{576}\beta^{12}\ln\beta + \frac{1+18\nu}{1152}\beta^{12} - \frac{3+20\nu}{288}\beta^{10}\ln^2\beta\right.\\
&\quad - \left(\frac{18-7\nu}{144}k - \frac{633-1469\nu}{13824}\right)\beta^{10}\ln\beta\\
&\quad + \left(\frac{59-27\nu}{144}k - \frac{3813-5177\nu}{27648}\right)\beta^{10} + \frac{1+51\nu}{96}\beta^8\ln^3\beta\\
&\quad - \left(\frac{89-93\nu}{72}k - \frac{925-1365\nu}{1152}\right)\beta^8\ln^2\beta\\
&\quad - \left(\frac{109-89\nu}{72}k^2 + \frac{2437+231\nu}{1728}k - \frac{7098-7225\nu}{1152}\right)\beta^8\ln\beta
\end{aligned}
$$

$$+ \left(\frac{643 - 393\nu}{144} k^2 + \frac{13943 - 8747\nu}{6912} k - \frac{299981 - 289049\nu}{82944}\right) \beta^8$$

$$- \frac{35}{144} \lambda_1 \beta^6 \ln^4 \beta + \lambda_1 \left(\frac{53}{8} k - \frac{683}{1152}\right) \beta^6 \ln^3 \beta$$

$$- \left(\frac{331}{36} \lambda_1 k^2 + \frac{1141 - 2525\nu}{288} k - \frac{49555 - 43363\nu}{13824}\right) \beta^6 \ln^2 \beta$$

$$- \left(\frac{37}{6} \lambda_1 k^3 + \frac{16673 - 14201\nu}{864} k^2 + \frac{47553 - 13361\nu}{6912} k\right.$$

$$\left. + \frac{1450375 - 1393351\nu}{331776}\right) \beta^6 \ln \beta + \left(\frac{159 - 155\nu}{9} k^3\right.$$

$$+ \frac{44355 - 41963\nu}{1728} k^2 + \frac{606815 - 604223\nu}{41472} k$$

$$\left. + \frac{3284315 - 3394619\nu}{663552}\right) \beta^6 - \frac{47}{9} \lambda_1 k \beta^4 \ln^4 \beta$$

$$+ \lambda_1 \left(\frac{76}{3} k^2 + \frac{803}{72} k + \frac{289}{288}\right) \beta^4 \ln^3 \beta$$

$$- \left(\frac{44}{3} \lambda_1 k^3 - \frac{1229 - 1389\nu}{36} k^2 - \frac{18287 - 15023\nu}{1728} k\right.$$

$$\left. + \frac{5227 - 6419\nu}{2304}\right) \beta^4 \ln^2 \beta - \left(\frac{80}{9} \lambda_1 k^4 + \frac{1679 - 1659\nu}{18} \lambda_1 k^3\right.$$

$$+ \frac{36151 - 38343\nu}{432} k^2 + \frac{263473 - 234337\nu}{20736} k$$

$$\left. - \frac{53989 - 78189\nu}{27648}\right) \beta^4 \ln \beta + \left(\frac{259}{9} \lambda_1 k^4 + \frac{41011 - 40315\nu}{432} k^3\right.$$

$$+ \frac{120725 - 129269\nu}{3456} k^2 - \frac{640589 - 654893\nu}{55296} k$$

$$\left. - \frac{436891 - 760123\nu}{663552}\right) \beta^4 - \tfrac{64}{9} \lambda_1 k(4k + 1) \beta^2 \ln^3 \beta$$

$$+ \tfrac{2}{9} \lambda_1 k(768k^2 + 604k + 103) \beta^2 \ln^2 \beta - \left(\frac{448}{3} \lambda_1 k^4\right.$$

$$+ \frac{3290 - 3192\nu}{9} k^3 + \frac{13847 - 14042\nu}{108} k^2$$

$$\left. + \frac{2765 - 3593\nu}{432} k - \frac{5405 - 3131\nu}{6912}\right) \beta^2 \ln \beta$$

$$+ \left(\frac{128}{9} \lambda_1 k^5 + \frac{1120 - 1104\nu}{9} k^4 + \frac{4321 - 4071\nu}{27} k^3 \right.$$

$$\left. - \frac{5539 - 4147\nu}{648} k^2 - \frac{18797 - 16511\nu}{1296} k - \frac{3193 - 1669\nu}{5184} \right) \beta^2$$

$$+ \left(24\lambda_1 k^4 + \frac{58 - 76\nu}{3} k^3 + \frac{23 - 32\nu}{3} k^2 + \frac{49 - 103\nu}{96} k \right.$$

$$\left. - \frac{353 - 191\nu}{3456} \right) \ln \beta - 16\lambda_1 k^5 - \frac{398}{9} \lambda_1 k^4 - \frac{281 - 326\nu}{9} k^3$$

$$\left. - \frac{1529 - 1889\nu}{144} k^2 - \frac{4495 - 6925\nu}{5184} k + \frac{1627 - 817\nu}{20736} \right\},$$

$$n_{-2,1} = -\frac{\beta^5 p_1}{8\lambda_1} \left\{ \frac{1 + 8\nu}{288} \beta^{10} + \frac{1}{8} \lambda_1 \beta^8 \ln^2 \beta \right.$$

$$+ \left(\frac{1}{2} \lambda_1 k - \frac{85 - 31\nu}{288} \right) \beta^8 \ln \beta + \left(\frac{61 - 25\nu}{144} k - \frac{17 - 15\nu}{128} \right) \beta^8$$

$$- \frac{7}{12} \lambda_1 \beta^6 \ln^3 \beta - \lambda_1 \left(k - \frac{95}{36} \right) \beta^6 \ln^2 \beta$$

$$+ \left(4\lambda_1 k^2 - \frac{217 - 163\nu}{72} k - \frac{155 - 161\nu}{48} \right) \beta^6 \ln \beta$$

$$+ \left(\frac{76 - 67\nu}{18} k^2 + \frac{649 - 703\nu}{288} k + \frac{2854 - 3259\nu}{2592} \right) \beta^6$$

$$- 3\lambda_1 k \beta^4 \ln^3 \beta - \lambda_1 \left(6k^2 - \frac{46}{3} k + \frac{2}{3} \right) \beta^4 \ln^2 \beta$$

$$+ \left(8\lambda_1 k^3 - \frac{40}{3} \lambda_1 k^2 - \frac{1637 - 1583\nu}{72} k + \frac{385 - 439}{288} \right) \beta^4 \ln \beta$$

$$+ \left(\frac{55}{3} \lambda_1 k^3 + \frac{223}{18} \lambda_1 k^2 - \frac{1163 - 1649\nu}{864} k - \frac{149 - 203\nu}{216} \right) \beta^4$$

$$- 3\lambda_1 k (16k^2 - 1) \beta^2 \ln \beta + \left[32\lambda_1 k^4 + (37 - 35\nu) k^3 \right.$$

$$\left. - \frac{164 - 155\nu}{6} k^2 - \frac{3041 - 2807\nu}{288} k - \frac{91. - 55\nu}{288} \right] \beta^2$$

$$- 8\lambda_1 k^4 - \frac{58 - 76\nu}{9} k^3 - \frac{23 - 32\nu}{9} k^2$$

$$\left. - \frac{49 - 103\nu}{288} k + \frac{353 - 191\nu}{10368} \right\},$$

$$n_{-2,2} = \frac{\beta^6}{4} \left[\frac{1}{4} \beta^4 - \frac{3}{2} \beta^2 \ln \beta + \left(3k + \frac{21}{8} \right) \beta^2 - 12k \ln \beta + 8k^2 + 21k \right],$$

$$n_{-2,3} = \frac{\beta^6}{3}\left(\frac{1}{4}\beta^2 + 2k\right),$$

$$n_{-2,4} = 0,$$

$$n_{0,0} = \frac{\beta^3 p_1}{16\lambda_1}\Bigg\{\frac{37 + 125v}{1728}\beta^{12}\ln\beta - \frac{37 + 125v}{3456}\beta^{12} + \frac{13 - 31v}{48}\beta^{10}\ln^3\beta$$

$$- \left(\frac{13 - 7v}{12}k + \frac{183 - 201v}{288}\right)\beta^{10}\ln^2\beta + \left(\frac{269 - 143v}{72}k\right.$$

$$\left.- \frac{35 - 71v}{576}\right)\beta^{10}\ln\beta - \left(\frac{1367 - 665v}{864}k - \frac{305 - 62v}{1728}\right)\beta^{10}$$

$$- \frac{25 - 79v}{24}\beta^8\ln^4\beta + \left(\frac{23 - 17v}{4}k + \frac{2051 - 3293v}{288}\right)\beta^8\ln^3\beta$$

$$- \left(\frac{35 - 29v}{3}k^2 + \frac{1735 - 1105v}{72}k + \frac{6451 - 7477v}{576}\right)\beta^8\ln^2\beta$$

$$+ \left(\frac{1405 - 1171v}{36}k^2 + \frac{1789 - 1357v}{72}k\right.$$

$$\left.+ \frac{113497 - 140551v}{20736}\right)\beta^8\ln\beta - \left(\frac{403 - 373v}{24}k^2\right.$$

$$\left.+ \frac{406 - 469v}{144}k + \frac{73105 - 78775v}{41472}\right)\beta^8 - \frac{5}{6}\lambda_1\beta^6\ln^5\beta$$

$$- \lambda_1\left(4k - \frac{223}{108}\right)\beta^6\ln^4\beta + \left(\frac{110}{3}\lambda_1 k^2 + \frac{1109 - 515v}{36}k\right.$$

$$\left.+ \frac{11 + 25v}{9}\right)\beta^6\ln^3\beta - \left(\frac{136}{3}\lambda_1 k^3 + \frac{3367 - 3169v}{18}k^2\right.$$

$$\left.+ \frac{23693 - 12083v}{432}k + \frac{2615 + 8023v}{1728}\right)\beta^6\ln^2\beta$$

$$+ \left(\frac{4114 - 4006v}{27}k^3 + \frac{7837 - 7621v}{36}k^2 - \frac{9641 - 16283v}{1296}k\right.$$

$$\left.- \frac{77779 - 124921v}{15552}\right)\beta^6\ln\beta - \left(\frac{3733 - 3571v}{54}k^3\right.$$

$$+ \frac{380 - 425v}{12}k^2 - \frac{190597 - 194971v}{10368}k$$

$$\left.- \frac{397807 - 519793v}{124416}\right)\beta^6 - \lambda_1\left(24k^2 + \frac{5}{3}k - \frac{13}{12}\right)\beta^4\ln^4\beta$$

$$+ \left(96\lambda_1 k^3 + \frac{445 - 463v}{3}k^2 + \frac{2297 - 2135v}{108}k\right.$$

$$
\begin{aligned}
&- \frac{3173 - 3659\nu}{864}\bigg) \beta^4 \ln^3 \beta - \bigg(64\lambda_1 k^4 + \frac{1684 - 1660\nu}{3} k^3 \\
&+ \frac{2977 - 3193\nu}{9} k^2 + \frac{1541 - 3431\nu}{216} k - \frac{6025 - 5431\nu}{864}\bigg) \beta^4 \ln^2 \beta \\
&+ \bigg(\frac{752}{3} \lambda_1 k^4 + \frac{1939 - 1897\nu}{3} k^3 + \frac{17807 - 22397\nu}{216} k^2 \\
&- \frac{62341 - 53755\nu}{648} k - \frac{98303 - 59585\nu}{20736}\bigg) \beta^4 \ln \beta - \bigg(\frac{3520}{27} \lambda_1 k^4 \\
&+ \frac{5063 - 4847\nu}{54} k^3 - \frac{89131 - 83461\nu}{1296} k^2 - \frac{42377 - 9815\nu}{7776} k \\
&- \frac{20255 + 141583\nu}{124416}\bigg) \beta^4 - 6\lambda_1 k(4k + 1)\beta^2 \ln^4 \beta \\
&+ 6\lambda_1 k(32k^2 + 28k + 5)\beta^2 \ln^3 \beta - \bigg[448\lambda_1 k^4 + \frac{2242 - 2134\nu}{3} k^3 \\
&+ (278 - 268\nu)k^2 + \frac{3655 - 3673\nu}{144} k - \frac{565 - 187\nu}{1728}\bigg] \beta^2 \ln^2 \beta \\
&+ \bigg[128\lambda_1 k^5 + (568 - 552\nu)k^4 + \frac{2675 - 2297\nu}{9} k^3 \\
&- \frac{451 - 601\nu}{12} k^2 - \frac{5503 - 5809\nu}{288} k + \frac{8741 - 9227\nu}{10368}\bigg] \beta^2 \ln \beta \\
&- \bigg[96\lambda_1 k^5 + (58 - 46\nu)k^4 - \frac{4337 - 4355\nu}{36} k^3 + \frac{561 - 403\nu}{16} k^2 \\
&+ \frac{10463 - 8801\nu}{384} k + \frac{81289 - 55207\nu}{41472}\bigg] \beta^2 \\
&+ \bigg(32\lambda_1 k^4 + \frac{232 - 304\nu}{9} k^3 + \frac{92 - 128\nu}{9} k^2 + \frac{49 - 103\nu}{72} k \\
&- \frac{353 - 191\nu}{2592}\bigg) \ln^2 \beta - \bigg(128\lambda_1 k^5 + \frac{1888}{9} \lambda_1 k^4 \\
&+ \frac{1204 - 1240\nu}{9} k^3 + \frac{701 - 737\nu}{18} k^2 + \frac{5021 - 5507\nu}{1296} k \\
&- \frac{77 + 85\nu}{5184}\bigg) \ln \beta + 64\lambda_1 k^5 + \frac{512}{9} \lambda_1 k^4 + \frac{254 - 164\nu}{9} k^3 \\
&+ \frac{149 + 31\nu}{36} k^2 + \frac{2375 + 55\nu}{2592} k + \frac{2041 - 1231\nu}{10368}\bigg\},
\end{aligned}
$$

$$
\begin{aligned}
n_{0,1} = & -\frac{\beta^3 p_1}{8\lambda_1}\left\{\frac{1+8\nu}{144}\beta^{12} + \frac{1}{4}\lambda_1\beta^{10}\ln^2\beta - \left(\frac{13-7\nu}{24}k\right.\right. \\
& \left. + \frac{83+7\nu}{288}\right)\beta^{10}\ln\beta + \left(\frac{79-37\nu}{48}k - \frac{59}{144}\lambda_1\right)\beta^{10} - \frac{7}{6}\lambda_1\beta^8\ln^3\beta \\
& + \left(\frac{77-59\nu}{12}k + \frac{469-523\nu}{144}\right)\beta^8\ln^2\beta - \left(\frac{35-29\nu}{6}k^2\right. \\
& \left. + \frac{601-391\nu}{48}k + \frac{1589-1571\nu}{576}\right)\beta^8\ln\beta + \left(\frac{53-44\nu}{3}k^2\right. \\
& \left. + \frac{1393-1303\nu}{288}k + \frac{5291-8045\nu}{10368}\right)\beta^8 - \frac{\lambda_1}{6}(32k-1)\beta^6\ln^3\beta \\
& + \left(\frac{128}{3}\lambda_1 k^2 + \frac{119-101\nu}{6}k - \frac{61-115\nu}{48}\right)\beta^6\ln^2\beta \\
& - \left(\frac{68}{3}\lambda_1 k^3 + \frac{1711-1657\nu}{18}k^2 + \frac{2759-1553\nu}{216}k\right. \\
& \left. - \frac{5045-7367\nu}{1728}\right)\beta^6\ln\beta + \left(\frac{631-613\nu}{9}k^3 + \frac{459-477\nu}{8}k^2\right. \\
& \left. - \frac{13055-12569\nu}{1296}k - \frac{1985-3929\nu}{5184}\right)\beta^6 - 6\lambda_1 k(4k+1)\beta^4\ln^3\beta \\
& + \lambda_1\left(96k^3 + \frac{292}{3}k^2 + \frac{38}{3}k - \frac{2}{3}\right)\beta^4\ln^2\beta - \left(32\lambda_1 k^4\right. \\
& + \frac{718-706\nu}{3}k^3 + \frac{943-979\nu}{9}k^2 - \frac{355+203\nu}{144}k \\
& \left. - \frac{1513-1063\nu}{576}\right)\beta^4\ln\beta + \left(\frac{352}{3}\lambda_1 k^4 + \frac{1403-1349\nu}{9}k^3\right. \\
& \left. - \frac{587-425\nu}{54}k^2 - \frac{3109-2758\nu}{216}k - \frac{296-53\nu}{864}\right)\beta^4 \\
& - \left(96\lambda_1 k^4 + \frac{248-224\nu}{3}k^3 + \frac{95-80\nu}{3}k^2 + \frac{181-145\nu}{36}k\right. \\
& \left. + \frac{445-283\nu}{1728}\right)\beta^2\ln\beta + \left[64\lambda_1 k^5 + (84-76\nu)k^4\right. \\
& + \frac{223-169\nu}{9}k^3 + \frac{169-91\nu}{24}k^2 + \frac{143-107\nu}{36}k \\
& \left. + \frac{1759-1435\nu}{2592}\right]\beta^2 - 64\lambda_1 k^5 - \frac{800}{9}\lambda_1 k^4 - 2(27-26\nu)k^3 - \\
& \left. - \frac{517-481\nu}{36}k^2 - \frac{4139-3653\nu}{2592}k - \frac{629-467\nu}{10368}\right\},
\end{aligned}
$$

$$n_{0,2} = \frac{\beta^3 p_1}{16\lambda_1}\left\{\frac{11+7\nu}{48}\beta^{10}\ln\beta - \frac{29-11\nu}{288}\beta^{10} - \frac{43+11\nu}{24}\beta^8\ln^2\beta\right.$$

$$+\left(\frac{85-67\nu}{12}k + \frac{67+203\nu}{288}\right)\beta^8\ln\beta - \left(\frac{7-25\nu}{18}k\right.$$

$$\left.+\frac{317-227\nu}{576}\right)\beta^8 + \frac{1}{3}\lambda_1\beta^6\ln^3\beta + \lambda_1\left(\frac{10}{3}k-1\right)\beta^6\ln^2\beta$$

$$+\left(\frac{146}{3}\lambda_1 k^2 - \frac{139+239\nu}{36}k - \frac{164+25\nu}{108}\right)\beta^6\ln\beta +$$

$$+\left(\frac{193-103\nu}{18}k^2 + \frac{557-179\nu}{144}k + \frac{4259-3125\nu}{5184}\right)\beta^6$$

$$-12\lambda_1 k\beta^4\ln^3\beta + 4\lambda_1\left(6k^2 + \frac{23}{3}k - \frac{1}{3}\right)\beta^4\ln^2\beta$$

$$+\left(96\lambda_1 k^3 + \frac{19-\nu}{3}k^2 - \frac{287-233\nu}{36}k + \frac{311-473\nu}{288}\right)\beta^4\ln\beta$$

$$+\left(\frac{320}{3}\lambda_1 k^3 + \frac{1157-1319\nu}{18}k^2 - \frac{5005-4519\nu}{216}k\right.$$

$$\left.+\frac{925-331\nu}{1728}\right)\beta^4 - 6\lambda_1 k(16k^2+8k+1)\beta^2\ln\beta$$

$$+\left[256\lambda_1 k^4 + (318-298\nu)k^3 + \frac{266}{3}\lambda_1 k^2 + \frac{47-353\nu}{144}k\right.$$

$$\left.-\frac{485-251\nu}{576}\right]\beta^2 - 32\lambda_1 k^4 - \frac{232-304\nu}{9}k^3 - \frac{92-128\nu}{9}k^2$$

$$\left.-\frac{49-103\nu}{72}k + \frac{353-191\nu}{2592}\right\},$$

$$n_{0,3} = \beta^4\left[\frac{1}{8}\beta^4 + \frac{1}{4}\beta^2\ln\beta - \left(\frac{5}{2}k + \frac{1}{48}\right)\beta^2 + 2k\ln\beta - 12k^2 - \frac{13}{2}k\right],$$

$$n_{0,4} = -\beta^4\left(\frac{1}{8}\beta^2 - k\right)$$

$$n_{2,0} = -\frac{\beta p_1}{32\lambda_1}\left\{\frac{1-19\nu}{48}\beta^{12}\ln^2\beta - \frac{17-143\nu}{576}\beta^{12}\ln\beta - \frac{17+253\nu}{1152}\beta^{12}\right.$$

$$-\frac{1-7\nu}{2}\beta^{10}\ln^3\beta - \left(\frac{11+19\nu}{12}k - \frac{59-93\nu}{16}\right)\beta^{10}\ln^2\beta$$

$$-\left(\frac{1-43\nu}{48}k + \frac{135-345\nu}{128}\right)\beta^{10}\ln\beta - \left(\frac{73-33\nu}{16}k\right.$$

$$\left.-\frac{3263-3029\nu}{2304}\right)\beta^{10} + \frac{19-73\nu}{12}\beta^8\ln^4\beta - \left(\frac{41-95\nu}{6}k\right.$$

$$+ \frac{641 - 767\nu}{48}\Big) \beta^8 \ln^3 \beta - \left(\frac{52 - 40\nu}{3} k^2 - \frac{607 - 544\nu}{18} k\right.$$

$$- \frac{6895 - 4465\nu}{864}\Big) \beta^8 \ln^2 \beta + \left[(2 - 6\nu) k^2 + \frac{425 - 119\nu}{36} k\right.$$

$$+ \frac{29585 - 54047\nu}{10368}\Big] \beta^8 \ln \beta - \left(\frac{1165 - 1063\nu}{24} k^2\right.$$

$$+ \frac{1333 - 1543\nu}{192} k + \frac{51283 - 72181\nu}{20736}\Big) \beta^8$$

$$+ \lambda_1 \left(\frac{8}{3} k + \frac{49}{9}\right) \beta^6 \ln^4 \beta - \left(\frac{40}{3} \lambda_1 k^2 + \frac{499 - 283\nu}{9} k\right.$$

$$+ \frac{1306 - 631\nu}{108}\Big) \beta^6 \ln^3 \beta - \left(\frac{256}{3} \lambda_1 k^3 - \frac{791 - 449\nu}{9} k^2\right.$$

$$+ \frac{1319 - 2075\nu}{108} k + \frac{19007 - 21599\nu}{2592}\Big) \beta^6 \ln^2 \beta$$

$$- \left(\frac{194 - 50\nu}{9} k^3 - \frac{5849 - 5759\nu}{36} k^2 - \frac{88891 - 74797\nu}{1296} k\right.$$

$$+ \frac{10511 - 46151\nu}{10368}\Big) \beta^6 \ln \beta - \left(\frac{1603}{9} \lambda_1 k^3 + \frac{4520 - 4799\nu}{36} k^2\right.$$

$$- \frac{33083 - 34541\nu}{1296} k - \frac{25697 - 45218\nu}{10368}\Big) \beta^6 + 4\lambda_1 k \beta^4 \ln^5 \beta$$

$$- \lambda_1 \left(8k^2 - 6k + \frac{3}{2}\right) \beta^4 \ln^4 \beta - \left[(90 - 102\nu) k^2 + \frac{25}{2} \lambda_1 k\right.$$

$$- \frac{57 - 63\nu}{16}\Big] \beta^4 \ln^3 \beta - \left[128\lambda_1 k^4 - (212 - 228\nu) k^3\right.$$

$$+ (104 - 74\nu) k^2 + \frac{3307 - 2809\nu}{24} k + \frac{249 - 143\nu}{32}\Big] \beta^4 \ln^2 \beta$$

$$- \left[240\lambda_1 k^4 - (378 - 346\nu) k^3 - \frac{4751 - 4685\nu}{12} k^2\right.$$

$$+ \frac{365 - 301\nu}{16} k + \frac{2227 - 2137\nu}{384}\Big] \beta^4 \ln \beta - \left[224\lambda_1 k^4\right.$$

$$+ (375 - 350\nu) k^3 - \frac{541 - 475\nu}{12} k^2 - \frac{1438 - 811\nu}{48} k$$

$$- \frac{1133 + 694\nu}{1152}\Big] \beta^4 + 8\lambda_1 k(4k + 1) \beta^2 \ln^4 \beta$$

$$- 8\lambda_1 k(32k^2 + 12k + 1) \beta^2 \ln^3 \beta + \left[640\lambda_1 k^4 + \frac{856 - 712\nu}{3} k^3\right.$$

$$- (70 - 86\nu)k^2 - \frac{505 - 523\nu}{18}k + \frac{41 - 95\nu}{432}\Big]\beta^2 \ln^2\beta$$

$$- \Big[512\lambda_1 k^5 + (160 - 96\nu)k^4 - \frac{3196 - 3160\nu}{9}k^3$$

$$- \frac{79 - 130\nu}{6}k^2 + \frac{2707 - 2275\nu}{72}k + \frac{32147 - 21941\nu}{10368}\Big]\beta^2 \ln\beta$$

$$- \Big(208\lambda_1 k^4 + \frac{22 + 14\nu}{9}k^3 + \frac{163 - 115\nu}{4}k^2$$

$$+ \frac{203 - 165\nu}{8}k + \frac{13619 - 10217\nu}{5184}\Big)\beta^2$$

$$- \Big(32\lambda_1 k^4 + \frac{232 - 304\nu}{9}k^3 + \frac{92 - 128\nu}{9}k^2 + \frac{49 - 103\nu}{72}k$$

$$- \frac{353 - 191\nu}{2592}\Big)\ln^2\beta + \Big(128\lambda_1 k^5 + \frac{1024}{9}\lambda_1 k^4 + \frac{508 - 328\nu}{9}k^3$$

$$+ \frac{149 + 31\nu}{18}k^2 + \frac{2375 + 55\nu}{1296}k + \frac{2041 - 1231\nu}{5184}\Big)\ln\beta$$

$$+ 160\lambda_1 k^5 + \frac{1820}{9}\lambda_1 k^4 + \frac{1070 - 980\nu}{9}k^3 + \frac{2125 - 1765\nu}{72}k^2$$

$$+ \frac{9245 - 6815\nu}{2592}k + \frac{2455 - 1645\nu}{10368}\Big\},$$

$$n_{2,1} = \frac{\beta p_1}{32\lambda_1}\Big\{\frac{1 - 19\nu}{48}\beta^{12}\ln\beta - \frac{7 - 25\nu}{288}\beta^{12} + \frac{7 + 47\nu}{24}\beta^{10}\ln^2\beta$$

$$- \Big(\frac{11 + 19\nu}{12}k - \frac{815 - 1265\nu}{288}\Big)\beta^{10}\ln\beta$$

$$- \Big(\frac{81 - 45\nu}{36}k + \frac{289 - 451\nu}{576}\Big)\beta^{10} - \frac{1}{3}\lambda_1\beta^8\ln^3\beta$$

$$- \Big[(17 - 23\nu)k + \frac{21 - 51\nu}{8}\Big]\beta^8\ln^2\beta$$

$$- \Big(\frac{52 - 40\nu}{3}k^2 - \frac{1333 - 1063\nu}{36}k - \frac{8809 - 8431\nu}{864}\Big)\beta^8\ln\beta$$

$$- \Big(\frac{149 - 95\nu}{6}k^2 + \frac{1567 - 973\nu}{144}k + \frac{10253 - 9443\nu}{5184}\Big)\beta^8$$

$$- \frac{9}{2}\lambda_1\beta^6\ln^3\beta - \Big(\frac{416}{3}\lambda_1 k^2 + \frac{508 - 454\nu}{9}k - \frac{26 - 53\nu}{12}\Big)\beta^6\ln^2\beta$$

$$- \Big(\frac{256}{3}\lambda_1 k^3 - \frac{409 - 319\nu}{3}k^2 - \frac{2803 + 275\nu}{108}k$$

$$+ \frac{1243 - 2755\nu}{288}\bigg) \beta^6 \ln \beta - \bigg[\frac{1196 - 1052\nu}{9} k^3 + (87 - 82\nu)k^2$$

$$- \frac{25223 - 26681\nu}{648} k - \frac{2927 - 3710\nu}{864}\bigg] \beta^6 + 6\lambda_1 k(8k + 3)\beta^4 \ln^3 \beta$$

$$- 12\lambda_1 k(24k^2 + 25k + 6)\beta^4 \ln^2 \beta - \bigg[128\lambda_1 k^4 - (168 - 184\nu)k^3$$

$$- (61 - 83\nu)k^2 + \frac{535 - 427\nu}{12} k + \frac{2093 - 1679\nu}{288}\bigg] \beta^4 \ln \beta$$

$$- \bigg[400\lambda_1 k^4 + (424 - 436\nu)k^3 - \frac{32 + 58\nu}{3} k^2 - \frac{443 - 366\nu}{8} k$$

$$- \frac{101 + 142\nu}{288}\bigg] \beta^4 + \bigg(\frac{64 - 16\nu}{3} k^3 + \frac{46 - 16\nu}{3} k^2 + \frac{73 - 37\nu}{18} k$$

$$+ \frac{445 - 283\nu}{864}\bigg) \beta^2 \ln \beta - \bigg[512\lambda_1 k^5 + (800 - 736\nu)k^4$$

$$+ \frac{5036 - 4532\nu}{9} k^3 + \frac{614 - 566\nu}{3} k^2 + \frac{2197 + 2035\nu}{72} k$$

$$+ \frac{1669 - 1507\nu}{1296}\bigg] \beta^2 + 128\lambda_1 k^5 + \frac{1744}{9} \lambda_1 k^4 + \frac{1088}{9} \lambda_1 k^3$$

$$+ \frac{203}{6} \lambda_1 k^2 + \frac{1145}{324} \lambda_1 k + \frac{23}{432} \lambda_1 \bigg\}$$

$$n_{2,2} = - \frac{\beta p_1}{32\lambda_1} \bigg\{ \frac{19 - 37\nu}{24} \beta^{10} \ln \beta - \frac{59 - 95\nu}{144} \beta^{10}$$

$$- \frac{5 - 59\nu}{12} \beta^8 \ln^2 \beta - \bigg(\frac{61 - 43\nu}{6} k - \frac{401 - 527\nu}{48}\bigg) \beta^8 \ln \beta$$

$$+ \bigg(\frac{143 - 179\nu}{36} k - \frac{193 - 211\nu}{72}\bigg) \beta^8 - 3\lambda_1 \beta^6 \ln^3 \beta$$

$$- \bigg(\frac{86}{3} \lambda_1 k + \frac{7}{36} \lambda_1\bigg) \beta^6 \ln^2 \beta - \bigg(\frac{376}{3} \lambda_1 k^2 - \frac{8 + 46\nu}{3} k$$

$$- \frac{2725 - 997\nu}{432}\bigg) \beta^6 \ln \beta + \bigg(\frac{193 - 265\nu}{9} k^2 + \frac{257 - 365\nu}{18} k$$

$$+ \frac{784 - 2161\nu}{1296}\bigg) \beta^6 + 12\lambda_1 k \beta^4 \ln^3 \beta - 24\lambda_1 k(3k + 2)\beta^4 \ln^2 \beta$$

$$- \bigg[288\lambda_1 k^3 + (198 - 186\nu)k^2 + \frac{123}{2} \lambda_1 k + \frac{51 - 57\nu}{16}\bigg] \beta^4 \ln \beta$$

$$- \bigg[164\lambda_1 k^3 + (39 - 57\nu)k^2 - \frac{679 - 598\nu}{12} k - \frac{317 - 272\nu}{144}\bigg] \beta^4$$

$$- \left[640\lambda_1 k^4 + (744 - 712v)k^3 + \frac{860 - 842v}{3} k^2 \right.$$

$$\left. + \frac{278 - 287v}{9} k - \frac{121 - 31v}{288} \right] \beta^2 + 32\lambda_1 k^4 + \frac{232 - 304v}{9} k^3$$

$$\left. + \frac{92 - 128v}{9} k^2 + \frac{49 - 103v}{72} k - \frac{353 - 191v}{2592} \right\},$$

$$n_{2,3} = \beta^2 \left[\frac{3}{16} \beta^4 - \frac{3}{8} \beta^2 \ln \beta - \left(\frac{13}{4} k - \frac{17}{32} \right) \beta^2 - 16k^2 - 7k \right],$$

$$n_{2,4} = 2k\beta^2,$$

$$n_{4,0} = -\frac{1}{24\lambda_1} \left\{ \frac{1 - 3v}{4} \beta^8 \ln \beta - \frac{7 + 55v}{96} \beta^8 - \frac{3 - 17v}{4} \beta^6 \ln^2 \beta \right.$$

$$- \left(4vk - \frac{61 - 3v}{16} \right) \beta^6 \ln \beta - \left(\frac{57 + 5v}{12} k - \frac{660 - 823v}{144} \right) \beta^6$$

$$+ \frac{5 - 11v}{2} \beta^4 \ln^3 \beta - \left[(1 - 15v)k + \frac{33 - 15v}{8} \right] \beta^4 \ln^2 \beta$$

$$- \left[(28 - 20v)k^2 - \frac{133 - 59v}{4} k + \frac{277 - 587v}{96} \right] \beta^4 \ln \beta$$

$$- \left(\frac{331 - 269v}{6} k^2 - \frac{2209 - 2163v}{72} k - \frac{8063 - 6113v}{1152} \right) \beta^4$$

$$- 2\lambda_1 \beta^2 \ln^4 \beta + 2\lambda_1 \beta^2 \ln^3 \beta + \left[56\lambda_1 k^2 - (34 - 30v) k \right.$$

$$\left. + \frac{15 - 21v}{4} \right] \beta^2 \ln^2 \beta - \left[160\lambda_1 k^3 - (108 - 100v) k^2 \right.$$

$$\left. + 30\lambda_1 k - \frac{1 - 15v}{8} \right] \beta^2 \ln \beta - \left(232\lambda_1 k^3 - \frac{101 - 8v}{9} k^2 \right.$$

$$\left. - \frac{583 - 117v}{72} k - \frac{1487 - 1415v}{432} \right) \beta^2 + 8\lambda_1 k \ln^4 \beta$$

$$- 16\lambda_1 k(4k - 1) \ln^3 \beta + \left[192\lambda_1 k^3 - (84 - 92v) k^2 \right.$$

$$\left. + (17 - 15v)k + \frac{9 - 7v}{8} \right] \ln^2 \beta - \left[256\lambda_1 k^4 - (144 - 176v) k^3 \right.$$

$$\left. + 56\lambda_1 k^2 - \frac{9 - 7v}{2} k - \frac{9 - 7v}{8} \right] \ln \beta - \frac{992}{3} \lambda_1 k^4 - \frac{458 - 334v}{3} k^3$$

$$\left. - \frac{2983 - 2517v}{36} k^2 - \frac{1453 - 1243v}{108} k - \frac{423 - 329v}{1152} \right\},$$

$$n_{4,1} = \frac{1}{24\lambda_1}\left\{\frac{1-3\nu}{4}\beta^8 + \frac{3}{2}\lambda_2\beta^6\ln\beta - \left(4\nu k - \frac{255-329\nu}{48}\right)\beta^6\right.$$

$$+ \lambda_2\left(6k + \frac{19}{8}\right)\beta^4\ln\beta - \left[(28-20\nu)k^2 - \frac{295-321\nu}{12}k\right.$$

$$\left. - \frac{541-531\nu}{32}\right]\beta^4 - \left[160\lambda_1 k^3 + \frac{224-248\nu}{3}k^2 - \frac{163-101\nu}{6}k\right.$$

$$\left. - \frac{2069-1763\nu}{144}\right]\beta^2 - 256\lambda_1 k^4 - 32(11-10\nu)k^3$$

$$\left. - \frac{533-471\nu}{3}k^2 - \frac{1259-1109\nu}{36}k - \frac{171-133\nu}{96}\right\},$$

$$n_{4,2} = -\frac{1}{24\lambda_1}\left\{\frac{9-11\nu}{4}\beta^6 + \frac{3}{2}\lambda_2\beta^4\ln\beta + \left[(7-9\nu)k\right.\right.$$

$$\left. + \frac{107-113\nu}{8}\right]\beta^4 - \left[56\lambda_1 k^2 - (15-11\nu)k - \frac{35-32\nu}{2}\right]\beta^2$$

$$\left. - 192\lambda_1 k^3 - 4(41-39\nu)k^2 - \frac{119-113\nu}{3}k - \frac{9-7\nu}{8}\right\},$$

$$n_{4,3} = \frac{1}{6}\left(\beta^4 + \frac{55}{24}\beta^2 - 16k^2 - \frac{19}{3}k\right),$$

$$n_{4,4} = -\frac{1}{3}\left(\frac{1}{4}\beta^2 - k\right),$$

$$n_{6,0} = -\frac{1}{48}\left[\frac{1}{6}\beta^6\ln\beta + \frac{71}{288}\beta^6 - \frac{1}{2}\beta^4\ln^2\beta + \left(2k - \frac{1}{2}\right)\beta^4\ln\beta\right.$$

$$+ \left(\frac{71}{24}k + \frac{49}{128}\right)\beta^4 + \frac{2}{3}\beta^2\ln^3\beta - (4k-1)\beta^2\ln^2\beta$$

$$+ \left(8k^2 - 4k + \frac{1}{2}\right)\beta^2\ln\beta + \left(\frac{71}{6}k^2 + \frac{49}{16}k + \frac{5243}{6912}\right)\beta^2$$

$$- \frac{1}{3}\ln^4\beta + \left(\frac{8}{3}k - \frac{2}{3}\right)\ln^3\beta - \left(8k^2 - 4k + \frac{1}{2}\right)\ln^2\beta$$

$$+ \left(\frac{32}{3}k^3 - 8k^2 + 2k - \frac{1}{6}\right)\ln\beta + \frac{142}{9}k^3 + \frac{49}{8}k^2$$

$$\left. + \frac{5243}{1728}k + \frac{23921}{82944}\right]$$

$$n_{6,1} = \frac{1}{48}\left[\frac{1}{6}\beta^6 + \left(2k + \frac{47}{48}\right)\beta^4 + \left(8k^2 + \frac{47}{6}k + \frac{65}{32}\right)\beta^2\right.$$

$$\left. + \frac{32}{3}k^3 + \frac{47}{3}k^2 + \frac{65}{8}k + \frac{4667}{3456}\right],$$

$$n_{6,2} = -\frac{1}{48}\left[\frac{1}{2}\beta^4 + \left(4k + \frac{47}{24}\right)\beta^2 + 8k^2 + \frac{47}{6}k + \frac{65}{32}\right],$$

$$n_{6,3} = \frac{1}{144}\left(2\beta^2 + 8k + \frac{47}{12}\right),$$

$$n_{6,4} = -\frac{1}{144}. \tag{3.19}$$

4. Major results

From the solution in the previous section, we obtain the relation between the load and the centre deflection as

$$P = p_1 W_0 + p_3 W_0^3 \tag{4.1}$$

and the formula for the radial stress

$$S_r = S_2 W_0^2 + S_4 W_0^4. \tag{4.2}$$

Substituting (4.2) into (2.13), we obtain the formula for the tangential stress

$$S_\theta = S_2' W_0^2 + S_4' W_0^4, \tag{4.3}$$

where

$$\begin{aligned} S_2' = &-\frac{\beta^2 p_1^2}{32}\Bigg\{3\alpha^2 y^2 - \frac{1}{\lambda_1}\Bigg[\frac{1}{4}\lambda_2\beta^6 + \frac{1-7v}{2}\beta^4 \ln\beta \\ &+ \left(\lambda_2 k - \frac{13-19v}{8}\right)\beta^4 - 2\lambda_1\beta^2\ln^2\beta + 4\lambda_1(2k+1)\beta^2\ln\beta \\ &- \left(2(3-v)k - \frac{1-7v}{4}\right)\beta^2 + 4(3-v)k^2 + (5-3v)k + \frac{9-7v}{8}\Bigg] \\ &+ H(y-\beta)\Bigg[3y^2\ln^2 y - \left(3\beta^2 + 12k + \frac{5}{2}\right)y^2\ln y \\ &+ \left(3\beta^2\ln\beta + \frac{11}{4}\beta^2 - 3\ln^2\beta + 3(4k-1)\ln\beta + 11k + \frac{3}{8}\right)y^2 \\ &+ 2\beta^2\ln^2 y - 2\beta^2(\beta^2 + 4k - 1)\ln y + 2\beta^4\ln\beta - 3\beta^4 \\ &- 2\beta^2\ln^2\beta + 4(2k+1)\beta^2\ln\beta - (12k+1)\beta^2 + \beta^4 y^{-2}\ln y \\ &+ \beta^4\left(\frac{1}{4}\beta^2 - \frac{3}{2}\ln\beta + k + \frac{5}{8}\right)y^{-2}\Bigg]\Bigg\}, \end{aligned} \tag{4.4a}$$

$$\begin{aligned} S_4' = &\frac{\beta^4 p_1^4}{512}\Bigg\{\sum_{i=0}^{3} m_{2i}(2i+1)y^{2i} - H(y-\beta)\sum_{i=-2}^{3}\sum_{j=0}^{4} n_{2i,j}\Big[(2i+1)\ln^j y \\ &+ j\ln^{j-1} y\Big]y^{2i}\Bigg\}. \end{aligned} \tag{4.4b}$$

For a particular position at the centre of the plate, substituting $y = 0$ in expressions (4.2) and (4.3), we obtain the stresses at the centre of the plate

$$S_r(0) = S_\theta(0) = S_2(0)\,W_0^2 + S_4(0)\,W_0^4 \tag{4.5}$$

where

$$S_2(0) = \frac{\beta^2 p_1^2}{32\lambda_1}\left\{\frac{1}{4}\lambda_2\beta^6 + \frac{1-7\nu}{2}\beta^4\ln\beta + \left(\lambda_2 k - \frac{13-19\nu}{8}\right)\beta^4 - 2\lambda_1\beta^2\ln^2\beta + 4\lambda_1(2k+1)\beta^2\ln\beta - \left[2(3-\nu)k - \frac{1-7\nu}{4}\right]\beta^2 + 4(3-\nu)k^2 + (5-3\nu)k + \frac{9-7\nu}{8}\right\} \tag{4.6a}$$

$$S_4(0) = \frac{\beta^4 p_1^4}{512} m_0. \tag{4.6b}$$

For a particular position at the edge of the plate, substituting $y = 1$ in expressions (4.2) and (4.3), we obtain the stresses at the edge of the plate

$$\begin{aligned} S_r(1) &= S_2(1)\,W_0^2 + S_4(1)\,W_0^4, \\ S_\theta(1) &= \nu S_r(1), \end{aligned} \tag{4.7}$$

where

$$S_2(1) = \frac{\beta^2 p_1^2}{32\lambda_1}\left[\frac{1}{2}\beta^6 - 3\beta^4\ln\beta + \left(2k+\frac{3}{4}\right)\beta^4 - \left(4k+\frac{3}{2}\right)\beta^2 + 8k^2 + 2k + \frac{1}{4}\right] \tag{4.8}$$

$$S_4(1) = \frac{\beta^4 p_1^4}{512}\left(\sum_{i=0}^{3} m_{2i} - \sum_{i=-2}^{3} n_{2i,0}\right).$$

Numerical values of the coefficients p_1, p_3, $S_2(0)$, $S_4(0)$, $S_2(1)$ and $S_4(1)$ have been computed for various dimensionless characteristic parameter k and dimensionless radius β by assuming Poisson's ratio $\nu = 0.3$. The calculations have been completed by PRIME computer and are given in Tables 1–6.

Knowing the coefficients, we can easily calculate the deflection and the stresses of the circular sandwich plate from formulae (4.1), (4.5) and (4.7). The results of these calculations are represented graphically in Figs. 2–4.

Figure 2 indicates the relation between the dimensionless load P and the dimensionless centre deflection W_0 for several values of k and β. It is obvious that the curves rise monotonically. When β is smaller, for the same value of P, the centre deflection induced in a plate with large β is large. But when β is larger, the result turns out to be contrary to the above conclusions.

Table 1. Numerical values of p_1. $\nu = 0.3$.

k \ β	0.05	0.20	0.35	0.50	0.65	0.80
0	162.84	48.121	36.849	39.660	57.660	134.45
0.01	156.47	45.911	34.617	36.083	48.553	87.431
0.02	150.59	43.896	32.639	33.097	41.936	64.777
0.03	145.12	42.050	30.876	30.568	36.906	51.447
0.04	140.04	40.353	29.293	28.398	32.953	42.666
0.05	135.30	38.788	27.865	26.515	29.765	36.446
0.06	130.88	37.340	26.569	24.867	27.140	31.809
0.07	126.73	35.996	25.389	23.411	24.940	28.219
0.08	122.84	34.745	24.309	22.117	23.070	25.356
0.09	119.18	33.578	23.317	20.958	21.461	23.022
0.10	115.73	32.487	22.403	19.915	20.062	21.080

Table 2. Numerical values of p_3. $\nu = 0.3$.

k \ β	0.05	0.20	0.35	0.50	0.65	0.80
0	13.617	4.5554	3.8972	5.7758	47.776	4111.5
0.01	15.096	4.9248	4.0735	5.3604	26.778	743.25
0.02	16.416	5.2543	4.2349	5.1222	17.287	229.64
0.03	17.611	5.5528	4.3848	4.9890	12.487	95.721
0.04	18.705	5.8266	4.5254	4.9198	9.8449	48.792
0.05	19.716	6.0803	4.6582	4.8908	8.2918	28.913
0.06	20.657	6.3174	4.7843	4.8872	7.3292	19.290
0.07	21.541	6.5405	4.9042	4.9000	6.7061	14.146
0.08	22.375	6.7516	5.0187	4.9233	6.2880	11.174
0.09	23.167	6.9522	5.1282	4.9533	5.9988	9.3454
0.10	23.922	7.1436	5.2329	4.9872	5.7934	8.1610

Table 3. Numerical values of $S_2(0)$. $\nu = 0.3$.

k \ β	0.05	0.20	0.35	0.50	0.65	0.80
0	2.3957	2.1757	2.0441	1.9708	1.9338	1.9208
0.01	2.3254	2.1173	1.9905	1.9173	1.8782	1.8709
0.02	2.2642	2.0679	1.9480	1.8803	1.8507	1.8671
0.03	2.2106	2.0260	1.9140	1.8543	1.8370	1.8709
0.04	2.1636	1.9903	1.8868	1.8358	1.8307	1.8760
0.05	2.1222	1.9597	1.8648	1.8228	1.8284	1.8808
0.06	2.0856	1.9335	1.8469	1.8136	1.8284	1.8849
0.07	2.0532	1.9108	1.8325	1.8073	1.8297	1.8886
0.08	2.0245	1.8913	1.8207	1.8030	1.8318	1.8917
0.09	1.9988	1.8744	1.8111	1.8002	1.8343	1.8944
0.10	1.9760	1.8597	1.8034	1.7986	1.8370	1.8968

Table 4. Numerical values of $S_4(0)$. $\nu = 0.3$.

k \ β	0.05	0.20	0.35	0.50	0.65	0.80
0	0.010777	−0.0053845	0.0036663	1.0433	162.01	86327
0.01	−0.0025789	−0.013954	−0.030390	−0.10583	42.329	10552
0.02	−0.013588	−0.020560	−0.048084	−0.54886	7.9979	2477.8
0.03	−0.022645	−0.025619	−0.056489	−0.68391	−2.1399	823.80
0.04	−0.030075	−0.029448	−0.059398	−0.68571	−4.7358	340.03
0.05	−0.036142	−0.032285	−0.059090	−0.63313	−4.9110	162.65
0.06	−0.041060	−0.034320	−0.056928	−0.56209	−4.3606	86.577
0.07	−0.045009	−0.035694	−0.053718	−0.48881	−3.6630	49.972
0.08	−0.048133	−0.036527	−0.049956	−0.42023	−3.0095	30.733
0.09	−0.050559	−0.036908	−0.045922	−0.35882	−2.4522	19.890
0.10	−0.052386	−0.036914	−0.041807	−0.30506	−1.9944	13.423

Table 5. Numerical values of $S_2(1)$. $\nu = 0.3$.

k \ β	0.05	0.20	0.35	0.50	0.65	0.80
0	0.72901	0.82269	0.93330	1.0472	1.1617	1.2761
0.01	0.72965	0.82127	0.92981	1.0419	1.1561	1.2804
0.02	0.73219	0.82243	0.93033	1.0435	1.1633	1.3013
0.03	0.73620	0.82555	0.93365	1.0491	1.1751	1.3194
0.04	0.74137	0.83014	0.93893	1.0571	1.1882	1.3335
0.05	0.74743	0.83586	0.94555	1.0664	1.2012	1.3447
0.06	0.75419	0.84240	0.95309	1.0763	1.2136	1.3536
0.07	0.76147	0.84956	0.96123	1.0864	1.2251	1.3609
0.08	0.76914	0.85715	0.96973	1.0965	1.2357	1.3669
0.09	0.77709	0.86505	0.97843	1.1065	1.2454	1.3719
0.10	0.78525	0.87314	0.98720	1.1161	1.2543	1.3762

Table 6. Numerical values of $S_4(1)$. $\nu = 0.3$

k \ β	0.05	0.20	0.35	0.50	0.65	0.80
0	−0.0045683	0.00089932	0.019810	1.4897	246.67	132708
0.01	−0.024509	0.0019225	−0.0010191	0.41535	94.585	20043
0.02	−0.00017936	0.0035071	−0.010424	−0.054294	40.872	5508.7
0.03	0.0022370	0.0055180	−0.013229	−0.24659	19.248	2068.8
0.04	0.0047843	0.0078532	−0.012211	−0.30950	9.6591	940.78
0.05	0.0074469	0.010440	−0.0089951	−0.31241	5.0818	486.76
0.06	0.010212	0.013220	−0.0045519	−0.28864	2.7690	276.37
0.07	0.013063	0.016153	0.00054737	−0.25446	1.5481	168.29
0.08	0.015991	0.019202	0.0059514	−0.21774	0.88165	108.24
0.09	0.018980	0.022344	0.011466	−0.18211	0.50906	72.737
0.10	0.022024	0.025555	0.016958	−0.14915	0.29779	50.670

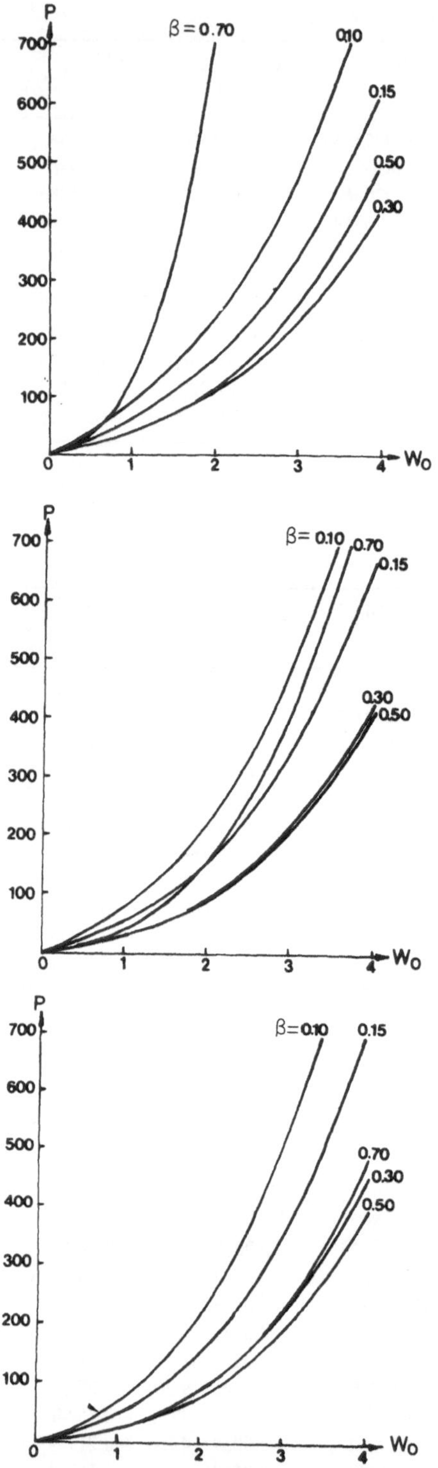

Figure 2. Variation of the load with centre deflection ($\nu = 0.3$). (a) $k = 0.01$. (b) $k = 0.05$. (c)$k = 0.10$.

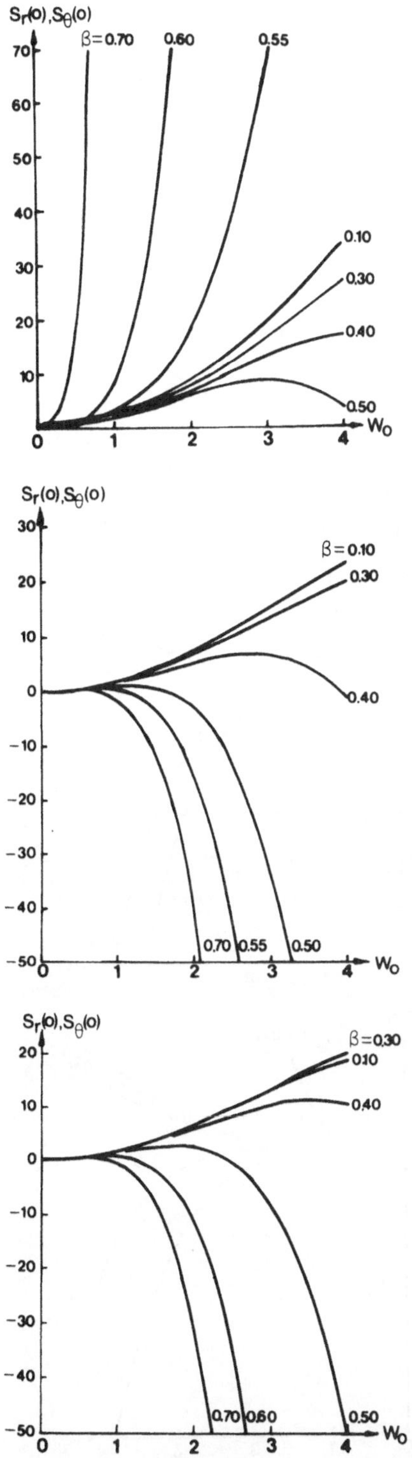

Figure 3. The radial stress and the tangential stress at the centre of the circular sandwich plate (ν = 0.3). (a) k = 0.01. (b) k = 0.05. (c) k = 0.10.

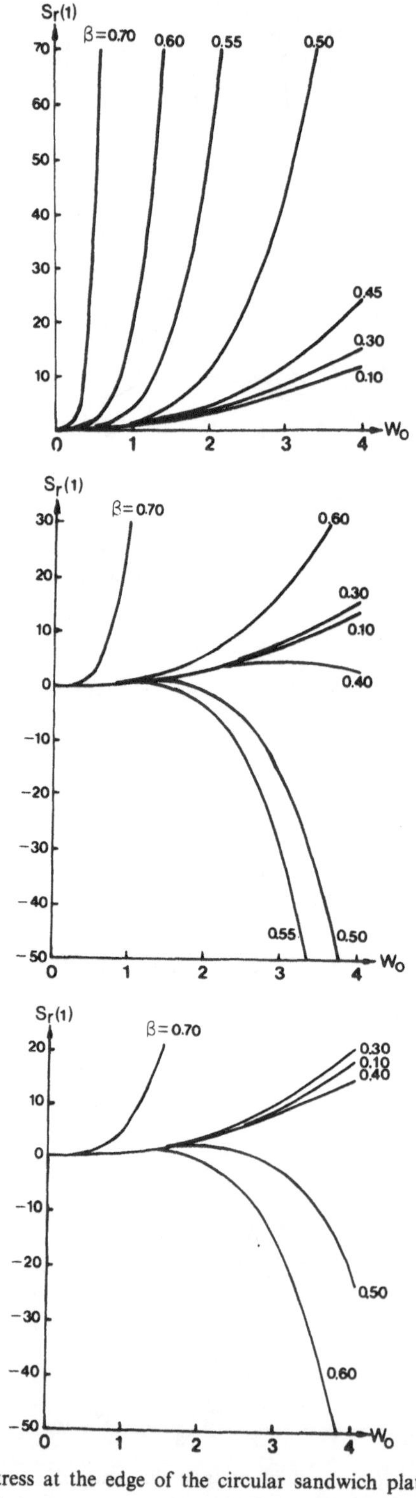

Figure 4. The radial stress at the edge of the circular sandwich plate ($\nu = 0.3$). (a) $k = 0.01$. (b) $k = 0.05$. (c) $k = 0.10$.

In Fig. 3, curves are given for the dimensionless centre radial stress $S_r(0)$ and the dimensionless centre tangential stress $S_\theta(0)$. It can be seen from the figures that, for small k, the stresses are positive. However, for larger k, when the centre deflection is larger, the stresses induced in a plate with large β are negative.

Figure 4 shows curves for the dimensionless edge radial stress $S_r(1)$. From the figures we see that, for small k the curves rise monotonically, and the stresses are positive. However for larger k, when the centre deflection is larger, the stresses in a plate with middle β are negative.

In the limiting case where β is infinitely small, from the above solution we can obtain the solution for a concentrated load acting at the centre of the circular sandwich plate with a rigidly clamped edge. But the solution does not hold near the point of application of a concentrated load, since the deflection and the stresses approach infinity as r approaches zero. The accurate theory in [2] must be considered if we want the deflection and the stresses near the point of application of the load.

Acknowledgement

I would like to thank Mr. Hu Si-yi for his help in the numerical calculations.

References

1. Liu Ren-huai, "Nonlinear axisymmetrical bending of circular sandwich plates under the action of uniform edge moment" (in Chinese), *Journal of the China University of Science and Technology* **10**, 2 (1980), 56.
2. Liu Ren-huai, "Nonlinear bending of circular sandwich plates", *Applied Mathematics and Mechanics* **2**, 2 (1981), 189.
3. Liu Ren-huai and Shi Yun-fang, "Exact solution for circular sandwich plates with large deflection", *Applied Mathematics and Mechanics* **3**, 1 (1982), 11.
4. Chien Wei-zang, "Large deflection of a circular clamped plate under uniform pressure", *Chinese Journal of Physics* **7**, 2 (1947), 102.

Fracture Mechanics

The physical mechanism of acoustic emission and the difficulty of fracture mechanics

TSAI SHU-TANG

Shanghai University of Technology, Shanghai, PRC

Abstract. The problems frequently met and disputed concerning acoustic emission and fracture mechanics, and the problems of the difficulty of fracture mechanics shall be discussed in the following article. The whole contents, including seven problems, will be discussed as follows: (1) What is meant by acoustic emission? (2) The process of acoustic emission and the balance of energy. (3) The frequency and amplitude of acoustic emission and number of signals; the relation between slit stretching and acoustic emission. (4) The receiving detector of acoustic emission and the mechanism of signal receiving. (5) The analysis of the stretching experiment of acoustic emission. (6) The problem of location of acoustic emission. (7) From the property of acoustic emission, an investigation of the difficulty of fracture mechanics.

1. Introduction

Utilization of acoustic emission to detect the cracks of the device [1] is a valuable technique with wide application, and which has been applied in a lot of fields in many countries. But during application of this technique, some difficult problems have frequently been met. For example, the cracking signals of acoustic emission are very different for different materials. For some materials, the spectra of the signals appears to be very little different before fracture, so that we cannot get the exact criterion of fracture to judge when the fracture will occur. Another example is how to locate the positions of cracks as the signals of acoustic emſssion are received. Thus we can select favourable means to avoid serious incident. Sometimes, there are some external signals, appearing on the record, which help us to distinguish the real signals from the external signals, etc. In this article, we will discuss these problems in relation to the physical mechanism.

2. What is meant by acoustic emission?

Acoustic emission is a kind of signal or elastic wave of decreasing stress emitting locally from a body or material as fracture occurs. This local region may be macroscopic or microscopic. Because it is the signal of decreasing stress, it is impossible for it to be a plastic wave and must therefore be an elastic wave. Either crack opening for stretching, or dislocation for shearing at the position of fracture,

Yeh, K. Y. (ed) Progress in Applied Mechanics
 ISBN-13: 978-94-010-8061-3

the stress decreases from a high value to a certain definite value (in most of the cases, the value is zero).

According to its essential property, acoustic emission belongs to the macroscopic physical phenomenon. It is incorrect to say that the signal of acoustic emission with least energy only corresponds to the energy of some atomic bonds. Because the energy level of some atomic bonds is only 10^{-12} erg, and the mass of the crystal of the detector of acoustic emission is in grams, we cannot, in general cases, measure the vibration with so little energy. It may be clearer to estimate the order of magnitude only.

$$m \sim 1\,\text{gm}, \quad \tfrac{1}{2}mv^2 \sim 10^{-12}\,\text{erg}, \quad v \sim a\omega \sim 10^{-6}\,\text{cm/sec},$$

$$f \sim 100\,\text{kc}, \quad \omega = 2\pi f \sim 10^6/\text{sec}, \quad a \sim 10^{-12}\,\text{cm} \sim 10^{-4}\,\text{A}.$$

The ordinary detector of acoustic emission cannot measure the vibration with so small an amplitude. By the above estimation, we may conclude that the ordinary receiver cannot detect the signals of breaking of some atomic bonds or even of dislocation.

The ordinary signals of acoustic emission may be put into two classes. The first class is produced by macroscopic slits. Macroscopic means that the body or material may be considered an homogeneous isotropic medium locally. In detail, the size of crystals of steel of 6–8 class is about 5 μ. The slits which produce acoustic emission must be much greater than 5 μ; in which cases the neighbourhood of the slits may be considered as homogeneous isotropic mediums locally. The special requirements of this kind of signal are that the number of signals must be small, the frequency must be low and the amplitude, large. Another class of signals is produced by microscopic slits. Here, microscopic does not mean that the cases of quantum effect are relatively large, but that the linear dimension of the cracks need to be comparable with the size of the crystals. At this instant, the body or material cannot be considered as an homogeneous isotropic medium, unless we consider the distribution of crystals and discuss the stress distribution of each crystal in particular. This kind of signal of acoustic emission is mostly dominated by a statistical law. For example, the signal is related to the size distribution and position distribution of crystals, etc. The characteristic of this kind of signal is that the number of signals is large, the frequency is high and the amplitude, small.

3. The process of acoustic emission and energy balance

When the fracture phenomenon has been produced at a certain point of a body or material, the elastic potential energy of the surrounding medium of this point is divided into three parts. The first part remains in the form of potential energy, and the second part transforms to work, i.e., the surface energy. The third part transforms to vibration energy, i.e., the energy of acoustic emission or elastic wave. When the signal of decreasing stress has propagated to a certain region, the medium of the region is relaxed. When the signal has propagated to the fixture, the fixture is also relaxed. Likewise, when the sample is fastened, the fixture must be moved. Thus, the fixture has done a certain amount of work. For a pressure vessel, each element of the area of the vessel is under a non-equilibrium condition when the signal of acoustic emission propagates to the whole vessel. At this instant, the internal pressure is not in equilibrium with the elastic stresses. Then the elements of the area of the vessel will move out, until they are finally fastened. In this case, the liquid or gas in the vessel has done some amount of work. This is the cyclical process of stress condition when the signals of acoustic emission are emitted.

Some investigation conclude that acoustic emission is a release of elastic energy, and, through the process of acoustic emission, that the fixture does not move. For signals starting from acoustic emission to signals propagating to the fixture, this statement is correct. But this statement is correct only to the time when the signals propagate to the fixture. In the next stage, the fixture moves to do work, and the sample is fastened again. The work done is transformed into elastic energy, and therefore the cyclical process is carried on. Therefore, we cannot conclude that the process of acoustic emission is only the process of releasing elastic energy. Because of the above statement we may sometimes be given the incorrect impression that the elastic energy is released straight-forward, but the actual facts are not so straightforward. Furthermore, under the equilibrium condition and in the case of larger cracks, not only the surface energy is larger, but also the elastic energy. If we simply say that acoustic emission is only the release of elastic energy, then where does the increasing surface energy and potential energy come from? Where does the vibration energy of signals of acoustic emission come from?

Some textbooks and papers, referring to fracture mechanics, so far as to state that the fixture does not move and the slits are stretching; but as to calculating the energy, the energy under equilibrium condition is considered. Such statements are undoubtedly erroneous. As we consider the energy of the equilibrium condition after the crack is stretched, the surface energy increases the elastic potential energy also increases, the fixture does not move so that no work has been done; and at the same time there is some energy lost in vibration of acoustic emission. Apparently the above statement is contrary to the law of the conservation of energy. There are two ways of correct calculation:

(1) The calculated condition before the elastic waves of acoustic emission propagate to the fixture. In such a case, the initial elastic potential energy is equal to the elastic potential energy after relaxing, plus the work done (i.e., surface energy) at slits stretching, plus the vibration energy of elastic waves, etc. During the process, the fixture does not move, so that no work is done.

(2) The calculated condition after the fixture is fastened and the system tending toward equilibrium again. At this moment, the work done due to fixture moving is equal to the difference of surface energy, plus the difference of elastic potential energy for cracks stretching, plus vibrational energy of elastic waves, etc. (If there is some plastic work, we must also add this part of the energy.)

4. The frequency and amplitude of acoustic emission, and number of signals: acoustic emission related to slits stretching

The frequency of elastic waves of acoustic emission depends upon the property of materials, the original length of the crack and the length of the crack after stretching, etc. Its spectrum is very wide, (usually, from many kilo-Hertz to many tens of mega-Hertz). Generally, in the stretching test of acoustic emission, the variation of amplitude of elastic waves of acoustic emission is also very large. For example, in a study undertaken by the Shenyang Institute of Metals of the Academy of Science of China, utilizing crystals ϕ 13.5 × 10 and ϕ 10 × 10 in experiments of acoustic emission in order to detect the signals, the measured voltage range was found to be some microvolts to hundred volts, and the ratio of maximum volt to minimum volt was 10^8 in order of magnitude. The circumstance is close to the following: As the sample starts to yield, the received signals of acoustic emission are of the order of magnitude of several micro-volts. As the sample starts to crack, the received signals of acoustic emission are about a hundred micro-volts in order of magnitude, and as the sample is before breaking down, the received signals of acoustic emission attain several tens of volts to a hundred volts. The maximum point of the measured spectrum of frequency is about hundred kilo-Hertz (80–120 kilo-Hertz).

The counters in the test of acoustic emission usually give two numerical values. One is the total amount of the number of signals of acoutic emission and the other is the number of acoustic emission per unit time, i.e., the rate of counts. In general, when the test material reaches the yielding point, the weak links in crystals and intercrystal material will cause it to be cracked everywhere, and the peak rate of the counts will finally appear. As the material reaches the crack points due to the material being the continuous medium to be cracked, the number of signals of acoustic emission per unit time increases again. But this time the cracking is due to the stretching of macroscopic slits along the surface of maximum macroscopic mean stress, and not along the weak surfaces of the crystal or intercrystal materials. The intensity of these kind of signals of acoustic emission is determined by the original length of the macroscopic slit and its increment, which is much greater than the former kind. Furthermore, at the final instant, the original length and its increment become still larger and the signals of acoustic emission more intensive, until the sample breaks down. But there are some exceptional cases, as at 15 MnV, where the signals of acoustic emission produced under different stress conditions, are distributed relatively uniformly. At the yield and cracking points, there are no apparent peaks of the rate of counts of signals of acoustic emission, the cause of which is not clear yet. For this kind of material, utilizing acoustic emission to judge the yield point and the crack point is comparatively difficult.

The frequency of acoustic emission related to the length of slit and its increment may be explained by utilizing a simplified model qualitatively. There is a slit of length $2a$ in an infinite plane as shown in Fig. 1. After cracking, its increment is δa [4]. It is well-known that the displacement about each end of the slit in the infinite plane, which is under stress, is the following:

$$\begin{aligned} W_x &= \frac{\sqrt{2}\sigma}{E}\sqrt{ra}\left[(1-\mu)+(1+\mu)\sin^2\frac{\theta}{2}\right]\cos\frac{\theta}{2}, \\ W_y &= \frac{\sqrt{2}\sigma}{E}\sqrt{ra}\left[2-(1+\mu)\cos^2\frac{\theta}{2}\right]\sin\frac{\theta}{2}, \end{aligned} \tag{4.1}$$

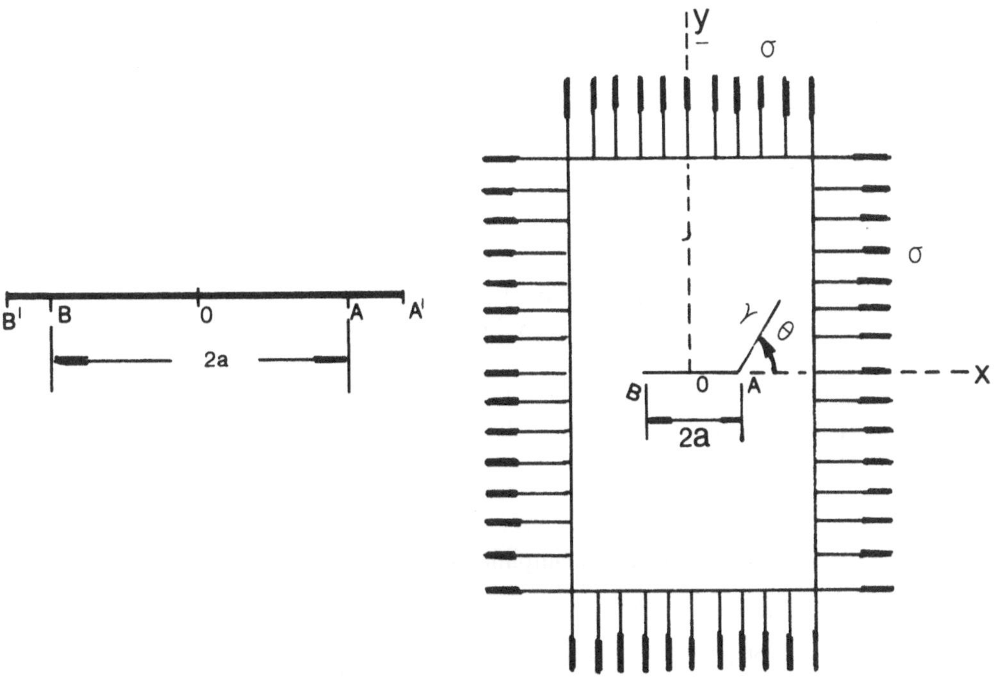

Figure 1

where σ is the uniform stress; r is the distance from the end of the slit to the point; θ is the angle between the radial vector and the x-axis; μ is Poisson's ratio; E is Young's modulus; W_x is the displacement along the x direction; and W_y is the displacement along the y direction. All these symbols are commonly used, and may be seen in [1]. Now we consider the vibration of point A, after the crack stretches from A to A'. We assume that the initial position of A is the original position and its initial velocity is zero, after cracking. The equilibrium position after cracking may be obtained by the above formulae for displacements with $(a + \delta a)$, instead of a, δa instead r, and $\theta = \pi$. After substitution, we get

$$W_{xA} = 0; \quad W_{yA} = 2\frac{\sigma}{E}\sqrt{2}\sqrt{(a + \delta a)\,\delta a}. \tag{4.2}$$

Due to the fact that A is not at the equilibrium position at the initial moment, vibration with amplitude W_{yA} is produced. During vibration, only the elastic deformation can be recovered (i.e., only the elastic potential energy is playing a role). At the initial moment, the elastic deformation is the elastic deformation for yielding ε_e. Because in general cases, the deformation is larger about the crack point than in elastic deformation, only the maximum elastic deformation ε_e can be recovered. Thus the maximum vibrational velocity is ωW_{yA} and the maximum kinetic energy density is $\frac{1}{2}\varrho\omega^2 W_{yA}^2$, where ϱ is the density of the material and $\omega = 2\pi f$ is the circular frequency. The elastic potential energy is $\frac{1}{2}E\varepsilon_e^2$. During vibration, elastic potential energy and kinetic energy are usually interchanged. Thus we have

$$\tfrac{1}{2}\varrho\omega^2 W_{yA}^2 \sim \tfrac{1}{2}E\varepsilon_e^2, \quad \omega^2 W_{yA}^2 \sim \frac{E}{\varrho}\varepsilon_e^2 \sim V_s^2\varepsilon_e^2,$$

$$\omega W_{yA} \sim V_s\varepsilon_e$$

$$f = \frac{\omega}{2\pi} \sim \frac{1}{2\pi}\frac{V_s\varepsilon_e}{W_{yA}} \sim \frac{V_s}{2\pi}\frac{E\varepsilon_e}{2\sqrt{2}\,\sigma\,\sqrt{(a+\delta a)\,\delta a}} = \frac{V_s\sigma_y}{4\sqrt{2}\,\pi\sigma\,\sqrt{(a+\delta a)\,\delta a}}, \tag{4.3}$$

where V_s is the velocity of sound, and σ_y is the yielding stress. Therefore the vibrational frequency f is proportional to the velocity of sound and the yielding stress and is inversely proportional to the external stress σ and the square root of $(a + \delta a)$ and δa.

The square of the amplitude A_v of the elastic wave of acoustic emission is evidently proportional to the energy change during cracking, and is proportional to the area A of the crack, i.e.,

$$A_v^2 \propto \delta A.$$

From the above-mentioned and simplified model, we have

$$A_v^2 \propto b\delta a.$$

The number of signals of acoustic emission is a statistical distribution function, which depends upon the property of the material, the external load, the distribution function of size and position of crystals, etc. As an example, the number N of signals of acoustic emission whose amplitude is greater than A_v is:

$$N = f(A_v, \sigma, \sigma_B, \sigma_y, a, \delta a, \ldots), \tag{4.4}$$

where f is a certain function, σ_B is the stress breaking down, and the other symbols the same as above. The simplest statistical distribution law is the normal distribution law

$$y = c \exp\left[-(x - x_0)^2/2\sigma_0^2\right], \tag{4.5}$$

where c is a certain constant, x_0 is the mean value of x, σ_0 is the root mean square value of $x - x_0$. There is a peak at $x = x_0$, and the value is very small at the other positions. Sometimes, we may roughly utilize the normal distribution law to represent the approximate distribution law at the neighbourhood of the peak, i.e., at $x = x_0$. There occurs a maximum of the rate of counts, and the rate of counts is very small elsewhere. The rate of counts of acoustic emission has a maximum at the yield point or cracking point, and is very small elsewhere. The cause is the statistical distribution law playing a role. The maximum value of rate of counts at the yield point or cracking point does not reject the possibility of some signals appearing elsewhere.

5. The detector of acoustic emission and receiving mechanism

The detector of acoustic emission is made up of a crystal and a pedestal, etc. Due to selecting a more favourable size of crystal and pedestal, we discuss the natural frequency [3] of the crystal and detector. For the natural frequency of crystal along the symmetry axis, we have

$$f = V_a/2L, \tag{5.1}$$

where V_a is the sound velocity of crystal along a symmetrical axis, and L is the thickness of the crystal. For the natural frequency of the crystal alone the radial direction, we have

$$f = \frac{\eta V_r}{2\pi r} \approx \frac{V_r}{\pi r}, \tag{5.2}$$

where V_r is the velocity of sound along a radial direction, r is the radius of the crystal, and η is the root of the following equation

$$\eta \frac{\mathrm{d}J_1(\eta)}{\mathrm{d}\eta} + \mu J_1(\eta) = 0, \quad \text{(thin crystal)}, \tag{5.3}$$

or

$$\eta \frac{\mathrm{d}J_1(\eta)}{\mathrm{d}\eta} - \frac{\mu}{1-\mu} J_1(\eta) = 0, \quad \text{(thick crystal)}, \tag{5.4}$$

where J_1 is a Bessel function of the first order. The value of μ varies from 1.8 to 2.2, and by taking the average, we get the final approximate formula.

After adding the pedestal to the crystal, the frequency will be decreased. For vibration along the direction of a symmetrical axis, the ratio [3] of the natural frequency f, after adding the pedestal to the natural frequency f_0 of the original crystal, is approximately as follows:

$$\frac{f}{f_0} \approx \frac{L_1}{L_1 + \beta\,(\varrho_2/\varrho_1)\,L_2}, \quad \beta = \frac{2}{1 + \sqrt{1 + [(\varrho_2/\varrho_1)(\pi L_2/L_1)]^2}}, \tag{5.5}$$

where L_1 is the thickness of the crystal; L_2 is the thickness of the pedestal; ϱ_1 is the density of crystal; and ϱ_2 is the density of the pedestal.

For radial vibration, we have the approximate formula [3]

$$\frac{f}{f_0} \approx \left[1 - \left(1 + \frac{R_1}{R_2} - 2\frac{V_1}{V_2}\right) \mathrm{e}^{-\alpha}\right],$$

$$\alpha \approx \frac{\varrho_2}{\varrho_1}\left(\frac{V_2}{V_1}\right)^2 \frac{L_1}{L_2} \frac{[(V_1/V_2) - (R_1/R_2)]}{[1 + (R_1/R_2) - 2(V_1/V_2)]}, \tag{5.6}$$

where R_1 is the radius of the crystal; R_2 is the radius of the pedestal; V_1 is the radial velocity of sound of the crystal; V_2 is radial sound velocity of the pedestal; and the other symbols are the same as above. This expression is obtained under the conditions that the thickness of the crystal is not too small in comparison with its radius, and the thickness of the crystal is not too large, etc. When we compare these two expressions with the experimental data, we obtain errors within 10%. When the pedestal is added on, the peak of radial vibration will usually be diminished in these cases, which we are incapable of measuring; but the peak of vibration along the symmetrical axis does not decrease much.

There are two problems which must be stated as follows:

(1) We discover that the velocity of sound of the crystal of the detector is almost isotropic. This conclusion is very different from the ordinary products of piezoelectric ceramics in catalogue [2]. For ordinary products, the velocity of sound along a symmetrical axis is much greater than the radial sound velocity. By utilizing the natural frequency (along a symmetrical axis and along the radial direction), we find the ratio of the two sound velocity to be about 2. It is possible that the processes of technique for thick crystals and thin crystals may be quite different, i.e., the degree of pressing are quite different.

(2) For a thin crystal, only the frequency of radial vibration is in the required range (about 100 kc). Due to the addition of the pedestal, the vibration in the radial direction will be diminished slightly, yet the vibration along a symmetrical axis will not decrease so much. It appears therefore that the principal vibration takes place along the symmetrical axis. Some authors have stated that the principal vibration of crystal is a radial vibration. This is possible for the properties of thin crystals but is not true for thick crystals.

We consider that the mechanism of receiving signals of acoustic emission by the crystal is mainly due to the normal pressure acting on the lower surface of the crystal. It is well-known that, if the crystal is under pressure in the direction of the thickness, the radius will stretch. As the vibrational pressure has passed, the crystal vibrates with its natural frequencies along the radial direction and thickness direction respectively. The contribution of shearing stress is found to be about 10^{-3} in order of magnitude of the contribution of normal pressure through estimation. Let ε be the strain of the sample along the thickness direction, and p be the normal pressure acting on the sample and the film of oil on the sample. Through transmission by the oil film, p is also the normal pressure acting on the surface of the crystal. Then, we have

$$p \approx E\varepsilon,$$

where E is the Young's modulus. The maximum vibrational velocity V_v may be given by the following expression in order of magnitude:

$$\varrho V_v^2 \approx E\varepsilon^2.$$

Since $E/\varrho \approx V_s^2$, where V_s is the sound velocity of the sample, we have

$$\begin{aligned} V_v^2 &= V_s^2\varepsilon^2, \\ V_v &\approx V_s\varepsilon = V_s\frac{p}{E}\frac{p}{\varrho V_s}. \end{aligned} \tag{5.7}$$

The shearing stress acting on the oil film is also the shearing stress acting on the crystal. Its value is

$$\tau = \eta_0\frac{V_v}{d} \approx \eta_0\frac{p}{\varrho V_s d}, \tag{5.8}$$

where d is the thickness of the oil film and η_0 is the viscosity coefficient of the oil. In this expression, we have used the maximum vibrational velocity to estimate the order of magnitude of the flowing velocity of the oil. Hence, we have finally

$$\frac{\tau}{p} \approx \eta_0/\varrho V_s d.$$

In an ordinary case, $\eta_0 \sim 10^2$, $V_s \sim 5 \times 10^5$, $\varrho \sim 8$, $d \sim 10^{-2}$, we have

$$\tau/p \sim 2 \times 10^{-3}.$$

Therefore, the contribution of the shearing stress is about 10^{-3} of the contribution of the normal pressure in order of magnitude.

6. The stress analysis during tests of acoustic emission

The details of this problem may be found in [3] and here we introduce it simply.

6.1. *Simplified model*

The specimens are under the stress condition as shown in Fig. 2. All specimens are cut with a slit. The length of the slit is "a" which is about one half of the width W. It is well-known that the bending rigidity is proportional to $Eh^3/12$. Hence, for

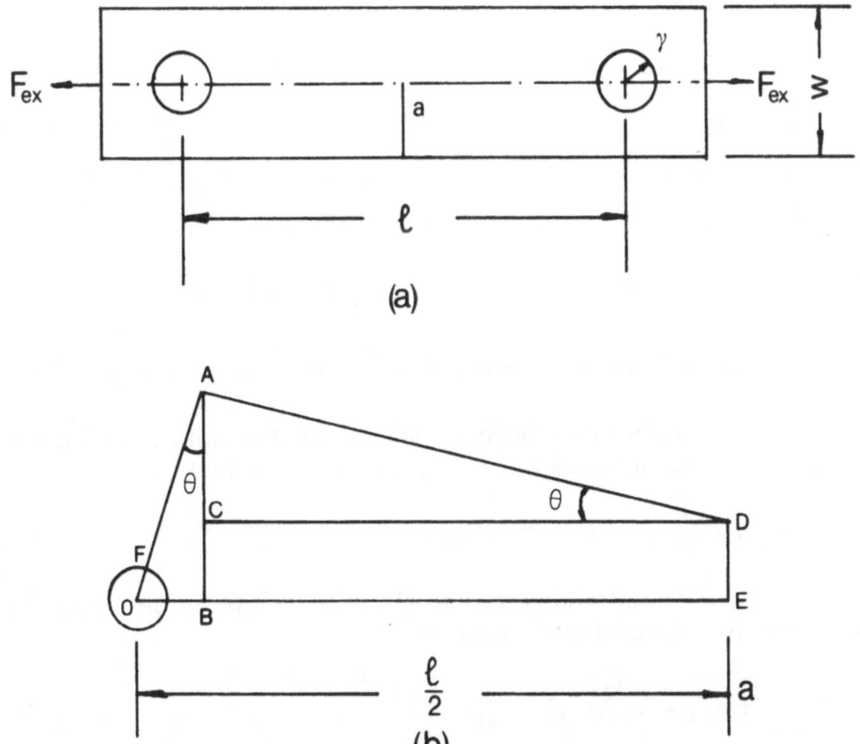

Figure 2

the position of cut, the bending ridigity is $\frac{1}{12}E(W - a)^3$; and for the position without a cut, the bending rigidity is $EW^3/12$. The ratio of these two values is $[(W - a)/W)]^3 \sim \frac{1}{10}$. Therefore, as the same bending moment is applied to these two positions, the degrees of bending are very different, i.e., the ratio of these two radii of curvatures is approximately 10. On the other hand, due to the fact that the slit is very narrow, it may be considered as a line. Thus we may obtain a simplified model. The two wings without cuts do not bend but rotate about a certain point on the extended line of the slit. This point is the so called centre of rotation.

6.2. *Geometrical relation*

Suppose that the specimen is cut off along the slit. We will discuss the condition of the stress acting on the left half. As in Fig. 2, D and E lie on the central line along the slit. We take the centre of the fixture as the reference point. Then the horizontal coordinate of the central line is $x = l/2$, where l is the distance between the centre of the two fixtures. For the specimen without any stress, the ordinate of any fibre is $(r + z)$, where r is the radius of the fixture. In Fig. 2

$$\mathrm{OA} = r + z, \quad \mathrm{OF} = r, \quad \mathrm{AF} = z,$$

$$\mathrm{OB} = (r + z)\sin\theta, \quad \mathrm{OE} = \frac{l}{2}, \quad \mathrm{CD} = \mathrm{BE} = \frac{l}{2} - (r + z)\sin\theta,$$

$$\mathrm{AD} = \mathrm{CD}\sec\theta = \left[\frac{l}{2} - (r + z)\sin\theta\right]\sec\theta$$

$$= \left[\frac{l}{2}\sec\theta - (r + z)\,\mathrm{tg}\,\theta\right]$$

$$\mathrm{DE} = \mathrm{AB} - \mathrm{AC} = (r + z)\cos\theta - \mathrm{AD}\sin\theta$$

$$= (r + z)\cos\theta - \left[\frac{l}{2}\sec\theta - (r + z)\,\mathrm{tg}\,\theta\right]\sin\theta$$

$$= (r + z)(\cos\theta + \sin\theta\,\mathrm{tg}\,\theta) - \frac{l}{2}\mathrm{tg}\,\theta = (r + z)\sec\theta - \frac{l}{2}\mathrm{tg}\,\theta.$$

Let $z = z_0$ be the ordinate of the fibre whose length does not extend. Then the position of the rotating centre is at $(r + z_0)\sec\theta - (l/2)\,\mathrm{tg}\,\theta$. Let

$$z_0 = a - \frac{W}{2} - r + \gamma\,(W - a), \tag{6.1}$$

where γ is the so-called rotating factor. The relation between displacement of opening and the measured displacement is

$$\phi = \frac{V_g}{1 + (z^* + a)/\gamma(W - a)}, \tag{6.2}$$

where ϕ is the displacement of the opening, V_g is the measured displacement, and z^* is the distance from the blade to the surface of the specimen. At this moment the extension of each fibre is $(z_0 - z)$ tg θ, when we assume that, within the neighbourhood of a central cut, the length of the curve of the fibre is approximately equal to the length of the broken line, the strain is

$$\varepsilon = \frac{(z_0 - z)\,\mathrm{tg}\,\theta}{(l/2)\sec\theta - (r + 2)\,\mathrm{tg}\,\theta} = \frac{(z_0 - z)\sin\theta}{(l/2) - (r + z)\sin\theta} \approx \frac{2\theta}{l}(z_0 - z). \tag{6.3}$$

The range of variation of z is from $(-(W/2) - r)$ to $((W/2) - r)$.

6.3. *The equilibrium relation of forces*

The forces acting on the specimen are the tensile external forces F_{ex} acting on each end. If we consider the left half of the specimen as above, and assume that the resultant of the external forces acting on the left half of the specimen is passing through the centre of the fixture, then we have

$$F_{ex} = b\int_{a-\left(\frac{W}{2}+r\right)}^{\frac{W}{2}-r} \sigma_{xx}\,\mathrm{d}z,$$

$$M_{ex} = 0 = b\int_{a-\left(\frac{W}{2}+r\right)}^{\frac{W}{2}-r} \sigma_{xx}\left[(r + z)\sec\theta - \frac{l}{2}\,\mathrm{tg}\,\theta\right]\mathrm{d}z, \tag{6.4}$$

where b is the thickness of the specimen. In this case, there are two unknown values z_0 and θ, and there are equations of equilibrium. Therefore, the problem is soluble, and we will now discuss it by dividing it into two cases.

(a) *Elastic case.* Within the whole region we utilize the relation $\sigma_{xx} = E\varepsilon$, and put it into the integrals, from which we get two equations for z_0 and θ. In this case the end of the slit starts to yield,

$$\varepsilon_y = k\left[\frac{2[z_0 - a + (W/2) + r]\theta}{l}\right] \tag{6.5}$$

where k is the stress concentration factor. In this case, F_{ex} is an unknown value.

(b) *Elastic and plastic case.* Firstly, we must determine the yield line (i.e. the fibre which starts to yield). We may use c to represent its value of z, which may be determined by the relation:

$$\varepsilon_y = \frac{2(z_0 - c)\theta}{l}. \tag{6.6}$$

Then the whole region may be divided into two subregions. For the elastic region, we use the relation $\sigma_{xx} = E\varepsilon$. For the plastic region, we use the relation

$\sigma/\sigma_1 = (\varepsilon/\varepsilon_1)^n$, where σ_1, ε_1 and n are the constant values. We divide the integrals into two parts

$$F_{ex} = b\int_{a-\left(\frac{W}{2}+r\right)}^{c} \sigma_{xx}\,\mathrm{d}z + b\int_{c}^{\frac{W}{2}-r} \sigma_{xx}\,\mathrm{d}z, \tag{6.7}$$

$$M_{ex} = 0 = b\int_{a-\left(\frac{W}{2}+r\right)}^{c} \sigma_{xx}\left[(r + z)\sec\theta - \frac{l}{2}\operatorname{tg}\theta\right]\mathrm{d}z$$
$$+ b\int_{c}^{\frac{W}{2}-r} \sigma_{xx}\left[(r + z)\sec\theta - \frac{l}{2}\operatorname{tg}\theta\right]\mathrm{d}z. \tag{6.8}$$

Thus the three equations may determine three unknowns z_0, θ and c. As the end of slit cracks, F_{ex} is undermined. At this moment,

$$k\left[\frac{2[z_0 - a + (W/2) + r]\theta}{l}\right] = \varepsilon_b \tag{6.9}$$

where k is a factor related to stress concentration, and ε_b is the strain for breaking. Also $2[z_0 - a + (W/2) + r]$ is the critical displacement of opening.

7. The location problem of acoustic emission

We will divide this problem into several sub-problems.

7.1. *The velocity of propagation of elastic waves in plates and shells*

The elastic waves in plates and shells can be affected by two lateral boundary surfaces. The velocity of propagation of the longitudinal wave is $\sqrt{E/\varrho(1 - \mu^2)}$. The velocity of propagation of the transverse wave is still $\sqrt{E/2\varrho(1 + \mu)}$. These two values are independent of the radius of curvature, but the amplitude and the form of the waves are distorted. From the standpoint of location by acoustic emission, it is only necessary to receive the signals of the same emitting source with the same propagation velocity. The distortion of wave form is not worth consideration. Thus, we select signals of the longitudinal wave comparatively conveniently. But, after receiving a longitudinal signal, the transverse signal or the reflected signal need not receive again.

7.2. *Determination of a single emitting source*

Firstly, we will examine the determination of a single source on a plane. To determine a single emitting source it is necessary to determine three numbers, i.e., the coordinates x, y and the time t of emitting. Thus, we either need three equations, or we must, at least, receive signals of acoustic emission of this point source at three points. If we use V_s to represent the sound velocity, these three equations will be

$$(x - x_1)^2 + (y - y_1)^2 = V_s^2(t - t_1)^2, \tag{7.1a}$$

$$(x - x_2)^2 + (y - y_2)^2 = V_s^2(t - t_2)^2, \tag{7.1b}$$

$$(x - x_3)^2 + (y - y_3)^2 = V_s^2(t - t_3)^2, \tag{7.1c}$$

where (x_1, y_1, t_1), (x_2, y_2, t_2) and (x_3, y_3, t_3) are the coordinates and the receiving time of three measuring points respectively. The number of equations greater than three is also permissible, but they must be compatible and have a common solution, or, the received signals cannot be emitted from a single source of acoustic emission. For the case of a curved surface, we will use the following three equations

$$F(q_x, q_y, t; q_{x_1}, q_{y_1}, t_1) = 0, \tag{7.2a}$$

$$F(q_x, q_y, t; q_{x_2}, q_{y_2}, t_2) = 0, \tag{7.2b}$$

$$F(q_x, q_y, t; q_{x_3}, q_{y_3}, t_3) = 0, \tag{7.2c}$$

where q_x, q_y are the generalized coordinates. In this case, the sound waves propagate along the geodesic lines.

7.3. *The case of several emitting sources*

In principle, if we want to determine n unknown numbers, we must obtain at least n equations. In other words, we must receive sound waves at n points. But the difficulty of this problem is how to distinguish the signals of longitudinal waves or transversal waves from the same source; the signals of longitudinal waves or transversal waves from different sources; and the reflected waves from different boundary surfaces. Due to the existence of different emitting sources, different types of waves, and the reflected waves from different types of waves by different boundary surfaces, the problem becomes very much complicated and it is difficult to determine the positions of sources. We must increase many additional points to judge the position of the source of acoustic emission and only then can we determine the positions of the emitting sources correctly.

8. The difficulty of fracture mechanics from the property of acoustic emission

From the above statements, we may conclude that the position and time of the appearing micro-slit depend upon the distribution of grain size and position of crystals in the polycrystal and the inter-crystal material, etc. Since the linear dimension of the ends of the slit is of the same order of magnitude as the grain size of the crystals, the position and time of the micro-slit depend closely upon these properties of the crystals in the neighbourhood of the ends of the slit. These properties do not belong to the domain of solid mechanics in which the material is considered to be a continuous medium; rather are they objects for research within the domain of solid state physics. Therefore, we cannot solve the problem of the production of a micro-slit within the realm of solid mechanics. On the other hand, the stress distribution and the conditions arising from the forces on the body are the problems of solid mechanics because the body is considered as a continuous medium. To determine the initiation of micro-slits on the body, we must connect the stress distribution and the distribution of position and grain size of the crystals. Therefore the production of micro-slits cannot be solved within the domain of solid state physics solely. Furthermore, due to the distribution of grain size and the distribution of the position of crystals etc. with some random property, the production of micro-slits is also a random property. It is different from the ordinary determinate problem. On the other hand, the principal cause of the production of micro-slits is the stress distribution which is largely determined by the methods of solid mechanics using a continuous medium. Hence, though the production of micro-slits has some random property, it is far different in the case of a completely random state.

During stretching of the slit, the speed of stretching depends upon the condition which is related to the slits and the defects, the length of the slit and the property of the crystal about the end of the slit. Therefore, this is also a problem related to solid mechanics and solid state physics and with a certain random property. At the moment of breaking down, the condition of breaking is also related to the mechanics of solid and physics of solid. But at this moment, the distribution of the average stress dominates the problem, the local fluctuation of stress is worthless, and therefore the body broken down is almost along the surface of the maximum stress as the body is considered as a continuous medium. Hence, except for the existence of the initial and historical condition, the effect of other random properties is very small.

In short, using the property of acoustic emission as ground work to discuss the problem of fracture mechanics, we may apparently observe that fracture mechanics is not simply the problem of solid mechanics considering a continuous medium, but is also a problem of solid state physics and metal physics. From the mathematical point of view, it is not only the problem of differential equations but also the problem of probability and mathematical statistics. Therefore it is the science at the intersectional edge, and cannot simply be solved by any one side.

References

1. *Advances of External Technique of Pressure Vessels* (in Chinese), Institute of Commonly Used Machine (May, 1974), 155–156, 191–192.
2. *Catalogue of the Products of Shandong Zibo Factory of Piezoelectric Ceramics* (in Chinese).
3. Tsai Shu-tang, "Some problems of acoustic emission and fracture mechanics I–II" (in Chinese), *Journal of the Anhui University*, No. 1 (1979), 1–30.
4. Tsai Shu-tang, "Some problem of acoustic emission and fracture mechanics V" (in Chinese), *Journal of the Anhui University*, No. 1 (1980), 1–12.

An elastic-plastic analytical analysis near a crack tip

HSUEH DAH-WEI

Teaching Group of Strength of Materials, Beijing Institute of Technology, Beijing, PRC

Abstract. The elastic-plastic solutions of mode I and mode II under plane strain deformation, near a crack tip, are given in this paper. The special case is the well-known Irwin's solution. An estimate of the effect of pressure acting on the crack surfaces through the internal wall of a circular cylinder is also included.

1. Introduction

The elastic-plastic analytical solutions near a crack tip are very useful in practice and have been studied by many scientists. One of them is the successful elastic-plastic solution of mode III made by Rice [1]. In the current paper, based on the well-known HRR theory [2–4], the author tries to obtain elastic-plastic analytical solutions of mode I and mode II, under plane strain deformation, near a crack tip, and Irwin's solution of mode I for elastic material is treated here as a special case. An estimate of the effect of pressure acting on crack surfaces through the internal wall of a circular cylinder is also made.

2. The method for solving the problem

According to [4], the following equations and conditions must be satisfied when solving an elastic-plastic crack problem:

$$\sigma_r = (1 + n)r^{-n/(1+n)}\left[f(\theta) + \frac{1+n}{2+n}f''(\theta)\right], \tag{2.1}$$

$$\sigma_\theta = r^{-n/(1+n)}f(\theta), \tag{2.2}$$

$$\tau_{r\theta} = -\frac{1+n}{2+n}r^{-n/(1+n)}f'(\theta), \tag{2.3}$$

$$u_r = -r^{n/(1+n)}g'(\theta), \tag{2.4}$$

$$u_\theta = \frac{1+2n}{1+n}r^{n/(1+n)}g(\theta), \tag{2.5}$$

Yeh, K. Y. (ed) Progress in Applied Mechanics
 ISBN 90-247-3249-2.

$$\varepsilon_\theta = -\varepsilon_r = \frac{n}{1+n} r^{-1/(1+n)} g'(\theta), \tag{2.6}$$

$$\varepsilon_{r\theta} = -\tfrac{1}{2} r^{-1/(1+n)} \left[\frac{1+2n}{(1+n)^2} g(\theta) + g''(\theta)\right], \tag{2.7}$$

$$\left[f'' + \frac{n(2+n)}{(1+n)^2} f\right]\left[g'' + \frac{1+2n}{(1+n)^2} g\right] = -\frac{4n}{(1+n)^2} g'f' \tag{2.8}$$

$$\left[f'' + \frac{n(2+n)}{(1+n)^2} f\right]^2 + \frac{4}{(1+n)^2} f'^2$$

$$= \frac{4\tau_0^2}{\gamma_0^{2n}} \frac{(2+n)^2}{(1+n)^4} \left\{\frac{4n^2}{(1+n)^2} g'^2 + \left[g'' + \frac{1+2n}{(1+n)^2} g\right]^2\right\}^n, \tag{2.9}$$

$$f(0) = 1, \tag{2.10}$$

$$f'(0) = f'''(0) = 0, \tag{2.11}$$

$$f'(\pi) = f(\pi) = 0, \tag{2.12}$$

where, τ_0 and γ_0 are the yield shear stress and shear strain of the material respectively; n is the exponent of hardening; (r, θ) are the polar coordinates (Fig. 1) and $f(\theta)$ and $g(\theta)$ are unknown functions.

It is easy to see that the scheme suggested above by HRR is applicable not only to crack problems of mode I but also to crack problems of mode II, as long as one introduces conditions (2.13), (2.14) and (2.15), instead of conditions (2.10), (2.11) and (2.12):

$$\text{at } \theta = 0: \ \sigma_r = \sigma_\theta = 0, \quad \text{and so } f(0) = f''(0) = 0. \tag{2.13}$$

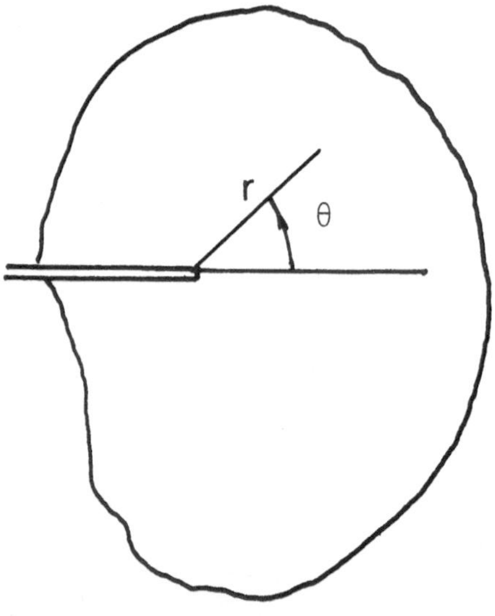

Figure 1

$$\begin{aligned} &\text{at}\quad \theta = \pm\pi:\quad \sigma_\theta = \tau_{r\theta} = 0, \\ &\text{and so}\quad f(\pm\pi) = f'(\pm\pi) = 0, \end{aligned} \tag{2.14}$$

and

$$f'(0) = -\frac{2+n}{1+n} \tag{2.15}$$

Because of

$$\frac{\sigma_r - \sigma}{\varepsilon_r} = \frac{\sigma_\theta - \sigma}{\varepsilon_\theta} = \frac{\tau_{r\theta}}{\varepsilon_{r\theta}}, \tag{2.16}$$

where σ is the average normal stress, we have

$$\sigma = \tfrac{1}{2}(\sigma_r + \sigma_\theta)$$

under plane strain deformation.

Substituting formulae (2.1), (2.2), (2.3), (2.6) and (2.7) into formula (2.16) we may obtain

$$\frac{nf + \dfrac{(1+n)^2}{2+n}f''}{2\dfrac{1+n}{2+n}f'} = \frac{-\dfrac{2n}{1+n}g'}{\dfrac{1+2n}{(1+n)^2}g + g''} = k(\theta), \tag{2.17}$$

where $k(\theta)$ is an unknown proportional factor chosen for different crack problems. From (2.17) we have

$$f'' - \frac{2k}{1+n}f' + \frac{n(2+n)}{(1+n)^2}f = 0, \tag{2.18}$$

$$g'' + \frac{2n}{k(1+n)}g' + \frac{1+2n}{(1+n)^2}g = 0. \tag{2.19}$$

It is obvious that no matter how $k(\theta)$ is chosen, equation (2.8) will be satisfied if one substitutes two unknown functions $f(\theta)$ and $g(\theta)$, obtained from (2.18) and (2.19), into (2.8). Equations (2.18) and (2.19) are differential equations of the second order, but $f(\theta)$ and $g(\theta)$ can satisfy four conditions.

Substituting (2.18) and (2.19) into (2.9), we obtain

$$g'(\theta) = \pm\frac{(1+n)^{(1+n)/n}\gamma_0 k(1+k^2)^{(1-n)/2n} f'^{1/n}}{2n(2+n)^{1/n}\tau_0^{1/n}}. \tag{2.20}$$

This is the relation between the two functions $f(\theta)$ and $g(\theta)$ and it is obviously simpler than (2.9).

The main point in this paper is to choose the unknown function $k(\theta)$ as appropriately as possible for the solving of elastic-plastic crack problems.

3. An estimate of the effect of pressure acting on crack surfaces

As an example of the utilization of the suggested method in this paper, let us consider an important problem in practice, that is to estimate the effect of

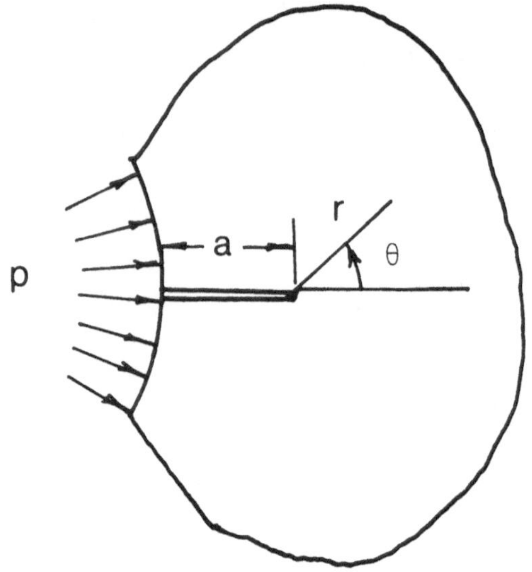

Figure 2

pressure acting on crack surfaces through the inner wall of a circular cylinder (Fig. 2). Let

$$k(\theta) = -\operatorname{tg}\frac{1}{1+n}\theta$$

then one can obtain a solution of equation (2.18)

$$f(\theta) = \frac{n(2+n)}{2(1+n)}\left[\frac{1}{2+n}\cos\frac{2+n}{1+n}\theta + \frac{1}{n}\cos\frac{n}{1+n}\theta\right].$$

This formula satisfies the conditions (2.10), (2.11) and the first condition of (2.12). If one introduces a multiplier D into equation (2.2) in such a manner that

$$\sigma_\theta = Dr^{-n/(1+n)}f(\theta),$$

then one can see that σ_θ does not satisfy the condition of uniform inner pressure p acting on the crack surfaces through the inner wall of a circular cylinder. So we use the method of least squares and let

$$\int_0^a [\sigma_\theta(\pi) - p]^2\, dr$$

be the minimum, where a is the length of the inner crack. From this one can obtain

$$D = p(1-n)a^{n/(1+n)}\Big/\cos\left(\frac{n}{1+n}\right)\pi.$$

Multiplying all the items on the right side of formulae (2.1), (2.2) and (2.3) by the multiplier D which is obtained above, all the stress components can also then be obtained. Thus one can see that:

(1) Comparing the above with Irwin's solution of elastic material, one can come to the conclusion that the effect of the internal pressure p on crack surfaces through the internal wall of a circular cylinder is negligible.

(2) If the length of a crack is very small, or if one supposes that the expressions obtained in this paragraph can be used in the entire vicinity region of the crack length, then one can get

$$\int_0^a \sigma_\theta \, \mathrm{d}r = p(1 - n^2)a,$$

$$\int_0^a \sigma_\theta r \, \mathrm{d}r = p(1 - n^2)a^2/(2 + n).$$

These values approach the main vector pa, and main moment $\frac{1}{2}pa^2$ respectively because the values of n of elastic-plastic materials are usually very small. Therefore, it may be appropriate to conclude that there are no forces acting on the crack surfaces, and only the inner pressure p acting on the inner wall of the circular cylinder is under consideration.

4. Crack problems of mode I

It can be proved by the method of substitution that if

$$\begin{aligned} k_1 = \Bigg[& 2(2 + n)A_1 \cos \frac{2 + n}{1 + n}\theta \\ & + (1 + 2n + 3n^2)A_2 \cos \frac{1 + 2n}{1 + n}\theta - 2nA_3 \cos \frac{n}{1 + n}\theta \\ & + (1 - 2n - n^2)A_4 \cos \frac{1}{1 + n}\theta \Bigg] \Bigg/ 2\Bigg[A_1(2 + n) \sin \frac{2 + n}{1 + n}\theta \\ & + A_2(1 + 2n) \sin \frac{1 + 2n}{1 + n}\theta + A_3 n \sin \frac{n}{1 + n}\theta + A_4 \sin \frac{1}{1 + n}\theta \Bigg], \end{aligned} \tag{4.1}$$

where

$$\left.\begin{aligned} A_1 &= \frac{5 + 7n}{32(1 + n)}; & A_2 &= \frac{3 + n}{32(1 + n)}; \\ A_3 &= \frac{11 + 9n}{32(1 + n)}; & A_4 &= \frac{13 + 15n}{32(1 + n)}, \end{aligned}\right\} \tag{4.2}$$

and

$$\begin{aligned} f_1 = {} & A_1 \cos \frac{2 + n}{1 + n}\theta + A_2 \cos \frac{1 + 2n}{1 + n}\theta \\ & + A_3 \cos \frac{n}{1 + n}\theta + A_4 \cos \frac{1}{1 + n}\theta, \end{aligned} \tag{4.3}$$

then formula (4.3), together with formula (4.1), satisfies equation (2.18), and f_1 satisfies all the conditions (2.10), (2.11) and (2.12).

Substituting formula (4.3) into (2.1), (2.2) and (2.3) and introducing a multiplier D_1, one can obtain the expression for the stresses as follows:

$$\begin{aligned}
\sigma_r &= D_1(1+n)r^{-n/(1+n)}\left\{A_1\left[1-\frac{2+n}{1+n}\right]\cos\frac{2+n}{1+n}\theta\right.\\
&\quad + A_2\left[1-\frac{(1+2n)^2}{(1+n)(2+n)}\right]\cos\frac{1+2n}{1+n}\theta\\
&\quad + A_3\left[1-\frac{n^2}{(1+n)(2+n)}\right]\cos\frac{n}{1+n}\theta\\
&\quad \left. + A_4\left[1-\frac{1}{(1+n)(2+n)}\right]\cos\frac{1}{1+n}\theta\right\},\\
\sigma_\theta &= D_1 r^{-n/(1+n)}\left[A_1\cos\frac{2+n}{1+n}\theta + A_2\cos\frac{1+2n}{1+n}\theta\right.\\
&\quad \left. + A_3\cos\frac{n}{1+n}\theta + A_4\cos\frac{1}{1+n}\theta\right],\\
\tau_{r\theta} &= D_1\frac{1}{2+n}r^{-n/(1+n)}\left\{A_1(2+n)\sin\frac{2+n}{1+n}\theta\right.\\
&\quad + A_2(1+2n)\sin\frac{1+2n}{1+n}\theta + A_3 n\sin\frac{n}{1+n}\theta\\
&\quad \left. + A_4\sin\frac{1}{1+n}\theta\right\},
\end{aligned} \tag{4.4}$$

where [5]

$$D_1 = \left[\frac{1}{(1+n)\pi}\right]^{n/(1+n)}(\sigma_y^*)^{(1-n)/(1+n)}K_1^{2n/(1+n)}. \tag{4.5}$$

σ_y^* is the yield stress under plane strain deformation, and K_1 is the stress intensity factor of mode I.

Substituting formula (4.3) into equation (2.9), and taking (2.18) and (2.19) into consideration, one can obtain

$$g_1'(\theta) = \frac{\pm(1+n)^{(1+n)/n}\gamma_0 k_1(1+k_1^2)^{(1-n)/2n}f_1^{1/n}}{2n(2+n)^{1/n}\tau_0^{1/n}}, \tag{4.6}$$

where "$\pm$" results from the extraction of a root.

One can get g_1 and g_1'' from equation (4.6) respectively by integration and differentiation. If one substitutes them into (2.4), (2.5), (2.6) and (2.7) and determines the multiplier D_2 from equation (2.9), then one can get the displacement fields and strain fields as shown below:

$$
\begin{aligned}
u_r &= -D_2 r^{n/(1+n)} g_1'(\theta), \\
u_\theta &= D_2 \frac{1+2n}{1+n} r^{n/(1+n)} g_1(\theta), \\
\varepsilon_\theta &= -\varepsilon_r = D_2 \frac{n}{1+n} r^{-1/(1+n)} g_1'(\theta), \\
\varepsilon_{r\theta} &= -\frac{D_2}{2} r^{-1/(1+n)} \left[\frac{1+2n}{(1+n)^2} g_1(\theta) + g_1''(\theta)\right],
\end{aligned}
\tag{4.7}
$$

where

$$
D_2 = D_1^{1/n} = \left[\frac{1}{(1+n)\pi}\right]^{1/(1+n)} (\sigma_y^*)^{(1-n)/n(1+n)} K_{\mathrm{I}}^{2/(1+n)}. \tag{4.8}
$$

It is worth noting that when $n = 1$ is used for elastic material, one can get Irwin's solution from (4.4) and (4.7) while taking formula (4.8) into consideration.

Although we have derived the important relation shown in formula (4.6), it is still too complex, and only a few values of n can be integrated to obtain the unknown function g_{I}.

From formula (4.4), and according to the von Mises criterion, the distinguishing surfaces between elastic regions and plastic regions of the body can be determined as follows:

$$
\begin{aligned}
& D_1^2 r^{-2n/(1+n)} \left\{ -2A_1 \cos\frac{2+n}{1+n}\theta - \frac{1+2n+3n^2}{2+n} A_2 \cos\frac{1+2n}{1+n}\theta \right. \\
& \left. + \frac{2n}{2+n} A_3 \cos\frac{n}{1+n}\theta + \frac{n^2+2n-1}{2+n} A_4 \cos\frac{1}{1+n}\theta \right\}^2 \\
& + 4D_1^2 \frac{1}{(2+n)^2} r^{-2n/(1+n)} \left\{ A_1(2+n)\sin\frac{2+n}{1+n}\theta \right. \\
& + A_2(1+2n)\sin\frac{1+2n}{1+n}\theta + A_3 n \sin\frac{n}{1+n}\theta \\
& \left. + A_4 \sin\frac{1}{1+n}\theta \right\}^2 = \tfrac{4}{3}\sigma_s^2.
\end{aligned}
\tag{4.9}
$$

5. Crack problems of mode II

It can be proved by the method of substitution that if

$$
\begin{aligned}
k_{\mathrm{II}}(\theta) = & \left[-2(2+n)\sin\frac{2+n}{1+n}\theta \right. \\
& -(1+2n+3n^2)\sin\frac{1+2n}{1+n}\theta + 2n\sin\frac{n}{1+n}\theta
\end{aligned}
$$

$$+(n^2 + 2n - 1)\sin\frac{1}{1+n}\theta\Big]\Big/2\Big[(2+n)\cos\frac{2+n}{1+n}\theta$$

$$+(1+2n)\cos\frac{1+2n}{1+n}\theta + n\cos\frac{n}{1+n}\theta + \cos\frac{1}{1+n}\theta\Big], \tag{5.1}$$

then one can get

$$f_{\mathrm{II}} = B\left(\sin\frac{2+n}{1+n}\theta + \sin\frac{1+2n}{1+n}\theta + \sin\frac{n}{1+n}\theta + \sin\frac{1}{1+n}\theta\right), \tag{5.2}$$

where "II" represents the problems of mode II and

$$B = -\frac{2+n}{4(1+n)}. \tag{5.3}$$

Equation (5.2) already satisfies equation (2.18) and all the conditions (2.13), (2.14) and (2.15) which must be satisfied by $f_{\mathrm{II}}(\theta)$. Substituting (5.2) into formula (2.1), (2.2), (2.3) and introducing a multiplier D_3, one can obtain the expression for stresses as follows:

$$\sigma_r = D_3 B(1+n)r^{-n/(1+n)}\left\{\left(1 - \frac{2+n}{1+n}\right)\sin\frac{2+n}{1+n}\theta\right.$$

$$+\left[1 - \frac{(1+2n)^2}{(1+n)(n+2)}\right]\sin\frac{1+2n}{1+n}\theta$$

$$+\left[1 - \frac{n^2}{(1+n)(2+n)}\right]\sin\frac{n}{1+n}\theta$$

$$\left.+\left[1 - \frac{1}{(1+n)(2+n)}\right]\sin\frac{1}{1+n}\theta\right\},$$

$$\sigma_\theta = D_3 B r^{-n/(1+n)}\left[\sin\frac{2+n}{1+n}\theta + \sin\frac{1+2n}{1+n}\theta\right. \tag{5.4}$$

$$\left.+\sin\frac{n}{1+n}\theta + \sin\frac{1}{1+n}\theta\right],$$

$$\tau_{r\theta} = -D_3 B\frac{1}{2+n}r^{-n/(1+n)}\left[(2+n)\cos\frac{2+n}{1+n}\theta\right.$$

$$+(1+2n)\cos\frac{1+2n}{1+n}\theta$$

$$\left.+n\cos\frac{n}{1+n}\theta + \cos\frac{1}{1+n}\theta\right],$$

where

$$D_3 = \left[\frac{1}{(1+n)\pi}\right]^{n/(1+n)} \tau_s^{(1-n)/(1+n)} K_{\rm II}^{2n/(1+n)} \tag{5.5}$$

and τ_s is the yielding shear stress of material; $K_{\rm II}$ is the stress intensity factor of mode II.

Substituting formula (5.2) into equation (2.9) and taking equations (2.18) and (2.19) into consideration one can get

$$g'_{\rm II}(\theta) = \frac{\pm(1+n)^{(1+n)/n}\gamma_0 k_{\rm II}(1+k_{\rm II}^2)^{(1-n)/2n} f_{\rm II}^{1/n}}{2n(2+n)^{1/n}\tau_0^{1/n}}. \tag{5.6}$$

If

$$D_4 = D_3^{1/n} = \left[\frac{1}{(1+n)\pi}\right]^{1/(1+n)} \tau_s^{(1-n)/n(1+n)} K_{\rm II}^{2/(1+n)} \tag{5.7}$$

and substituting $g_{\rm II}(\theta)$ together with $g''_{\rm II}(\theta)$, obtained from formula (5.6), into formula (2.4), (2.5), (2.6), and (2.7), one can obtain the displacement fields and the strain fields as follows:

$$\begin{aligned}
u_r &= -D_4 r^{n/(1+n)} g'_{\rm II}(\theta),\\
u_\theta &= D_4 \frac{1+2n}{1+n} r^{n/(1+n)} g_{\rm II}(\theta),\\
\varepsilon_\theta &= -\varepsilon_r = D_4 \frac{n}{1+n} r^{-1/(1+n)} g'_{\rm II}(\theta),\\
\varepsilon_{r\theta} &= -\frac{D_4}{2} r^{-1/(1+n)} \left[\frac{1+2n}{(1+n)^2} g_{\rm II}(\theta) + g''_{\rm II}(\theta)\right].
\end{aligned} \tag{5.8}$$

It is worthwhile to note that (5.4) and (5.8), when $\nu = \frac{1}{2}$, are the same as Irwin's solutions of elastic material.

According to von Mises criterion and from (5.4), it can be shown that the boundary surfaces between elastic regions and plastic regions of a body are as follows:

$$\begin{aligned}
&D_3^2 B^2 r^{-2n/(1+n)} \left\{-2\sin\frac{2+n}{1+n}\theta - \frac{1+2n+3n^2}{2+n}\sin\frac{1+2n}{1+n}\theta\right.\\
&\quad \left. + \frac{2n}{2+n}\sin\frac{n}{1+n}\theta + \frac{n^2+2n-1}{2+n}\sin\frac{1}{1+n}\theta\right\}^2\\
&\quad + \frac{4D_3^2B^2}{(2+n)^2} r^{-2n/(1+n)} \left\{(2+n)\cos\frac{2+n}{1+n}\theta + (1+2n)\cos\frac{1+2n}{1+n}\theta\right.\\
&\quad \left. + n\cos\frac{n}{1+n}\theta + \cos\frac{1}{1+n}\theta\right\}^2 = \tfrac{4}{3}\sigma_s^2.
\end{aligned} \tag{5.9}$$

6. Discussion

(1) The unknown function $g(\theta)$ must satisfy the following conditions:
For problems of mode I:

$$\begin{aligned} &\text{At } \theta = 0\colon\ u_\theta = \varepsilon_{r\theta} = 0, \\ &\text{and so } g_{\mathrm{I}}(0) = g_{\mathrm{I}}''(0) = 0. \end{aligned} \tag{6.1}$$

$$\begin{aligned} &\text{At } \theta = \pi\colon\ \varepsilon_{r\theta} = u_r = 0, \\ &\text{and so } \frac{1 + 2n}{(1 + n)^2} g_{\mathrm{I}}(\pi) + g_{\mathrm{I}}''(\pi) = g_{\mathrm{I}}'(\pi) = 0. \end{aligned} \tag{6.2}$$

For problems of mode II:

$$\begin{aligned} &\text{At } \theta = 0\colon\ u_r = u_\theta = 0, \\ &\text{and so } g_{\mathrm{II}}(0) = g_{\mathrm{II}}'(0) = 0. \end{aligned} \tag{6.3}$$

$$\begin{aligned} &\text{At } \theta = \pi\colon\ u_\theta = \varepsilon_{r\theta} = 0, \\ &\text{and so } g_{\mathrm{II}}(\pi) = g_{\mathrm{II}}''(\pi) = 0. \end{aligned} \tag{6.4}$$

As the expressions of $g(\theta)$ cannot be obtained easily, checking of the degree of precision of the expressions in this paper is not discussed. Consequently, the analysis made in this paper is only an approximation.

(2) The method of getting solutions ofcrack problems as suggested in this paper, is feasible. The conclusion is that the existence and uniqueness of the correct function $k(\theta)$ is beyond doubt.

(3) It is necessary to note that when an external force is loaded on the crack surface, the invariance of J-integration is still true if some additional conditions are satisfied. For example, in the problem of this paper shown in Fig. 2, if one takes four points A, B and A′, B′ which are situated on the upper and lower crack

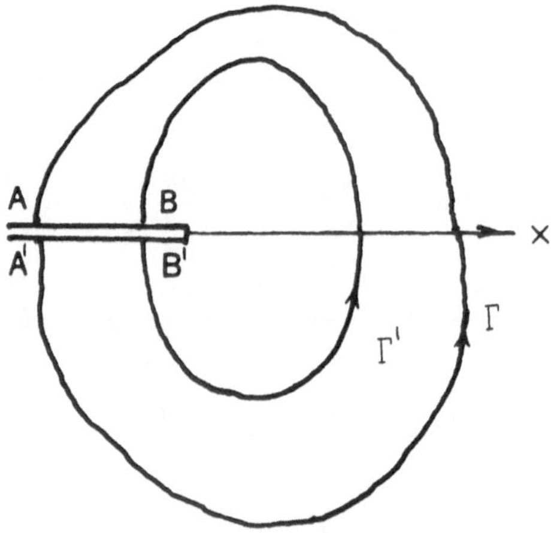

Figure 3

surfaces respectively, in such a manner that A and A′, B and B′ are symmetric points respectively (Fig. 3), then since

$$\int_{\overline{AB}+\overline{B'A'}} \left(W\,\mathrm{d}y - T_i \frac{\partial u_i}{\partial x}\,\mathrm{d}s \right) = 0$$

the conclusion as mentioned above is obtained.

References

1. Rice J. R., "Stresses due to a sharp notch in a workhardening elastic-plastic material loaded by longitudinal shear", *J. Appl. Mech.* **34**, 2 (June, 1967), 287–298.
2. Hutchison, J. W., "Singular behaviour at the end of a tensile crack in a hardening material", *J. Mech. Phys. Sol.* **16**, 1 (Jan., 1968), 13–31.
3. Hutchinson, J. W., "Plastic stress and strain fields at a crack tip", *J. Mech. Phys. Sol.* **16**, 5 (Sep., 1968), 337–347.
4. Rice, J. R., and Rosengren, G. F., "Plane strain deformation near a crack tip in a power-law hardening material", *J. Mech. Phys. Sol.* **16**, 1 (Jan., 1968), 1–12.
5. Chi Chen, "A modification of the plastic zone of hardening material" (in Chinese), *Mechanics* (April, 1975), 78–81.

On path-independent integrals in fracture dynamics for visco-elastic plastic media

OUYANG CHANG

Department of Mechanics, Fudan University, Shanghai, PRC

Abstract. In this paper we extend the results in [1] to fracture dynamics of visco-elastic plastic media. Some new path-independent integrals are worked out for this case, and an integral for the crack extension force in such media is given. It is shown that the path-independent integral given equals the crack extension force. It is possible to propose a dynamic fracture criterion on the basis of such integrals.

1. Introduction

In nonlinear fracture statics, J. R. Rice [1], proposed a path-integral J, which has been the theoretical basis of one kind of fracture criterion, J_c. In [2], the present author extended the research to nonlinear fracture dynamics, where the dynamic effect and crack propagation phenomenon were included. In this paper we extend the results in [2] to the nonlinear visco-elastic plastic case. New path-independent integrals are given for this case, and integral forms of the crack extension force in such media are shown. Thus, it is possible to construct a nonlinear fracture criterion in fracture dynamics for visco-elastic plastic media.

2. Path-independent integrals in fracture dynamics for visco-elastic plastic media

Now we will consider the notched crack propagation in certain kinds of visco-elastic plastic media as shown in Fig. 1. Here Γ is an arbitrary path around the crack tip, V is the volume bounded by Γ and the crack surface; ν is the outer normal of Γ, and u_i, T_i are the displacement and traction vectors, respectively.

Then we have the following:

THEOREM 1. The integral

$$Y_1 = \int_{t_0}^{t_1} \left(\int_\Gamma (W_e - F_i u_i - K)\, \mathrm{d}y - T_i \partial u_i/\partial x\, \mathrm{d}s \right) \mathrm{d}t + \int_{t_0}^{t_1} \int_V \sigma_{ij} \partial e_{ij}^{cp}/\partial x\, \mathrm{d}x\, \mathrm{d}y\, \mathrm{d}t + \int_V \varrho v_i \partial u_i/\partial x\, \mathrm{d}V \Big|_{t_0}^{t_1} \tag{2.1}$$

Yeh, K. Y. (ed) Progress in Applied Mechanics

ISBN-13: 978-94-010-8061-3

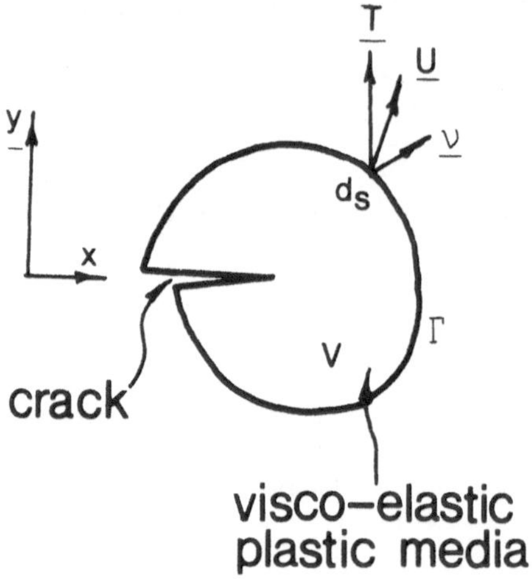

Figure 1. Dynamic crack extension in visco-elastic plastic media.

is path-independent for any path Γ around the crack tip and $t_1 > t > t_0 \geqslant 0$. Here W_e is the elastic strain energy density, σ_{ij}, e^e_{ij} are the stress tensors and elastic strain tensors, and the kinetic energy density is

$$K = 1/2\varrho v_i v_i. \tag{2.2}$$

v_i is the velocity $\partial u_i/\partial t$. The body force F_i is independent of x. Also, e^{cp}_{ij} is the creep and plastic strain. The proof is similar to that in [1] and is omitted here for reasons of simplicity.

In crack propagation problems, it is sometimes convenient to choose the path $\Gamma(t)$ to vary with time t. For example, for the problem of steady state crack propagation, we could choose paths moving with the crack tip to obtain simplification of treatment. Then we could obtain the following:

THEOREM 2: The integral

$$\begin{aligned} Y_2 = {} & \int_{t_0}^{t_1}\left(\int_{\Gamma(t)}(W_e + (\varrho a_i - F_i)u_i)\,\mathrm{d}y - T_i\,\partial u_i/\partial x\,\mathrm{d}s\right)\mathrm{d}t \\ & + \int_{t_0}^{t_1}\int_V \sigma_{ij}\partial e^{cp}_{ij}/\partial x\,\mathrm{d}x\,\mathrm{d}y\,\mathrm{d}t - \int_{t_0}^{t}\int_V \varrho u_i \partial a_i/\partial x\,\mathrm{d}V\,\mathrm{d}t \end{aligned} \tag{2.3}$$

or more simply

$$\begin{aligned} Y_3 = {} & \int_{\Gamma(t)}(W_e + (\varrho a_i - F_i)u_i)\,\mathrm{d}y - T_i\,\partial u_i/\partial x\,\mathrm{d}s \\ & + \int_V \sigma_{ij}\partial e^{cp}_{ij}/\partial x\,\mathrm{d}V - \int_V \varrho u_i \partial a_i/\partial x\,\mathrm{d}V \end{aligned} \tag{2.4}$$

is path-independent for any path $\Gamma(t)$ and $t_1 > t > t_0 \geqslant 0$. Here a_i is the acceleration.

THEOREM 3: The integral

$$Y_4 = \int_{t_0}^{t_1} \left(\int_{\Gamma(t)} (W_e + (\varrho a_i - F_i)u_i^e)\, \mathrm{d}y - T_i\, \partial u_i^e/\partial x\, \mathrm{d}s \right) \mathrm{d}t$$

$$- \int_{t_0}^{t_1} \int_V \varrho u_i^e\, \partial a_i/\partial x\, \mathrm{d}V\, \mathrm{d}t \tag{2.5}$$

or more simply

$$Y_5 = \int_{\Gamma(t)} (W_e + (\varrho a_i - F_i)u_i^e)\, \mathrm{d}y - T_i\, \partial u_i^e/\partial x\, \mathrm{d}s$$

$$- \int_V \varrho u_i^e\, \partial a_i/\partial x\, \mathrm{d}V \tag{2.6}$$

is path-independent for any path $\Gamma(t)$ and $t_1 > t > t_0 \geqslant 0$. Here e_{ij}^e is the elastic displacement.

3. The dynamic crack extension force and integral *Y*

In the following, we will consider the integral form of dynamic crack extension force in visco-elastic plastic media.

For this purpose, we consider a notched crack with crack length a as shown in Fig. 2. Suppose the crack extends a small amount Δa, and we evaluate the energy variation in this crack extension process.

The work done by the traction T_i is

$$\Delta A_T = \int_{S_T} T_i \Delta u_i\, \mathrm{d}s. \tag{3.1}$$

Here Δu_i is the displacement increment during crack extension.

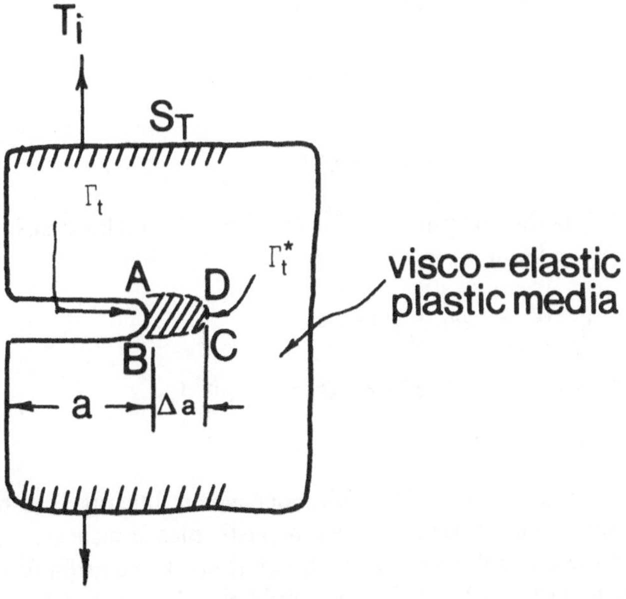

Figure 2

The work done by the body forces consists of two parts: the one is that released in the shaded region ΔV

$$\Delta A_{B1} = - \int_{\Delta V} (F_i - \varrho a_i) u_i^e \, \mathrm{d}V \tag{3.2}$$

the other is equal to

$$\Delta A_{B2} = \int_{V-\Delta V} (F_i - \varrho a_i) \Delta u_i \, \mathrm{d}V. \tag{3.3}$$

The work done by the internal force also consists of two parts: the one is that in ΔV, equal to the elastic strain energy released

$$\Delta A' = - \int_{\Delta V} W_e \, \mathrm{d}V \tag{3.4}$$

the other is that in $(V - \Delta V)$:

$$\Delta A'' = \int_{V-\Delta V} \sigma_{ij} \Delta e_{ij} \, \mathrm{d}V. \tag{3.5}$$

The dynamic crack extension force G is now determined by the following relation:

$$\tilde{G} \Delta a = \Delta A_T + \Delta A_{B1} + \Delta A_{B2} - (\Delta A' + \Delta A'') \tag{3.6}$$

or

$$\begin{aligned} \tilde{G} &= \int_{S_T} T_i \, \partial u_i / \partial a \, \mathrm{d}S + \int_V (F_i - \varrho a_i) \partial u_i / \partial a \, \mathrm{d}V \\ &\quad - \int_V \sigma_{ij} \partial e_{ij} / \partial a \, \mathrm{d}V + \lim_{\Delta a \to 0} 1/\Delta a \int_{\Delta V} W_e \, \mathrm{d}y \\ &\quad - \lim_{\Delta a \to 0} \int_{\Delta V} (F_i - \varrho a_i) u_i^e \, \mathrm{d}V \\ &= \int_{\Gamma_t} W_e \, \mathrm{d}y - \int_{\Gamma_t} (F_i - \varrho a_i) u_i^e \, \mathrm{d}y. \end{aligned} \tag{3.7}$$

Equation (3.5) is the integral form for the dynamic crack extension force.

From (2.6) we know

$$Y_5 = \int_{\Gamma_t} (W_e + (\varrho a_i - F_i) u_i^e) \, \mathrm{d}y. \tag{3.8}$$

Thus, combining (3.3) and (3.6) we get the equality

$$\tilde{G} = Y_5.$$

Therefore, we have shown the path-independent integral Y_5 to be equal to the dynamic crack extension force in a visco-elastic plastic media.

Note that the integral form for the dynamic crack extension force is first given here for visco-elastic plastic media, and could be considered as the theoretical basis for some kind of dynamic fracture criterion.

References

1. Rice, J. R., "A path-independent integral and the approximate analysis of strain concentration by notches and cracks", *Journal of Applied Mechanics, Transactions of the ASME*, Ser. e, **35**, 2 (1968), 378–386.
2. Ouyang Chang, "On path-independent integrals and fracture criteria in non-linear fracture dynamics", *Intern. Journal of Non-linear Mechanics* **18**, 1 (1983), 79–86.

A fundamental solution for flexural problems of a circular cylinder with cracks

TANG REN-JI

Department of Mathematics and Mechanics, Lanzhou University, Lanzhou, Gansu 730001, PRC

Abstract. In this paper, the method proposed by the author [1] is extended to the Saint-Venant's flexural problem of a circular cylinder with cracks. A fundamental solution expressed by the screw dislocation density function is obtained which can be used for the flexural problems of a circular cylinder with an arbitrary radial crack system. To conclude, an example is calculated and the expressions for the stress intensity factor are finally derived.

1. Introduction

The Saint-Venant's flexural problem of a circular cylinder with several radial cracks was investigated by Wigglesworth [2], Sih [3] and others. The method used by them is the complex variable technique, which cannot be used to solve the arbitrary radial crack system. Using a Mellin transform, and adopting the mixed boundary condition of the stress and displacement differences after having solved a set of triple integral equations and after a Neumann's problem exactly, we obtained a fundamental solution for a single crack. Because the fundamental solution is expressed by the screw dislocation density function defined on a crack line, it can eliminate the difficulty of the multiply-connected region and can be used to solve the arbitrary radial crack system by making use of the superposition principle. Here a general method is suggested for solving the Saint-Venant's flexural problem of an arbitrary radial crack system. In order to illustrate the use of the fundamental solution, a detailed discussion is given for solving the flexural problem of two non-colinear equal radial cracks and the expressions for the stress intensity factor are thereby derived.

2. General relationships

According to Saint-Venant's flexural theory for prisms, in the case of a cylindrical coordinate system (ϱ, θ, z), the stress components $(\sigma_{z\varrho}, \sigma_{z\theta}, \sigma_{zz})$ and the displacement components $(u_\varrho, u_\theta, u_z)$ can be expressed by the flexural function $f(\varrho, \theta)$ and

Yeh, K. Y. (ed) Progress in Applied Mechanics
 ISBN-13: 978-94-010-8061-3

the torsional function $\varphi(\varrho, \theta)$. Both functions are defined on the cross section Ω of the prism as follows [5]

$$\sigma_{z\varrho} = \alpha\mu \frac{\partial \varphi}{\partial \varrho} - \beta\mu \frac{\partial \chi}{\partial \varrho} - \frac{\beta\mu}{2}(\nu + 6 \sin^2\Theta\varrho^2 \cos \Theta, \tag{2.1}$$

$$\sigma_{z\theta} = \frac{\alpha\mu}{\varrho}\left(\frac{\partial \varphi}{\partial \theta} + \varrho^2\right) - \frac{\beta\mu}{\varrho}\left[\frac{\partial \chi}{\delta \theta} + \frac{(4 + \nu - 6 \sin^2\Theta)\varrho^3 \sin \Theta}{2}\right], \tag{2.2}$$

$$\sigma_{zz} = -\beta E(l - z)\varrho \cos \Theta, \tag{2.3}$$

$$u_\varrho = \frac{\beta \cos \Theta}{6}[3\nu(l - z)\varrho^2 + 3lz^2 - z^3], \tag{2.4}$$

$$u_\theta = \alpha\varrho z + \frac{\beta \sin \Theta}{6}[3\nu(\lambda - z)\varrho^2 - 3lz^2 + z^3], \tag{2.5}$$

$$u_z = \alpha\varphi - \beta\chi - \frac{\beta}{2}[(2lz - z^2)\varrho \cos \Theta + 2\varrho^3 \sin^2\Theta \cos \Theta], \tag{2.6}$$

where $\Theta = \Theta_0 + \theta$ is the polar angle measured from the principal axis OX of the cross section, $\Theta_0 =$ const., E is Young's modulus, ν is Poisson's ratio, l the length of the prism, $\beta = Q/EJ$ the factor of the lateral force Q applied along the axis OX, J the movement of inertia of the cross section Ω about the principal axis OY, and α the twisting ratio, determined by the following expression.

$$\alpha = \frac{\beta \int_\Omega \left[\frac{\partial \chi}{\partial \theta} + \frac{(4 + \nu - 6 \sin^2\Theta)\varrho^3 \sin \Theta}{2}\right] d\Omega}{\int_\Omega \left(\frac{\partial \varphi}{\partial \theta} + \varrho^2\right) d\Omega}. \tag{2.7}$$

When the cylinder contains an arbitrary radial crack system as shown in Fig. 1, the torsional function $\varphi(\varrho, \theta)$ caused by the disturbance of the crack is not equal to zero, so that we call it the disturbed torsional function. Similarly, the

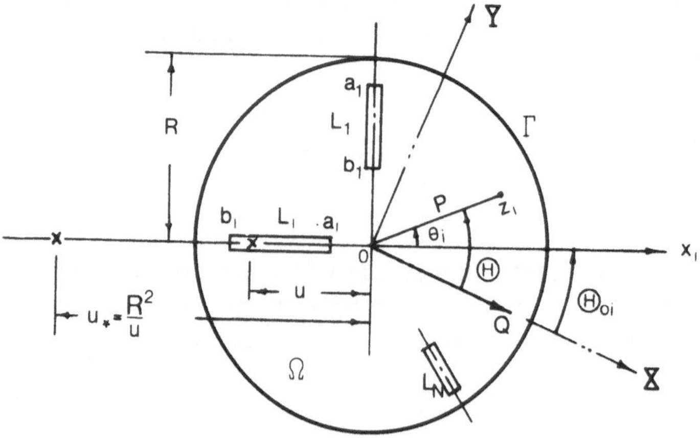

Figure 1

corresponding displacements and stresses are called the disturbed displacements and stresses which are denoted by

$$\bar{u}_z = \alpha\varphi(\varrho, \theta), \tag{2.8}$$

$$\bar{\sigma}_{z\varrho} = \alpha\mu \frac{\partial\varphi}{\partial\varrho}, \qquad \bar{\sigma}_{z\theta} = \frac{\alpha\mu}{\varrho}\frac{\partial\varphi}{\partial\theta}. \tag{2.9}$$

The flexural function $\chi(\varrho, \theta)$, which is different from the torsional function, consists of two parts as follows

$$\chi(\varrho, \theta) = \chi_0(\varrho, 0) + \omega(\varrho, \theta) \tag{2.10}$$

in which $\chi_0(\varrho, \theta)$ is an original flexural function of the non-crack cylinder

$$\chi_0(\varrho, \theta) = \frac{-(3 + 2v)R^2\varrho \cos \Theta + \varrho^3 \cos 3\Theta}{4}, \tag{2.11}$$

where R is the radius of the cylinder and the function $\omega(\varrho, \theta)$ is caused by the crack's disturbance, so that we name it the disturbed flexural function. The corresponding displacements and stresses are called the disturbed displacements and stresses respectively, which are

$$\tilde{u}_z = -\beta\omega(\varrho, \theta) \tag{2.12}$$

$$\tilde{\sigma}_{z\varrho} = -\beta\mu \frac{\partial\omega}{\partial\varrho}, \qquad \tilde{\sigma}_{z\theta} = -\frac{\beta\mu}{\varrho}\frac{\partial\omega}{\partial\theta}. \tag{2.13}$$

Then the total stresses and displacements in the cylinder can be written as

$$\sigma_{z\varrho} = \bar{\sigma}_{z\varrho} + \tilde{\sigma}_{z\varrho} + \frac{\beta\mu}{4}(3 + 2v)(R^2 - \varrho^2) \cos \Theta, \tag{2.14}$$

$$\sigma_{z\theta} = \bar{\sigma}_{z\theta} + \alpha\mu\varrho + \tilde{\sigma}_{z\theta} + \frac{\beta\mu}{4}[4\varrho^2 - (3 + 2v)(R^2 + \varrho^2)] \sin \Theta, \tag{2.15}$$

$$\sigma_{zz} = -\beta E(l - z)\varrho \cos \Theta, \tag{2.16}$$

$$u_\varrho = \frac{\beta \cos \Theta}{6}[3v(l - z)\varrho^2 + 3lz^2 - z^3], \tag{2.17}$$

$$u_\theta = \alpha\varrho z + \frac{\beta \sin \Theta}{6}[3v(l - z)\varrho^2 - 3lz^2 + z^3], \tag{2.18}$$

$$u_z = \bar{u}_z + \tilde{u}_z + \frac{[(3 + 2v)R^2 - 2(2lz - z^2) - \varrho^2]\beta\varrho \cos \Theta}{4}. \tag{2.19}$$

From this it can be concluded that the flexural problem of a cracked cylinder is reduced to finding two disturbed functions, i.e., $\varphi(\varrho, \theta)$, $\omega(\varrho, \theta)$, which are plane harmonic functions and satisfy the following identical equation inside the cross section, or domain Ω

$$\nabla^2(\,) = \frac{\partial^2(\,)}{\partial\varrho^2} + \frac{1}{\varrho}\frac{\partial(\,)}{\partial\varrho} + \frac{1}{\varrho^2}\frac{\partial^2(\,)}{\partial\theta^2} = 0. \tag{2.20}$$

The boundary conditions of the problem are in the following form

$$\left.\frac{\partial \omega}{\partial \varrho}\right|_{\Gamma} = 0, \quad \left.\frac{\partial \omega}{\partial \theta}\right|_{L} = \frac{[4\varrho^2 - (3 + 2\nu)(R^2 + \varrho^2)]\varrho \sin \Theta}{4}, \tag{2.21}$$

$$\left.\frac{\partial \varphi}{\partial \varrho}\right|_{\Gamma} = 0, \quad \left.\frac{\partial \varphi}{\partial \theta}\right|_{L} = -\varrho^2, \tag{2.22}$$

where Γ is the outer boundary of the domain Ω, and $L = L_1 + L_2 + \cdots + L_n$ denotes the union of all the crack surfaces, or inner boundary of the domain Ω.

Because the statement of the above two problems (2.21) and (2.22) is the same, we will only discuss the first dusturbed flexural function $\omega(\varrho, \theta)$, which may be separated into two parts, as follows

$$\omega(\varrho, \theta) = \omega^*(\varrho, \theta) + \omega^{**}(\varrho, \theta), \tag{2.23}$$

where $\omega^*(\varrho, \theta)$ is an harmonic function, caused by the disturbance of the crack L and defined on the cracked infinite plane Ω^∞; the limit case of the domain Ω shown in Fig. 1 as its outer boundary Γ goes to infinite. Note that this function has a jump on the crack line and is regular at infinite, so that it can be determined by the Mellin transform technique. The Mellin transform and its reverse are denoted respectively by

$$\Omega^*(S, \theta) = M[\omega^*(\varrho, \theta); S] = \int_0^\infty \omega^*(\varrho, \theta)\varrho^{S-1}\,d\varrho, \tag{2.24}$$

$$\omega^*(\varrho, \theta) = M^{-1}[\Omega^*(S, \theta); \varrho] = \frac{1}{2\pi i}\int_{L_0} \Omega^*(S, \theta)\varrho^{-S}\,dS, \tag{2.25}$$

where L_0 is an infinite straight line and parallels the imaginary axis in the plane of complex variable S. The stresses and displacements caused by the function $\omega^*(\varrho, \theta)$ are denoted respectively by

$$\tilde{\sigma}^*_{z\varrho} = -\beta\mu\frac{\partial \omega^*}{\partial \varrho}, \quad \tilde{\sigma}^*_{z\theta} = -\frac{\beta\mu}{\varrho}\frac{\partial \omega^*}{\partial \theta}, \tag{2.26}$$

$$\tilde{u}^*_z = -\beta\omega^*(\varrho, \theta). \tag{2.27}$$

Their Mellin transforms are

$$M[\varrho\tilde{\sigma}^*_{z\varrho}; S] = -S\Omega^*(S, \theta), \quad M[\varrho\tilde{\sigma}^*_{z\theta}; S] = -\beta\mu\frac{\partial \Omega^*(S, \theta)}{\partial \theta} \tag{2.28}$$

$$M[\tilde{u}^*_z; S] = -\beta\Omega^*(S, \theta). \tag{2.29}$$

The second function $\omega^{**}(\varrho, \theta)$ in (2.23) is also a plane harmonic function defined on a non-crack finite domain Ω–L, or the domain Ω without the crack L. Since the function $\Omega(\varrho, \theta)$ satisfies the first of the boundary conditions (2.21), $\omega^{**}(\varrho, \theta)$ is a solution of the Neumann's problem on the domain Ω–L, which can be determined by using the complex variable method. The stresses and displacements caused by this functiona are expressed respectively by

$$\tilde{\sigma}^{**}_{z\varrho} = -\beta\mu\frac{\partial \omega^{**}}{\partial \varrho}, \quad \tilde{\sigma}^{**}_{z\theta} = -\frac{\beta\mu}{\varrho}\frac{\partial \omega^{**}}{\partial \theta}, \tag{2.30}$$

$$\tilde{u}_z^{**} = -\beta\omega^{**}(\varrho, \theta). \tag{2.31}$$

When there is only one crack L_1 in the domain Ω, the disturbed flexural function $\omega(\varrho, \theta)$ of the crack L_1 may be determined without any difficulties by using the above expression (2.23). When the function is obtained, the general radial crack system may be solved easily.

3. $\omega_1^*(\varrho, \theta_1)$ of the single crack L_1

When only a radial crack L_1 appears on Fig. 1, the disturbed flexural function caused by L_1 is

$$\omega_1(\varrho, \theta_1) = \omega_1^*(\varrho, \theta_1) + \omega_1^{**}(\varrho, \theta_1), \tag{3.1}$$

where $\omega_1^*(\varrho, \theta_1)$ and $\omega_1^{**}(\varrho, \theta_1)$ correspond to those in (2.23), and the polar angle θ_1 is measured from the axis OX_1. Then the mixed boundary value problem can be stated as

$$\tilde{\sigma}_{z\theta}^*(\varrho, \pi) - \tilde{\sigma}_{z\theta}^*(\varrho, -\pi) = 0, \qquad (0 < \varrho < \infty), \tag{3.2}$$

$$\tilde{u}_z^*(\varrho, \pi) - \tilde{u}_z^*(\varrho, -\pi) = 0, \qquad (0 < \varrho < a_1, b_1 < \varrho < \infty), \tag{3.3}$$

$$\tilde{\sigma}_{z\theta}^*(\varrho, \pi) + \tilde{\sigma}_{z\theta}^*(\varrho, -\pi) = 2q^*(\varrho), \quad (a_1 < \varrho < b_1), \tag{3.4}$$

in which $q^*(\varrho)$ is the sum of the circumferential loads applied on the upper and lower crack surfaces. Since $\omega_1^*(\varrho, \theta_1)$ is a plane harmonic function on the infinite plane Ω^∞, is regular at infinite, and uses the Mellin transform for the boundary condition (3.2), it can easily be determined as follows

$$\Omega_1^*(S, \theta_1) = M[\omega_1^*(\varrho, \theta_1); S] = \frac{\psi_1(S)}{S \sin S\pi} \sin S\theta_1. \tag{3.5}$$

Here $\psi_1(S)$ is an unknown function which may be determined from the following triple integral equations

$$M^{-1}\left[\frac{\psi_1(S)}{S}; \varrho\right] = 0, \qquad (0 < \varrho < a_1, b_1 < \varrho < \infty), \tag{3.6}$$

$$M^{-1}\left[\frac{\psi_1(S)}{\operatorname{tg} S\pi}; \varrho\right] = \frac{\varrho q^*(\varrho)}{\beta\mu}, \quad (a_1 < \varrho < b_1), \tag{3.7}$$

which may be solved by the method shown in [1] and whose solution is

$$\psi_1(S) = \int_{a_1}^{b_1} f_1(\varrho)\varrho^S \, d\varrho. \tag{3.8}$$

Here, $f_1(\varrho)$ is the screw dislocation density function, i.e., the derivative of the jump of the flexural displacement of the crack L_1, and defined as follows

$$f_1(\varrho) = \frac{1}{2\beta}\frac{\partial}{\partial\varrho}[\tilde{u}_z^*(\varrho, \pi) - \tilde{u}_z^*(\varrho, -\pi)], \tag{3.9}$$

which satisfies the following integral condition

$$\int_{a_1}^{b_1} \tilde{f}_1(\varrho)\,\mathrm{d}\varrho \;=\; \tilde{A}_1. \tag{3.10}$$

Here $\tilde{A}_1$ is a constant and equal to zero for an internal crack. Then from (3.5) we obtain

$$\omega_1^*(\varrho,\theta_1) \;=\; M^{-1}\left[\frac{\int_{a_1}^{b_1}\tilde{f}_1(\varrho)\varrho^S\,\mathrm{d}\varrho}{S\sin S\pi}\sin S\theta_1;\,\varrho\right]. \tag{3.11}$$

4. Solution $\omega_1^{**}(\varrho, \theta_1)$ of the Neumann problem

According to the foregoing discussion, $\omega_1^{**}(\varrho, \theta_1)$ is a solution of the following Neumann problem

$$\nabla^2\omega_1^{**}(\varrho,\theta_1) \;=\; 0, \qquad \omega_1^{**}(\varrho,\theta_1)\in\Omega - L_1, \tag{4.1}$$

$$\left.\frac{\partial\omega_1^{**}}{\partial\varrho}\right|_\Gamma \;=\; -\left.\frac{\partial\omega_1^{*}}{\partial\varrho}\right|_\Gamma. \tag{4.2}$$

From [5], its solution is

$$\omega_1^{**}(\varrho,\theta_1) \;=\; \mathrm{Re}\,\frac{1}{\pi i}\int\frac{\mathrm{d}z}{z}\oint_{t=Re^{i\theta_1}}\frac{-(\partial\omega_1^*/\partial\varrho)|_\Gamma}{t-z}\,\mathrm{d}t, \tag{4.3}$$

where

$$\left.\frac{\partial\omega_1^*}{\partial\varrho}\right|_\Gamma \;=\; \frac{i}{\pi}\int_{a_1}^{b_1}\frac{1}{2u}\left[\frac{u}{R}-\frac{Ru}{tu-R^2}-\frac{u^2}{(t-u)R}\right]\tilde{f}_1(u)\,\mathrm{d}u. \tag{4.4}$$

Substituting it in (4.3), we have

$$\omega_1^{**}(\varrho,\theta_1) \;=\; \frac{1}{\pi}\int_{a_1}^{b_1}\arcsin\frac{\varrho\sin\theta_1}{\sqrt{\varrho^2+2\varrho u_*\cos\theta_1+u_*^2}}\,\tilde{f}_1(u)\,\mathrm{d}u, \tag{4.5}$$

where $u_* = R^2/u$.

5. Fundamental solution

Summing up the above two solutions the disturbed flexural function caused by the crack L_1 is

$$\omega_1(\varrho,\theta_1) \;=\; M^{-1}\left[\frac{\int_{a_1}^{b_1}\tilde{f}_1(\varrho)\varrho^S\,\mathrm{d}\varrho}{S\sin S\pi}\sin S\theta_1;\,\varrho\right]$$

$$+\,\frac{1}{\pi}\int_{a_1}^{b_1}\arcsin\frac{\varrho\sin\theta_1}{\sqrt{\varrho^2+2\varrho u_*\cos\theta_1+u_*^2}}\,\tilde{f}_1(u)\,\mathrm{d}u. \tag{5.1}$$

Obviously this function has already satisfied the boundary condition on the outer boundary Γ of the domain Ω, i.e., the first of equations (2.21), and has a jump on the crack line L_1. Then the disturbed flexural stresses caused by the above function are

$$\tilde{\sigma}_{z\varrho}(\varrho, \theta_1) = -\frac{\beta\mu}{\pi}\int_{a_1}^{b_1}\frac{1}{2u}\zeta_{z\varrho}\left(\frac{\varrho}{u}, \theta_1\right)\tilde{f}_1(u)\, du, \tag{5.2}$$

$$\tilde{\sigma}_{z\theta}(\varrho, \theta_1) = -\frac{\beta\mu}{\pi}\int_{a_1}^{b_1}\frac{1}{2u}\zeta_{z\theta}\left(\frac{\varrho}{u}, \theta_1\right)\tilde{f}_1(u)\, du, \tag{5.3}$$

where

$$\zeta_{z\varrho}(\varrho, \theta) = -\sin\theta + \frac{\sin\theta\cos\theta + \sin\theta \operatorname{sh}\ln\varrho}{\cos\theta + \operatorname{ch}\ln\varrho} + \frac{2R^2\sin\theta}{u^2\varrho^2 + 2\varrho R^2\cos\theta + u_*^2}, \tag{5.4}$$

$$\zeta_{z\theta}(\varrho, \theta) = -\cos\theta - \frac{\sin^2\theta - \cos\theta \operatorname{sh}\ln\varrho}{\cos\theta + \operatorname{ch}\ln\varrho} + \frac{2(\varrho u^2 + R^2\cos\theta)}{u^2\varrho^2 + 2\varrho R^2\cos\theta + u_*^2}. \tag{5.5}$$

In a similar manner, the disturbed torsional function is

$$\varphi_1(\varrho, \theta_1) = M^{-1}\left[\frac{\int_{a_1}^{b_1}\bar{f}_1(\varrho)\varrho^S\, d\varrho}{S\sin S\pi}\sin S\theta_1; \varrho\right] + \frac{1}{\pi}\int_{a_1}^{b_1}\arcsin\frac{\varrho\sin\theta_1}{\sqrt{\varrho^2 + 2\varrho u_*\cos\theta_1 + u_*^2}}\bar{f}_1(u)\, du, \tag{5.6}$$

where $\bar{f}_1(\varrho)$ is the screw dislocation density function, i.e. the derivative of the jump of the torsional displacement of the crack L_1, which is defined as follows

$$\bar{f}_1(\varrho) = -\frac{1}{2\alpha}\frac{\partial}{\partial\varrho}[\bar{u}_z^*(\varrho, \pi) - \bar{u}_z^*(\varrho, -\pi)], \tag{5.7}$$

where $\bar{u}_z^*$ corresponds to $\tilde{u}_z^*$, and $\bar{f}_1(\varrho)$ satisfies the following integral condition

$$\int_{a_1}^{b_1}\bar{f}_1(\varrho)\, d\varrho = \bar{A}_1, \tag{5.8}$$

where $\bar{A}_1$ is a constant and equal to zero for an internal crack.

The disturbed torsional stresses caused by $\bar{f}_1(\varrho)$ are

$$\bar{\sigma}_{z\varrho}(\varrho, \theta_1) = \frac{\alpha\mu}{\pi}\int_{a_1}^{b_1}\frac{1}{2u}\zeta_{z\varrho}\left(\frac{\varrho}{u}, \theta_1\right)\bar{f}_1(u)\, du, \tag{5.9}$$

$$\bar{\sigma}_{z\theta}(\varrho, \theta_1) = \frac{\alpha\mu}{\pi}\int_{a_1}^{b_1}\frac{1}{2u}\zeta_{z\theta}\left(\frac{\varrho}{u}, \theta_1\right)\bar{f}_1(u)\, du, \tag{5.10}$$

where the integral kernels are expressed by (5.4) and (5.5), and α is the torsional ratio, determined by (2.7). For the cracked cylinder as shown in Fig. 1, it may be calculated by

$$\alpha = \frac{\mu\beta \int_\Omega \frac{\partial\omega}{\partial\theta}\, d\Omega}{\mu \int_\Omega \left(\frac{\delta\varphi}{\partial\theta} + \varrho^2\right) d\Omega} = \frac{\tilde{M}}{\bar{D} + D_0}, \tag{5.11}$$

where

$$\tilde{M} = \mu\beta \int_\Omega \frac{\partial\omega}{\partial\theta}\, d\Omega, \tag{5.12}$$

$$\bar{D} = \mu \int_\Omega \frac{\partial\varphi}{\partial\theta}\, d\Omega, \qquad D_0 = \frac{\mu\pi R^4}{2}. \tag{5.13}$$

Note that $\tilde{M}$ is a disturbed twisting moment caused by a crack which only depends on the disturbed flexural function $\omega(\varrho, \theta)$, $\bar{D}$ is the loss caused by the cracks L_1 of the twisting rigidity and depends on the disturbed torsional function $\varphi(\varrho, \theta)$, and D_0 is the original twisting rigidity of the uncracked cylinder.

In order to evaluate the twisting ratio α, we must calculate the total disturbed twisting moment $\tilde{M}$ and the total loss of the twisting rigidity, so that the calculation for a single crack is important. Obviously in the case of single crack L_1 these values are

$$\tilde{M}_1 = \mu\beta \int_\Omega \frac{\partial\omega_1}{\partial\theta_1}\, d\Omega = \mu\beta \int_{a_1}^{b_1} (u^2 - b_1^2)\tilde{f}_1(u)\, du, \tag{5.14}$$

$$\bar{D}_1 = \mu \int \frac{\partial\varphi_1}{\partial\theta_1}\, d\Omega = \mu \int_{a_1}^{b_1} (u^2 - b_1^2)\bar{f}_1(u)\, du. \tag{5.15}$$

Summing up the results contributed by each crack, we then get the total values $\tilde{M}$, $\bar{D}$, from which the twisting ratio α is

$$\alpha = \frac{\sum_1^n \tilde{M}_i}{\sum_1^n D_i + D_0}. \tag{5.16}$$

Finally, all the above results are expressed in terms of the screw dislocation density functions $\tilde{f}_1(\varrho)$ and $\bar{f}_1(\varrho)$, which can be used to solve the Saint-Venant flexural problem of an arbitrary radial crack system. We will therefore call this result the fundamental solution.

6. Application of the fundamental solution

Now we will consider the case of two equal non-colinear radial cracks. The crack configuration and the lateral force Q are shown in Fig. 2, in which angle α_1 can very arbitrarily, Q is symmetrical to the crack, and $a_1 = a_2$, $b_1 = b_2$. According to the

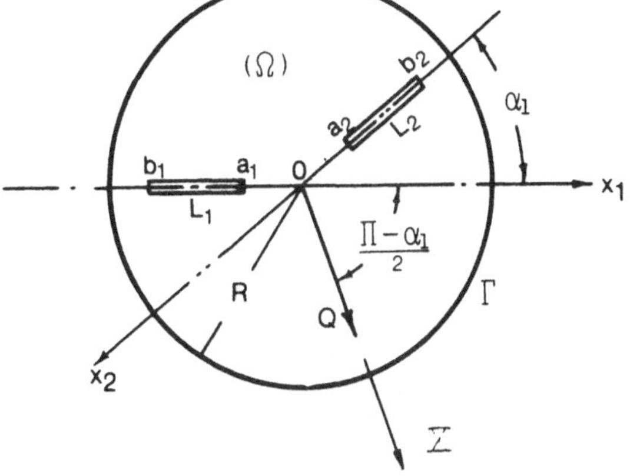

Figure 2

symmetry of the problem, we have $\tilde{f}_1 = -\tilde{f}_2$ and $\bar{f}_1 = \bar{f}_2$. Using the fundamental solution, we can find the total circumferential shearing stress $\sigma_{z\theta}$ from (2.15) and let it equal zero on the upper and lower crack surfaces of L_1 and L_2. Then the integral equations are obtained as follows:

$$\int_{a_1}^{b_1} \frac{\tilde{f}_1(u)}{u - \varrho} \, du + \int_{a_1}^{b_1} \zeta(\varrho, u, \alpha_1) \tilde{f}_1(u) \, du$$

$$= -\frac{\pi[4\varrho^2 - (3 + 2\nu)(R^2 + \varrho^2) \cos(\alpha_1/2)]}{4}, \tag{6.1}$$

$$\int_{a_1}^{b_1} \tilde{f}_1(u) \, du = \tilde{A}_1, \tag{6.2}$$

$$\int_{a_1}^{b_1} \frac{\bar{f}_1(u)}{u - \varrho} \, du + \int_{a_1}^{b_1} \zeta(\varrho, u, \alpha_1) \bar{f}_1(u) \, du = -\pi\varrho \tag{6.3}$$

$$\int_{a_1}^{b_1} \bar{f}_1(u) \, du = \bar{A}_1, \tag{6.4}$$

where the kernel is

$$\zeta(\varrho, u, \alpha_1) = \frac{1}{\varrho - u_*} - \frac{\varrho + u \cos \alpha_1}{\varrho^2 + 2\varrho u \cos \alpha_1 + u^2},$$

$$+ \frac{\varrho + u_* \cos \alpha_1}{\varrho^2 + 2\varrho u_* \cos \alpha_1 + u_*^2}, \tag{6.5}$$

where $u_* = R^2/u$, which is conjugate with point u.

The above two sets of integral equations are the Cauchy-type singular integral equations which can be solved by analytical or numerical methods [4, 7]. When the two unknown screw dislocation functions $\tilde{f}_1$ and $\bar{f}_1$ are obtained, then, using the fundamental solution, the total stresses caused by the lateral force Q can be determined by

$$\sigma_{z\varrho}(\varrho, \Theta) = \frac{\beta\mu}{\pi} \sum_{1}^{2} \int_{a_n}^{b_n} \frac{1}{2u} \zeta_{z\varrho}\left(\frac{\varrho}{u}, \theta_n\right)\left[\frac{\alpha}{\beta}\tilde{f}_n(u) - \tilde{f}_n(u)\right] du$$

$$+ \frac{\beta\mu}{4}(3 + 2\nu)(R^2 - \varrho^2)\cos\Theta, \tag{6.6}$$

$$\sigma_{z\theta}(\varrho, \Theta) = \frac{\beta\mu}{\pi} \sum_{1}^{2} \int_{a_n}^{b_n} \frac{1}{2u} \zeta_{z\theta}\left(\frac{\varrho}{u}, \theta_n\right)\left[\frac{\alpha}{\beta}\tilde{f}_n(u) - \tilde{f}_n(u)\right] du$$

$$+ \frac{\beta\mu}{4}[4\varrho^2 - (3 + 2\nu)(R^2 + \varrho^2)]\sin\Theta + \alpha\mu\varrho. \tag{6.7}$$

Note that the stress $\sigma_{zz}(\varrho, \Theta)$ is given by the original formula, and the integral kernels above are expressed by (5.4) and (5.5), in which θ must be replaced by the angle θ_n of n-th crack.

In the same way as stated in [8], and using the singularity of the circumferential stress at the crack tips, the expression for the stress intensity factor are respectively given below

$$k(a_n) = \beta\mu \lim_{\varrho \to a_n} \sqrt{2(\varrho - a_n)}\left[\frac{\alpha}{\beta}\tilde{f}_n(\varrho) - \tilde{f}_n(\varrho)\right], \tag{6.8}$$

$$k(b_n) = -\beta\mu \lim_{\varrho \to b_n} \sqrt{2(b_n - \varrho)}\left[\frac{\alpha}{\beta}\tilde{f}_n(\varrho) - \tilde{f}_n(\varrho)\right]. \tag{6.9}$$

The twisting ratio is

$$\alpha = \frac{\beta \sum_{n=1}^{2} \int_{a_n}^{b_n} (u^2 - b_n^2)\tilde{f}_n(u)\, du}{\sum_{n=1}^{2} \int_{a_n}^{b_n} (u^2 - b_n^2)\tilde{f}_n(u)\, du + D_0}. \tag{6.10}$$

From the viewpoint of fracture mechanics, the problem mentioned above is solved.

Next, owing to the symmetry of the problem the screw dislocation density function $\tilde{f}_1 = -\tilde{f}_2$, so that the twisting ratio $\alpha = 0$. Thus, the problem can be further simplified, and the stress intensity factor only depends on the function $\tilde{f}_n$, which is the derivative of the jump of the flexural displacement. For this reason we will call it the flexural stress intensity factor and denote it by

$$k_f(a_n) = -\beta\mu \lim_{\varrho \to a_n} \sqrt{2(\varrho - a_n)}\tilde{f}_n(\varrho), \tag{6.11}$$

$$k_f(b_n) = -\beta\mu \lim_{\varrho \to b_n} \sqrt{2(b_n - \varrho)}\tilde{f}_n(\varrho). \tag{6.12}$$

Therefore, in order to calculate the stresses and the stress intensity factors of the cracked cylinder, it is sufficient to solve th first set of the singular integral equations (6.1) and (6.2), which are similar to those given in [8] and can also be solved in a similar way. Due to the limitations of space it is omitted here.

References

1. Tang Ren-ji and Wang Kai, "On the Griffith crack, whose surfaces are loaded asymmetrically", *Acta Mechanica Sinica* **3** (1980), 269–278 (in Chinese).
2. Wigglesworth, L. A., "Flexture and torsion of a circular shaft with two cracks", *Proc. London Math. Soc.* **47** (1940), 20–37.
3. Sih, G. C., "Strength of stress singularities at crack tips for flexural and torsional problems", *J. Appl. Mech.* **33** (1963), 000–000.
4. Erdogan F. and G. D. Gupta, "On the numerical solution of singular integral equations", *Quart. Appl. Math.* **29** (1972), 523–534.
5. Muskhelishvili, N. I., *Some Basic Problems of the Mathematical Theory of Elasticity*, Noordhoff (1953).
6. Tang Ren-ji and Jiang Zhu-zhong, "An analysis of radial crack systems on a finite circular plate", *Acta Mechanics Solida Sinica* **4** (1981), 445–458 (in Chinese).
7. Muskhelishvili, N. I., *Singular Integral Equations*, Noordhoff (1953).
8. Tang Ren-ji, "Saint-Venant's torsion problem for a circular cylinder with cracks", *Acta Mechanica Sinica* **4** (1982), 332–340 (in Chinese).

Computational Mechanics

The whiplash effect on tall buildings

ZHONG WAN-XIE and LIN JIA-HAO

Research Institute of Engineering Mechanics, Dalian Institute of Technology, Dalian, PRC

Abstract. The prominent parts of the tops of tall buildings will vibrate drastically under strong earth tremors which may cause these parts to be damaged or even collapse. In this paper, the mechanism of the so-called whiplash effect is explained by means of an analysis of the characteristics of coupled vibrators. The correctness of this theoretical analysis has been justified by numerous examples. It has been shown that increasing the stiffness does not in all cases result in a reduction of the whiplash effect. It is suggested that in order to reduce the whiplash effect, dynamic calculations should be executed so as to adjust the stiffness and mass distribution of the buildings.

Introduction

Many tall buildings have prominent parts on their tops. Under the action of strong earth tremors, these parts may be subject to severe damage or even collapse while the lower parts (main structures) remain in good condition. In order to cope with this whiplash effect, the relevant code [1] stipulates that for the prominent parts such as the water tank cabin, and the elevator cabin, etc., the area of the earthquake-resistant walls must be increased up to three times that calculated for ordinary structures.

Although the code provides a way, without explanation, to cope with the whiplash effect or to increase the area of stiffness which leads to the strengthening of the prominent parts, it does not in fact reduce the whiplash effect in all cases. Moreover, engineering practice governs that it has often been difficult for engineers to follow this stipulation, for the stiffeners are sometimes too dense to arrange. It is necessary to find the reason for the whiplash effect so as to resolve this difficulty.

In this paper, the cause of the whiplash effect is explained in terms of the analysis of free vibration and forced vibration of coupled vibrators, and measurements are suggested for weakening the whiplash effect. It should also be pointed out that the whiplash effect will appear not only in tall buildings, but also in other structures such as offshore platforms, ship masts, and so on.

2. Free vibration analysis of coupled vibrators

Let us first consider a two-DOF system as that shown in Fig. 1. It is assumed that $K_1 \gg K_2, M_1 \gg M_2$. If the value of K_1/M_1 is far from that of K_2/M_2, then (K_1, M_1)

Yeh, K. Y. (ed) Progress in Applied Mechanics
 ISBN-13: 978-94-010-8061-3

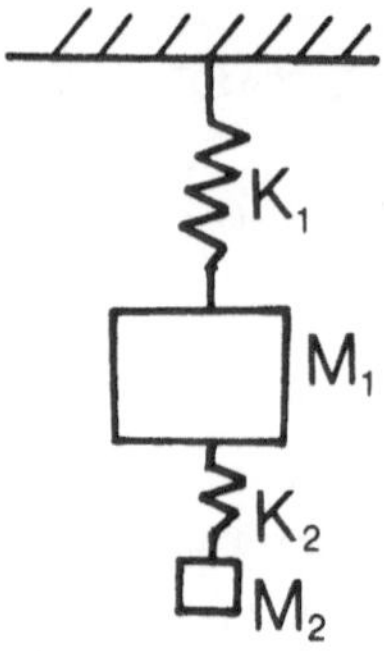

Figure 1

and (K_2, M_2) can be regarded as two vibrators independent of each other. The characteristic frequencies of the system consist of the characteristic frequencies of both vibrators, i.e.,

$$\omega_1^2 = \frac{K_1}{M_1}, \qquad \omega_2^2 = \frac{K_2}{M_2}. \tag{2.1}$$

However, if K_1/M_1 is identical or close to K_2/M_2, the two vibrators will be coupled. If the tension of each spring is of a generalized displacement, the stiffness matrix and mass matrix will then be

$$[K] = \begin{bmatrix} K_1 & \\ & K_2 \end{bmatrix}, \qquad [M] = \begin{bmatrix} M_1 + M_2 & M_2 \\ M_2 & M_2 \end{bmatrix}. \tag{2.2}$$

The frequency equation of the system is:

$$\|[K] - \omega^2[M]\| = \begin{vmatrix} K_1 - (M_1 + M_2)\omega^2 & -M_2\omega^2 \\ -M_2\omega^2 & K_2 - M_2\omega^2 \end{vmatrix} = 0. \tag{2.3}$$

Introducing the following symbols:

$$\omega_1^2 = \frac{K_1}{M_1}, \quad \omega_2^2 = \frac{K_2}{M_2}, \quad \lambda_1 = \frac{\omega_1^2}{\omega^2}, \quad \lambda_2 = \frac{\omega_2^2}{\omega^2}, \quad \varepsilon = \frac{M_2}{M_1}, \tag{2.4}$$

Equation (2.3) can then be transformed into

$$(\lambda_1 - 1)(\lambda_2 - 1) = \varepsilon\lambda_2. \tag{2.5}$$

When ε is very small, $\lambda_1 = 1$, and $\lambda_2 = 1$, i.e., the characteristic frequencies of the system can be expressed by the first two expressions of (2.4). Furthermore, if $\omega_1^2 = \omega_2^2$, by denoting $\lambda = \lambda_1 = \lambda_2$, the solution of (2.5) is approximately

$$\lambda = 1 \mp \sqrt{\varepsilon}. \tag{2.6}$$

It can be seen that a system consisting of two vibrators with identical frequencies possess two closely spaced characteristic frequencies. The explanation lies in the coupling between the two vibrators, and the coupling extent is characterized by $\varepsilon = M_2/M_1$. The corresponding closely spaced modes are:

$$\left.\begin{aligned} &\text{mode 1:}\quad X_1 = 1,\quad X_2 = 1/\sqrt{\varepsilon}\quad (\text{for } \lambda_a = 1 - \sqrt{\varepsilon}),\\ &\text{mode 2:}\quad X_1 = 1,\quad X_2 = -1/\sqrt{\varepsilon}\quad (\text{for } \lambda_b = 1 + \sqrt{\varepsilon}).\end{aligned}\right\} \tag{2.7}$$

It is known that the simultaneous vibration of two coupled vibrators with identical or close characteristic frequencies will produce a beating phenomenon which results in the transfer of the energy from one vibrator to the other [2]. Now, let M_1 be very large, and the fixed end of K_1 have an initial displacement D_0 when $t = 0$, which means that the initial condition is as follows:

$$X_1 = D_0,\quad X_2 = 0,\quad \dot{X}_1 = 0,\quad \dot{X}_2 = 0. \tag{2.8}$$

The general solution of the equation of free vibration is

$$\left.\begin{aligned} X_1 &= A_1 \cos \omega_a t + A_2 \sin \omega_a t + A_3 \cos \omega_b t + A_4 \sin \omega_b t,\\ X_2 &= -\frac{A_1}{\sqrt{\varepsilon}} \cos \omega_a t - \frac{A_3}{\sqrt{\varepsilon}} \sin \omega_a t + \frac{A_2}{\sqrt{\varepsilon}} \cos \omega_b t + \frac{A_4}{\sqrt{\varepsilon}} \sin \omega_b t.\end{aligned}\right\} \tag{2.9}$$

Using the initial condition, one obtains $A_2 = A_4 = 0$, $A_1 = A_3 = D_0/2$, so that

$$\left.\begin{aligned} X_1 &= \frac{D_0}{2}\{\cos \omega_a t + \cos \omega_b t\} = D_0 \cos \frac{\omega_a + \omega_b}{2} t \cos \frac{\omega_b - \omega_a}{2} t,\\ X_2 &= \frac{D_0}{2\sqrt{\varepsilon}}\{-\cos \omega_a t + \cos \omega_b t\} = \frac{D_0}{\sqrt{\varepsilon}} \sin \frac{\omega_a + \omega_b}{2} t \sin \frac{\omega_a - \omega_b}{2} t.\end{aligned}\right\} \tag{2.10}$$

As ω_1 and ω_2 are very close, the frequency $(\omega_b - \omega_a)/2$ is very low, and $(\omega_a + \omega_b)/2$ is the frequency of the primal vibrators; it can be seen that the beating phenomenon occurs. When $t = 0$, vibrator X_2 has no energy. When $t = \pi/(\omega_b - \omega_a)$, almost all the energy of the system has been transferred to X_2, and its amplitude is $D_0/\sqrt{\varepsilon}$ which is very large as ε is very small. The vibrator X_1 consists of a large mass and a large spring which implies the main structure of a tall building. When an earthquake takes place, this vibrator absorbs a great amount of energy from the ground. The vibrator X_2 consists of a small mass and a small spring which implies the prominent part on the top of the tall building. When the energy absorbed by the vibrator X_1 completely transfers to the vibrator X_2, this small vibrator X_2 will vibrate drastically, and the whiplash effect then occurs.

3. Forced vibration of coupled vibrators

Let us consider a SDOF system as shown in Fig. 2(a). Under the action of an horizontal ground acceleration $\ddot{X}_g$, the displacement X of mass M_1 relative to the ground, is governed by the following equation

$$M_1\ddot{X} + K_1 X = -\ddot{X}_g. \tag{3.1}$$

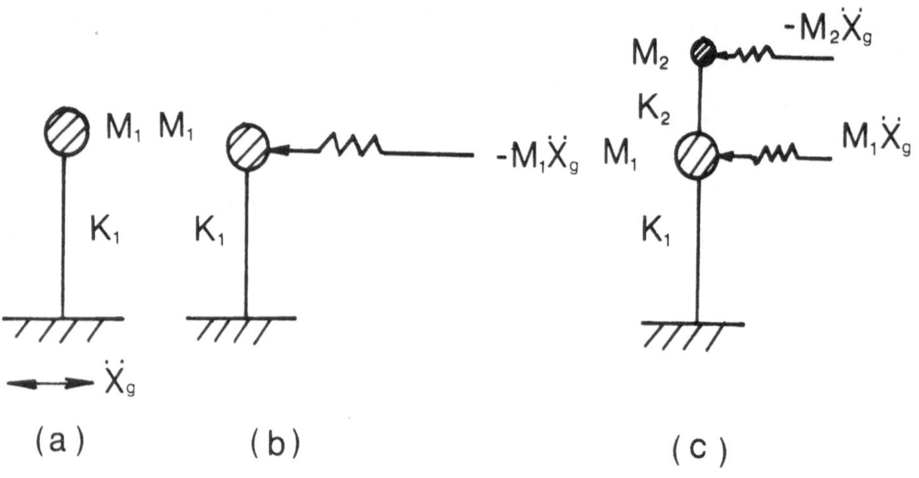

Figure 2

The action of the ground acceleration can be regarded as an inertia force acting on M_1 while the ground is assumed immovable, cf. Fig. 2(b). In appearance, the variation of $\ddot{X}_g$ with time, seems orderless. In fact, if $\ddot{X}_g(t)$ is taken to execute Fourier's decomposition in a limited time region, a prominently strong Fourier component will be found at a certain frequency which reflects the dynamic characteristics of the site itself. For sites with medium hardness the strongest Fourier component appears generally near the frequency of $\omega = 16\,\text{sec}^{-1}$ or period $T = 2\pi/\omega = 0.4$ sec. Here T is known as the predominant period of the site, which is small for hardrock sites and large for soft sites. In general, T varies between 0.2 sec and 0.6 sec. For buildings with three or more storeys, it is often found that at least one of the characteristic frequencies is within this region.

If the characteristic period T of the above SDOF system happens to be identical to the prominent period of the site, the system will get into resonance with the Fourier component of ground motion, which leads to a considerable amplitude of M_1. Furthermore, if a system with much smaller stiffness K_2 and much smaller mass M_2 is attached to M_1 as shown in Fig. 2(c), and $K_1/M_1 = K_2/M_2$, it is known that the subsidiary system (M_2, K_2) turns out to be a dynamic damper. When the frequency of a sinusoidal force applied to M_1 is $\omega = \sqrt{K_2/M_2}$, the system (M_2, K_2) will cancel or reduce the force acting on M_1. In cases where the inertia force acting on M_2 is neglected, M_1 will remain static and become an anti-resonant point [4]. The system (M_2, K_2), when reducing the dynamic response of M_1, is subjected to an extremely great load, and is therefore likely to be damaged.

If the inertia force on M_2 is not neglected as shown in Fig. 2(c), the motion equation of the whole system becomes

$$\begin{bmatrix} M_1 & \\ & M_2 \end{bmatrix} \begin{bmatrix} \ddot{X}_1 \\ \ddot{X}_2 \end{bmatrix} + \begin{bmatrix} K_2 + K_1 & -K_2 \\ -K_2 & K_2 \end{bmatrix} \begin{bmatrix} X_1 \\ X_2 \end{bmatrix}$$

$$= -\begin{bmatrix} M_1 \\ M_2 \end{bmatrix} \ddot{X}_g = \omega^2 \begin{bmatrix} M_1 \\ M_2 \end{bmatrix} \sin \omega t. \tag{3.2}$$

Assume

$$\begin{bmatrix} X_1 \\ X_2 \end{bmatrix} = \begin{bmatrix} B_1 \\ B_2 \end{bmatrix} \sin \omega t \tag{3.3}$$

and

$$\frac{K_1}{M_1} = \frac{K_2}{M_2} = \omega^2, \qquad \frac{M_1}{M_2} = n \gg 1, \tag{3.4}$$

then (3.2) can be transformed into

$$-\omega^2 \begin{bmatrix} M_1 & \\ & M_2 \end{bmatrix} \begin{bmatrix} B_1 \\ B_2 \end{bmatrix} + \begin{bmatrix} K_1 + K_2 & -K_2 \\ -K_2 & K_2 \end{bmatrix} \begin{bmatrix} B_1 \\ B_2 \end{bmatrix} = \begin{bmatrix} M_1 \\ M_2 \end{bmatrix} \omega^2 \tag{3.5}$$

or

$$\begin{bmatrix} 1 & -1 \\ -1 & 0 \end{bmatrix} \begin{bmatrix} B_1 \\ B_2 \end{bmatrix} = \begin{bmatrix} n \\ 1 \end{bmatrix}. \tag{3.6}$$

The solution to (3.6) is

$$\begin{cases} B_1 = -1 \\ B_2 = -(n+1). \end{cases} \tag{3.7}$$

It can be seen that $|B_2| \gg |B_1|$, i.e., the amplitude of M_2 is much bigger than that of M_1 and this illustrates the whiplash effect.

The damping factor has not been taken into account in the above analysis. The deduced conclusion, however, is applicable to structures with weak damping such as steel structures or RC structures. However, let us take the structure shown in Fig. 2(c) as an example, and assume

$$M_1 = 100, \quad M_2 = 1, \quad K_1 = 10\,000, \quad K_2 = 100.$$

The structural damping ratio is taken as $\zeta_1 = \zeta_2 = 0.05$. A sinusoidal horizontal ground acceleration $\ddot{X}_g = \sin \omega t \; (\omega = 10 \sec^{-1})$ is adopted to execute step-by-step integration in terms of the Wilson-ϑ method with $\vartheta = 1.4$, and the step size is $\Delta t = 0.006 \sec^{-1}$; 3000 steps (amounting to 30 cycles) are calculated. Within this duration, the maximum displacements of X_1 and X_2 are respectively

$$\bar{X}_1 = 0.0609, \qquad \bar{X}_2 = 0.5226;$$

so that the whiplash effect appears. It is noted from the calculation that X_1 and X_2 reach their maxima $\bar{X}_1$ and $\bar{X}_2$ alternatively with a phase lag $T/4$.

One way to reduce the whiplash effect is to increase the stiffness of the upper part, say, by 20%, i.e., let $K_2 = 120$ while other conditions remain unchanged. Repeating the above 3000 steps of calculation produces the maxima as follows

$$\bar{X}_1 = 0.0789, \qquad \bar{X}_2 = 0.3304.$$

It can be seen that the whiplash effect of the upper part is considerably reduced though the response of the lower part is now a little bit stronger (as the dynamic damper loses its effect).

Now, let us decrease the stiffness of the upper part by 20%, i.e., put $K_2 = 80$ while the other conditions remain unchanged. Repetition of the above calculation produces

$$\bar{X}_1 = 0.0812, \qquad \bar{X}_2 = 0.3825.$$

The whiplash effect is also reduced herein by means of decreasing the stiffness of the upper part. For the present system (with $K_2 = 80$), further reduction of the whiplash effect requires a strengthening of the stiffness of K_1, or a weakening instead of a strengthening of K_2. However, the whiplash effect will get even stronger if the stiffness of K_2 is strengthened blindly. Nevertheless, this blindness is not yet fully understood in engineering practice.

The above deduced conclusion can be extended to deal with MDOF systems. Figure 3(a) shows the structural mode of a four-storey shear-type building of which the basic frequency is $\omega_1 = 16.44\,\text{sec}^{-1}$, so $T_1 = 2\pi/\omega_1 = 0.3822\,\text{sec}^{-1}$. Here the dimensions of mass and stiffness follow the T.M.S. metric system. If a two DOF system is built on top of the building as shown in Fig. 3(b), and $M_5 = M_6 = 0.05$, $K_5 = K_6 = 35.38$, the characteristic frequencies of this system (fixed to immovable M_4) will be $\omega_1 = 16.44\,\text{sec}^{-1}$, $\omega_2 = 43.4\,\text{sec}^{-1}$. According to the above analysis, this structure is likely to produce whiplash effect under the action of earthquakes. In fact, if a horizontal ground acceleration $\ddot{X}_g = \sin 16.44t$ M/sec^2 is applied to the base of the structure whose Rayleigh damping matrix is consistuted according to the specified damping ratios $\zeta_1 = \zeta_2 = 0.05$, the step size is $\Delta t = 0.003822\,\text{sec}^{-1}$. The Wilson-$\vartheta$ method is adopted to calculate 1000 integration steps (about 10 cycles of the ground motion), and within

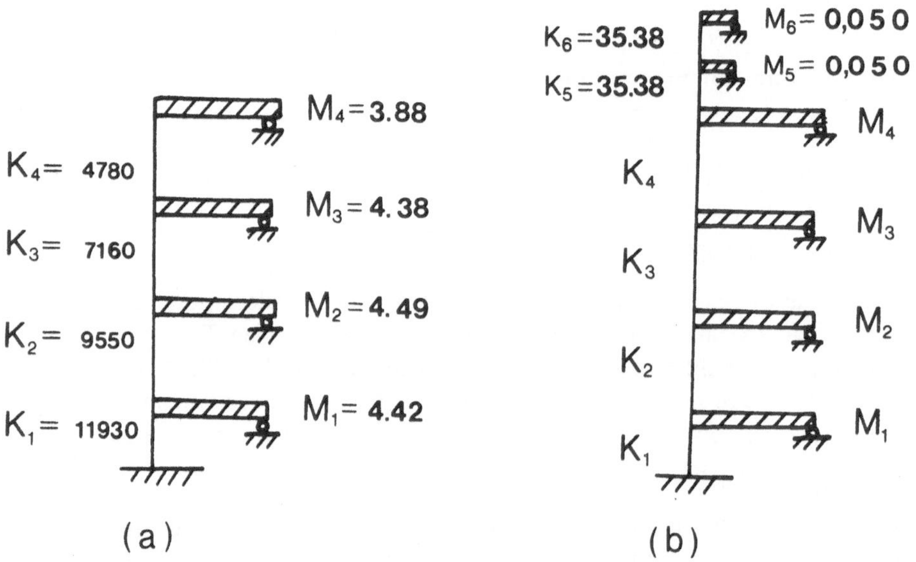

Figure 3

this time duration, the maximum value of each interfloor displacement is successively (M):

$$0.0069,\ 0.0077,\ 0.0079,\ 0.0064,\ 0.1791,\ 0.1061. \tag{3.8}$$

Thereby, the whiplash effect is found on the added part.

If the stiffness of the added part is increased by 40%, i.e., assume $K_5 = K_6 = 50$ with the other conditions kept unchanged; the above maximum values then become

$$0.0098,\ 0.0112,\ 0.0116,\ 0.0096,\ 0.0693,\ 0.0400. \tag{3.9}$$

Here, the whiplash effect is considerably reduced (the difference is about 2.5 times).

If the stiffness values of the lower part K_1 to K_4 are all reduced by 10%, i.e., set $K_1 = 10\,700$, $K_2 = 8600$, $K_3 = 6440$, $K_4 = 4300$, $K_5 = K_6 = 50$; then the whiplash effect is further weakened, and the six maxima become

$$0.0077,\ 0.0089,\ 0.0093,\ 0.0078,\ 0.0484,\ 0.0279. \tag{3.10}$$

In the above calculations, only the sinusoidal ground motion is taken into account. As an alternative, if a real time-history record of ground acceleration is taken as the load, then analogous conclusions can be obtained. Now, let a horizontal ground acceleration record of the El-centro earthquake (1940, N–S) act on the structure shown in Fig. 3(b), and take $\Delta t = 0.02$ sec, and then the earthquake duration is 10 sec. Thus, the six maxima corresponding to the above case (3.8) are now (cm)

$$0.6635,\ 0.7194,\ 0.7181,\ 0.6705,\ 8.2171,\ 4.9596. \tag{3.8a}$$

The maxima corresponding to the above case (3.9) are

$$0.6903,\ 0.7409,\ 0.7617,\ 0.7129,\ 4.7952,\ 2.8901. \tag{3.9a}$$

And those corresponding to the above (3.10) are

$$0.7349,\ 0.8105,\ 0.8870,\ 0.8157,\ 4.6361,\ 2.7707. \tag{3.10a}$$

It can be seen that (3.9a) and (3.10a) show an analogous law as (3.9) and (3.10) to reduce the whiplash effect.

4. Conclusion

So far, in engineering, the whiplash effect has not yet been properly studied and explained. This paper shows that when a certain characteristic period of the prominent part of the top of a building is identical to one of the characteristic periods of the building (the main structure), and this period is identical to or near the predominant period of the site, the whiplash effect will most likely take place under the action of earthquakes. Increasing the stiffness of the prominent part blindly does not necessarily lead to reduction of the whiplash effect. The correct method is to adjust the stiffness and mass distribution of the whole structure based on structural dynamic analysis, so as to remove the factor causing the whiplash effect as explained in this paper. This suggestion provides a supplement to the revision of code [1], currently in effect. It might also be helpful to structural engineers.

References

1. *Earthquake resistant Design Code for Industrial and Civil Architecture*, TJ-11-78, Press of China Architectural Industry (1978).
2. French, A. P., *Vibrations and Waves*, W. W. Norton & Company, Inc. (1971).
3. Wang Guang-yuan, *Vibration of Architectural Structures*, Science Press (1978).
4. Dalian Institute of Technology, *Computational Structural Dynamics* (textbook) (1983).

Solutions of geometrical non-linear problems of stiffened plates and shells by the finite element method

LIU ZHENG-XING, FENG TAI-HUA and LI DING-XIA

Address ?

Abstract. In this paper, the geometrical non-linear problems of stiffened plates and shells is studied by using the finite element method. The discrete elements used are triangular plate elements and beam elements. Both stresses due to the axial action of the beam and the membrane action of the plate are considered. A comoving coordinate system and an incremental process are adopted to analyse deformation behaviour and stress distribution. The effect on loads due to large deflections is also discussed. The inverse iteration method is adopted for calculation of lower critical stresses.

1. Introduction

When deflections under lateral loads develop in thin plates, and thin shells and beams which exceed the limit permitted by linear theory, geometric non-linearity must be considered. Geometric non-linearity is caused by large displacements and large rotations. Thus, in the strain deflection relationship, the quadratic terms must be considered along with the effect of deformation on the configuration and dimension of the structures, and on equilibrium equations.

The comoving coordinate system suggested by Chien Wei-zang [1] is adopted in this paper. Here, the local coordinate systems of each element are fixed to the element so that it changes with the deformation of the structure. Thus, the following basic equation for analysis of a geometrically non-linear structure can be given:

$$([K_E] + [K_G])\{d\} = \{R\}, \qquad (1.1)$$

where $[K_G]$ is the geometric stiffness matrix, and $[K_E]$ is the elastic stiffness matrix of a structure. Here, $[K_E]$ and $[K_G]$ are calculated by the behaviour after deformation. When an incremental process is adopted for each step, the equation (1.1) can be written as follows:

$$([K_E]_{i-1} + [K_G]_{i-1})\{d_i\} = \{\Delta R_i\}, \qquad (1.2)$$

where $[K_E]_{i-1}$ is the elastic stiffness matrix and $[K_G]_{i-1}$ is the geometric stiffness matrix for a deformed structure obtained in the $(i - 1)$-th step, i.e., the preceeding step. $\{\Delta R_i\}$ is the increment of the load in the i-th step. In this paper the load $\{R\}$

Yeh, K. Y. (ed) Progress in Applied Mechanics
 ISBN-13: 978-94-010-8061-3

is divided equally into n parts, i.e., $\{\Delta R_i\} = \{\Delta R\} = \{R\}/n$ for each step. $\{\Delta d_i\}$ is the increment of displacements found in the present step.

For stiffened structures the discrete elements used are triangular plate elements and beam elements. A fictitious rotational stiffness is added when plate elements are coplanar. For a beam element, a correction for eccentric effects is made as the nodes are not placed on the neutral axis of the beam element. The effect of deflection on nodal loads is also considered. The problem is very complicated. In this paper a preliminary analysis is given for further discussion.

2. Element stiffness matrices

2.1. *Triangular plate elements*

In Fig. 1 $x' - y' - z'$ is a local coordinate system where the origin coincides with the geometric centre.

(a) *Elastic stiffness matrix.* Membrane stiffness matrix [2]

$$[K_1] = [B]^T[D][B]t\Delta. \tag{2.1}$$

Bending stiffness matrix [2]

$$[K_2] = \frac{1}{64\Delta^5}[T]^T[C]^T[I][C][T]. \tag{2.2}$$

Fictitious rotational stiffness matrix [3]

$$[K_3] = 0.03Et\Delta \begin{bmatrix} 1 & -0.5 & -0.5 \\ -0.5 & 1 & -0.5 \\ -0.5 & -0.5 & 1 \end{bmatrix} \tag{2.3}$$

and the elastic stiffness matrix is

$$[K_E] = [K_1] + [K_2] + [K_3]. \tag{2.4}$$

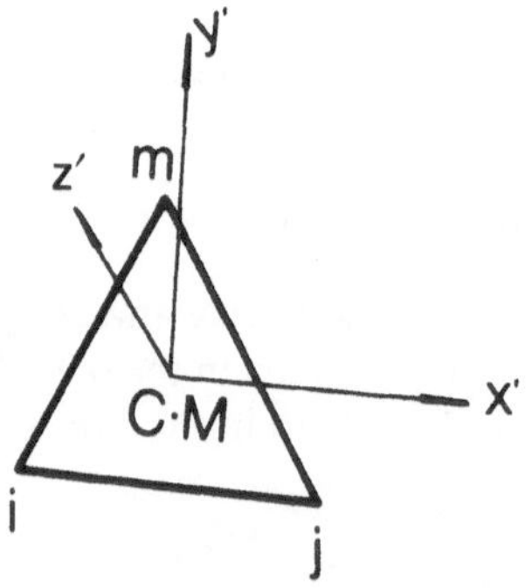

Figure 1. Triangular plate element.

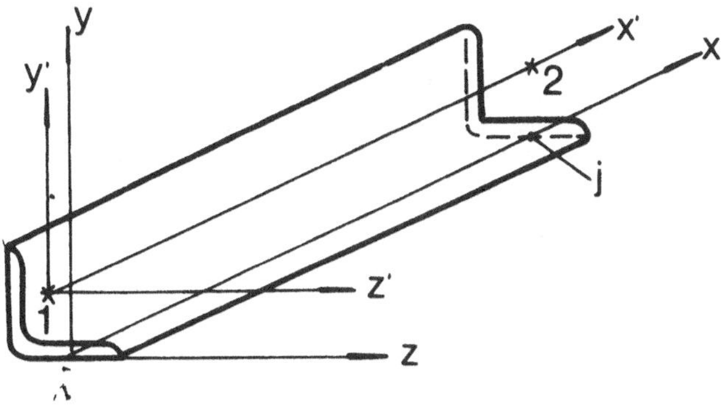

Figure 2. Space beam element.

(b) *Geometric stiffness matrix* [4]

$$[K_G] = [K_{Gx}] + [K_{Gy}] + [K_{Gxy}] = \frac{t}{2\Delta}\{\sigma_x[b_i^2 A + b_i b_j B + b_j C]$$
$$+\sigma_y[c_i^2 A + c_i c_j B + c_j^2 C] + \tau_{xy}[2b_i c_i A$$
$$+(b_j c_i + b_i c_j)B + 2b_j c_j C]\} \tag{2.5}$$

where σ_x, σ_y, τ_{xy} are membrane stresses in each step. They are taken as constants.

2.2 *Space beam elements*

In Fig. 2, points 1 and 2 are the centroid of the cross section, i and j are the nodes of the beam element, x–y–z is a local coordinate system passing through the nodes, x'–y'–z' is local coordinate system with the origin passing through the centroid.

We use the elastic stiffness matrix $[K_E]$ of a straight beam element with equal cross section area, with reference to its neutral axis [5].

The geometric stiffness matrix with reference to its neutral axis is given [6] as:

$$[K_G] = \frac{\sigma_x A}{30L} S. \tag{2.6}$$

When the nodes are not placed on the neutral axis of the beam element, the stiffness matrix

$$[K_T] = [TT]^T[K][TT] \tag{2.7}$$

is used, where $[TT]$ is the correcting matrix suggested in [7].

3. Geometric non-linear analysis

3.1. *Analysis of a circular plate*

Figure 3 shows a circular plate with the circumference clamped under uniformly distributed transverse loads.

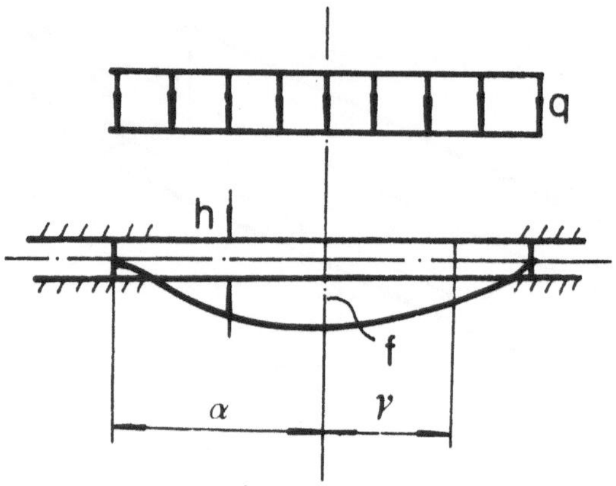

Figure 3. Circular thin plate.

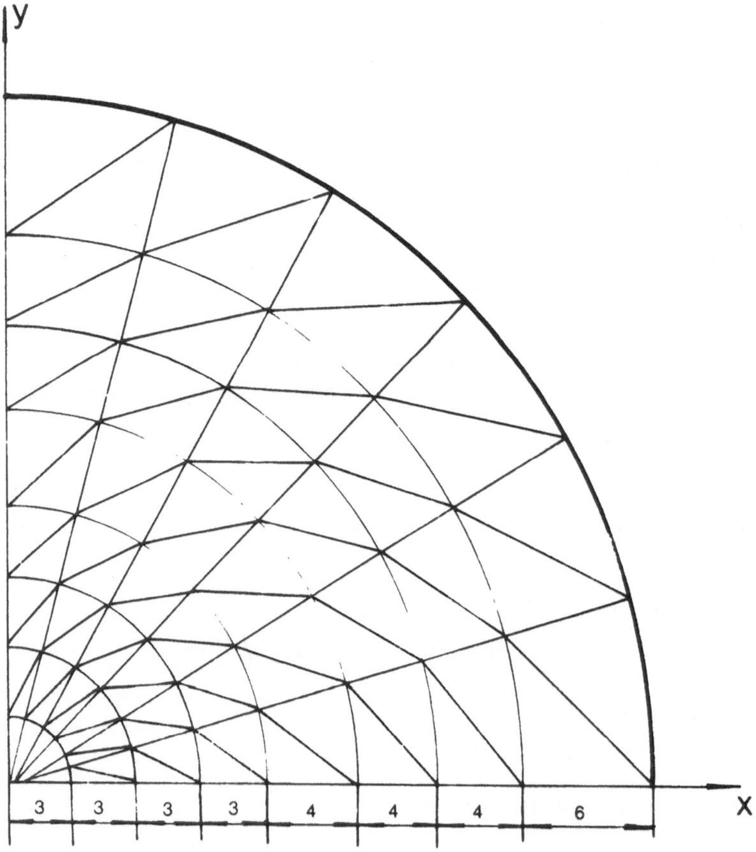

Figure 4. The calculation model of the circular plate.

Table 1. Final results by the finite element method.

Load	Deflection	Membrane stress				Bending stress			
q^*	ξ	$\sigma^*_{rm}(0)$	$\sigma^*_{\phi m}(0)$	$\sigma^*_{rm}(a)$	$\sigma^*_{\phi m}(a)$	$\sigma^*_{rb}(0)$	$\sigma^*_{\phi b}(0)$	$\sigma^*_{rb}(a)$	$\sigma^*_{\phi b}(a)$
2	0.3583	0	0	0	0	1.2408	0.5789	−1.6371	−0.2439
4	0.6720	0.2184	−0.0018	0.1104	0.0295	2.2680	1.0608	−3.1295	−0.4662
6	0.8848	0.3274	0.1065	0.1836	0.0509	2.2075	1.0528	−3.8724	−0.5769
8	1.0057	0.4974	0.2745	0.2881	0.0786	2.3214	1.1214	−4.6220	−0.7001
10	1.0966	0.7066	0.4202	0.3968	0.1065	2.3072	1.1305	−5.2540	−0.8103
12	1.1872	0.8072	0.5821	0.5069	0.1397	2.3163	1.1499	−5.9569	−0.9150
14	1.2679	0.9668	0.7408	0.6180	0.1738	2.2963	1.1553	−6.6368	−1.0132
16	1.3444	1.1307	0.9038	0.7334	0.2095	2.3522	1.1603	−7.2711	−1.1078
18	1.4157	1.2948	1.0670	0.8505	0.2463	2.2510	1.1613	−7.8780	−1.1983
20	1.4832	1.4600	1.2314	0.9696	0.2840	2.2244	1.1613	−8.4621	−1.2854
22	1.5471	1.6254	1.3961	1.0899	0.3226	2.1968	1.1601	−9.0248	−1.3693
24	1.6079	1.7911	1.5610	1.2114	0.3618	2.1698	1.1585	−9.5681	−1.4504
26	1.6659	1.9566	1.6259	1.3336	0.4017	2.1438	1.1568	−10.0935	−1.5287
28	1.7215	2.1220	1.8905	1.4564	0.4420	2.1194	1.1553	−10.6023	−1.6047
30	1.7748	2.2870	2.0548	1.5795	0.4828	2.0970	1.1541	−11.0957	−1.6783

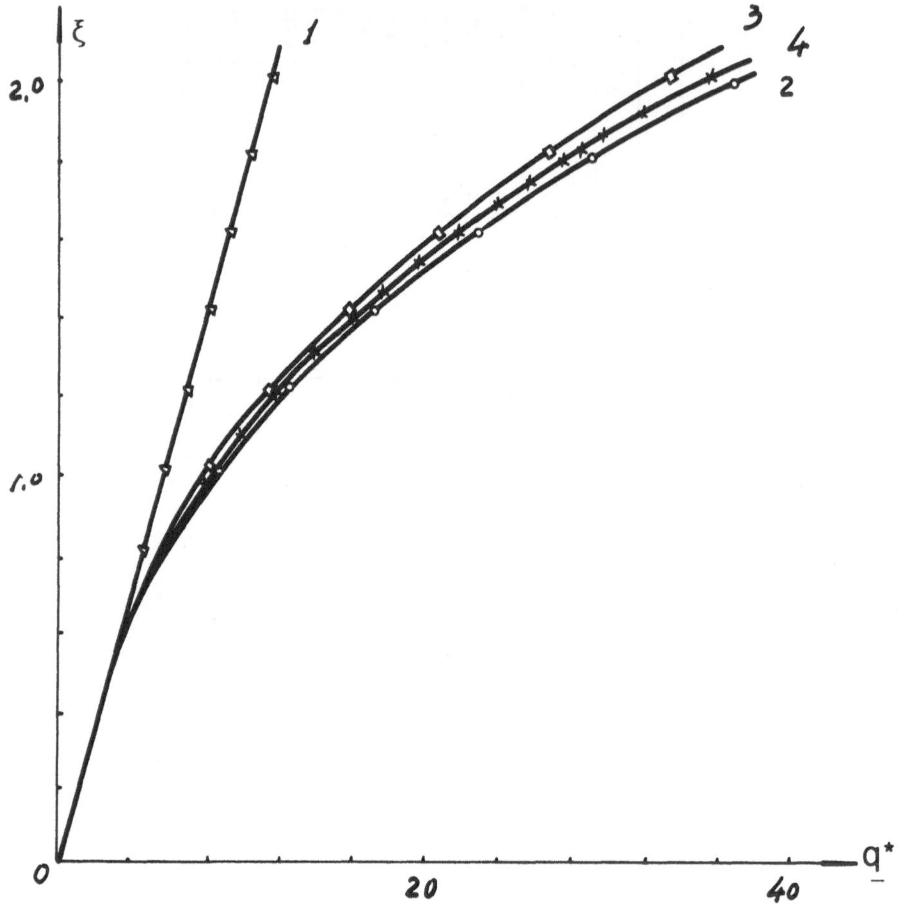

Figure 5. Load, deflection at centre.

In the calculation, the following data are taken: $a = 30\,\text{cm}$, $h = 0.3\,\text{cm}$, $E = 7.2 \times 10^5\,\text{Kg/cm}^2$, $\mu = 0.3$, and dimensionless parameters are used.

$\xi = f/h$ is the deflection at centre

$q^* = \dfrac{qa^4}{Eh^4}$ is the load

$\sigma^* = \dfrac{\sigma a^2}{Eh^2}$ is the stress

$\sigma^*_{rm}(r)$ is the radial membrane stress

$\sigma^*_{\phi m}(r)$ is the circumferential membrane stress

$\sigma^*_{rb}(r)$ is the radial bending stress

$\sigma^*_{\phi b}(r)$ is the circumferential bending stress.

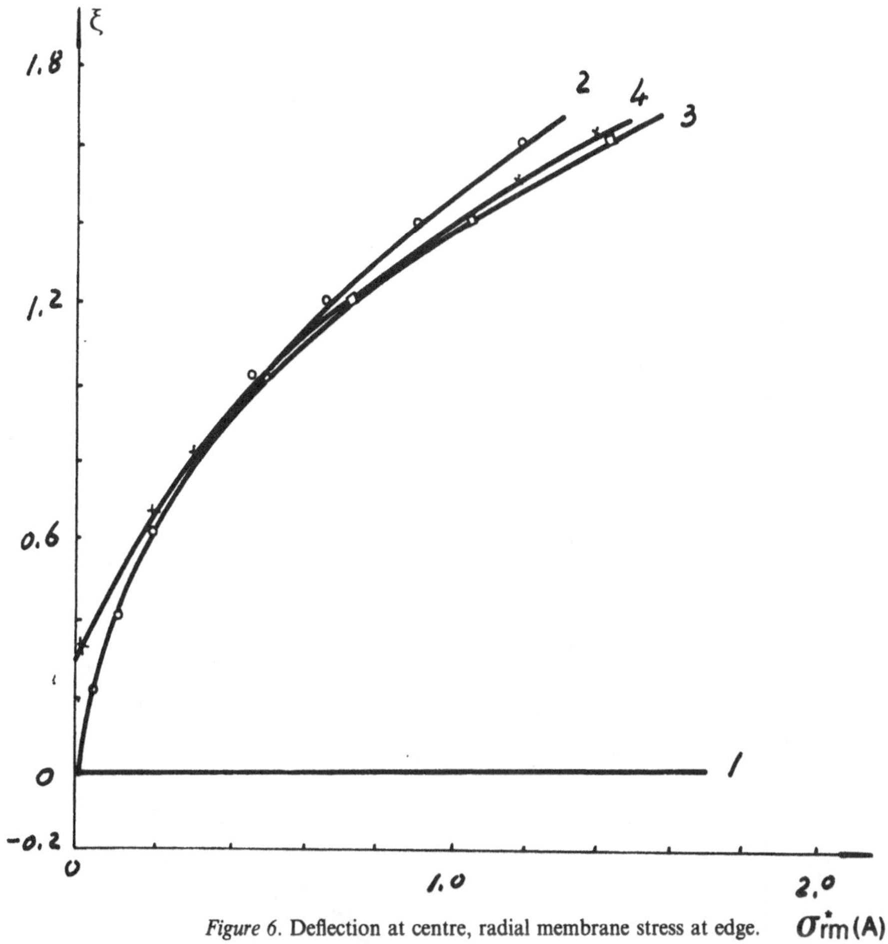

Figure 6. Deflection at centre, radial membrane stress at edge.

The calculation model of the finite element method is shown in Fig. 4, where we take a quarter of the circular plate. The calculation is carried out with linear steps $n = 15$ and load increment $\Delta q^* = 2$.

The final results are listed in Table 1. A comparison between the solution of this paper and the solutions found by using other methods is made in Figs 5–9, where:

1 —△—△—△— is the Timoshenko linear solution [8]

2 —○—○—○— is the solution obtained by Chien wei-zang and Yeh Kai-yuan perturbation method [11]

3 —□—□—□— is the A. S. Volmir first-order approximation solution [9]

4 —×—×—×— is the finite element method solution

It can be seen that the solutions of this paper are much closer to the solutions found by Chien and Yeh.

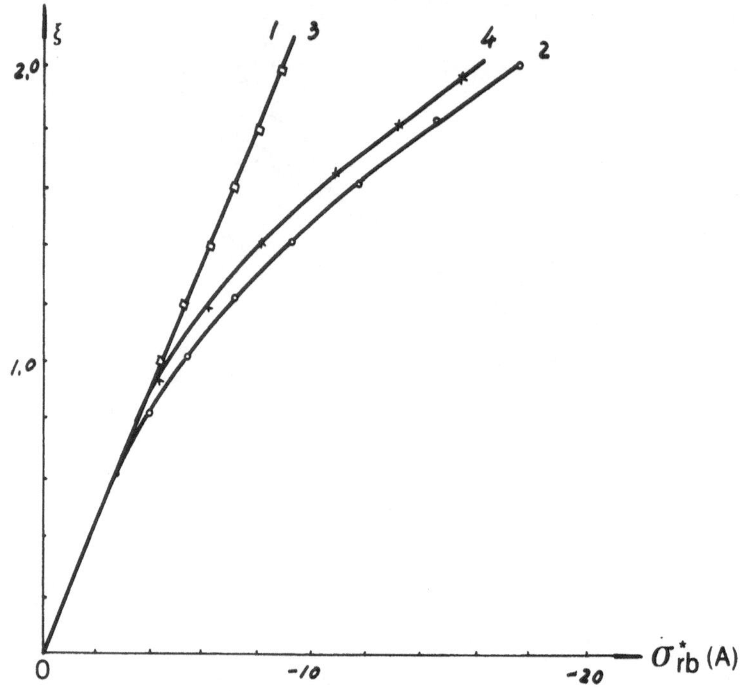

Figure 7. Deflection at centre, radial bending stress at edge.

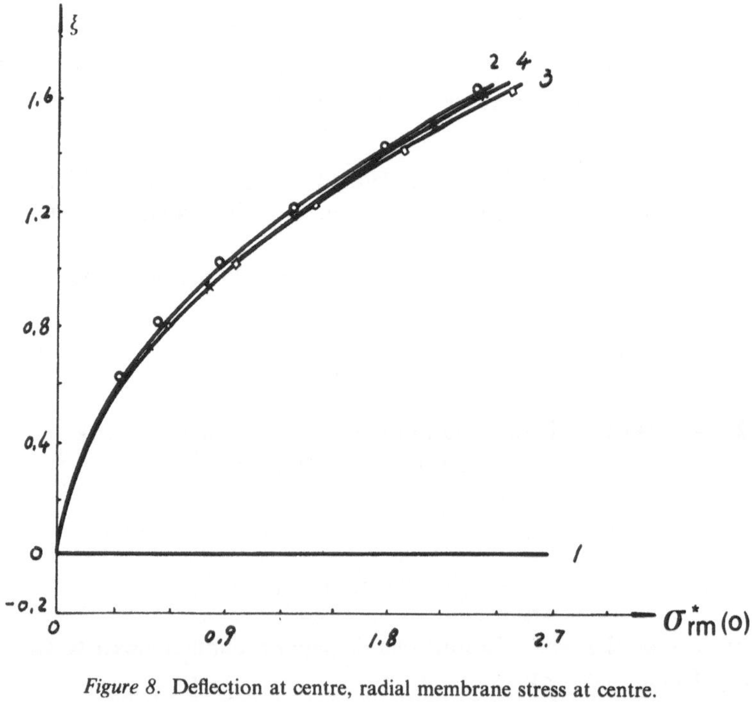

Figure 8. Deflection at centre, radial membrane stress at centre.

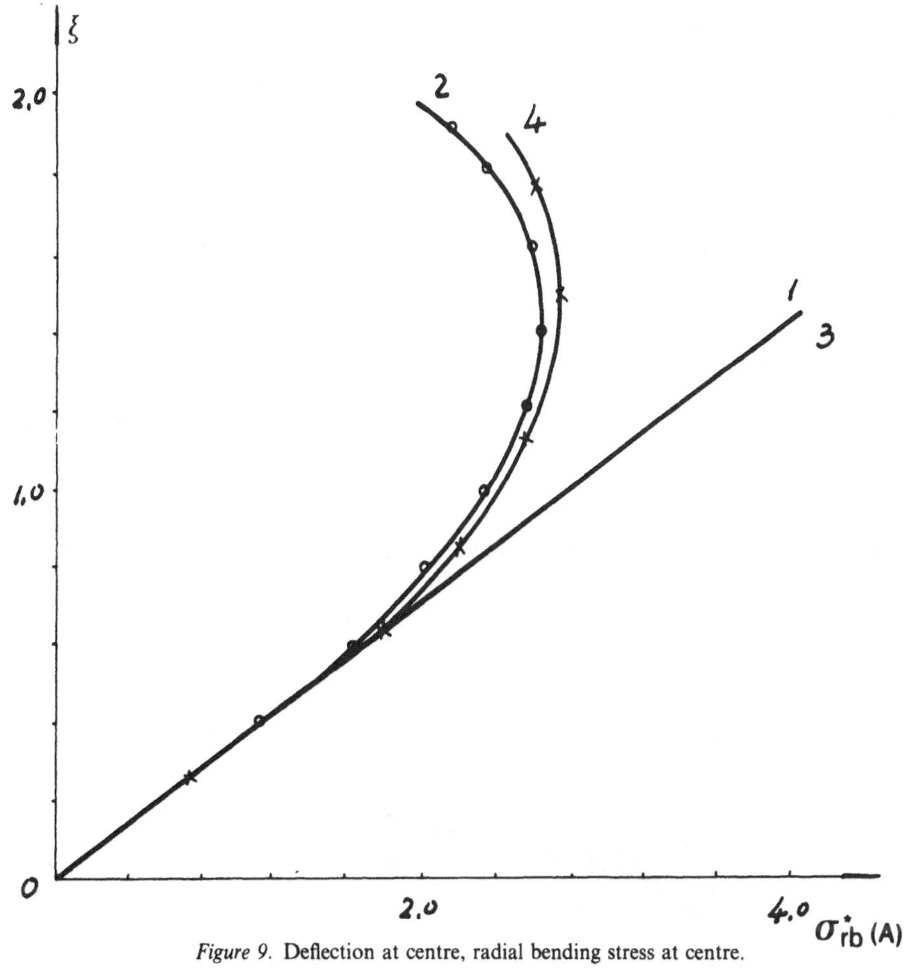

Figure 9. Deflection at centre, radial bending stress at centre.

3.2. *Rectangular plate*

Figure 10 shows a simply supported rectangular plate under a uniformly distributed lateral load, with the following data: $t = 1\,\text{cm}$, $a = 100\,\text{cm}$, $b = 50\,\text{cm}$, $\mu = 0.3$, $E = 7.2 \times 10^5\,\text{Kg/cm}^2$ and $q = 3.5\,\text{Kg/cm}^2$.

As the deflected surface of the plate is symmetrical with respect to both axes, only a quarter of the plate with 24 nodes and 30 triangular elements are taken in the analysis. Calculations are made with respect to $n = 1$, 3, 5, 8.

Table 2 lists the maximum deflection, membrane stresses and bending stresses of the plate when $n = 8$. Results calculated by the linear theory [8] and large deflection theory [9] are also listed in Table 2. Curves of deflections versus loads made from these results are shown in Fig. 11.

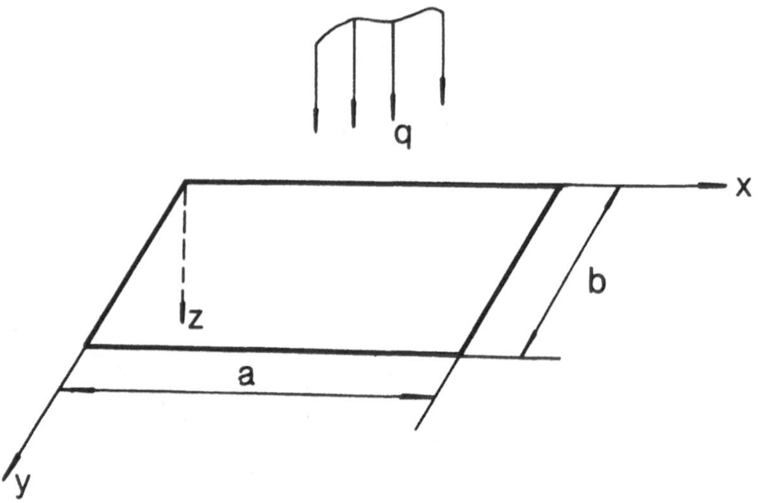

Figure 10. Simply supported rectangular plate.

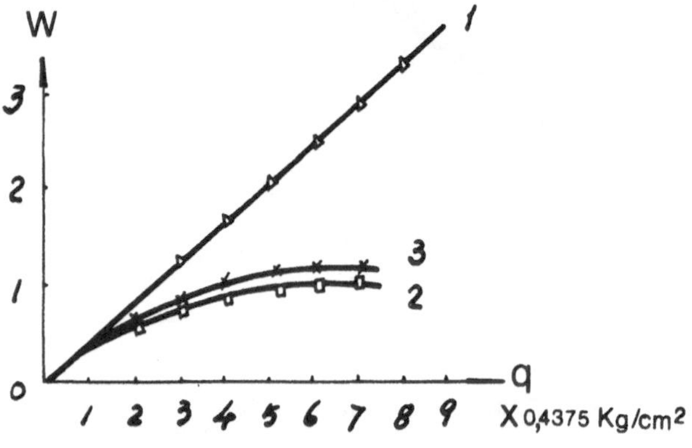

Figure 11. Load, maximum deflection.

The data listed and the curves show that:

(1) The linear theory can be used when the maximum deflection of the plate is less than 40% of the thickness of the plate.

(2) The membrane stress increases with the deflection, and geometrical non-linearity should be considered when the deflection is larger than 40 percent of the thickness.

Figures 12, 13 and 14 show the relationship of steps n and the results of the previous problem. If sufficient steps are taken, then satisfactory results will be obtained.

3.3. *The effect on loads due to large deflection*

In the above calculation we have neglected the effect of large deflection on nodal forces. But in some cases the effect must be considered. S. Timoshenko [10] has

Table 2. Maximum deflection and stress $q_0 = 0.4375\,\text{Kg/cm}^2$.

n		1	2	3	4	5	6	7	8
Load q (Kg/cm)		q_0	$2q_0$	$3q_0$	$4q_0$	$5q_0$	$6q_0$	$7q_0$	$8q_0$
Maximum deflection W (cm)	Linear theory	0.418	0.835	1.252	1.670	2.088	2.505	2.922	3.340
	Large deflection theory	0.380	0.570	0.680	0.780	0.860	0.922	0.980	1.090
	Finite element method	0.428	0.681	0.807	0.908	0.992	1.064	1.130	1.180
Membrane stress (Kg/cm^2)	Linear theory	0	0	0	0	0	0	0	0
	Large deflection theory	113	254	372	475	579	662	751	912
	Finite element method	0	168	302	431	552	667	777	875
Bending stress (Kg/cm^2)	Linear theory	662	1325	1987	2650	3312	3975	4657	5300
	Large deflection theory	594	890	1078	1218	1343	1437	1531	1687
	Finite element method	640	997	1165	1299	1409	1503	1586	1658

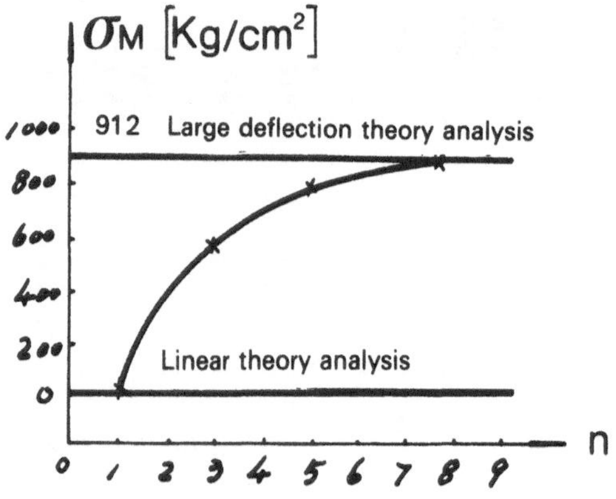

Figure 12. Membrane stress, *n*.

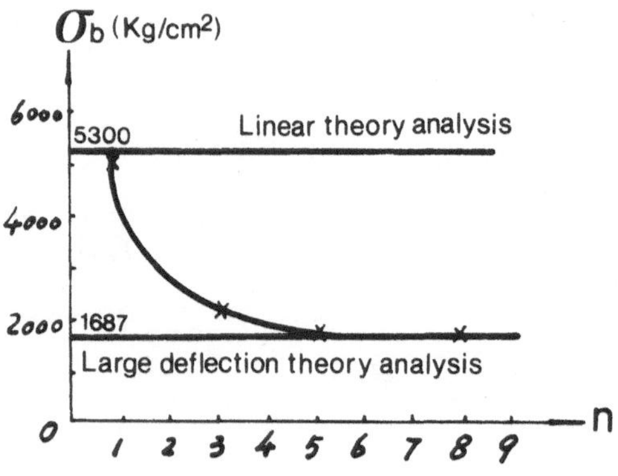

Figure 13. Bending stress, *n*.

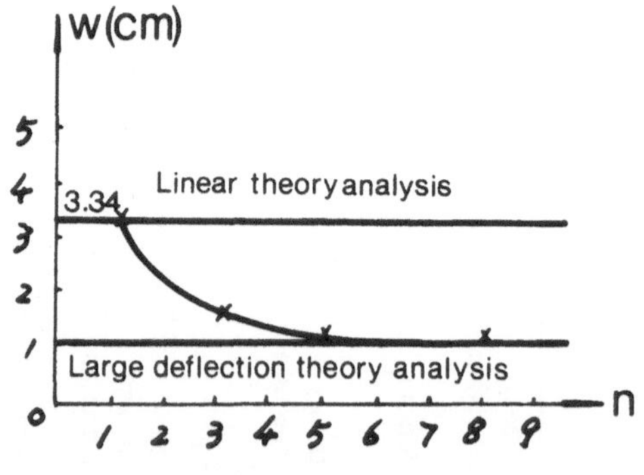

Figure 14. Maximum deflection, *n*.

depicted the change of loads due to the large deflection of a beam. We will refer to this case as the ponding problem.

For vessels under the action of pressure, the behaviour of the wall may be changed due to deflection, e.g. a flat part may behave like a shallow shell. When the incremental process of the finite element method is used, the change of the area

Table 3. Maximum deflections for the two cases.

Case	Linear step					
	1	2	3	4	5	Σ
Without the effect on load	0.686	0.238	0.123	0.103	0.088	1.238
With the effect on load	0.686	0.477	0.228	0.228	0.213	1.832

of the elements, and the direction and value of the nodal force can be considered after each step. The equation used in the condition becomes

$$([K_{E(i-1)}] + [K_{G(i-1)}])\{\Delta d_{(i)}\} = \{\Delta R_i(d_{i-1})\}, \tag{3.1}$$

where $\{\Delta R_i(d_{i-1}\}$ is the increment of load vector in the i-th step. It is determined according to the displacement vector $\{d_{i-1}\}$. For an element the nodal force vector is

$$\begin{aligned}\{\Delta R_i\} = {} & \Delta q^e_{(i)}\Delta^e_{(i)}\Delta^e_{(i-1)}[\tfrac{1}{3}, -\tfrac{1}{8}y'_{i(i-1)}, \tfrac{1}{8}x'_{i(i-1)}, \tfrac{1}{3}, -\tfrac{1}{8}y'_{j(i-1)}, \tfrac{1}{8}x'_{j(i-1)}, \tfrac{1}{3},\\ & -\tfrac{1}{8}y'_{m(i-1)}, \tfrac{1}{8}x'_{m(i-1)}]^T\end{aligned} \tag{3.2}$$

where $\Delta q^e_{(i)}$ is the pressure acting on the element for i step, $x'_{i(i-1)}, y'_{i(i-1)}, \ldots, x'_{m(i-1)}, y'_{m(i-1)}$ are coordinates of the nodal i, j and m of the triangular element in the local coordinate system x'–y'–z' for $i - 1$ step and $\Delta^e_{(i-1)}$ is the area of the element after the previous deflection. Then

$$\Delta^e_{(i-1)} = \frac{1}{2}\begin{vmatrix} 1 & x'_{i(i-1)} & y'_{i(i-1)} \\ 1 & x'_{j(i-1)} & y'_{j(i-1)} \\ 1 & x'_{m(i-1)} & y'_{m(i-1)} \end{vmatrix}.$$

As an example, the deflections of the above plate (Fig. 10) under $q = 3.5\,\mathrm{Kg/cm^2}$ are listed in Table 3 for both cases, i.e., with and without the consideration of the effect on load. There are obvious differences between the results of the two cases.

4. Stability analysis

4.1. *Basic equation*

The usual formulation for solving linear buckling problems by the finite element method is [5]

$$([K_E] + [K_G])\{d\} = 0. \tag{5.1}$$

We assume that for an elastic structure under certain kinds of loads, the stresses of each element exist with one and only one proportional relationship before buckling [6, 13], i.e.,

$$\begin{aligned}&\sigma_{x_i} : \sigma_{y_i} : \tau_{xy_i} : \sigma_{x_j} = R_i : m_i : n_i : mb_j\\ &(i = 1, 2, \ldots, Ne, \quad j = 1, 2, \ldots, Nb)\end{aligned} \tag{4.2}$$

where σ_{x_i}, σ_{y_i}, and τ_{xy_i} denote the membrane stress components of the i-th plate element; σ_{x_j} denotes the axial stress of the j-th beam element; R_i, m_i, n_i and mb_j are proportion constants; and Ne and Nb denote the total number of plate elements and beam elements, respectively.

We introduce a proportion factor λ, then

$$\frac{\sigma_{x_i}}{\lambda} = R_i; \quad \frac{\sigma_{y_i}}{\lambda} = m_i; \quad \frac{\tau_{xy_i}}{\lambda} = n_i, \quad (i = 1, 2, \ldots, Ne),$$
$$\frac{\sigma_{x_j}}{\lambda} = mb_j, \quad (j = 1, 2, \ldots, Nb). \tag{4.3}$$

The geometric stiffness matrices of the plate element and the beam element can be written as

$$\begin{aligned}[K_G^e]_{p(i)} &= \frac{\lambda t}{2\Delta}\{R_i[b_i^2 A + b_i b_j B + b_j^2 C] + m_i[c_i^2 A + c_i c_j B + c_j^2 C] \\ &\quad + n_i[2b_i c_i A + (b_j c_i + b_i c_j)B + 2b_j c_j C]\} \\ &= \lambda[K_G^e]_p^*\end{aligned} \tag{4.4}$$

$$[K_G^e]_{b(j)} = \lambda\frac{mb_j A}{30l}[S] = \lambda[K_G^e]_b^*, \tag{4.5}$$

where $[K_G^e]_p^*$ and $[K_G^e]_b^*$ are the geometric stiffness matrices of the plate and the beam element, when $\lambda = 1$, respectively.

The equation for the analysis of the structure can be written as

$$([K_E] + \lambda[K_G]^*)\{d\} = 0. \tag{4.6}$$

Thus, the problem becomes one of determining the lower eigenvalues λ. When λ is determined, the state of stress is found by (4.3) and the lower critical value is also found.

4.2. *Determination of eigenvalue and critical stress*

In terms of stability problems, we are only interested in the lowest critical stress in engineering. Therefore, the inverse iteration method [13] is used to solve the eigenvalue of (4.6). The procedure is as follows:

(a) Rewrite (4.6) so that

$$[K_E]\left(-\frac{1}{\lambda}\{d\}\right) = [K_G]^*\{d\}. \tag{4.7}$$

(b) Take a non-zero vector $\{d.\}$ as the initial vector. Set

$$[K_G]^*\{d\} = \{PK_1\}, \qquad -\frac{1}{\lambda}\{d\} = \{\delta\lambda_1\}.$$

Then (4.7) becomes

$$[K_E]\{\delta\lambda_1\} = \{PK_1\}. \tag{4.8}$$

(c) Find $\{\delta\lambda_1\}$ from (4.8).

(d) Normalization. Take the maximum absolute value of $\{\delta\lambda_1\}$ as $d_{1\max}$, and calculate $\{d_1\}$ from the following equation

$$\{\delta\lambda_1\} = d_{1\max}\left\{\frac{\delta\lambda_1}{d_{1\max}}\right\} = d_{1\max}\{d_1\}.$$

(e) Checking of convergence. Calculate $\{\Delta d_1\}$ according to

$$\{\Delta d_1\} = \{d_1\} - \{d_0\}.$$

Take $\{\Delta d_{m1}\}$ from $\{\Delta d_1\}$, that is, the component of $\{\Delta d_1\}$ with the maximum absolute value. If $|\Delta d_{m1}| > \varepsilon$, where ε is a prescribed value, then the above process should be repeated until $|\Delta dm| < \varepsilon$. If in the i-th iteration $\{\Delta d_i\} = \{d_i\} - \{d_{i-1}\}$ and $|\Delta dm_i| < \varepsilon$, then $\{d_i\}$ is the required eigenvector with respect to the configuration of the buckled plate. The eigenvalue is determined by the following formula

$$\lambda = -1/d_{(i-1)\max}$$

the critical stresses of the plate element are

$$\sigma_{xcri} = \lambda R_i,$$

$$\sigma_{ycri} = \lambda m_i, \quad (i = 1, 2, \ldots, Ne)$$

$$\tau_{xycri} = \lambda n_i,$$

and the critical stress of the beam element is

$$\sigma_{xcrj} = \lambda mb_j, \quad (j = 1, 2, \ldots, Nb).$$

EXAMPLE 1. A rectangular stiffened plate with four sides simply supported and with the following data under axial compression (Fig. 15):

$$a = 100\,\text{cm}, \quad b = 50\,\text{cm}, \quad t = 1\,\text{cm},$$

Cross section area of stiffener: $A = 3.875\,\text{cm}^2$,

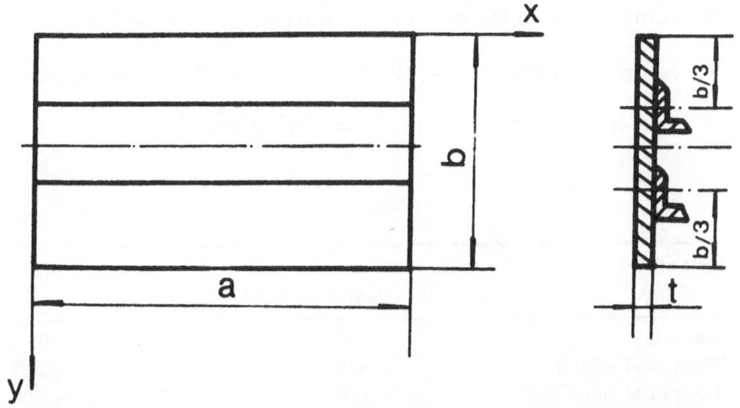

Figure 15. Rectangular stiffened plate with four sides simply supported.

Moment of inertia of stiffener: $J_y = 1.665\,\text{cm}^4$, $J_z = 17.949\,\text{cm}^4$,

$E = 7.2 \times 10^5\,\text{Kg/cm}^2$, $\mu = 0.3$.

In the calculation we take $\varepsilon = 0.03$. By five iterations we find

$$\sigma_{xcr} = 1621\,\text{Kg/cm}^2$$

while the exact solution is given in [15]

$$\sigma_{xcr} = 1749\,\text{Kg/cm}^2.$$

EXAMPLE 2. An integral cylindrical plate with longitudinal and transverse stiffeners, of the following parameters under axial compression (Fig. 16):

$t = 0.18\,\text{cm}$, $t_{ws} = 0.8\,\text{cm}$, $t_{wr} = 0.7\,\text{cm}$, $h_s = h_r = 0.77\,\text{cm}$,

$L = 96.5\,\text{cm}$, $S = 52.2\,\text{cm}$, $R = 96.5\,\text{cm}$, $l = 19\,\text{cm}$, $d = 9\,\text{cm}$,

$r_w = 0.77\,\text{cm}$, $E = 7.2 \times 10^5\,\text{Kg/cm}^2$, $\mu = 0.3$

and internal pressure $p = 0$.

For the whole structure we take 77 nodes, 120 plate elements and 74 beam elements. In the calculation, the correction of the eccentric effect of stiffeners is made, and ε is taken to be 0.05. By four iterations we find the critical compression $p_{xcr} = 23\,680\,\text{Kg}$. The theoretical value and tested value are given in [17] and [14], respectively. These values are listed in Table 4 where the error is given with respect to the test value.

EXAMPLE 3. Integral cylindrical panel with longitudinal and transverse stiffeners, of the following parameters subjected to axial compression and internal pressure (Fig. 16):

$L = 91.5\,\text{cm}$, $S = 52.2\,\text{cm}$, $R = 96.5\,\text{cm}$, $l = 12.5\,\text{cm}$,

$d = 7\,\text{cm}$, $t = 0.16\,\text{cm}$, $t_{ws} = 0.6\,\text{cm}$, $t_{wr} = 0.5\,\text{cm}$,

$h_s = h_r = 0.74\,\text{cm}$, $r_w = 0.74\,\text{cm}$, $E = 7.2 \times 10^5\,\text{Kg/cm}$,

$\mu = 0.3$ and internal pressure $p = 0.6\,\text{Kg/cm}$.

Table 4

Method	Critical compression (Kg)	Error %
Test value	25 300	0
Theoretical value 1	33 910	34.03
Theoretical value 2	31 640	25.06
F.E.M. solution	23 680	−6.40

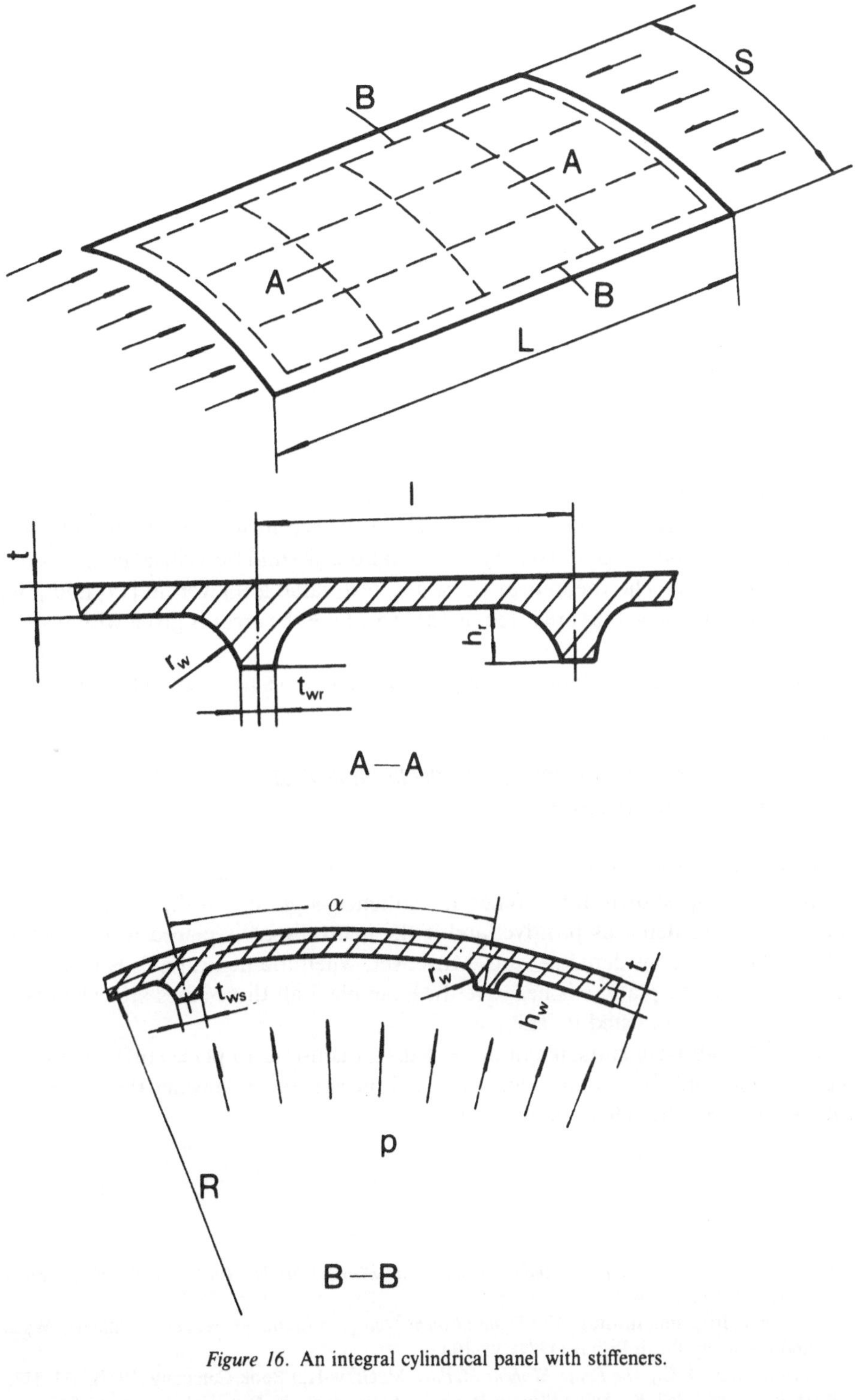

Figure 16. An integral cylindrical panel with stiffeners.

Table 5

Method	Critical compression (Kg)	Error %
Test value	34 700	0
Theoretical value 1	33 830	−2.51
Theoretical value 2	32 290	−6.95
F.E.M. solution	35 474	2.23

Table 6

b (cm)	Critical stress (Kg/cm^2)
0.45	3411
0	3489
−0.45	3567

For the whole structure we take 72 nodes, 99 beam elements and 112 plate elements. In the calculation, the correction of the eccentric effect of the stiffeners is made and ε is taken to be 0.02. By four iterations we find the critical compression $p_{xcr} = 35\,474$ Kg. The theoretical value and test value are given in [17] and [14], respectively. These values are listed in Table 5 where the error is given with respect to test value.

It can be seen that the finite element analysis gives a result closer to the test value.

4.3. *The effect of eccentricity between the beam neutral axis and the plate middle surface on critical stress*

The eccentricity between the stiffener neutral axis and plate middle surface is denoted by b as shown in [7]. When the stiffener is placed on the concave of the plate, b is considered as positive; and if the neutral axis is placed in the middle plane, then b is considered as zero. Moreover, when the neutral axis is placed on the convex of the plate, b is negative. In Example 3 all these cases are calculated, and the results are listed in Table 6.

From the above results, it can be seen that it is useful to place stiffeners on the convex side, and that in the calculation, it is necessary to consider the correction due to the eccentric effect of stiffeners.

References

1. Chien Wei-zang, *Variational Method of the Finite Element Method* (in Chinese), Science Press, Beijing (1980), 491–493.
2. Haudong Irrigation Institute, *The Finite Element Method in Elastic Mechanics* (in Chinese), Water and Electricity Press, Beijing (1974), 95–100.
3. Zienkiewicz, O. C., *The Finite Element Method*, McGraw-Hill Book Company (1977), 331–337.
4. Hollad, I. and Bell, K., *Finite Element Method in Stress Analysis*, Technical University of Norway (1972), 475–485.

5. Przemieniecki, J. S., *Theory of Matrix Structural Analysis*, McGraw-Hill Book Company (1968), 70–82.
6. Kabaila, A. P., *Bifurcation of Space Frames*, AFFDL-TR-36 (1970).
7. Liu Zheng-xing and Feng Tai-Hua, *The Correction of Eccentric Effect for Space Beam Elements* (in Chinese), Scientific and Technical Material of Nanjing Aeronautical Institute, No. 494 (1979).
8. Timoshenko, S. and Woinowsky-Krieger, S., *Theory of Plates and Shells* (second edition), McGraw-Hill Book Company (1959), 113–124.
9. Volmir, A. S., *Flexible Plates and Shells* (in Chinese), Science Press, Beijing (1963), 171–176.
10. Timoshenko, S. and Gere, J., *Mechanics of Materials*, Van Nostrand Reinhold Company (1972), 195–199.
11. Chien Wei-zang and Yeh Kai-yuan, "On the large deflection of circular plate" (in Chinese), *Acta Physica Sinica* **10**, 3 (1954), 209–211.
12. Chen Shan-lin and Kuang Ji-chang, "The perturbation parameter in clamped circular plates" (English Edition), *Applied Mathematics and Mechanics* **2**, 1 (1981), 137–154.
13. Cook, R., *Concepts and Applications of Finite Element Analysis*, John Wiley (1974), 264–266.
14. Liu Zheng-xing and Feng Tai-hua, *Solution of Stability Problems of Stiffened Plates and Shells by the Finite Element Method* (in Chinese), Scientific and Technical Material of Nanjing Aeronautical Institute, No. 807 (1980).
15. Timoshenko, S. P. and Gere, J. M., *Theory of Elastic Stability* (second edition), McGraw-Hill (1961), 360–370.
16. Singer, J., Barauch, M., and Harari, O., "On the stability of eccentrically stiffened cylindrical shells under axial compression", *Int. J. Sol. Struc.* **34** (1967), 445–470.
17. *The Stability Analysis of Integral Cylindrical Plates with Stiffnesses* (in Chinese), The Mechanical Research Institute of the Chinese Academy of Sciences, Science Reference Press (1976).

Author Index

Subject Index